ALLERGEN MANAGEMENT IN THE FOOD INDUSTRY

ALLERGEN MANAGEMENT IN THE FOOD INDUSTRY

Edited by

JOYCE I. BOYE
Agriculture & Agri-Food Canada
Saint-Hyacinthe, Quebec, Canada

SAMUEL BENREJEB GODEFROY
Health Canada
Ottawa, Ontario, Canada

WILEY

A JOHN WILEY & SONS, INC., PUBLICATION

Published by John Wiley & Sons, Inc., Hoboken, New Jersey
Published simultaneously in Canada

Library of Congress Cataloging-in-Publication Data:
Allergen management in the food industry / edited by Joyce I. Boye, Samuel Benrejeb Godefroy.
p. cm.
Includes index.
Summary: "This book comprehensively addresses the sources of allergenic contaminants in foods, their fate during processing, and the specific measures that need to be taken to minimize their occurrence in foods. The book provides up-to-date information on the nine major allergens (as well as other emerging allergens) and practical guidelines on how these allergens can be identified and controlled during production and processing. Starting with an introduction to food allergens, the book follows with sections on food allergen management during production and processing, guidelines for the processing of specific allergen-free foods, techniques for hypo-allergenization and allergen detection, and allergen-free certification"—Provided by publisher.
Summary: "This book comprehensively addresses the sources of allergenic contaminants in foods, their fate during processing, and the specific measures that need to be taken to minimize their occurrence in foods"—Provided by publisher.
ISBN 978-0-470-22735-0 (hardback)
1. Food allergy—Prevention. 2. Food industry and trade. I. Boye, Joyce I. I. Godefroy, Samuel Benrejeb.
RC596.A557 2010
616.97′5—dc22
2010024927

Printed in Singapore

10 9 8 7 6 5 4 3 2 1

When food is your enemy, knowledge is your best weapon of defence.

Anonymous, 2009

CONTENTS

PREFACE

Food allergic reactions have emerged as a growing challenge to the food industry. The statistics are quite remarkable: ~4–8% of children and ~2–4% of the adult population are believed to have food allergies. Allergic reactions are particularly bothersome because in some unfortunate situations, they have resulted in anaphylaxis and even death. For example, the death of a young girl in Ontario, the most populated province in Canada, after consumption of french fries that had been inadvertently in contact with a dairy product in her school cafeteria resulted in increased awareness about food allergies, leading legislators in the province to enact what is now known as Sabrina's law. This legislation requires schools in Ontario to be proactive about allergy education and preparedness. The death of Sabrina is not unique, and unfortunately, other incidents have occurred around the world. Of the over 160–180 foods known to be allergenic, some are considered priority allergens. These include eggs, milk, soy, peanuts, tree nuts, fish, shellfish, and wheat (gluten). The health burden of allergies and allergy-related diseases still remains unclear.

To protect food-allergic consumers, several countries around the world have put in place food allergen labeling regulations requiring food industries to label the priority allergens when they are present in foods. This has posed some challenges to the food industry and has resulted in discussions on the need of allergen management programs to allow the identification of allergens in the food supply chain and their avoidance in foods targeted toward food-allergic consumers. The objective of food safety regulators and the food industry, ultimately, is to provide safe foods to allergic consumers and their families while maximizing choice.

The market for allergen-free foods has grown astronomically as the food industry has awakened to the potential for growth of this sector. These developments are exciting and provide a glimmer of hope in a world that has proven to be hazardous for allergic individuals. The eagerness of the food industry to embrace this market and the readiness of regulatory agencies to provide guidelines to support the sector is palpable. It is in this context that this book has been compiled, which focuses on providing practical and timely information on the manufacturing of foods targeted for allergic consumers. Renowned experts from around the world working in the area of food allergy have contributed excellent reviews covering the spectrum of allergen management from "farm to fork." The target audience for the proposed book includes food scientists and food processors in academia (professors, researchers, students) and the food industry. The book's comprehensive nature will also appeal to scientists and researchers in general involved in allergy research, food inspection, food safety certification, and policy making.

The book is divided into four parts. Part I begins with Chapter 1, which provides an overview of food allergy and food intolerance and the mechanisms involved in sensitization and elicitation. A clear distinction between food allergy and food intolerance is provided as well as a definition of gluten hypersensitivity. Chapters 2, 3, and 4 provide a summary of the major issues that need to be addressed in allergen management and the criteria for determining priority allergens, using tree nuts and mustard as an example.

Part II contains three chapters that address, respectively, allergen management in agricultural practices (on the farm), food processing, and the foodservice industry.

In Part III, Chapters 8–13, respectively, provide in-depth reviews of each of the following priority allergens: dairy, eggs, fish and crustaceans, peanuts and tree nuts, gluten, and soy. Critical information is provided for the processing of foods targeted for consumers suffering from theses specific allergies. Chapter 14, the last chapter in this part, provides a practical example of a food product formulated to be "free from" milk, eggs, and soybeans.

Chapters 15–17 of Part IV, address, respectively, allergen risk assessment and risk management, precautionary labeling, and allergen-free certification. As more "allergen-free" products are introduced into the market, consumers, as well as the food industry, need to consider the risks and benefits of auto-certification versus third-party certification. Chapter 18 discusses in detail some of the emerging allergens (e.g., lupin, sesame, and mustard) and their properties and the need to monitor these foods as global movement and trade exposes consumers to an ever-increasing array of new and novel foods. The last chapter, Chapter 19, provides practical tools for managing food allergen risks.

One of the important areas of food allergy requiring further investigation is the threshold dose required to provoke allergic reactions. Zero tolerance, while desirable, poses serious challenges to the food industry and could increase processing costs considerably. Knowledge of the amount of allergens required to elicit allergic reactions may help to provide realistic targets for the industry. While this subject is beyond the scope of this book, results of current studies such as those being conducted by the European Union Network (EuroPREVALL, http://www.europrevall.org) should shed more light on this important area. Methods for allergen detection are not addressed in this book, as they have been effectively discussed in the book *Detecting Allergens in Food* (edited by S.J. Koppelman and S.L. Hefle, CRC Press and Woodhead Publishing, 2006). There continues to be a need, however, for improved methods for allergen detection that take into consideration the effect of processing and matrix.

The reader will sometimes find the use of multiple terminologies in the text and different statistics for allergy prevalence rates. This is expected in a field that is transitioning from infancy into adolescence. An example that requires mentioning is the use of the terms "cross-contact" and "cross-contamination." As "cross-contamination" has been traditionally used in the food safety context to refer to contamination by pathogenic microorganisms, "cross-contact" has evolved as a more adequate terminology to describe the inadvertent presence of allergenic substances in the food chain. These two terms are used interchangeably throughout the book. Additionally, various numbers have been reported in the literature for allergy prevalence rates, ranging from 4% to 8% for children and 1% to 4% of the adult population. This variability is reflected in some of the literature cited in the text.

We would also like to mention that while effort has been made to minimize duplication, in a volume that attempts to cover a spectrum as large as this, some repetition is unavoidable. There may also be some errors in the text especially in a field such as this one, which is quickly evolving. We take responsibility for any such errors and would kindly request that readers inform us of any such errors so appropriate corrections and additions can be made in any future editions. Finally, we would like to thank the contributors for doing such a stellar job of compiling the most up-to-date information for the management of food allergens. It is a remarkable piece of work, and we sincerely thank them for their hard work and dedication.

Joyce I. Boye
Samuel Benrejeb Godefroy

CONTRIBUTORS

Allaoua Achouri, Food Research and Development Centre, Agriculture and Agri-Food Canada, Saint-Hyacinthe, Quebec, Canada

Anton J. Alldrick, BRI Campden, Chipping Campden, UK

Elke K. Arendt, Department of Food and Nutritional Sciences, University College Cork, Ireland

Joyce I. Boye, Food Research and Development Centre, Agriculture and Agri-Food Canada, Saint-Hyacinthe, Quebec, Canada, J2S 8E3; Email: joyce.boye@agr.gc.ca

Michel Britten, Food Research and Development Centre, Agriculture and Agri-Food Canada, 3600 Casavant Blvd West, Saint-Hyacinthe, Quebec, Canada

Vernon D. Burrows, Eastern Cereal and Oilseed Research Centre, Agriculture and Agri-Food Canada, Central Experimental Farm, Ottawa, Ontario, Canada

René W.R. Crevel, Safety & Environmental Assurance Centre, Unilever, Sharnbrook, Bedford, UK

Angelina O. Danquah, Department of Home Science, University of Ghana, Legon, Ghana

Philippe Delahaut, CER Groupe—Laboratoire d'Hormonologie, Marloie, Belgium

Lilla Dömötör, Sződliget, Budapest, Hungary

Sheila Dubois, Food Directorate, Health Products and Food Branch, Health Canada, Ottawa, Ontario, Canada

Valéry Dumont, CER Groupe—Laboratoire d'Hormonologie, Marloie, Belgium

Audrey DunnGalvin, College of Medicine and Health, Department of Paediatrics and Child Health, University College Cork, Ireland

Christine Dupuis, Bureau de Normalisation du Québec (BNQ), Montreal, Quebec, Canada

Fiona Fleming, George Weston Foods, Sydney, New South Wales, Australia

Zoë Gillespie, Bureau of Chemical Safety, Food Directorate, Health Canada, Ottawa, Ontario, Canada

Samuel Benrejeb Godefroy, Food Directorate, Health Canada, Ottawa, Ontario, Canada

M. Hazel Gowland, Allergy Action, St Albans, UK

Kirsten Grinter, Nestle Australia Ltd., Sydney, New South Wales, Australia

Sandra Kerbach, Eurofins Analytik GmbH, Hamburg, Germany

Sébastien La Vieille, Food Directorate, Health Products and Food Branch, Health Canada, Ottawa, Ontario, Canada

Kim Leighton, Australian Food and Grocery Council, Canberra, Australian Capital Territory, Australia

Lamia L'Hocine, Food Research and Development Centre, Agriculture and Agri-Food Canada, Saint-Hyacinthe, Quebec, Canada

E.N. Clare Mills, Institute of Food Research, Colney, Norwich, UK

Kevin Norman, Peanut Company of Australia, Kingaroy, Queensland, Australia

Maria Helena B. Nunes, Department of Food and Nutritional Sciences, University College Cork, Ireland

Takahide Obata, Research Institute, Morinaga and Co. Ltd., Yokohama, Japan

Sylvia Pfaff, Food Information Service Europe, Bad Bentheim, Germany

Roland E. Poms, International Association for Cereal Science and Technology (ICC), Vienna, Austria

Bert Popping, Eurofins Scientific Group, Yorkshire, UK

Chris Preston, Legal Finesse, Sydney, New South Wales, Australia

Olga M. Pulido, Bureau of Food Policy and Science Integration, Food Directorate, Health Products and Food Branch, Health Canada, Ottawa, Ontario, Canada; and Department of Pathology and Laboratory Medicine, Faculty of Medicine, University of Ottawa, Ontario, Canada

Sahul H. Rajamohamed, Food Research and Development Centre, Agriculture and Agri-Food Canada, Saint-Hyacinthe, Quebec, Canada

Maria Said, Anaphylaxis Australia, Sydney, New South Wales, Australia

Masahiro Shoji, Morinaga Institute of Biological Science, Inc., Yokohama, Japan

Benjamin K. Simpson, Department of Food Science and Agricultural Chemistry, Macdonald Campus, McGill University, Sainte-Anne-De-Bellevue, Quebec, Canada

Warren E. Stone, Grocery Manufacturers Association, Washington, DC

Ferdinand Tchounkeu, Québec Food Allergies Association (QFAA), St-Léonard, Montréal, Québec, Canada

Sandor Tömösközi, BUTE, Department of Applied Biotechnology and Food Science, Budapest, Hungary

Jupiter M. Yeung, Nestlé Nutrition, Nestlé Product Technology Center, Fremont, MI

GLOSSARY OF TERMS

Adaptive (acquired) immunity (specific immunity). This is the response of the immune system to a specific immune stimulus (antigen). The immune system "remembers" that it has encountered a specific antigen and reacts more rapidly on subsequent exposure (immune surveillance). Critical to adaptive immunity are antigen-presenting cells (APCs) including macrophages and dendritic cells, antigen-dependent stimulation of T-cell subtypes, B-cell activation leading to antibody production, and the activation of macrophages and natural killer (NK) cells.

Allergen. A substance that causes an inappropriate reaction by the immune system "an allergic reaction."

Allergic proctocolitis. A benign disorder manifesting with blood-streaked stools in otherwise healthy-appearing infants who are breast- or formula-fed. Its clinical features and laboratory results are often nonspecific. Symptoms resolve within 48–72 hours following elimination of dietary cow's milk protein. The underlying mechanism is not known, though IgE is clearly not implicated. Endoscopy shows focal or diffuse colitis, with edema and erosions. The biopsy reveals eosinophilic infiltration with focal distribution.

Allergy. A hypersensitivity reaction initiated by immunologic mechanisms.

Anaphylaxis. A systemic IgE-mediated allergic reaction that can be fatal within minutes, by compromising the airways or through a dramatic drop in blood pressure. In a sensitized, susceptible person, contact with or ingestion of an allergen may elicit an IgE-mediated adverse immune response leading to airway obstruction, hypotension, and loss of consciousness, resulting in anaphylactic shock. In anaphylaxis, several systems are usually affected simultaneously, including the respiratory tract, cardiovascular system, and gastrointestinal tract.

Angioedema. Refers to locally diffuse and painful soft-tissue swelling that may be asymmetric, especially on the eyelids, lips, face, and tongue, but also on the back of hands or feet and on the genitals. If angioedema affects the throat, the person's airway could be blocked, which could be life threatening.

Antibodies (Abs). Also called immunoglobulins (Igs) that are released by plasma cells. When a B cell encounters an antigen, it is stimulated to mature into a plasma cell or a memory B cell. Each antibody molecule has two parts. One part varies and is specialized to attach to a specific antigen. The other part is one of five structures, which determines the antibody's class—IgM, IgG, IgA, IgE, or IgD. This part is the same within each class and determines the function of the antibody.

Antigen-presenting cells (APCs). T-cell-dependent acquired immune responses typically require antigen-presenting cells to present Ag-derived peptides within major histocompatibility complex (MHC) molecules.

Antigens (Ag). A substance, usually proteins or polysaccharides, capable of stimulating the immune system to produce antibodies.

Asthma. An allergic-mediated response in the bronchial airways and a common disorder characterized by chronic inflammation of the bronchial tree with consequent reduction of airflow and symptomatic wheezing and dyspnea. Asthmatics are more responsive than nonasthmatics to a wide range of triggers capable of initiating an asthmatic episode. The narrowing of the bronchial tree (bronchi) is usually reversible, but in some patients with chronic asthma, there may be an element of irreversible airflow obstruction. Asthma involves only the bronchi and does not affect the air sacs (alveoli) or the lung parenchyma itself.

Atopic dermatitis (AD). A pruritic, chronic inflammatory skin disease of unknown origin that usually starts in early infancy; it is characterized by eczematous lesions, dryness, and thickening of the skin. AD may be associated to acute allergic reactions to foods. Genetic factors are important in the development of AD and are often associated with a personal or family history of other atopic diseases. The association of food allergy with atopic dermatitis has been demonstrated and IgE and non-IgE cellular mechanisms have been implicated. AD is considered to be an inherited genetic disorder with an allergic diathesis.

Atopy. A personal or familial tendency to produce IgE antibodies in response to low doses of allergens, confirmed by a positive skin prick test, and typical symptoms such as asthma, rhinoconjunctivitis, or eczema/dermatitis.

Autoimmune disorders. Refers to medical conditions that occur when the immune system mistakenly attacks "itself" and destroys healthy body tissue. There are more than 80 different types of autoimmune disorders.

Blood pressure. The pressure of the blood within the arteries produced primarily by the contraction of the heart muscle. Its measurement is recorded by two numbers. The first (systolic pressure) is measured after the heart contracts and is recorded by the highest number. The second (diastolic pressure) is measured before the heart contracts and is recorded by the lowest number. Elevation of blood pressure is called "hypertension."

Bronchial asthma. Refers to the definition of asthma.

Bronchospasm. Spasmodic contraction of the muscular walls of the bronchial air passages as observed in asthma; it is associated with breathing difficulty.

Cardiovascular. The heart and the blood vessels as a unified body system.

Conjunctivitis. Inflammation of the mucous membrane lining the inner surface of the eyelids and covering the front part of the eyeball.

Cross-contact/contamination. Refers to a food contaminating or entering in contact with another unrelated food leading to a "hidden" source of allergenic proteins.

Cross-reactivity. The concept of cross-reactivity concerns two allergens and an antibody. The term is used to describe a relation between two allergens and a cross-reactive antibody. The closer the similarity between the two allergens, the more likely it is to find a cross-reactive antibody. A variety of cross-reacting allergens are present among foods and aeroallergens. Allergen cross-reactivity can be detected when tested *in vitro*, but clinical correlation of the cross-reactivity is more variable. For example, cow's milk allergy is a common disease of infancy and childhood. Goat's milk cross-reacts with cow's milk. Cow's milk allergic patients may also react to goat's and/or sheep's milk.

Cytokines. Polypeptides secreted by immune and other cells when the cell interacts with a specific antigen, endotoxin, or other cytokines.

Dermatitis. An umbrella term for local inflammation of the skin.

Diaphoresis. Perspiration, especially when profuse.

Dyspnea. Shortness of breath.

Eczema. A general term for many types of skin inflammation, also known as dermatitis, and is characterized by itching and the formation of scales. It is a very common condition and can affect all races and ages, including young infants. The most common form of eczema is atopic dermatitis.

Eosinophilic esophagitis. A primary clinicopathologic disorder of the esophagus, characterized by (1) symptoms including, but not restricted to, food impaction and dysphagia in adults, and feeding intolerance and gastroesophageal reflux disease (GERD) symptoms in children; (2) biopsy with ≥15 eosinophils/high power field; (3) exclusion of other disorders associated with similar clinical, histological, or endoscopic features, especially GERD. Appropriate treatments include dietary approaches based on eliminating exposure to food allergens, or topical corticosteroids.

Erythema. An abnormal redness of the skin caused by various agents, as sunlight, drugs, and so on, that irritate and congest the capillaries.

Exercise-induced anaphylaxis. Exercise can induce an allergic reaction to food. The usual scenario is that of a person eating a specific food and then exercising. As the individual exercises and their body temperature increases, they begin to itch, get lightheaded, and soon develop the characteristic allergic reactions of hives, asthma, abdominal symptoms, and even anaphylaxis. Refer to the definition of anaphylaxis or systemic reaction.

Favism. Disease that develops in genetically predisposed individuals when they ingest broad beans or fava beans or inhale the flower pollen. The condition is a result of a deficiency of glucose-6-phosphate dehydrogenase in the red blood cells and of reduced glutathione, which is needed for red blood cell integrity. Fava beans contain substances that oxidize glutathione, which results in acute hemolytic anemia. The condition is not IgE mediated and is therefore regarded as food intolerance. Areas of the world most affected by this disease are the Mediterranean, Asia, Middle East, and Formosa. In the United States, favism is reported to affect 1–2% of Caucasian-Americans and 10–15% of African-Americans.

Flushing. Sudden redness of the skin, especially of the head and neck.

Gastrointestinal. Refers collectively to the stomach and small and large intestines.

Glottis. The middle part of the larynx; the area where the vocal cords are located.

Gluten sensitivity. This term as used in the literature is confusing and has been used to express various types of adverse reactions to dietary gluten. **Gluten sensitivity enteropathy** is a term used inter-changeably with celiac disease (CD). The term gluten sensitivity has also been used to encompass other autoimmune conditions associated with gluten exposure but that often present without gastrointestinal symptoms and bowel pathology, e.g., gluten ataxia and gluten neuropathy. On the other hand, emerging research has used the term gluten sensitivity to differentiate between the autoimmune enteropathy CD and those individuals who may present with symptoms similar to CD, but without anti-tTG autoantibodies or the autoimmune comorbidities (Hadjivassiliou et al. 2010; Sapone et al. 2010). The specific genetic profiles and mechanisms involved in distinguishing the different conditions associated with gluten toxicity are just beginning to be elucidated and the terminology will need further definition and clarification. Further investigation is also needed to assess the gluten threshold among individuals with various gluten toxicity profiles.

Haplotypes. A haplotype is the set of "single nucleotide polymorphism (SNP)," alleles along a region of a chromosome. Some of the segments of the ancestral chromosomes occur as regions of DNA sequences that are shared by multiple individuals. These segments are the haplotypes that enable geneticists to search for genes involved in diseases and other medically important traits (http://www.hapmap.org/originhaplotype.html.en; http://www.genome.gov/10001665).

Heiner syndrome (HS). A food immune-mediated pulmonary disease that affects primarily infants and is mostly caused by cow's milk.

Histamine and histamine intolerance. Histamine intolerance results from disequilibrium of accumulated histamine and the capacity for histamine degradation. Histamine is a biogenic amine that occurs to various degrees in many foods. In healthy persons, dietary histamine can be rapidly detoxified by amine oxidases, whereas persons with low amine oxidase activity are at risk of histamine toxicity. Diamine oxidase (DAO) is the main enzyme for the metabolism of ingested histamine.

HLAs (human leukocyte antigens). Proteins found on the surface of nearly every cell in the human body. HLAs are found in large amounts on the surface of white blood cells. They help the immune system tell the difference between "self" body tissues and foreign substances.

HLA system. Major histocompatibility complex (MHC) in humans, controlled by genes located on chromosome 6. It encodes cell surface molecules specialized to present antigenic peptides to the T-cell receptor (TCR) on T cells.

Hypotension. Blood pressure that is below the normal expected for an individual in a given environment. Hypotension is the opposite of hypertension (abnormally high blood pressure). It is a relative term because the blood pressure normally varies greatly with activity, age, medications, and underlying medical conditions. Unlike high blood pressure, low blood pressure is defined primarily by signs and symptoms of low blood flow and not by a specific blood pressure number. A sudden fall in blood pressure can also be dangerous. A change of just 20 mmHg, a drop from 130 systolic to 110 systolic, for example, can cause dizziness and fainting when the brain fails to receive an adequate supply of blood. Big blood pressure plunges, especially those caused by uncontrolled bleeding, severe infections, or allergic reactions, can be life threatening.

Hypotension for children. Defined as systolic blood pressure <70 mmHg from 1 month to 1 year, [<70 mmHg + (2 age)] from 1 to 10 years, and <90 mmHg from 11 to 17 years.

Hypoxia. An abnormal condition resulting from a decrease in the oxygen supplied to or utilized by body tissue.

Immune response. The action taken by the body's immune system to defend itself from pathogens. The immune system must be able to determine what is a normal part of the body or "self," from what is foreign or "non-self." The immune response can be roughly divided into two broad categories: innate (natural) immunity and adaptive (acquired) immunity.

Immunoglobulins (Ig). Glycoprotein molecules that are produced by plasma cells in response to an immunogen and which function as antibodies. All immunoglobulins have a four-chain structure as their basic unit. They are composed of two identical light chains (23 kD) and two identical heavy chains (50–70 kD).

Incontinence. Inability to restrain a natural discharge or feces from the body.

Infantile colic. Refers to colicky babies who cry constantly and hard at about the same time each day at least 3 days a week. It is more common in boys and in firstborn children. It usually begins at about 2 weeks of age and goes away by the fourth month. Infants who are experiencing symptoms of cow's milk allergy have a high rate (44%) of colic. However, the role of allergy as opposed to other causes among those with colic and without other symptoms of food allergy remains controversial and in need of additional study.

Innate immunity. This is a nonspecific, fast-acting response and is not directed against one type of pathogen/antigen, but is capable of destroying many different invaders.

Lymphocytes. These are types of white blood cell responsible for acquired immunity and may be T cells or B cells. *T cells* are produced in the thymus, where they learn to distinguish self from non-self. Only the T cells that ignore self-antigen molecules are allowed to mature and leave the thymus. *B cells* are formed in the bone marrow. They have particular receptor sites on their surface where antigens can attach.

Oral allergy syndrome (OAS). Oral allergy syndrome is an allergic (immunologic) reaction to certain proteins in a variety of fruits, vegetables, and nuts, which develops in some people with pollen allergies, particularly birch pollen allergies, but it can also affect people with allergies to the pollens of grass, ragweed (more common in North America), and mugwort (more common in Europe). These reactions can occur at any time of the year but are often worse during the pollen season. Oral and/or pharyngeal pruritus appears within minutes after the intake of the food and may be the first symptom of generalized anaphylactic reactions or the only manifestation (http://www.inspection.gc.ca/english/fssa/concen/tipcon/orale.shtml).

Rhinitis. Allergic symptoms involving the nose (e.g., itching, sneezing) with increased secretion and blockage.

Rhinoconjunctivitis. Allergic conjunctivitis is also called "rhinoconjunctivitis." It is the most common allergic eye disorder. The condition is usually seasonal and is associated with hay fever. The main cause is pollens, although indoor allergens such as dust mites, molds, and dander from household pets such as cats and dogs may affect the eyes year-round. Typical complaints include itching, redness, tearing, burning, watery discharge, and eyelid swelling. To a large degree, the acute (initial) symptoms appear related to histamine release.

Sensitivity and specificity. These terms are used as measures of how good a medical test, sign, or symptom is. The sensitivity of a test refers to how many cases of a disease a particular test can find. On the other hand, specificity of a test refers to how accurately it diagnoses a particular disease without giving false-positive results. See also **Gluten sensitivity**.

Shock. Medically, shock is a critical condition brought on by a sudden drop in blood flow through the body. There is failure of the circulatory system to maintain adequate blood flow, curtailing the delivery of oxygen and nutrients to vital organs. The signs and symptoms of shock include low blood pressure (hypotension), over breathing (hyperventilation), a weak rapid pulse (tachycardia), cold clammy grayish-bluish (cyanotic) skin, decreased urine flow (oliguria), and mental changes (a sense of great anxiety and confusion).

Stridor. A harsh, high-pitched whistling sound, produced in breathing by an obstruction in the bronchi, trachea, or larynx.

Syncope. The temporary loss of consciousness followed by the return to full wakefulness; fainting.

Systemic reactions. Several systems within the body are affected simultaneously, including the upper and lower respiratory tracts, cardiovascular system, and gastrointestinal tract. In the context of allergy, this refers to anaphylaxis.

Threshold. Allergen threshold refers to the levels (exposure amounts) below which it is unlikely that a food-allergic individual would experience an adverse effect. It also applies to the establishment of a limit by statute, below which no regulatory action will be taken.

Toxic peptide. The term "toxic peptide" has been used to describe any gluten sequence able to induce damage of the intestinal mucosa in celiac individuals. However, this term is now more specifically used to refer to those gluten peptides affecting *in vitro* cells and intestinal preparations, producing damage *in vivo*, eliciting the innate response. Whereas the peptide fragments eliciting the mucosal adaptive immune response are termed as "immunostimulatory," "immunogenic," or "immunodominant."

Urticaria. A common allergic skin condition, transient in nature, characterized by erythematous (red) edematous plaques or wheals within the superficial dermis, usually pruritic (itching), burning, or stinging. The lesions typically result from an inflammatory reaction that induces localized transudation of fluid from dilated small blood vessels and capillaries in the superficial dermis. It can be acute (6 weeks' duration or less), whereas urticaria recurring frequently for longer than 6 weeks is referred to as chronic.

REFERENCES

Brown, S.G. (2004). Clinical features and severity grading of anaphylaxis. *The Journal of Allergy and Clinical Immunology*, *114*, 371–376.

Chapman, J., Bernstain, J., Lee, R., Oppenheimer, J. (2006). Food allergy: a practice parameter. *Annals of Allergy, Asthma & Immunology*, *96*, S1–S68.

Eichenfield, L.F. (2004). Consensus guidelines in diagnosis and treatment of atopic dermatitis. *Allergy*, *59*, 86–92.

FDA Threshold Working Group. (2008). Approaches to establish thresholds for major food allergens and for gluten in food. *Journal of Food Protection*, *71*, 1043–1088.

Furuta, G.T., Liacouras, C.A., Collins, M.H., Gupta, S.K., Justinich, C., Putnam, P.E., Bonis, P., Hassall, E., Straumann, A., Rothenberg, M.E. (2007). Eosinophilic esophagitis in children and adults: a systematic review and consensus recommendations for diagnosis and treatment. *Gastroenterology*, *133*, 1342–1363.

Hadjivassiliou, M., Sanders, D.S., Grünewald, R.A., Woodroofe, N., Boscolo, S., Aeschlimann, D. (2010). Gluten sensitivity: from gut to brain. *The Lancet Neurology*, *9*(3), 318–330. Review.

Jesenak, M., Rennerova, Z., Babusikova, E., Havlicekova, Z., Jakusova, L., Villa, M.P., Ronchetti, R., Banovcin, P. (2008). Food allergens and respiratory symptoms. *Journal of Physiology and Pharmacology*, *59*, 311–320.

Johansson, S.G., Bieber, T., Dahl, R., Friedmann, P.S., Lanier, B.Q., Lockey, R.F., Motala, C., Ortega Martell, J.A., Platts-Mills, T.A., Ring, J., Thien, F., Van Cauwenberge, P., Williams, H.C. (2004). Revised nomenclature for allergy for global use: report of the Nomenclature Review Committee of the World Allergy Organization. *The Journal of Allergy and Clinical Immunology*, *113*, 832–836.

Maintz, L., Novak, N. (2007). Histamine and histamine intolerance. *The American Journal of Clinical Nutrition*, *85*, 1185–1196.

MedicineNet.com © Website. (1996–2009). Available at http://www.medicinenet.com/script/main/hp.asp

Merk Manuals: Online Medical Library. (2008). September. Available at http://www.merck.com

Montalto, M., Santoro, L., D'Onofrio, F., Curigliano, V., Gallo, A., Visca, D., Cammarota, G., Gasbarrini, A., Gasbarrini, G. (2008). Adverse reactions to food: allergies and intolerances. *Digestive Diseases*, *2*, 96–103.

Muraro, A., Roberts, G., Clark, A., Eigenmann, P.A., Halken, S., Lack, G., Moneret-Vautrin, A., Niggemann, B., Rance, F. (2007). The management of anaphylaxis in childhood: position paper of the European Academy of Allergology and Clinical Immunology. *Allergy*, *62*, 857–871.

National Institutes of Health. (2005). Consensus development conference statement on celiac disease, June 28-30, 2004. *Gastroenterology*, *128*, S1–S9.

Nowak-Wegrzyn, A., Sampson, H.A. (2006). Adverse reactions to foods. *Medical Clinics of North America*, *90*, 97–127.

Ortolani, C., Pastorello, E.A. (2006). Food allergies and food intolerances. *Best Practice and Research Clinical Gastroenterology*, *20*, 467–483.

Rubio-Tapia, A., Kelly, D.G., Lahr, B.D., Dogan, A., Wu, T.T., Murray, J.A. (2009). Clinical staging and survival in refractory celiac disease: a single center experience. *Gastroenterology*, *136*, 99–107.

Sampson, H.A., Munoz-Furlong, A., Bock, S.A., Schmitt, C., Bass, R., Chowdhury, B.A., Decker, W.W., Furlong, T.J., Galli, S.J., Golden, D.B., Gruchalla, R.S., Harlor, A.D. Jr., Hepner, D.L., Howarth, M., Kaplan, A.P., Levy, J.H., Lewis, L.M., Lieberman, P.L., Metcalfe, D.D., Murphy, R., Pollart, S.M., Pumphrey, R.S., Rosenwasser, L.J., Simons, F.E., Wood, J.P., Camargo, C.A. Jr. (2005). Symposium on the definition and management of anaphylaxis: summary report. *The Journal of Allergy and Clinical Immunology*, *115*, 584–591.

Sapone, A., Lammers, K.M., Mazzarella, G., Mikhailenko, I., Cartenì, M., Casolaro, V., Fasano, A. (2010). Differential mucosal IL-17 expression in two gliadin-induced disorders: gluten sensitivity and the autoimmune enteropathy celiac disease. *International Archives of Allergy and Immunology*, *152*(1), 75–80.

Silano, M., Vincentini, O., De, V.M. (2009). Toxic, immunostimulatory and antagonist gluten peptides in celiac disease. *Current Medicinal Chemistry*, *16*, 1489–1498.

Sollid, L.M. (2000). Molecular basis of celiac disease. *Annual Review of Immunology*, *18*, 53–81.

YourDictionary.com © Website. (2005). *Webster's New World College Dictionary*. Cleveland, OH: Wiley Publishing. Available at http://www.yourdictionary.com

Zarkadas, M., Scott, F., Salminen, J., Ham Pong, A. (1999). Common allergenic foods and their labelling in Canada—a review. *Canadian Journal of Allergy and Immunology*, *4*, 118–141.

PART I

FOOD ALLERGY AND THE CONSUMER

1

IMMUNE-MEDIATED ADVERSE REACTIONS TO DIETARY PROTEINS

Olga M. Pulido

1.1. INTRODUCTION

An adverse reaction to food is a general term applied to the clinically abnormal response to an ingested food, food ingredient, or food additive. Adverse reactions to food may or may not be mediated by the immune system [1–6]. Nonimmune-mediated adverse reactions to food mimicking food allergy are termed food intolerances and can be the result of toxicity, for example, histamine in scromboid fish poisoning or *nonallergic food hypersensitivity* (Fig. 1.1) [4–7]. In turn, *nonallergic food hypersensitivity* (Fig. 1.1) can result from (1) chemical/pharmacological action of food ingredients (e.g., caffeine in coffee, tyramine in aged cheese, sulfites in wine, phenylethylamine in chocolate, or the flavor enhancer monosodium glutamate [8–10]; (2) physiological factors associated with specific characteristics of the host (e.g., lactase deficiency leading to lactose intolerance and deficiency of glucose-6-phosphate dehydrogenase in favism) [5, 11–13]; or (3) others such as psychogenic causes (e.g., eating disorders may present clinical symptoms suggestive of an adverse reaction to food) [6, 14]. Conversely, an adverse reaction to food (Table 1.1) may be mediated by an immunologic response and should be distinguished from food intolerances that do not have an immune basis, but may be similar in clinical presentation [2, 7, 15–17].

Allergen Management in the Food Industry, Edited by Joyce I. Boye and Samuel Benrejeb Godefroy

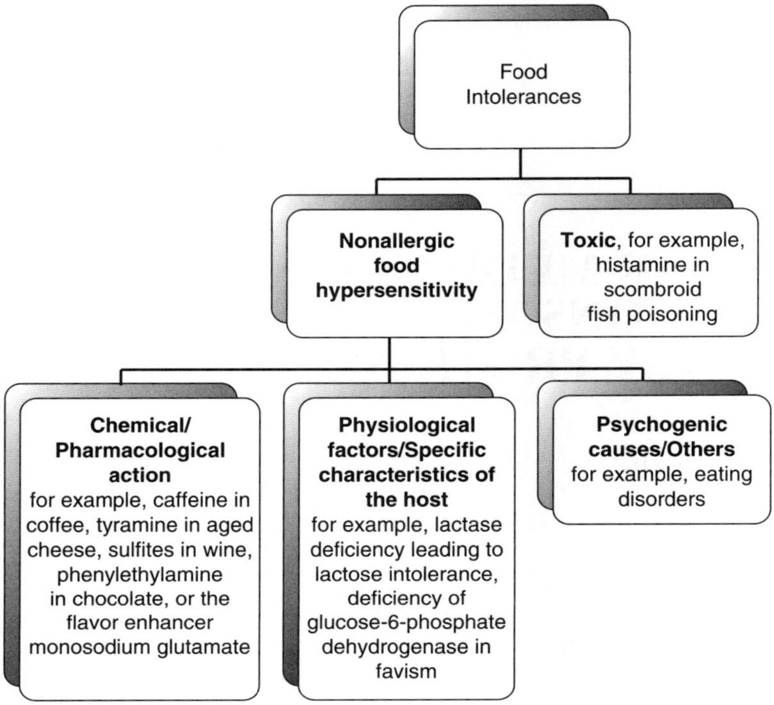

Fig. 1.1 Nonimmune-mediated adverse reaction to food "Food Intolerances." See text and Glossary of Terms for further explanation and references.

Allergy is defined as a *hypersensitivity* reaction to intrinsically harmless antigens, most of which are environmental, and the process is initiated by specific immunologic mechanisms [3]. The term *food allergy* has been recommended when an adverse reaction to food is mediated by immunologic mechanisms [1, 3, 5, 18]. Food allergens are defined as the antigenic molecules giving rise to the immunologic response [3, 19–21]. Proteins are the food constituents responsible for eliciting immune-mediated adverse responses to food [3, 19–21]. Hence, the eliciting dietary proteins are known as *allergens*. The term *IgE-mediated food allergy* is used when immunoglobulin E (IgE) is involved in the reaction [3, 5, 6]. Allergies to food contaminants such as dust mites, mold, and parasites should also be distinguished from food allergy elicited by dietary proteins.

Under the above definitions, all immune-mediated adverse reactions to dietary proteins are considered *food allergy*. Together, *food allergy* encompasses a wide range of clinical disorders, which are grouped in Table 1.1 as IgE, non-IgE, and mixed IgE/non-IgE [2, 5, 6, 22]. These include IgE-mediated food allergy, celiac disease, dermatitis herpetiformis (DH), and clinical conditions such as allergic eosinophilic esophagitis and food protein-induced

TABLE 1.1 Food Allergies: Immune-Mediated Mechanisms, Associated Clinical Presentations, and Most Often Offending Foods

Target Body System	IgE Mediated (Foods: Peanut, Milk, Soy, Egg, Tree Nuts, Wheat, Fish/Shellfish)	Non-IgE and Mixed IgE and Cell Mediated (Foods: Milk, Soy)	IgE and Non-IgE Reactions to Gluten (Foods: Wheat, Rye, Barley, and Related Cereals)
Generalized/systemic reactions	Anaphylaxis Immediate onset Late-onset reactions Food-dependent, exercise-induced anaphylaxis	—	IgE mediated Wheat allergy Wheat-dependent, exercise-induced anaphylaxis
Skin/cutaneous reactions	Urticaria/angioedema/flushing/acute contact urticaria	Atopic dermatitis Contact dermatitis	Non-IgE Dermatitis herpetiformis (skin form of celiac disease [CD])
Digestive system/gastrointestinal (GI)	Oral allergy syndrome, immediate Immediate gastrointestinal reactions Vomiting/diarrhea/abdominal pain	Allergic eosinophilic esophagitis Allergic eosinophilic gastroenteritis Allergic proctocolitis Food protein-induced enterocolitis syndrome Infantile colic	Non-IgE Celiac disease and related conditions Non-IgE adverse reactions to dietary gluten include a wide scope of conditions with and without GI involvement, for which scientific information is fast evolving (see Section 1.4)
Respiratory system	Acute rhinoconjunctivitis/bronchospasm	Asthma	Pulmonary hemosiderosis (Heiner's syndrome)

TABLE 1.2 Food Allergies

- Food allergens are generally glycoproteins with molecular weights ranging from 10 to 70 kDa.
- Innate allergenic capacity of foods may be determined by a combination of factors, for example, solubility, resistance to pH, heat, and proteolysis by digestive enzymes.
- Allergenic capacity of food allergens may be modified (increase or decrease) by food processing and manipulation, for example, heating.
- The individual must first be sensitized by exposure to the allergenic protein.
- The route of initial exposure and sensitization can be oral, respiratory (aeroallergens), or dermal (skin) contact.
- In infants, the route of initial exposure and sensitization can occur in utero or through breast milk.
- Food allergy occurs when a sensitized individual is exposed to the same or a cross-reactive protein through food ingestion.
- Food allergies are often encountered by infants or children during a developmental window of immunologic immaturity.
- IgE-mediated food allergies are characterized by the rapid release of powerful cellular chemicals such as histamine.
- In IgE-mediated food, allergy reactions can occur within minutes or up to 4 hours after ingestion.
- Clinical disorders associated to non-IgE-mediated mechanisms or to mixed IgE and non-IgE, typically have delayed onset of symptoms (>2 hours) and a chronic, relapsing course.
- Severity of reactions and presenting symptoms may vary with time and exposure.

enterocolitis syndrome (FPIES) [14, 23–25]. Many factors are implicated in the basic pathophysiological mechanisms of *food allergy*, such as host genetics, biochemical characteristics of the proteins, exposure, changes induced through food processing, or genetic engineering in "genetically modified foods" (Table 1.2) [20, 21, 26–28].

During the last two decades, there has been an increasing trend in the prevalence of food allergy in Western countries. It is estimated that food allergy affects between 5% and 8% of infants and young children and approximately 2–4% of adults [2, 7, 15, 17, 29]. Today, food allergies, both IgE and non-IgE mediated, are important health concerns from the point of view of risk management, policy setting, public health, diagnosis, and treatment for the consumers, their families, and the communities where they live, and for the food industry at large [13, 18, 30–32].

An understanding of the basic mechanisms underlying adverse reactions to foods and an enhanced awareness of the various clinical presentations is important for the overall management of food allergies. To this extent, this chapter presents an overview of the current understanding of the basic immune mechanisms mediating adverse reactions to food proteins and their various clinical presentations. For further clarification, refer to the Glossary of Terms on pages xix–xxvii.

Discussion of other aspects relevant to *food allergy*, such as allergen thresholds dose, clinical diagnostic tests, and methods used to detect specific allergenic proteins in food, are beyond the scope of this chapter. The chapter is organized in sections based on the implicated immune-mediated mechanism and associated clinical conditions. With the exception of celiac disease, which is discussed separately, a brief description of symptoms and medical conditions associated with food allergies is presented under each category throughout the text or in the Glossary of Terms.

1.2. ORAL IMMUNE TOLERANCE

The gut is responsible for the digestion and absorption of nutrients while acting as the first line of immune defense against pathogenic microbes within the gastrointestinal tract. The gut mucosal immune system accomplishes this task partly by establishing a tolerance to macronutrients [28, 33]. The gastrointestinal tract is the largest immune organ in the body and is constantly exposed to dietary proteins from ingested food. Immune tolerance to dietary proteins is maintained by active suppressive mechanisms involving antigen-specific regulatory T cells. In the first few years of life, humans gradually develop an intricate balance between tolerance and immune reactivity in the gut mucosa, along with a tremendous expansion of gut-associated lymphoid tissue (GALT). GALT is comprised predominantly of clusters of organized lymphoid tissue in the terminal ileum (Peyer's patches), appendix, and isolated lymphoid follicles located beneath the epithelium throughout the gut [34].

Several factors can cause disturbances at different steps in the process of developing oral tolerances, disrupting intestinal barrier function, and contributing to disease pathogenesis [35]. The factors implicated in the development and/or altering the risk of adverse immune reactions to dietary proteins can include genetic susceptibility, gastrointestinal infection, age, exposure (route, dose, and time), timing and length of initial exposure, association with breast-feeding, gastric pH, and type of protein [2, 13, 36–38]. Food allergies may be the result of a breach in oral tolerance to ingested food or from cross-reactivity between food and nonfood allergens. For example, individuals with allergies to fruits and vegetables may have been sensitized by pollen exposure known as pollen-food allergy syndrome or oral allergy syndrome (OAS) [5, 13].

1.3. FOOD ALLERGY

Food allergy is defined as an exaggerated immune response (hypersensitivity) to dietary proteins [1–3]. Allergies to food develop when exposure to a food protein is mistakenly identified as harmful by the human body. Failure of the development of gut tolerance for a specific food protein leads to hypersensitivity to that protein [21, 28, 33]. Food allergies are often seen during the early period of life that coincides with the critical period of development of immune tolerance and typically occurs during this period of immunologic immaturity [2, 15, 17, 28, 39].

Host factors, for example, genetics, age, gut flora, asthma, history of atopy, exercise, and extrinsic factors such as characteristics and dose of the protein (threshold), all influence the potential allergic reaction [2, 5, 33, 36, 40, 41]. The allergenic capacity of the protein may be modified by food processing and manipulation (e.g., heating) [27, 42, 43]. Food proteins that are resistant to digestion are considered to be the most allergenic. The ability of the allergenic protein to trigger direct oral sensitization is modulated by gastric acidity [37].

Although any food protein can potentially provoke an immune reaction, relatively few food proteins are responsible for the vast majority of significant food-induced allergic reactions [21]. The most common food allergens in the pediatric population include cow's milk, eggs, peanuts, tree nuts, soy, wheat, fish, and shellfish, whereas peanuts, tree nuts, fish, and shellfish predominate in adults [2, 5, 13, 14, 25, 44]. Gluten from wheat, barley, and rye are the proteins of most concern for celiac disease, DH, and other gluten-induced conditions (Section 1.4), while rice is emerging as a food of concern for FPIES [45, 46].

The most common food allergies in a given population and the criteria for identification of priority allergenic proteins will vary based on world regions, individual countries, dietary habits, and regulatory systems [30].

1.3.1. IgE-Mediated Food Allergy

IgE-mediated food allergies constitute the majority of food allergic reactions and are the best studied. An IgE-mediated reaction develops when an allergenic protein binds with specific IgE antibodies on mast cells and basophils activating the release of potent compounds such as histamine. The first step in the development of IgE food allergies is sensitization. The first time the susceptible individual is exposed to the specific food allergen, the body's immune system misidentifies the protein as harmful and responds by creating specific antibodies (IgE) to that allergen. Repeat exposures to the same food protein trigger an immune reaction with the release of IgE antibodies [2, 5, 17, 33]. The conjugation of the IgE antibody with the allergens triggers a stimulus to mast cells and basophils, which degranulate, releasing mediators (e.g., histamine) and promoting the synthesis of prostaglandins, leukotrienes, and cytokines [2, 18, 33]. This reaction represents an effort by the immune system to reject/remove the protein, mistakenly identified as harmful, from the body. In turn, the chemicals released have powerful effects on the respiratory system, gastrointestinal tract, skin, and cardiovascular system.

Histamine is a powerful biogenic amine, released during IgE-mediated allergic reactions. It is synthesized by mast cells, basophils, platelets, and other cells such as histaminergic neurons and enterochromaffin cells, where it is stored in the cytoplasm in vesicles and released upon stimulation. Conversely, histamine exerts its effects by binding to a family of receptors on target cells in various tissues mediating numerous biological reactions. These biological reactions include smooth muscle contraction, vasodilation, increased vascular permeability, mucus secretion, tachycardia, alterations of blood pressure, arrhythmias, and stimulation of gastric acid secretion [12]. This mechanism explains the fast onset of symptoms and potential severity of clinical symptoms observed with IgE-mediated food allergies.

IgE-mediated reactions can start within minutes to 1 hour (rarely past 2 hours) of exposure. Reactions often affect the skin (urticaria, angioedema, morbilliform eruptions, flushing, pruritus) [47] and can involve the respiratory tract (sneezing, rhinorrhea, congestion, cough, wheezing, difficulty breathing) [48], the gastrointestinal tract (OAS, nausea, vomiting, diarrhea, cramping, abdominal pain) [17, 49], and the cardiovascular system (tachycardia, hypotension) [50, 51]. Severe systemic reactions can result in anaphylactic shock and death [52].

A late-phase response may follow the immediate reaction beginning 4–6 hours after contact with the allergen and continuing for several days. This response is caused by chemotactic mediators released at the same time as the immediate reaction, which promote selective recruitment of inflammatory cells, mainly eosinophils and neutrophils, which infiltrate the tissue producing an inflammation that can last for a few days [6].

1.3.1.1. Anaphylaxis. Anaphylaxis is a serious generalized allergic reaction that may cause death. Anaphylaxis represents the most severe form of IgE-mediated food allergy and is clinically defined as a food allergic reaction involving two or more organ systems [50–52]. It can include cutaneous (skin), respiratory, cardiovascular, and gastrointestinal symptoms. The onset of symptoms after exposure to food is usually abrupt. In extremely sensitive individuals, reactions may be triggered by minute amounts of food proteins [31]. Symptoms can start within seconds to 2 hours following allergen ingestion and can include feelings of impending doom, throat tightness, coughing or wheezing, abdominal pain, vomiting, diarrhea, and loss of consciousness. Skin symptoms such as flushing, urticaria, and angioedema are present in most anaphylactic reactions. However, the most rapidly progressive anaphylactic reactions may not have cutaneous manifestations. Severe anaphylaxis is characterized by life-threatening upper airway obstruction, bronchospasm, and/or hypotension. In children, bronchospasm is a common symptom, and a background of atopy and asthma is often present [52].

An international task force on anaphylaxis and the European Academy of Allergology and Clinical Immunology recommend the following working clinical definition of anaphylaxis in which the diagnosis is considered highly likely when any one of the following three criteria are met [50, 51]:

1. acute onset of an illness (minutes to several hours) with involvement of the skin, mucosal tissue, or both (e.g., generalized hives, pruritus or flushing, swollen lips, tongue–uvula), with respiratory (e.g., dyspnea,

bronchospasm, stridor, hypoxia) or/and cardiovascular compromise (e.g., hypotension, collapse); or

2. two or more of the following that occur rapidly after exposure to a likely allergen for that patient (minutes to several hours):
   a. involvement of the skin or mucosal tissue (e.g., generalized hives, itch, flushing, and swelling),
   b. respiratory compromise (e.g., dyspnea, bronchospasm, stridor, hypoxia),
   c. cardiovascular compromise (e.g., hypotension, collapse), and
   d. persistent gastrointestinal symptoms (e.g., crampy abdominal pain, vomiting);
3. hypotension after exposure to known allergen for that patient (minutes to several hours).

In the literature, grading systems for acute systemic hypersensitivity reactions vary considerably, have a number of deficiencies, and lack a consistent definition of anaphylaxis. Despite limitations, the following clinical criteria and grading system of anaphylaxis [52] provide general guidance:

I. Severe reactions include symptoms that are strongly associated with hypotension and hypoxia: confusion, collapse, unconsciousness, and incontinence. Preexisting asthma and lung disease are viewed as an increased risk of hypoxia.

II. Moderate reactions include diaphoresis, vomiting, pre-syncope, dyspnea, stridor, wheeze, chest/throat tightness, nausea, vomiting, and abdominal pain.

III. Mild reactions are limited to the skin (urticaria, erythema, and angioedema); however, when angioedema includes the face with involvement of the glottis, this is considered severe, since it is associated with hypoxia.

Anaphylaxis may have a biphasic course of onset in as many as 20–25% of cases, with initial improvement occurring with or without treatment followed by the recurrence of severe symptoms within 1–2 hours. The severity of late symptoms cannot be predicted based on the early symptoms; for instance, early mild symptoms may be followed by anaphylactic shock. Given the potential for late-phase reactions, an observation period of at least 4 hours is recommended following a reaction. Peanut, tree nuts (e.g., almond, cashew, hazelnut, pecan, and walnut), fish, and shellfish are most often responsible for food-induced anaphylaxis [2]. Cross-reactivity among food allergens or with aeroallergens may be the eliciting cause [2]. In rare circumstances, anaphylaxis may have a protracted course of onset, with symptoms lasting for days.

1.3.1.2. Skin (Cutaneous) Manifestations of Food Allergy. The skin is the target organ most often involved in food allergy. Both mixed IgE and non-IgE cell-mediated mechanisms have been implicated in various skin manifestations associated with food allergy [5–7, 47]. Acute generalized urticaria characterized by pruritus and hives, with or without angioedema, is the most common clinical presentation of IgE-mediated allergic reactions to ingested foods in both children and adults. Onset of symptoms may be rapid (e.g., within minutes of ingesting the responsible food). Skin involvement may be isolated or associated with other organ systems in food anaphylaxis. Acute IgE-mediated urticaria can also be induced by skin contact with cow's milk, raw egg white, raw meats, fish, vegetables, and fruits. Urticaria symptoms lasting longer than 6 weeks are rarely caused by food allergy.

Atopic dermatitis is the most common mixed IgE/cell-mediated skin manifestation of food allergy [5–7, 47]. It is characterized by eczema that generally begins in early infancy, often associated with extreme pruritus, and a chronically relapsing course [53, 54]. DH is a non-IgE-immune-mediated condition elicited by gluten in susceptible individuals and is discussed under celiac disease (Section 1.4).

1.3.1.3. Oral Allergy Syndrome (OAS). OAS is a very common manifestation of food allergy, especially in adults that are allergic to tree pollen (pollen-food allergy syndrome) and, to a lesser extent, among those who are allergic to grass, ragweed, and mugwort pollens [2, 5, 13, 17]. It is seen in response to contact to raw fruits and vegetables and is usually confined to the oral cavity. It affects approximately 50% of pollen-allergic adults and represents the most common adult food allergy. OAS can occur as a result of cross-reactivity between the allergenic proteins in pollens and plant foods such as birch (apple, cherry, peach, carrot), grass (tomato, kiwi), ragweed (melon, banana, tomato), and mugwort (carrot, celery) [2, 5, 13, 17]. Upon contact with an allergenic food protein, the susceptible individual develops a reaction characterized by oral itching, lip swelling, and oral angioedema. Symptoms can also involve other organs and become more severe. There are four levels of increasing severity: (1) oral mucosal symptoms only, (2) oral mucosal plus gastrointestinal symptoms, (3) oral mucosal plus systemic symptoms (urticaria, rhinoconjunctivitis, or asthma), and (4) oral mucosal symptoms plus life-threatening problems (glottis edema, anaphylactic shock) [6, 52].

1.3.1.4. Respiratory Manifestation of Food Allergy. Allergic rhinoconjunctivitis, bronchospasm, laryngeal edema, and asthma may follow the ingestion of food allergens in allergic individuals [2, 6, 48, 52]. It is rare that patients present with isolated respiratory symptoms. They usually present in association with clinical symptoms involving the skin or the gastrointestinal tract or in the context of food anaphylaxis. Food-induced upper respiratory tract symptoms seem to be more common in infants and young children. Allergic rhinoconjunctivitis is characterized by periocular pruritus, tearing and conjunctival erythema, nasal congestion, rhinorrhea, and sneezing shortly after the ingestion of the allergenic food [48]. Chronic serous otitis media may develop secondary to chronic rhinitis and eustachian tube dysfunction, or the middle ear itself can be the primary involved organ [48]. Food-induced asthma is more common in young children, particularly in association with atopic eczema. Acute bronchospasm is a feature of severe food-induced anaphylaxis. Food allergy is considered a risk factor for severe asthma.

The Heiner syndrome is a chronic pulmonary disease caused by non-IgE food allergy, particularly to cow's milk proteins during infancy. It is characterized by recurrent pneumonia, pulmonary infiltrates, iron deficiency anemia, and a failure to thrive in small children [48]. If the associated pulmonary vasculitis is severe, alveolar bleeding occurs and causes pulmonary hemosiderosis (iron deposits in the form of hemosiderin).

1.3.1.5. Adverse Reactions to Dietary Gluten. Cereals including wheat, barley, and rye are consumed in large quantities all over the world. Worldwide, cereal grains account for about 70% of protein consumption. The cereals form part of the Gramineae (grasses) family and are divided into four subfamilies: the Bambusoideae (rice), the Chloridoideae (including ragi and teff), the Panicoideae (most millets, maize, and sorghum), and the Pooideae, which are further divided into the Triticeae (wheat, barley, and rye) and the Aveneae (oats) [41].

The wheat grain comprises three major components: starch, protein, and fiber (cell wall polysaccharides), with proteins accounting for about 10–15% of the dry weight [41]. Gluten is a generic term used for the storage proteins from wheat, barley, and rye and is discussed in more detail under celiac disease (Section 1.4).

Dietary intake of gluten can cause distinct immunologically mediated adverse reactions manifesting with gastrointestinal symptoms. These include celiac disease, other non-IgE-mediated gluten induced clinical conditions, and IgE-mediated food allergy. The pathogenic mechanisms underlying these diseases are different. The coexistence of gluten-induced IgE and non-IgE-mediated reactions in one individual seems to be rare. Diagnosis and management are also different. Hence, establishing a differential diagnosis between cereal (e.g., wheat) induced IgE allergy, celiac disease, and other related non-IgE reactions (Section 1.4) is important for the management of these conditions [44, 55–57].

1.3.1.5.1. IgE-Mediated Wheat Allergy. Wheat is the cereal most often implicated in IgE cereal-induced food allergy. Dietary wheat allergy is observed in adults and children, and like other IgE-mediated food allergies, there is a risk of anaphylaxis [44, 55–57]. The best known IgE allergic response to wheat ingestion is wheat-dependent, exercise-induced anaphylaxis (WDEIA). WDEIA is the most common type of food-dependent, exercise-induced anaphylaxis (FDEIA) (Section 1.3.1.6). This syndrome is associated with one major type of wheat gluten protein known as ω-gliadins. Other IgE-mediated allergic responses to dietary wheat include atopic dermatitis, urticaria, and anaphylaxis. These reactions may vary between populations and may be related to a wider range of wheat proteins and to nonspecific lipid transfer proteins. The other known type of allergy to wheat is baker's asthma, which results from the inhalation of flour and dust during grain processing [41].

1.3.1.6. FDEIA. FDEIA is a rare, potentially life-threatening condition reported in young, athletic individuals, especially women in late teens to mid-30s [40]. FDEIA can occur in two ways: anaphylaxis may occur when exercise follows the ingestion of a particular food to which IgE sensitivity can be identified (e.g., wheat, shellfish, fish, and celery) or, less commonly, 2–4 hours after the ingestion of these foods (postprandial anaphylaxis) in association with physical exertion. When food intake and exercise are independent of each other, there are no allergic symptoms. Although the pathogenesis of FDEIA is not yet known, multiple factors may be involved in eliciting or mediating these adverse reactions [58, 59]. For example, affected patients frequently identify hot and humid weather as an aggravating factor, and afflicted patients generally have asthma and/or other atopic disorders. Wheat gluten is the most common dietary protein associated with FDEIA.

1.3.2. Non-IgE and Mixed Food Allergy

In non-IgE-mediated food allergy, multiple inflammatory cells and their mediators play a role in the immunopathogenesis [2, 5, 6, 22]. These include activation of lymphocytes and recruitment of eosinophils and mast cells. Other immune-mediated factors such as immune complexes formed by food and food antibodies or cell-mediated immunity have been suggested as the mediating mechanism. In non-IgE-mediated disorders (Table 1.1), clinical manifestations of adverse reactions usually become evident hours to days after exposure to the dietary protein (*allergen*). Symptoms and signs may include diarrhea, vomiting, protein-losing enteropathy, rectal bleeding, and enterocolitis [2, 5, 6, 25]. Growth retardation may also be seen is some children. Milk and soy are the most common eliciting foods [21, 39, 45].

Non-IgE-mediated gastrointestinal allergic conditions include food protein-induced enterocolitis, allergic proctocolitis, and enteropathy (Table 1.1) [2, 5, 6, 25]. Celiac disease and DH are also considered non-IgE-mediated adverse reactions and are discussed separately in Section 1.4. Clinical disorders associated with non-IgE cell-mediated mechanisms, or with mixed IgE and non-IgE reactions, typically have delayed onset of symptoms (>2 hours) and a chronic, relapsing course. Therefore, the allergen cause–effect relationship may be difficult to establish.

In conditions such as allergic eosinophilic gastroenteropathy (allergic eosinophilic esophago-gastroenteritis, allergic eosinophilic esophagitis, allergic eosinophilic enterocolitis, dietary protein enterocolitis), IgE-mediated food allergy often cannot be demonstrated. The presence of eosinophils alone is not conclusive evidence of food allergy. However, food has been incriminated as the cause in a subset of patients [2, 25, 60].

1.3.2.1. Eosinophilic Esophagitis. The American Gastroenterological Association Institute and North American Society of Pediatric Gastroenterology, Hepatology, and Nutrition sponsored a systematic review that provides consensus recommendations for diagnosis and treatment of eosinophilic esophagitis in children and adults [60]. These authors define eosinophilic esophagitis as a primary clinicopathologic disorder of the esophagus, characterized by symptoms such as food impaction and dysphagia in adults, and feeding intolerance and gastroesophageal reflux disease symptoms in children. Children can also present with epigastric abdominal pain, dysphagia, and failure to thrive. The esophageal biopsy is characterized by high eosinophil count (≥15 eosinophils/high power field). Other disorders associated with similar clinical, histological, or endoscopic features, need to be excluded. The differential diagnosis includes conditions such as gastroesophagel reflux disease, Crohn's disease, hypereosinophilic syndrome, and drug hypersensitivity response. Appropriate treatments include dietary approaches based on eliminating exposure to food allergens [60].

Most studies characterizing the allergic phenotype of this condition have been performed in children [60]. The allergic etiology of eosinophilic esophagitis is based on several lines of evidence. The majority of patients with eosinophilic esophagitis (50–80%) are atopic. Usually, there is coexistence of atopic dermatitis, allergic rhinitis, and/or asthma and the presence of allergic antigen sensitization. Importantly, most patients improve on allergen-free diets, providing supportive evidence that antigenic dietary protein is eliciting the disease.

Evidence suggests that eosinophilic esophagitis is associated with T helper cell (Th2)-type immune responses. Elevated levels of Th2 cytokines (e.g., interleukin IL-4, IL-5, and IL-13) as well as mast cells are present in the esophagus of these patients. This view is further supported by experimental systems that demonstrate an intimate connection between the development of eosinophilic inflammation in the respiratory tract and esophagus not only in response to external allergic triggers but also to intrinsic Th2 cytokines [60].

1.3.2.2. FPIES. FPIES is an uncommon, pediatric, non-IgE-mediated disorder [45, 46]. The adverse reaction is triggered by dietary proteins with rice, soy, and cow's milk being the most common eliciting foods [45, 46, 49].

The pathophysiology of FPIES remains poorly understood. However, the most likely implicated mechanism is stimulation of T cells by food proteins in the gastrointestinal mucosa. The clinical presentation includes profuse vomiting and/or diarrhea about 2 hours after ingestion of the eliciting protein. Associated features may include pallor, lethargy, cyanosis, metabolic acidosis, and neutrophilia. The cutaneous or respiratory symptoms seen in IgE food allergies are often absent. Most children recover within a few hours, but there is up to 20% of them that may present with a hypovolemic shock. The diagnosis and link to food adverse reaction is often missed, due to the delay in the presentation of symptoms after food intake. The moribund appearance that many children have at presentation is often attributable to sepsis, a metabolic disorder, or a surgical abdominal emergency. Most children develop tolerance to the triggering food by 3 years of age. Although the most common presentation is acute, some children may present a chronic form of the condition characterized by chronic vomiting, diarrhea, and failure to thrive when continuously exposed to the offending food [25, 45, 49].

1.4. CELIAC DISEASE (CD) AND RELATED CONDITIONS

Celiac disease is a complex, systemic, autoimmune-mediated disorder, observed in genetically susceptible individuals in response to exposure to dietary gluten. It has been regarded primarily as a disease of the gastrointestinal tract and is characterized by chronic inflammation of the small intestinal mucosa [61–66]. This inflammation may result in atrophy of intestinal villi, malabsorption, and a variety of clinical manifestations [62, 63, 67–71]. Celiac disease has also been referred to as celiac sprue and gluten-sensitive enteropathy. More recently, the term "gluten syndrome" has been suggested to cover the myriad of extra-intestinal symptoms and clinical conditions described in association with celiac disease and the absence of gastrointestinal involvement in some cases [72]. These include neurological dysfunctions such as gluten ataxia and gluten neuropathy [97, 135].

The worldwide prevalence of celiac disease has been estimated to be between 1 in 100–200 individuals [63, 64, 73, 74]. Certain groups of people have markedly elevated risks of developing celiac disease. First-degree relatives of individuals diagnosed with celiac disease have a 10–20% increased risk of developing celiac disease [29, 75, 76]. A high prevalence of celiac disease is also found in individuals with Down syndrome, diabetes type 1 and IgA deficiency [65, 77, 78].

Celiac disease can be present in both silent and symptomatic forms affecting survival and risk of complications [79]. Silent celiac disease is characterized by positive serology and limited involvement of the gastrointestinal tract. Latent celiac disease includes individuals who are positive for serological markers or genetic susceptibility to disease but no pathology on biopsy. These individuals are asymptomatic but may later develop symptoms and/or histological changes [64, 79]. The difference between the number of clinically diagnosed celiac disease individuals with the vast amount of undetected cases has been described as the celiac disease iceberg, with those undetected lying beneath the surface. Gluten toxicity encompasses a wide spectrum of end target organ pathology, clinical disorders, and mechanisms involved [80, 81, 135, 136].

The clinical manifestations of celiac disease are highly variable in both character and severity. They are influenced by factors such as age, immunologic status, exposure to gluten (amount, duration, or timing of introduction to gluten), and the extent and severity of damage caused to the gastrointestinal tract [23, 61, 69, 70, 82–86] and other organs, for example, the nervous system [97, 135, 136]. The wide spectrum of clinical presentation results in frequent delays in diagnosis, and/or misdiagnoses [71, 85, 87]. Common examples of misdiagnoses include irritable bowel syndrome, chronic fatigue syndrome, and fibromyalgia [85].

Celiac disease can present with gastrointestinal, or "classic," and nongastrointestinal manifestations. Infants and children are more frequently inflicted with gastrointestinal manifestations including diarrhea, abdominal distension, or symptoms of malnutrition such as anemia [63–66, 69, 70, 88, 89]. For adults, nonspecific gastrointestinal complaints are common and include abdominal pain, flatulence, diarrhea, and, in severe cases, steatorrhea [63–65, 69, 71, 85, 88, 90].

Celiac disease is associated with various extraintestinal disorders and complications in addition to gastrointestinal symptoms and is therefore considered a multisystem disorder [68, 82, 84, 91]. Patients with celiac disease also have an increased risk of developing other autoimmune diseases, such as type 1 diabetes mellitus [78, 84, 86].

Nongastrointestinal manifestations are more insidious, are highly variable, and are the common presenting symptoms in older children and adults. These manifestations are either the result of long-term nutrient malabsorption and/or are part of the autoimmune systemic spectrum [65, 69, 81, 85, 86, 92, 93]. In children, nongastrointestinal manifestations may include short stature, enamel defects, neurological developmental delays, and delayed puberty [63, 66, 70, 94–96]. Many celiac patients experience neurological symptoms, frequently associated with malfunction of the autonomic nervous system, cerebella ataxia, learning disorders, depression, migraine, and headache [97, 135]. The absence of an enteropathy should not preclude patients from treatment with a gluten-free diet [97, 135].

In addition to neurological symptoms, there are many long-term consequences and complications for individuals undiagnosed, untreated, or under-treated [83, 93, 98–100].

Celiac disease is a lifelong condition. If celiac disease is not diagnosed early and treated with a strict gluten-free diet, it can be associated with serious complications, including osteoporosis, increased risk of fractures, recurrent miscarriage and infertility in both sexes, malignancy such as small bowel lymphoma, and higher mortality rate [79, 85, 92, 101–103]. Of special concern is the association of long-term untreated celiac disease with malignancy. These malignancies include small bowel lymphoma and both Hodgkin's and non-Hodgkin's lymphoma. Refractory celiac disease, which occurs when both symptoms and intestinal damage persist or recur despite strict adherence to a gluten-free diet, is associated with increased risk of lymphoma and high mortality [103].

1.4.1. Dermatitis Herpetiformis (DH)

DH is a condition of the skin that is also triggered by the ingestion of gluten in genetically susceptible individuals and is considered the dermatological form of celiac disease [23, 24, 88, 104]. DH is a chronic papulovesicular skin disorder in which lesions are symmetrically distributed over the extensor surfaces of the elbows, knees, and buttocks [23, 24, 88, 104]. The disorder is associated with a specific non-IgE-mediated immune sensitivity to gluten. A majority of DH patients have IgA specific for epidermal transglutaminase (TGe) and closely related tissue transglutaminase (tTG), and both TGe and tTG are considered to be autoantigens [105]. The concentration of these antibodies in these patients is reported to be independent of the degree of villous atrophy [105]. The test for establishing the diagnosis of DH is a biopsy from uninvolved skin for the detection of IgA [106]. Classically, in DH, there is granular IgA deposition along the dermo-epidermal junction with concentration at the papillary tips.

Patients with DH often do not have associated gastrointestinal symptoms. The extent of involvement of the small bowel varies, and 20% show apparently normal mucosa, but inflammatory changes consistent with celiac disease are present in most cases [107, 108]. The treatment of this condition includes a gluten-free diet, which helps to recover the injured small bowel and controls the rash even in those who do not have an abnormal small bowel biopsy [107, 108]. Other skin disorders such as psoriasis or vitiligo can also be seen in celiac disease [23, 109].

1.4.2. Genetic Factors in Celiac Disease

Although the etiology of celiac disease is not fully understood, it is considered to be an autoimmune disease with tTG suggested as the major autoantigen. The current consensus is that celiac disease is associated with human leukocyte antigen (HLA) DQ2 and DQ8 haplotypes [63, 69, 110, 111]. Virtually all celiac individuals express HLA-DQ2 or HLA-DQ8. These two class II molecules are chiefly responsible for the presentation of gluten peptides to the gluten-specific T cells that are found only in the gut of celiac patients. These predisposing HLA-DQ molecules bind enzymatically modified gluten peptides, and these HLA-DQ peptide complexes trigger inflammatory T-cell responses in the small intestine. In addition, gluten induces innate immune responses that contribute to the tissue damage that is characteristic of this condition (Figs. 1.2 and 1.3) [69, 112–115]. Thus, a combination of adaptive and innate immune responses triggered by gluten has been implicated as the cause of the clinical presentation of the disease (Fig. 1.4) [69, 111, 113]. Continued gluten exposure makes the adverse immune reactions self-perpetuating and increases the risk of serious complications [69, 111, 113].

However, many people with similar risk factors do not develop celiac disease. This suggests a multifactorial etiology [63, 64, 69]. Other genetic and environmental factors have been implicated in playing a role in the manifestation of this disease, such as gastrointestinal infections and stress [116, 117]. Regardless of the possible confounding etiological factors, it is agreed in the literature that early diagnosis and dietary treatment can prevent severe, sometimes life-threatening, complications.

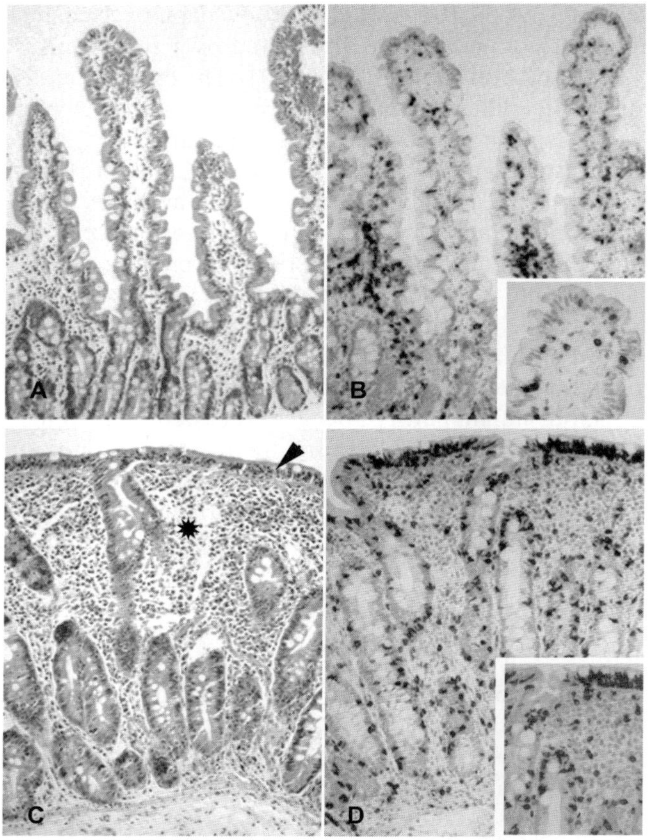

Fig. 1.2 Microscopic view of small intestinal biopsy stained with haematoxylin and eosin (A,C) and with CD3 immunohistochemistry (B,D). From Dr. Mohsin Rashid. A and B show the normal mucosa of the small bowel with well-maintained architecture, easily seen villi, and sparse CD3+ intraepithelial lymphocytes (IELs). Inset shows the tip of the villi. C and D are sections from the small bowel of an untreated celiac individual showing injured mucosa, characterized by villous atrophy (arrow), absence of villi, crypt hyperplasia, dense inflammatory infiltrate (asterisk), and marked increase of CD3+ IELs best seen at the upper portion of the mucosa D (insert).

1.4.3. Gluten and the Pathogenesis of Celiac Disease

1.4.3.1. Gluten Proteins. The major endosperm storage proteins of most cereal grains are prolamins [118]. Approximately 80% of the total grain protein is accounted for by this major storage protein fraction [41]. Early classification based on extraction in a series of solvents, defined four protein fractions, which are extracted sequentially in water (albumins), dilute saline (globulins), alcohol/water mixture (prolamins), and dilute acid (glutenins). Prolamins of

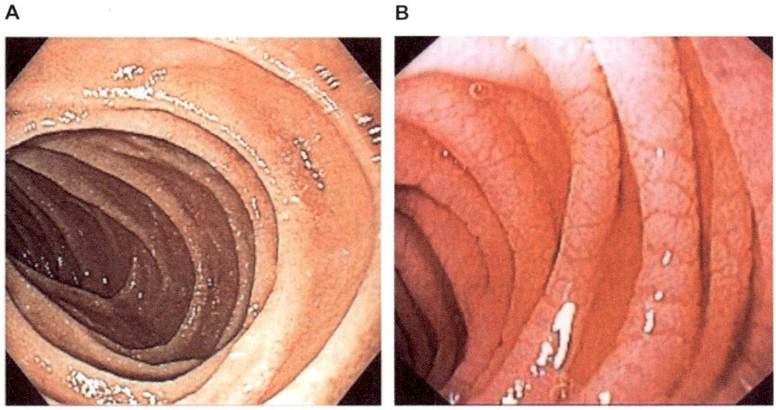

Fig. 1.3 Endoscopic view of the duodenum. (A) Normal (right). (B) Untreated celiac patient showing scalloping of the mucosal folds (left). From Dr. Mohsin Rashid.

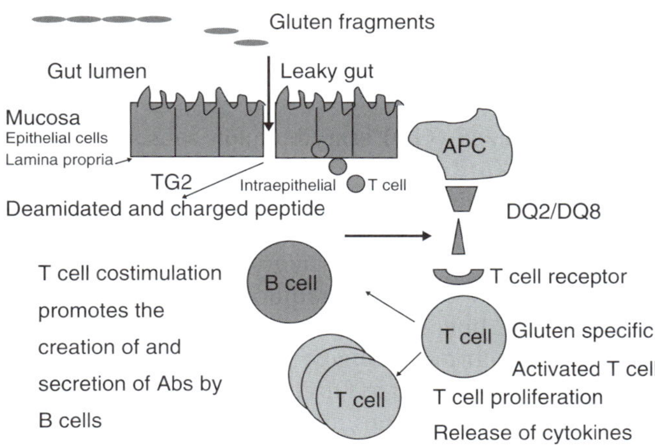

Fig. 1.4 Schematic representation of the immunopathology of celiac disease (CD). CD involves a complex interplay of many factors, including environmental, genetic, and immunologic. Under certain conditions, incompletely digested peptides of gluten from wheat, barley, and rye can cross the epithelium in the mucosa of the small intestine. Factors such as gastrointestinal infections may affect the permeability of the mucosa (leaky gut). After absorption, glutamine residues from gluten peptides are converted to negatively charged glutamic acids through deamidation by tissue transglutaminase (tTG2). Antigen-presenting cells (APCs) expressing the human leukocyte antigens HLA-DQ2 and HLA-DQ8 have an increased affinity for these deamidated peptides, resulting in peptide complexes that can activate a range of inappropriate immunogenic responses including reactivity against host tissues. Both innate and adaptive immune responses are involved, including antibodies (Abs) to gluten and to tissue proteins, for example, IgA anti-transglutaminase, T cell reactivity to gluten, increased number of intraepithelial T cells, and increased cytokines that can in turn promote inflammation and villous damage in the small intestine. tTG is pivotal in the pathogenesis of CD and is the main autoantigen.

the Triticeae (wheat, barley, and rye) comprise three broad groups: (1) sulfur-rich (S-rich): α/β-gliadins, γ-gliadins, B- and C-type low-molecular-weight (LMW) subunits of glutenin; (2) sulfur-poor (S-poor): ω-gliadins and D-type LMW subunits of glutenin; and (3) high-molecular-weight (HMW) subunits [41, 118].

Wheat proteins are cohesive with each other in wheat dough, giving elasticity to the dough called "gluten" [41, 118]. It is the gluten in wheat flour that binds and gives structure to bread, baked goods, and other foods, making it widely used in the production of many processed and packaged foods. The alcohol-soluble fractions from barley (hordein) and rye (secalin) have similar amino acid sequences to wheat (gliadin), but they lack the cohesive ability of wheat gluten [119]. Nonetheless, the term "gluten" encompasses the homologous prolamins of wheat, rye, barley, and related cereals [119]. Hence, gluten is a generic name given to storage proteins in wheat, barley, rye, and other closely related cereal grains.

The alcohol-soluble fractions (prolamins) of wheat (gliadin), rye (secalin), and barley (hordein) are considered to be the protein constituents of most concern to celiac individuals. The wheat gliadins are monomeric proteins and are classified on the basis of their electrophoretic mobility at low pH: these are α/β-gliadins (fast), γ-gliadins (intermediate), and ω-gliadins (slow). The glutenins are polymers of individual proteins linked by interchain disulfide bonds and include HMW and LMW groups after separation by gel electrophoresis [41]. The presence of amino acid sequences consisting of repeated blocks of peptide motifs is responsible for the high proportions of glutamine (Q), proline (P), and other specific amino acids in some prolamin groups. These proteins consist almost entirely of repetitive sequences rich in glutamine (Q) and proline (P) (e.g., PQQPFPQQ) [120]. The repetitive units of α/β-gliadins are dodecapeptides such as QPQPFPQQPYP, which are usually repeated five times and modified by the substitution of a single residue [120]. The typical unit of γ-gliadins is QPQQPFP, which is repeated 16 times and interspaced by additional residues. The total wheat gluten protein is characterized by ≈28–33% of α/β-gliadins units and ≈23–31% of γ-gliadins [120]. The distribution of total gliadins among different types is strongly dependent on wheat variety (genotype) and growing conditions (soil, climate, fertilization) [120].

The glutenin fraction comprises proteins linked by disulfide bonds and their molecular weight distribution has been recognized as one of the main determinants of dough properties and baking performance [120]. After the reduction of disulfide bonds, the resulting glutenin subunits show solubility in aqueous alcohols similar to gliadins. The predominant protein type are LMW glutenin subunits (LMW-GSs) representing ≈20% of the total gluten protein, whereas the HMW (HMW-GS) represents ≈10% of the gluten proteins [120]. The LMW-GSs are related to α/β- and γ-gliadins in molecular weight and amino acid composition. They consist of an N-terminal domain rich in repetitive glutamine (Q) and proline (P) units (e.g., QQQPPFS), and a C-terminal domain homologous to that of α/β- and γ-gliadins. Both wheat prolamins (gliadins) and glutenins subunits (LMW-GS and HMW-GS) are characterized by high glutamine and proline content, all having protein fragments of concern to individuals with celiac disease [119–122].

1.4.3.2. Gluten and Celiac Disease. Gluten is partially digested by humans. A 33-amino acid peptide molecule (33mer) and other immunogenic peptides remain after the action of gastric, duodenal, and pancreatic enzymes. It is this fragment (33mer) α_2-gliadin 57–89 (LQLQPFPQPQLPYPQPQLPYPQPQLPYPQPQPF) that is considered to elicit the immune response in genetically susceptible individuals [123] and can bind to the DQ molecule [124]. The 33mer contains partly overlapping copies of specific T-cell epitopes [125], which are PFPQPQLPY, PQPQLPYPQ (three copies), and PYPQPQLPY (two copies) [123]. HLA-DQ2 binds to the ligands via anchor residues that are at positions 1, 4, 6, and 9 [119]. Molecules acting as DQ2 ligand have selectively large hydrophobic residues both at the terminal positions 1 and 9. Glutamate residues (E) formed by tTG-mediated deamidation in positions P4 and P6, and rarely in P7, are required for T-cell recognition [119, 124]. The T-cell stimulatory gluten sequence is a 9-amino acid peptide, and the ideal sequence is $(PQ)_1$, X_2, X_3, Q_4, X_5, P_6, Q_7, P_8, X_9. The presence of at least two P in epitope sequence is necessary for the peptide to survive the gastrointestinal digestion [119]. Gluten sequence with 10 amino acid peptides and a P in position 1 can also bind the DQ2 molecule. The binding motif of DQ8 also displays a preference for binding negatively charged residues at several positions, for example, P1, P4, and P9 [126]. Hence, both DQ2 and DQ8 molecules share a preference for negatively charged residues at some of the anchor positions.

It is unclear how the toxic and immunogenic peptides enter the intestinal mucosa [119]. Many factors and mechanisms have been investigated [69, 91, 110, 111, 117, 126]. For individuals with celiac disease, undigested gluten protein fragments trigger an immune-mediated adverse response resulting in an inflammatory injury in the mucosal surface (site of absorption) of the small intestine [69, 112–114, 119]. This results in malabsorption of protein, fat, carbohydrate, fat-soluble vitamins, folate, and minerals, especially iron and calcium [69, 83, 111, 127].

Although the precise molecular mechanisms of the toxic reaction to gluten are not fully elucidated, the primary event requires that the gliadin oligopeptides gain access to their antigen-binding sites, presumably within the lamina propria region interior to the relatively impermeable surface of the intestinal epithelial layer (Fig. 1.4) [91, 126, 128]. Ordinarily, such oligopeptides, generated through the action of pancreatic proteases, are efficiently hydrolyzed into amino acids, di-, or tripeptides by peptidases located in the brush-border membrane of the intestinal enterocyte before they can be transported across the epithelial layer. Homologous gluten proteins from wheat, rye, and barley are rich in proline (P) and glutamine (Q). The large amount of P residues in the sequence of these proteins, blocks the cleavage of the polypeptide by gastrointestinal enzymes at positions immediately next to P [119, 126]. The size of the digested peptides and the position of the Q residues in the primary structure of the peptides, both play a pivotal role in their capacity to elicit an immune-mediated response in individuals with celiac disease [119, 126].

Under certain conditions, the partially digested peptides of gluten from wheat, or their homologous epitopes from barley and rye, can cross the epithelium in the mucosa of the small intestine [91, 113, 126]. Factors such as gastrointestinal infections may affect the permeability of the mucosa (leaky gut) [116, 117]. After absorption, glutamine (Q) residues from gluten peptides are converted to negatively charged glutamic acids (E) through deamidation by transglutaminase-TG2, which is mainly localized in the lamina propria [126, 128]. The affinity of gliadin peptides for tTG depends on the relative position of Q residues and on the amino acid located nearby the target amino acid, particularly P [119, 124, 126, 129]. Once the gluten-derived peptide enters into the immunologic compartment of the intestinal mucosa and gets deamidated by tTG, it binds to the DQ2/8 molecule on the antigen-presenting cell (APC) membranes (Fig. 1.4) [126]. These APCs expressing the HLAs HLA-DQ2 and HLA-DQ8 have an increased affinity for these deamidated peptides, resulting in peptide complexes that can activate a range of inappropriate immunogenic responses including reactivity against host tissues [68, 91, 110, 111, 126]. Both innate and adaptive immune responses are involved, including antibodies (Abs) to gluten and to tissue proteins, for example, IgA anti-transglutaminase, T-cell reactivity to gluten, increased number of intraepithelial T cells, and increased cytokines that can in turn promote inflammation and villous damage in the small intestine (Figs. 1.2–1.4) [111–113, 126].

1.4.3.3. Grains of Concern and Gluten Threshold. Presently, the only treatment of celiac disease is a strict lifelong exclusion of wheat, rye, barley, and other related cereal grains from the diet [87, 93, 130]. The amount of gluten that can be tolerated varies among people with celiac disease. Some individuals tolerate an average of 34–36 mg of gluten per day without any clinical manifestations of celiac disease, while others who consume approximately 10 mg of gluten per day developed mucosal abnormalities [131, 132]. Although there is no evidence to suggest a single definitive threshold for a tolerable gluten intake, there is evidence that a daily gluten intake of <10 mg is unlikely to cause significant histological abnormalities in the small bowel mucosa [131, 132]. Little is known about thresholds for those with other gluten-induced conditions [135, 136].

The taxonomy and biochemistry of the cereal is relevant to its potential toxicity [119, 121, 126, 127]. Cereal grains that are known to trigger celiac disease/DH reactions include the following: wheat (including durum wheat or "durum," spelt wheat or "spelt," kamut), barley, rye, triticale, bulgur, einkorn, emmer, and farro [85, 87, 130].

The introduction of oats in the diet of celiac individuals has been a source of controversy. However, we have recently conducted a systematic review on the safety of oats in celiac disease and concluded that most celiac individuals can tolerate moderate amounts of pure oats, not contaminated with gluten from wheat, barley, and rye [133]. Wheat, rye, and barley have a common ancestral origin in the grass family, whereas oats are more distantly related to wheat, rye, and barley. The oat prolamins (avenin) have substantially lower proline content. Avenin accounts for 5–15% of the total protein in oats, whereas in wheat, barley, and rye, prolamins constitute 40–50% of the total protein [120, 121].

1.5. CONCLUSIONS

The significance of food allergy to public health has being recognized by the World Health Organization (WHO) and other food safety authorities. An expert group appointed by the Food Allergy Task Force of the International Life Sciences Institute ILSI Europe proposed a revised set of criteria together with a framework to help to decide which allergenic foods are of sufficient public health importance to be included in allergen lists [30]. The criteria include the demonstration of an IgE-mediated adverse reaction, estimates of the prevalence, severity of reactions, allergenic potency of the food and the extent, pattern, and nature of exposure to the food. In their proposed framework, the first stage is to assess the available evidence to decide whether or not the allergen in question induces an IgE-mediated food allergy. The public health importance given to IgE-mediated food allergy is due to the high prevalence and potential acute severity of the condition. However, this set of proposed criteria (diagnosis, potency of allergenic protein, severity of reactions, prevalence, exposure, and modulating factors such as food processing) could also be applied when assessing the risk of dietary proteins triggering non-IgE and mixed adverse reactions. Of particular relevance are gluten proteins from wheat, barley, rye, and closely related cereals, which can elicit both an IgE-mediated (e.g., wheat allergy) and a non-IgE-mediated immune response (celiac disease, DH), depending on the genetic susceptibility of the population. Gluten proteins from wheat, barley, rye, and related cereals are also included in allergen priority lists by regulatory authorities, including Canada (http://www.hc-sc.gc.ca/fn-an/label-etiquet/allergen/index-eng.php).

Food allergies, both IgE- and non-IgE-mediated, are important health concerns to consumers who are predisposed to these illnesses. Food allergies are known to impose a significant impact on the daily life of those affected, their families, and communities [87, 134]. The present-day diagnosis of various disorders elicited by dietary proteins can be impeded by intrinsic limitations in generating accurate information from patient history, diagnostic criteria, and methods. Distinguishing among the various conditions elicited by adverse reaction to food proteins is important in the management of these disorders [2, 6, 11]. Food allergy encompasses a variety of clinical conditions with diverse underlying mechanisms presenting in many instances with common symptoms, for example, gastrointestinal symptoms. Increased awareness of the clinical presentations, and the foods eliciting these adverse reactions, is the first step in the identification of cause–effect and in the management of these conditions.

Employing food process technologies to eliminate food constituents, which can be harmful to susceptible individuals, is a potentially viable approach for reducing risk to food-related disorders. The development of practice guidelines and standardization of diagnostic tests, methods for food testing, approaches to establishing thresholds, food labeling policies, and regulations are all positive steps toward the diagnosis, prevention, and management of adverse food reactions in hypersensitive individuals [2, 13, 18]. A continued joint effort is needed from the public, the food industry, governments, healthcare providers, researchers, and others to support the needs of these consumers by minimizing risk while maximizing the availability of healthy foods.

ACKNOWLEDGMENTS

The author gives special thanks to Dr. Mohsin Rashid for contributing the endoscopy and histology photographs illustrating the intestinal damage observed in celiac disease. The author acknowledges the contribution of Ms. Zoe Gillespie in the preparation of the glossary of terms and reading of the manuscript. Thanks to Mr. Frank Geraghty, Mr. Michael Abbott, and Ms. Marion Zarkadas for reading of the manuscript. Special thanks to Dr. Costa Zarkadas for his expert opinion on protein biochemistry.

REFERENCES

1. Bruijnzeel-Koomen, C., Ortolani, C., Aas, K., Bindslev-Jensen, C., Bjorksten, B., Moneret-Vautrin, D., Wuthrich, B. (1995). Adverse reactions to food. *Allergy*, *51*, 623–635.
2. Chapman, J., Bernstain, J., Lee, R., Oppenheimer, J. (2006). Food allergy: a practice parameter. *Annals of Allergy, Asthma & Immunology*, *96*, S1–S68.
3. Johansson, S.G., Bieber, T., Dahl, R., Friedmann, P.S. Lanier, B.Q., Lockey, R.F., Motala, C., Ortega Martell, J.A., Platts-Mills, T.A., Ring, J., Thien, F., Van Cauwenberge, P., Williams, H.C. (2004). Revised nomenclature for allergy for global use: report of the Nomenclature Review Committee of the World Allergy Organization. *The Journal of Allergy and Clinical Immunology*, *113*, 832–836.
4. Montalto, M., Santoro, L., D'Onofrio, F., Curigliano, V., Gallo, A., Visca, D., Cammarota, G., Gasbarrini, A., Gasbarrini, G. (2008). Adverse reactions to food: allergies and intolerances. *Digestive Diseases*, *2*, 96–103.

5. Nowak-Wegrzyn, A., Sampson, H.A. (2006). Adverse reactions to foods. *Medical Clinics of North America*, *90*, 97–127.
6. Ortolani, C., Pastorello, E.A. (2006). Food allergies and food intolerances. *Best Practice and Research Clinical Gastroenterology*, *20*, 467–483.
7. Sampson, H.A. (2004). Update on food allergy. *The Journal of Allergy and Clinical Immunology*, *11*, 805–819.
8. Ring, J., Brockow, K., Behrendt, H. (2001). Adverse reactions to foods. *Journal of Chromatogrphy. B, Biomedical Science and Applications*, *756*, 3–10.
9. Simon, R.A. (1998). Update on sulfite sensitivity. *Allergy*, *53*, 78–79.
10. Simon, R.A. (2003). Adverse reactions to food additives. *Current Allergy and Asthma Report*, *3*, 62–66.
11. Kitts, D., Yuan, Y., Joneja, J., Scott, F., Szilagyi, A., Amiot, J., Zarkadas, M. (1997). Adverse reactions to food constituents: allergy, intolerance, and autoimmunity. *Canadian Journal of Physiology and Pharmacology*, *75*, 241–254.
12. Maintz, L., Novak, N. (2007). Histamine and histamine intolerance. *The American Journal of Clinical Nutrition*, *85*, 1185–1196.
13. Zarkadas, M., Scott, F., Salminen J., Ham Pong, A. (1999). Common allergenic foods and their labelling in Canada—a review. *Canadian Journal of Allergy and Immunology*, *4*, 118–141.
14. Justinich, C. (2006). Food allergies: a GP's guide. *The Canadian Journal of Continuing Medical Education*, *18*, 80–84.
15. Asero, R., Ballmer-Weber, B.K., Beyer, K., Conti, A., Dubakiene, R., Fernandez-Rivas, M., Hoffmann-Sommergruber, K., Lidholm, J., Mustakov, T., Oude Elberink, J.N., Pumphrey, R.S., Stahl, S.P., Van, R.R., Vlieg-Boerstra, B.J., Hiller, R., Hourihane, J.O., Kowalski, M., Papadopoulos, N.G., Wal, J.M., Mills, E.N., Vieths, S. (2007). IgE-mediated food allergy diagnosis: current status and new perspectives. *Molecular Nutrition and Food Research*, *51*, 35–147.
16. Lack, G. (2008). Clinical practice: food allergy. *The New England Journal of Medicine*, *359*, 1252–1260.
17. Sicherer, S.H., Sampson, H.A. (2006). Food allergy. *The Journal of Allergy and Clinical Immunology*, *117*, S470–S475.
18. FDA Threshold Working Group. (2008). Approaches to establish thresholds for major food allergens and for gluten in food. *Journal of Food Protection*, *71*, 1043–1088.
19. Aalberse, R.C. (2007). Assessment of allergen cross-reactivity. *Clinical and Molecular Allergy*, *5*, 2–6.
20. Breiteneder, H., Clare Mills, E.N. (2005). Plant food allergens—structural and functional aspects of allergenicity. *Biotechnology Advances*, *23*, 395–399.
21. Hefle, S.L. (1996). The molecular biology of food allergy. *Immunology and Allergy Clinics of North America*, *16*, 565–590.
22. Chehade, M. (2007). IgE and non-IgE-mediated food allergy: treatment in 2007. *Current Opinion in Allergy and Clinical Immunology*, *7*, 264–268.
23. Abenavoli, L., Proietti, I., Leggio, L., Ferrulli, A., Vonghia, L., Capizzi, R., Rotoli, M., Amerio, P.L., Gasbarrini, G., Addolorato, G. (2006). Cutaneous manifestations in celiac disease. *World Journal of Gastroenterology*, *12*, 843–852.

24. Gawkrodger, D.J., Blackwell, J.N., Gilmour, H.M., Rifkind, E.A., Heading, R.C., Barnetson, R.S. (1984). Dermatitis herpetiformis: diagnosis, diet and demography. *Gut*, *25*, 151–157.

25. Mansueto, P., Montalto, G., Pacor, M.L., Esposito-Pellitteri, M., Ditta ,V., Lo, B.C., Leto-Barone, S.M., Di, L.G. (2006). Food allergy in gastroenterologic diseases: review of literature. *World Journal of Gastroenterology*, *12*, 7744–7752.

26. Meredith, C. (2005). Allergenic potential of novel foods. *The Proceedings of the Nutrition Society*, *64*, 487–490.

27. Nowak-Wegrzyn, A., Fiocchi, A. (2009). Rare, medium, or well done? The effect of heating and food matrix on food protein allergenicity. *Current Opinion in Allergy and Clinical Immunology*, *9*, 234–237.

28. Strobel, S., Mowat, A.M. (2006). Oral tolerance and allergic responses to food proteins. *Current Opinion in Allergy and Clinical Immunology*, *6*, 207–213.

29. Fasano, A., Berti, I., Gerarduzzi, T., Not, T., Colletti, R.B., Drago, S., Elitsur,Y., Green, P.H., Guandalini, S., Hill, I.D., Pietzak, M., Ventura, A., Thorpe, M., Kryszak, D., Fornaroli, F., Wasserman, S.S., Murray, J.A., Horvath, K. (2003). Prevalence of celiac disease in at-risk and not-at-risk groups in the United States: a large multicenter study. *Archives of Internal Medicine*, *163*, 286–292.

30. Bjorksten, B., Crevel, R., Hischenhuber, C., Lovik, M., Samuels, F., Strobel, S., Taylor, S.L., Wal, J.M., Ward, R. (2008). Criteria for identifying allergenic foods of public health importance. *Regulatory Toxicology and Pharmacology*, *51*, 42–52.

31. Taylor, S.L., Hefle, S.L., Bindslev-Jensen, C., Bock, S.A., Burks, A.W. Jr., Christie, L., Hill, D.J., Host, A., Hourihane, J.O., Lack, G., Metcalfe, D.D., Moneret-Vautrin, D.A., Vadas, P.A., Rance, F., Skrypec, D.J., Trautman, T.A., Yman, I.M., Zeiger, R.S. (2002). Factors affecting the determination of threshold doses for allergenic foods: how much is too much? *The Journal of Allergy and Clinical Immunology*, *109*, 24–30.

32. Taylor, S.L., Hefle, S.L. (2006). Food allergen labeling in the USA and Europe. *Current Opinion in Allergy and Clinical Immunology*, *6*, 186–190.

33. Chehade, M., Mayer, L. (2005). Oral tolerance and its relation to food hypersensitivities. *The Journal of Allergy and Clinical Immunology*, *115*, 3–12.

34. Spencer, J., Barone, F., Dunn-Walters, D. (2009). Generation of immunoglobulin diversity in human gut-associated lymphoid tissue. *Seminars in Immunology*, *21*, 139–146.

35. Groschwitz, K.R., Hogan, S.P. (2009). Intestinal barrier function: molecular regulation and disease pathogenesis. *The Journal of Allergy and Clinical Immunology*, *124*, 3–20.

36. Hourihane, J.O., Knulst, A.C. (2005). Thresholds of allergenic proteins in foods. *Toxicology and Applied Pharmacology*, *207*, 152–156.

37. Untersmayr, E., Jensen-Jarolim, E. (2008). The role of protein digestibility and antacids on food allergy outcomes. *The Journal of Allergy and Clinical Immunology*, *121*, 1301–1308.

38. Poole, J.A., Barriga, K., Leung, D.Y., Hoffman, M., Eisenbarth, G.S., Rewers, M., Norris, J.M. (2006). Timing of initial exposure to cereal grains and the risk of wheat allergy. *Pediatrics*, *117*, 2175–2182.

39. Benhamou, A.H., Schappi Tempia, M.G., Belli, D.C., Eigenmann, P.A. (2009). An overview of cow's milk allergy in children. *Swiss Medical Weekly*, *139*, 300–307.
40. Beaudouin, E., Renaudin, J.M., Morisset, M., Codreanu, F., Kanny, G., Moneret-Vautrin, D.A. (2006). Food-dependent exercise-induced anaphylaxis—update and current data. *European Annals of Allergy and Clinical Immunology*, *38*, 45–51.
41. Tatham, A.S., Shewry, P.R. (2008). Allergens to wheat and related cereals. *Clinical and Experimental Allergy*, *38*, 1712–1726.
42. Hays, T., Wood, R.A. (2005). A systematic review of the role of hydrolyzed infant formulas in allergy prevention. *Archives of Pediatrics and Adolescent Medicine*, *159*, 810–816.
43. Tanabe, S. (2008). Analysis of food allergen structures and development of foods for allergic patients. *Bioscience, Biotechnology, and Biochemistry*, *72*, 649–659.
44. Jones, S.M., Magnolfi, C.F., Cooke, S.K., Sampson, H.A. (1995). Immunologic cross-reactivity among cereal grains and grasses in children with food hypersensitivity. *The Journal of Allergy and Clinical Immunology*, *96*, 341–351.
45. Mehr, S., Kakakios, A., Frith, K., Kemp, A.S. (2009). Food protein-induced enterocolitis syndrome: 16-year experience. *Pediatrics*, *123*, e459–e464.
46. Mehr, S.S., Kakakios, A.M., Kemp, A.S. (2009). Rice: a common and severe cause of food protein-induced enterocolitis syndrome. *Archives of Disease in Childhood*, *94*, 220–223.
47. Burks, W. (2003). Skin manifestations of food allergy. *Pediatrics*, *111*, 1617–1624.
48. Jesenak, M., Rennerova, Z., Babusikova, E., Havlicekova, Z., Jakusova, L., Villa, M.P., Ronchetti, R., Banovcin, P. (2008). Food allergens and respiratory symptoms. *Journal of Physiology and Pharmacology*, *59*, 311–320.
49. Sicherer, S.H. (2003). Clinical aspects of gastrointestinal food allergy in childhood. *Pediatrics*, *111*, 1609–1616.
50. Muraro, A., Roberts, G., Clark, A., Eigenmann, P.A., Halken, S., Lack, G., Moneret-Vautrin, A., Niggemann, B., Rance, F. (2007). The management of anaphylaxis in childhood: position paper of the European Academy of Allergology and Clinical Immunology. *Allergy*, *62*, 857–871.
51. Sampson, H.A., Munoz-Furlong, A., Bock, S.A., Schmitt, C., Bass, R., Chowdhury, B.A., Decker, W.W., Furlong, T.J., Galli, S.J., Golden, D.B., Gruchalla, R.S., Harlor, A.D. Jr., Hepner, D.L., Howarth, M., Kaplan, A.P., Levy, J.H., Lewis, L.M., Lieberman, P.L., Metcalfe, D.D., Murphy, R., Pollart, S.M., Pumphrey, R.S., Rosenwasser, L.J., Simons, F.E., Wood, J.P., Camargo, C.A. Jr. (2005). Symposium on the definition and management of anaphylaxis: summary report. *The Journal of Allergy and Clinical Immunology*, *115*, 584–591.
52. Brown, S.G. (2004). Clinical features and severity grading of anaphylaxis. *The Journal of Allergy and Clinical Immunology*, *114*, 371–376.
53. Eichenfield, L.F. (2004). Consensus guidelines in diagnosis and treatment of atopic dermatitis. *Allergy*, *59*, 86–92.
54. Sturgill, S., Bernard, L.A. (2004). Atopic dermatitis update. *Current Opinion in Pediatrics*, *16*, 396–401.
55. Battais, F., Richard, C., Jacquenet, S., Denery-Papini, S., Moneret-Vautrin, D.A. (2008). Wheat grain allergies: an update on wheat allergens. *European Annals of Allergy and Clinical Immunology*, *40*, 67–76.

56. Pourpak, Z., Mansouri, M., Mesdaghi, M., Kazemnejad, A., Farhoudi, A. (2004). Wheat allergy: clinical and laboratory findings. *International Archives of Allergy and Immunology*, *133*, 168–173.
57. Scibilia, J., Pastorello, E.A., Zisa, G., Ottolenghi, A., Bindslev-Jensen, C., Pravettoni, V., Scovena, E., Robino, A., Ortolani, C. (2006). Wheat allergy: a double-blind, placebo-controlled study in adults. *The Journal of Allergy and Clinical Immunology*, *117*, 433–439.
58. Lehto, M., Palosuo, K., Varjonen, E., Majuri, M.L., Andersson, U., Reunala, T., Alenius, H. (2003). Humoral and cellular responses to gliadin in wheat-dependent, exercise-induced anaphylaxis. *Clinical and Experimental Allergy*, *33*, 90–95.
59. Matsuo, H., Kohno, K., Morita, E. (2005). Molecular cloning, recombinant expression and IgE-binding epitope of omega-5 gliadin, a major allergen in wheat-dependent exercise-induced anaphylaxis. *The FEBS Journal*, *272*, 4431–4438.
60. Furuta, G.T., Liacouras, C.A., Collins, M.H., Gupta, S.K., Justinich, C., Putnam, P.E., Bonis, P., Hassall, E., Straumann, A., Rothenberg, M.E. (2007). Eosinophilic esophagitis in children and adults: a systematic review and consensus recommendations for diagnosis and treatment. *Gastroenterology*, *133*, 1342–1363.
61. Catassi, C., Fasano, A. (2008). Celiac disease. *Current Opinion in Gastroenterology*, *24*, 687–691.
62. Craig, D., Robins, G., Howdle, P.D. (2007). Advances in celiac disease. *Current Opinion in Gastroenterology*, *23*, 142–148.
63. Fasano, A., Araya, M., Bhatnagar, S., Cameron, D., Catassi, C., Dirks, M., Mearin, M.L., Ortigosa, L., Phillips, A. (2008). Federation of International Societies of Pediatric Gastroenterology, Hepatology, and Nutrition consensus report on celiac disease. *Journal of Pediatric Gastroenterology and Nutrition*, *47*, 214–219.
64. National Institutes of Health. (2005). Consensus development conference statement on celiac disease, June 28-30, 2004. *Gastroenterology*, *128*, S1–S9.
65. Presutti, R.J., Cangemi, J.R., Cassidy, H.D., Hill, D.A. (2007). Celiac disease. *American Family Physician*, *76*, 1795–1802.
66. Tully, M.A. (2008). Pediatric celiac disease. *Gastroenterology Nursing*, *31*, 132–140.
67. Armstrong, M.J., Robins, G.G., Howdle, P.D. (2009). Recent advances in coeliac disease. *Current Opinion in Gastroenterology*, *25*, 100–109.
68. Briani, C., Samaroo, D., Alaedini, A. (2008). Celiac disease: from gluten to autoimmunity. *Autoimmunity Reviews*, *7*, 644–650.
69. Green, P.H., Cellier, C. (2007). Celiac disease. *The New England Journal of Medicine*, *357*, 1731–1743.
70. Rashid, M., Cranney, A., Zarkadas, M., Graham, I.D., Switzer, C., Case, S., Molloy, M., Warren, R.E., Burrows, V., Butzner, J.D. (2005). Celiac disease: evaluation of the diagnosis and dietary compliance in Canadian children. *Pediatrics*, *116*, e754–e759.
71. Zarkadas, M., Case, S. (2005). Celiac disease and the gluten free diet: an overview. *Topics in Clinical Nutrition*, *20*, 127–138.
72. Ford, R.P. (2009). The gluten syndrome: a neurological disease. *Medical Hypotheses*, *73*, 438–440.

73. Casellas, F., Lopez, V.J., Malagelada, J.R. (2006). Current epidemiology and accessibility to diet compliance in adult celiac disease. *Revista Española de Enfermedades Digestivas*, *98*, 408–419.
74. Catassi, C. (2005). The world map of celiac disease. *Acta Gastroenterologica Latinoamericana*, *35*, 37–55.
75. Bardella, M.T., Elli, L., Velio, P., Fredella, C., Prampolini, L., Cesana, B. (2007). Silent celiac disease is frequent in the siblings of newly diagnosed celiac patients. *Digestion*, *75*, 182–187.
76. Book, L., Zone, J.J., Neuhausen, S.L. (2003). Prevalence of celiac disease among relatives of sib pairs with celiac disease in U.S. families. *The American Journal of Gastroenterology*, *98*, 377–381.
77. Book, L., Hart, A., Black, J., Feolo, M., Zone, J.J., Neuhausen, S.L. (2001). Prevalence and clinical characteristics of celiac disease in Downs syndrome in a US study. *American Journal of Medical Genetics Part A*, *98*, 70–74.
78. Frisk, G., Hansson, T., Dahlbom, I., Tuvemo, T. (2008). A unifying hypothesis on the development of type 1 diabetes and celiac disease: gluten consumption may be a shared causative factor. *Medical Hypotheses*, *70*, 1207–1209.
79. Rubio-Tapia, A., Kyle, R.A., Kaplan, E.L., Johnson, D.R., Page, W., Erdtmann, F., Brantner, T.L., Kim, W.R., Phelps, T.K., Lahr, B.D., Zinsmeister, A.R., Melton, L.J., Murray, J.A. (2009). Increased prevalence and mortality in undiagnosed celiac disease. *Gastroenterology*, *137*, 88–93.
80. West, J., Logan, R.F., Hill, P.G., Khaw, K.T. (2007). The iceberg of celiac disease: what is below the waterline? *Clinical Gastroenterology and Hepatology*, *5*, 59–62.
81. West, J. (2009). Celiac disease and its complications: a time traveller's perspective. *Gastroenterology*, *136*, 32–34.
82. Alaedini, A., Green, P.H. (2005). Narrative review: celiac disease: understanding a complex autoimmune disorder. *Annals of Internal Medicine*, *142*, 289–298.
83. Barton, S.H., Kelly, D.G., Murray, J.A. (2007). Nutritional deficiencies in celiac disease. *Gastroenterology Clinics of North America*, *36*, 93–108.
84. Barton, S.H., Murray, J.A. (2008). Celiac disease and autoimmunity in the gut and elsewhere. *Gastroenterology Clinics of North America*, *37*, 411–428.
85. Cranney, A., Zarkadas, M., Graham, I.D., Butzner, J.D., Rashid, M., Warren, R., Molloy, M., Case, S., Burrows, V., Switzer, C. (2007). The Canadian celiac health survey. *Digestive Diseases and Sciences*, *52*, 1087–1095.
86. Gianfrani, C., Troncone, R., La, C.A. (2008). Autoimmunity and celiac disease. *Mini Reviews in Medicinal Chemistry*, *8*, 129–134.
87. Zarkadas, M., Cranney, A., Case, S., Molloy, M., Switzer, C., Graham, I.D., Butzner, J.D., Rashid, M., Warren, R.E., Burrows, V. (2006). The impact of a gluten-free diet on adults with coeliac disease: results of a national survey. *Journal of Human Nutrition and Dietetics*, *19*, 41–49.
88. Losowsky, M.S. (2008). A history of coeliac disease. *Digestive Diseases*, *26*, 112–120.
89. Halfdanarson, T.R., Litzow, M.R., Murray, J.A. (2007). Hematologic manifestations of celiac disease. *Blood*, *109*, 412–421.

90. Lee, S.K., Green, P.H. (2006). Celiac sprue (the great modern-day imposter). *Current Opinion in Rheumatology*, *18*, 101–107.

91. Fasano, A. (2006). Systemic autoimmune disorders in celiac disease. *Current Opinion in Gastroenterology*, *22*, 674–679.

92. Briani, C., Zara, G., Alaedini, A., Grassivaro, F., Ruggero, S., Toffanin, E., Albergoni, M.P., Luca, M., Giometto, B., Ermani, M., De, L.F., D'Odorico, A., Battistin, L. (2008). Neurological complications of celiac disease and autoimmune mechanisms: a prospective study. *Journal of Neuroimmunology*, *195*, 171–175.

93. Murray, J.A., Rubio-Tapia, A., Van Dyke, C.T., Brogan, D.L., Knipschield, M.A., Lahr, B., Rumalla, A., Zinsmeister, A.R., Gostout, C.J. (2008). Mucosal atrophy in celiac disease: extent of involvement, correlation with clinical presentation, and response to treatment. *Clinical Gastroenterology and Hepatology*, *6*, 186–193.

94. Bhadada, S.K., Bhansali, A., Kochhar, R., Menon, A.S., Sinha, S.K., Dutta, P., Nain, C.K. (2008). Does every short stature child need screening for celiac disease? *Journal of Gastroenterology and Hepatology*, *23*, e353–e356.

95. Catassi, C., Fasano, A. (2002). New developments in childhood celiac disease. *Current Gastroenterology Reports*, *4*, 238–243.

96. Wierink, C.D., van Diermen, D.E., Aartman, I.H., Heymans, H.S. (2007). Dental enamel defects in children with coeliac disease. *International Journal of Paediatric Dentistry*, *17*, 163–168.

97. Hadjivassiliou, M., Grunewald, R.A., Davies-Jones, G.A. (2002). Gluten sensitivity as a neurological illness. *Journal of Neurology, Neurosurgery, and Psychiatry*, *72*, 560–563.

98. Freeman, H.J. (2007). Pancreatic endocrine and exocrine changes in celiac disease. *World Journal of Gastroenterology*, *13*, 6344–6346.

99. Giovenale, D., Meazza, C., Cardinale, G.M., Sposito, M., Mastrangelo, C., Messini, B., Citro, G., Delvecchio, M., Di, M.S., Bozzola, M. (2006). The prevalence of growth hormone deficiency and celiac disease in short children. *Clinical Medicine and Research*, *4*, 180–183.

100. Stazi, A.V., Trinti, B. (2007). Risk of osteoporosis in endocrine disorders and celiac disease. *Annali dell'Istituto Superiore di Sanità*, *43*, 430–433.

101. Anderson, L.A., McMillan, S.A., Watson, R.G., Monaghan, P., Gavin, A.T., Fox, C., Murray, L.J. (2007). Malignancy and mortality in a population-based cohort of patients with coeliac disease or gluten sensitivity. *World Journal of Gastroenterology*, *13*, 146–151.

102. Goddard, C.J., Gillett, H.R. (2006). Complications of coeliac disease: are all patients at risk? *Postgraduate Medical Journal*, *82*, 705–712.

103. Rubio-Tapia, A., Kelly, D.G., Lahr, B.D., Dogan, A., Wu, T.T., Murray, J.A. (2009). Clinical staging and survival in refractory celiac disease: a single center experience. *Gastroenterology*, *136*, 99–107.

104. Reunala, T., Kosnai, I., Karpati, S., Kuitunen, P., Torok, E., Savilahti, E. (1984). Dermatitis herpetiformis: jejunal findings and skin response to gluten free diet. *Archives of Disease in Childhood*, *59*, 517–522.

105. Marietta, E.V., Camilleri, M.J., Castro, L.A., Krause, P.K., Pittelkow, M.R., Murray, J.A. (2008). Transglutaminase autoantibodies in dermatitis herpetiformis and celiac sprue. *The Journal of Investigative Dermatology*, *128*, 332–335.

106. Fry, L. (2002). Dermatitis herpetiformis: problems, progress and prospects. *European Journal of Dermatology*, *12*, 523–531.
107. Kaukinen, K., Collin, P., Mäki, M. (2007). Latent coeliac disease or coeliac disease beyond villous atrophy? *Gut*, *56*, 1339–1340.
108. Savilahti, E., Reunala, T., Mäki, M. (1992). Increase of lymphocytes bearing the gamma/delta T cell receptor in the jejunum of patients with dermatitis herpetiformis. *Gut*, *33*, 206–211.
109. Humbert, P., Pelletier, F., Dreno, B., Puzenat, E., Aubin, F. (2006). Gluten intolerance and skin diseases. *European Journal of Dermatology*, *16*, 4–11.
110. Jabri, B., Sollid, L.M. (2006). Mechanisms of disease: immunopathogenesis of celiac disease. *Nature Clinical Practice Gastroenterology and Hepatology*, *3*, 516–525.
111. Koning, F. (2008). Celiac disease: sandwiched between innate and adaptive immune responses induced by gluten. *Journal of Pediatric Gastroenterology and Nutrition*, *46*, E8–E9.
112. Chang, F., Mahadeva, U., Deere, H. (2005). Pathological and clinical significance of increased intraepithelial lymphocytes (IELs) in small bowel mucosa. *Acta Pathologica, Microbiologica, et Immunologica Scandinavica*, *113*, 385–399.
113. Dewar, D., Pereira, S.P., Ciclitira, P.J. (2004). The pathogenesis of coeliac disease. *The International Journal of Biochemistry and Cell Biology*, *36*, 17–24.
114. Dickson, B.C., Streutker, C.J., Chetty, R. (2006). Coeliac disease: an update for pathologists. *Journal of Clinical Pathology*, *59*, 1008–1016.
115. Rostami, K., Villanacci, V. (2009). Microscopic enteritis: novel prospect in coeliac disease clinical and immuno-histogenesis: evolution in diagnostic and treatment strategies. *Digestive and Liver Disease*, *41*, 245–252.
116. Stene, L.C., Honeyman, M.C., Hoffenberg, E.J., Haas, J.E., Sokol, R.J., Emery, L., Taki, I., Norris, J.M., Erlich, H.A., Eisenbarth, G.S., Rewers, M. (2006). Rotavirus infection frequency and risk of celiac disease autoimmunity in early childhood: a longitudinal study. *The American Journal of Gastroenterology*, *101*, 2333–2340.
117. Troncone, R., Auricchio, S. (2007). Rotavirus and celiac disease: clues to the pathogenesis and perspectives on prevention. *Journal of Pediatric Gastroenterology and Nutrition*, *44*, 527–528.
118. Shewry, P.R., Halford, N.G. (2002). Cereal seed storage proteins: structures, properties and role in grain utilization. *Journal of Experimental Botany*, *53*, 947–958.
119. Silano, M., Vincentini, O., De, V.M. (2009). Toxic, immunostimulatory and antagonist gluten peptides in celiac disease. *Current Medicinal Chemistry*, *16*, 1489–1498.
120. Weiser, H. (2007). Chemistry of gluten proteins. *Food Microbiology*, *24*, 115–119.
121. Wieser, H., Koehler, P. (2008). The biochemical basis of celiac disease. *Cereal Chemistry*, *85*, 1–13.
122. Dewar, D.H., Amato, M., Ellis, H.J., Pollock, E.L., Gonzalez-Cinca, N., Wieser, H., Ciclitira, P.J. (2006). The toxicity of high molecular weight glutenin subunits of wheat to patients with coeliac disease. *European Journal of Gastroenterology and Hepatology*, *18*, 483–491.
123. Shan, L., Molberg, O., Parrot, I., Hausch, F., Filiz, F., Gray, G.M., Sollid, L.M., Khosla, C. (2002). Structural basis for gluten intolerance in celiac sprue. *Science*, *297*, 2275–2279.

124. Qiao, S.W., Bergseng, E., Molberg, O., Jung, G., Fleckenstein, B., Sollid, L.M. (2005). Refining the rules of gliadin T cell epitope binding to the disease-associated DQ2 molecule in celiac disease: importance of proline spacing and glutamine deamidation. *Journal of Immunology*, *175*, 254–261.

125. Arentz-Hansen, H., Korner, R., Molberg, O., Quarsten, H., Vader, W., Kooy, Y.M., Lundin, K.E., Koning, F., Roepstorff, P., Sollid, L.M., McAdam, S.N. (2000). The intestinal T cell response to alpha-gliadin in adult celiac disease is focused on a single deamidated glutamine targeted by tissue transglutaminase. *The Journal of Experimental Medicine*, *191*, 603–612.

126. Sollid, L.M. (2000). Molecular basis of celiac disease. *Annual Review of Immunology*, *18*, 53–81.

127. Ciclitira, P.J., Ellis, H.J., Lundin, K.E. (2005). Gluten-free diet—what is toxic? *Best Practice and Research Clinical Gastroenterology*, *19*, 359–371.

128. Skovbjerg, H., Anthonsen, D., Knudsen, E., Sjostrom, H. (2008). Deamidation of gliadin peptides in lamina propria: implications for celiac disease. *Digestive Diseases and Sciences*, *53*, 2917–2924.

129. Vader, L.W., de Ru, A., van der Wal, Y., Kooy, Y.M., Benckhuijsen, W., Mearin, M.L., Drijfhout, J.W., Van Veelen, P., Koning, F. (2002). Specificity of tissue transglutaminase explains cereal toxicity in celiac disease. *The Journal of Experimental Medicine*, *195*, 643–649.

130. Case, S. (2008). Gluten-free diet. In: *A Comprehensive Resource Guide*, Shelly Case (Ed.), expanded edition. Regina: Case Nutrition Consulting, pp. 19–72.

131. Akobeng, A.K., Thomas, A.G. (2008). Systematic review: tolerable amount of gluten for people with coeliac disease. *Alimentary Pharmacology and Therapeutics*, *27*, 1044–1052.

132. Catassi, C., Fabiani, E., Iacono, G., D'Agate, C., Francavilla, R., Biagi, F., Volta, U., Accomando, S., Picarelli, A., De Vitis, I., Pianelli, G., Gesuita, R., Carle, F., Mandolesi, A., Bearzi, I., Fasano, A. (2007). A prospective, double-blind, placebo-controlled trial to establish a safe gluten threshold for patients with celiac disease. *The American Journal of Clinical Nutrition*, *85*, 160–166.

133. Pulido, O.M., Gillespie, Z., Zarkadas, M., Dubois, S., Vavasour, E., Rashid, M., Switzer, C., Godefroy, S.B. (2009). Chapter 6 introduction of oats in the diet of individuals with celiac disease a systematic review. *Advances in Food and Nutrition Research*, *57*, 235–285.

134. Sicherer, S.H., Noone, S.A., Munoz-Furlong, A. (2001). The impact of childhood food allergy on quality of life. *Annals of Allergy, Asthma & Immunology*, *87*, 461–464.

135. Hadjivassiliou, M., Sanders, D.S., Grünewald, R.A., Woodroofe, N., Boscolo, S., Aeschlimann, D. (2010). Gluten sensitivity: from gut to brain. *The Lancet Neurology*, *9*(3), 318–330. Review.

136. Sapone, A., Lammers, K.M., Mazzarella, G., Mikhailenko, I., Cartenì, M., Casolaro, V., Fasano, A. (2010). Differential mucosal IL-17 expression in two gliadin-induced disorders: gluten sensitivity and the autoimmune enteropathy celiac disease. *International Archives of Allergy and Immunology*, *152*(1), 75–80.

2

PROTECTING FOOD-ALLERGIC CONSUMERS: MANAGING ALLERGENS ACROSS THE FOOD SUPPLY CHAIN

Sandra Kerbach, Anton J. Alldrick, Rene W.R. Crevel, Lilla Dömötör, Audrey DunnGalvin, E.N. Clare Mills, Sylvia Pfaff, Roland E. Poms, Sandor Tömösközi, and Bert Popping

2.1. INTRODUCTION

2.1.1. General Facts on Food Allergens

While in principle, any food could trigger an allergic reaction, in practice the majority of food allergic reactions are caused by a restricted number of foods. These include cow's milk, hen's egg, fish, peanuts, and a variety of tree nut species [1]. The patterns and prevalence of specific types of food allergies vary between different population groups. Thus, cow's milk and hen's egg allergies are more common in infants and young children, while other foods including various tree nuts (especially hazelnut and walnut), peanuts, fish, crustacean, and mollusks cause allergies more frequently in older children and adults [2]. The components in foods that trigger allergies are almost always proteins. Many proteins have been identified as being responsible for triggering immunoglobulin E (IgE)-mediated allergic reactions, and many share characteristics in common [3] including:

Allergen Management in the Food Industry, Edited by Joyce I. Boye and Samuel Benrejeb Godefroy

- a molecular weight greater than 3 kDa since this is the molecular weight threshold required to elicit an antibody response;
- stability against enzymatic and chemical degradation (e.g., in acid environments, like in the stomach) and processing procedures (e.g., heat);
- comprising both B-cell and T-cell epitopes, responsible for interacting with immune effector molecules including interactions with IgE antibodies (see Section 2.1.2); and
- abundance in foodstuffs (e.g., 54% of hen's egg protein comprises ovalbumin), although there are exceptions such as the soybean allergen Gly m 4.

In common parlance in the analytical community, the foods themselves rather than the molecules are often termed allergens. This latter terminology will be used in this review.

2.1.2. Food Allergy and Adverse Reactions

Adverse reactions to foods, including IgE-mediated food allergies are of major concern to consumers and the food industry alike, partly because of the severe nature of some reactions and the fact that the prevalence of such adverse reactions against food is increasing.

Adverse reactions to foods can take many forms, ranging from toxic reactions to components such as histamine and deficiencies in enzymes such as lactase (which mean individuals cannot digest the milk sugar lactose), as well as food allergies that have an immune component [4]. Of these there are two main types: the gluten intolerance syndrome known as celiac disease and IgE-mediated food allergies. Induction of gluten intolerance results from deamidation of glutamine residues in digested gluten peptides by gut mucosal tissue transglutaminase. In susceptible individuals, these modified peptides are able to bind to class II human histocompatibility leukocyte antigen (HLA) molecules DQ2 and DQ8, an event which initiates an immune response leading to inflammation. This in turn results in the flattened gut mucosa characteristic of celiac disease [5]. Such reactions develop relatively slowly and, while debilitating, are not immediately life threatening. In contrast, IgE-mediated allergies result in rapid reactions that usually occur between minutes and hours after ingestion of a problem food. During normal immune functioning, IgE molecules are developed to combat parasitic infections such as malaria [6]. For reasons that are not completely understood, certain susceptible individuals make IgE toward environmental agents such as pollens, dusts, and foods in a process known as sensitization. Once an individual becomes sensitized to a molecule (allergen) in subsequent encounters, IgE molecules bound to the surface of histamine-packed cells interact with the allergen, triggering the release of the histamine and other inflammatory mediators. It is these mediators that cause the symptoms we know as an allergic reaction. The symptoms can be manifested as a skin rash (eczema, urticaria), respiratory symptoms, like asthma, and gastrointestinal symptoms (vomiting, diarrhea). In some instances, these can be life threatening, involving breathing difficulties (obstruction of airways, asthma) and occasionally anaphylaxis, which, among many other symptoms, is characterized by a rapid drop in blood pressure. Studies in unselected populations indicate that around 2–4% of the total population suffer from IgE-mediated food allergy [7, 8], the prevalence being higher among children at around 5–8%, and apparently increasing [9]. There is currently no effective cure for either celiac disease or IgE-mediated allergies. Consequently, individuals suffering from these conditions have to avoid consuming problem foods, generally for the rest of their lifetime. This, along with provision of emergency rescue medication, is the only treatment currently available for food-allergic individuals. Some individuals are very sensitive to tiny amounts of allergen, and there are case reports of individuals having a reaction after being kissed by someone who has eaten an offending food [10] or from hidden food allergens [11]. As a consequence of both the prevalence and severity of allergic reactions triggered by such small amounts, it was agreed worldwide that the major allergenic foods and derived ingredients should be identified on food labels to help allergic consumers to make an informed choice [12].

2.1.3. Stakeholder Perspectives in the Context of MoniQA

In order to manage allergens in foods, several stakeholder groups need to be involved. Each of these groups, which include food manufacturers, controlling authorities, retailers and allergic consumers, has its own requirements regarding food allergens. Elaboration of reliable, reproducible, and sensitive methods for detecting and measuring the allergenic constituents in food makes a critical contribution to managing allergens but is not an end in itself. Instead, it is a tool for affective allergen management to reduce risk at industry level and enforcement level and to allow consumers to make an informed choice.

Risk can be defined as the probability of an adverse outcome and can be represented as a function of hazard and exposure. Analytical techniques address the exposure side of this function, but only have meaning if the hazard in question has been characterized, that is, if the amount of allergen can be related to the probability of a reaction occurring and, ideally, to its severity. Application of the legislation is complicated by an absence of consensus on what constitutes the amount of an allergen that renders a food "unsafe." An upper limit for noningredient allergenic food components needs to consider the no observed adverse effect level (NOAEL) reported for each of the important allergenic foods. A practical way to deal with unintentional allergenic "cross-contact" could be the adoption of an upper limit for noningredient allergenic food components, which minimizes risk to the allergic consumer. The discussion on threshold values is of major concern for all affected stakeholder groups: allergic consumers seek "safe" food, food industry would like to provide them, and enforcement has to ensure the compliance with food regulation, supported by testing labs.

The Working Group Food Allergens of the European Union (EU) project MoniQA (www.moniqa.org) and a project like EuroPrevall (www.europrevall.org), both contributing to this chapter, have to be seen in a wider context, where the findings of these projects can be used to develop risk assessment/risk management tools for improved food safety (e.g., threshold level assessment, frequency of improper declaration).

2.2. LIVING WITH FOOD ALLERGIES—CHALLENGES FOR THE ALLERGIC CONSUMER

Living with a disease that is triggered by foods, which pose no threat to others, presents particular problems for allergic consumers and those around them. While there is much anecdotal information about the impact of food allergies, there has been little objective data regarding its effects on quality of life [13] or its economic cost [14]. Recent research has led to the development of the first validated age-specific and disease-specific Quality of Life (QoL) study instruments [15, 16] providing the first objective information that food-allergic individuals are at risk of negative emotional and social outcomes, including anxiety, avoidance, or risky behavior. These studies have also shown that food allergy impacts directly on a child's normal trajectory of psychosocial development in a disease-specific manner [17].

Food-allergic children have different views of their allergy and different coping strategies that evolve in response to age-, gender-, and context-specific stressors. The impact of food allergy also extends to parents who appear to be extremely worried about their children and demonstrate high levels of stress and anxiety due to constant high levels of vigilance and feelings of guilt when children have a reaction. Some of this worry is maladaptive, inhibits normal socialization of their child, and may have a long-term impact on quality of life [17].

One age group particularly at risk is teenagers and young adults. Preadolescence is an important transition point when children must begin to gain autonomy and self-belief in their ability to control events in their lives. A recent paper by Sampson and colleagues [18] found that adolescents and young adults appear to be at an increased risk for fatal food allergic reactions, and suggested that they may adopt more risk-taking behaviors with regard to their food allergy. Such research has emphasized the importance of discerning pathways to risk from early childhood to adolescence. Although its roots may lie in childhood, risk behavior may evolve more quickly in the vital transition period from preteen to teen, as developing children attempt to adapt rigid rules to novel social contexts. The following passage of text provides a good example:

> M: Why do you think John said that his food allergy had gone away, even when it hadn't?
>
> He really wanted the ice cream. (Girl, aged 12)

It was his only chance to get it. (Boy, aged 11)

I would try a little bit and see what happens. (Boy, aged 12)

Risk events are appraised in the context in which they occur and an awareness of expected behavior. For example, whereas an allergic reaction that takes place in the home may be regarded as a low-anxiety event, one that takes place in school can be appraised as highly stressful because it impacts on the child's goal to "fit in." This goal confronts children with the difficult balancing act of protecting their ego and managing risk. He or she learns to appraise (and weigh) threats to personal safety with threats to social identity. The stress appraisal process may result in children *just chancing it will be ok* when in the company of others, whereas others protect their self-esteem by avoiding novelty as much as possible. In an attempt to ensure their safety and control their anxiety, around 40% of children and teenagers prefer to avoid social events that directly or indirectly involve food (Table 2.1). Furthermore,

TABLE 2.1 Quotations from Food-Allergic Children and Their Parents [18]

Issues	Examples
Allergic children's anxiety around food	*Nearly everything says "may contain," so what can you eat?* (Girl, 10 years)
	If a [chocolate] bar is made in one country, it's all right to eat, but if it's made in another country, it's not safe and you can't eat it then and that just drives me crazy. (Boy, 9 years)
	It's so confusing that you can eat some things and not others. On Christmas I ended up in the hospital and I couldn't speak. I couldn't tell them how I felt. (Boy, 9 years)
	There's always food around you know. It doesn't have to be a food party. It's hard to relax. (Boy, 10 years)
	It's like they say in a plane: you have to keep an eye on the exits. (Boy, 14 years)
	He won't try anything that he hasn't had a thousand times before; he says he doesn't really care about food. (Parent of multiple allergic child, 10 years)
Avoiding social situations involving food	*I prefer to go to parties where there is no food. It is better to not go to restaurants. You never know… the waiters don't know. I only go to places where I know I am safe. In the cinema you don't know what someone next to you is having… To be honest, I don't bother going.*
Stress in carers	*I am absolutely terrified that I would buy something with nuts in it by mistake. If anything happened, I would never get over it.* (Mother of peanut-allergic child, 5 years)
Reading food labels	*My heart sinks if I read "new recipe" on the packet.* (Mother of milk-allergic child, 10 years)
	It's not just the main ingredients you know. You have to scan for hidden stuff too. (Father of multiple allergic child, 12 years)
	The writing can be very small and you have to read it several times. (Mother of multiple allergic child, 7 years)

it appears that many children and teenagers have a difficult relationship with food. Food management gives rise to feelings of anxiety and attendant stress in carers, a responsibility that tends to fall on mothers. A key part of the management strategies adopted by carers of allergic children is to control what food comes into the home. In the majority of cases, food containing the problem allergen is not allowed into the home, particularly when children are young. Parents agreed that children had a restrictive diet with less variety than siblings.

As a consequence of these issues, food labeling, with regards to both content and quality of information, is of crucial importance to allergic consumers in managing their condition and the resulting stress that it imposes on them. Food labeling has been mentioned as a significant source of uncertainty and therefore stress for most children [19]; such uncertainty gives rise to feelings of fear and confusion. All parents were perpetual readers of labels when purchasing food in supermarkets and elsewhere. While ingredient lists on prepackaged foods provide reassurance and reduce uncertainty, parents perceived that foods such as "cook-in sauces" often contain "unsafe" ingredients. All parents agreed that shopping takes much longer and is more stressful. As families get older and nonallergic siblings demand their favorite food, parents often segregate areas for "safe" and "not safe" foods. Again, labeling plays an important part in ensuring that food is placed in the "correct" section, or that it can be quickly recognized if the segregation breaks down.

However, there are problems regarding the readability and comprehension of food labels with issues such as the font size being too small, the contrast between background and text being inappropriate, present in too many languages, and a lack of standard position for the text [19].

Allergic consumers also report problems with precautionary "may contain" labeling with phrases such as "may contain traces of nuts," or "made in a factory where nuts are processed" becoming commonplace. Studies have shown that many allergic consumers ignore precautionary statements although 10% of foods with precautionary statements contained peanut although often not at clinically significant levels [20, 21]. The phrasing of precautionary statements (e.g., "may contain" vs. "shared facility" statements) did not seem to affect the likelihood of containing peanut, but consumers did attach a comparative risk assessment to the different phrases. There are also issues of how useful such statements are in multicultural environments. Providing allergic consumers with the level of information they might need to help them manage their condition may preclude the inclusion of all this information on the printed label. One option is to use new technology such as mobile phones to read information from special bar codes. However, whichever system is adopted to communicate information to consumers about allergens in foods, the crucial factors are adequate hazard control procedures and the testing framework in which food manufacturers work that underpins the information in a label, a data matrix, bar code, or radio frequency identification (RFID) tag.

2.3. LEGISLATIVE AND REGULATORY CONSIDERATIONS

Clearly, food allergy is a public health issue, and the protection of food-allergic consumers falls under the remit of food safety legislation and regulation. Recent standards and regulations have defined food safety. For instance, International Organization for Standardization (ISO) 22000:2005 defines food safety as the concept that food will not cause harm to the (average) consumer when it is prepared and/or eaten according to its intended use. According to the Food and Agriculture Organization (FAO)/World Health Organization (WHO) (2003), "food safety refers to all those hazards, whether chronic or acute, that may make food injurious to the health of the consumer" and "is not negotiable." The EU's Food Law 2002/178/EC [22] elaborates the concept further. Under this regulation, food is deemed unsafe if it is either injurious to human health or unfit for human consumption. The regulation also recognizes that safety is not an absolute condition and thus mirrors the criterion of "reasonable certainty of no harm" used in U.S. law from 1996 [23]. However, this legislation has been put in place with the average healthy consumer in mind.

For the allergic consumer, a new legislation has been issued in recent years across the world to help allergic consumers avoid problem foods by regulating the labeling of major allergenic foods (Table 2.2), in response to changes to the Codex General Standard for the Labelling of Prepackaged Foods [12]. The EU thus brought in Directive 2000/13/EC [24], as amended by Directives 2003/89/EC [25] and 2007/68/EC [26], to govern allergen labeling. The directive and its amendments identify 13 foods or food groups and sulfur dioxide (listed in Annex IIIa, see Table 2.3) that are found in a wide variety of processed foods, which are considered to be important relevant triggers of allergic reactions.

The EU has recently introduced a specific regulation for gluten (41/2009/EC) [27]. This regulation is intended to protect people intolerant to gluten and is based on the Codex Standard 118-1979, revised in 2008 [28]. Both define "gluten-free" foods as having a gluten level not exceeding 20 mg/kg. Regarding the labeling of foods specially processed to reduce the gluten content to a level up to 100 mg/kg, the European Commission (EC) regulation stipulates these foodstuffs shall bear the term "very low gluten," since the complete removal of gluten from gluten-containing grains is technically difficult and economically expensive.

Ingredients that are exempted from allergen declaration are listed in Directive 2007/68/EC [26] amending the Directive 2000/13/EC [25]. Under this legislation, even highly processed products like refined oils or polydextrins require labeling if derived from any of these allergens, even though the offending protein or substance might no longer be present. Food manufacturers wishing for specific products or ingredients derived from allergenic ingredients to be exempt from these labeling requirements have to make an application, supported by a detailed dossier demonstrating the safety of ingredients for relevant food-allergic individuals, which is then evaluated by the European Food Safety Authority. As a result, a number of exemptions have been granted,

TABLE 2.2 Allergen Labeling Requirements

Allergenic Food or Food Group, Including Derived Products (Unless Specifically Exempted)	Codex Alimentarius	United States	Canada	Australia and New Zealand	South Africa (Draft)	Switzerland[a]	EU (Annex IIIa)	Japan[b]
Cereals containing gluten, that is, wheat, rye, barley, oats, spelt, kamut, or their hybridized strains	✓	✓ (1)	✓	✓	✓	✓	✓	✓ (1)
Crustaceans	✓	✓	✓	✓	✓	✓	✓	✓ (2, 3)
Mollusks			✓		✓		✓	✓ (2, 4)
Eggs	✓	✓	✓	✓	✓	✓	✓	✓
Fish	✓	✓ (5)	✓	✓	✓	✓	✓	✓ (2, 6, 7)
Peanuts	✓	✓	✓	✓	✓	✓	✓	✓
Soybeans	✓	✓	✓	✓	✓	✓	✓	✓ (2)
Milk and dairy products	✓	✓	✓	✓	✓	✓	✓	✓
Tree nuts	✓	✓ (8, 9)	✓ (8)	✓	✓	✓ (8, 10)	✓ (8)	✓ (2)
Sesame seeds			✓	✓	✓	✓	✓	
Mustard							✓	
Celery						✓	✓	
Lupin							✓	

Buckwheat								✓
Beef								✓ (2)
Chicken (poultry)								✓ (2, 7)
Pork								✓ (2, 7)
Mushrooms								✓ (2)
Apples								✓ (2)
Kiwifruit								✓ (2)
Oranges								✓ (2)
Peaches								✓ (2)
Yams								✓ (2)
Sulfites (>10 mg/kg11)	✓	✓	✓	✓	✓	✓	✓	

Conferring cross-contact and precautionary labeling:

[a]Switzerland: if a listed allergen is present by cross-contact in amounts exceeding 1 g/kg (1000 mg/kg), it must be declared as an ingredient.

[b]Japan: allergens for which declaration is mandatory (i.e., those without the suffix (2) in the table) must be declared even if present by cross-contact. The use of the term "may contain" is prohibited, but alternatives are possible such as "manufactured on a line," which also manufactures "allergen X."

(1) Wheat only; (2) recommended by notice; (3) shrimp/prawn, crab; (4) abalone, squid; (5) fish should be listed by their common or usual species name; (6) salmon, salmon roe, mackerel; (7) gelatin included; (8) almond, Brazil nut, cashew nut, hazelnut, macadamia nut, pecan, pine nut, pistachio, walnut; (9) U.S. Food and Drug Administration (FDA) has published an extensive advisory list, some items of which are still in contention; (10) Swiss list includes all tree nuts listed in Annex IIIa (Table 2.3), but wording implies labeling could be required for others too; (11) sulfites are not allergens in the strict sense of the term but are universally included in allergen lists.

TABLE 2.3 Annex IIIa of Directive 2007/68/EC [25]: **Allergenic Food or Ingredients to be Labeled and Exemptions from Allergen Labeling**

Allergenic Food or Ingredient	Products Thereof Excluded
Cereals containing gluten, that is, wheat, rye, barley, oats, spelt, kamut	• Wheat-based glucose syrups including dextrose[a] • Wheat-based maltodextrins[a] • Glucose syrups based on barley • Cereals used for making distillates or ethyl alcohol of agricultural origin for spirit drinks and other alcoholic beverages
Crustaceans and products thereof	
Eggs and products thereof	
Fish and products thereof	• Fish gelatin used as carrier for vitamin or carotenoid preparations • Fish gelatin or isinglass used as fining agent in beer and wine
Peanuts and products thereof	
Soybean and products thereof	• Fully refined soybean oil and fat[a] including interesterified and partially hydrogenated soybean oil and fat • Natural mixed tocopherols (E306), natural D-alpha tocopherol, natural D-alpha tocopherol acetate, natural D-alpha tocopherol succinate from soybean sources • Vegetable oils derived phytosterols and phytosterol esters from soybean sources • Plant stanol ester produced from vegetable oil sterols from soybean sources
Milk and products thereof	• Whey used for making distillates or ethyl alcohol of agricultural origin for spirit drinks and other alcoholic beverages • Lactitol
Nuts, that is, almonds (*Amygdalus communis* L.), hazelnuts (*Corylus avellana*), walnuts (*Juglans regia*), cashews (*Anacardium occidentale*), pecan nuts (*Carya illinoiesis* (Wangenh.) K. Koch), Brazil nuts (*Bertholletia excelsa*), pistachio nuts (*Pistacia vera*), macadamia nuts and Queensland nuts (*Macadamia ternifolia*), and products thereof	• Nuts used for making distillates or ethyl alcohol of agricultural origin for spirit drinks and other alcoholic beverages

TABLE 2.3 Continued

Allergenic Food or Ingredient	Products Thereof Excluded
Celery and products thereof	
Mustard and products thereof	
Sesame seeds and products thereof	
Sulfur dioxide and sulfites at concentrations of more than 10 mg/kg or 10 mg/L expressed as SO_2	
Lupin and products thereof	
Mollusks and products thereof	

[a]And products thereof, in so far as the process that they have undergone is not likely to increase the level of allergenicity assessed by the European Food Safety Authority (EFSA) for the relevant product from which they originated.

which include some wheat-based glucose syrups, fish gelatin used as a carrier for vitamins, and fully refined soybean oil, among others (Table 2.3).

A similar legislation was passed in the United States in 2004 with the Food Allergen Labeling and Consumer Protection Act (FALCPA) [29], which came into force on January 1, 2006. This act contains provisions that differ in detail rather than principle from those of EU Directive 2003/89/EC [25]. The act mandates a shorter list of allergenic food groups but requires indication of the species on the label in the case of fish, crustacean, and tree nuts. It also specifically exempts highly refined oils from allergen labeling. Subsequent guidance on tree nuts describes a much longer list than the EU's Annex IIIa (Table 2.3). FALCPA provides for a process to obtain exemption from labeling, but so far, no derived ingredients have been exempted. As with the directive, FALCPA requires labeling of allergenic ingredients irrespective of the amount ultimately present in the product.

Both Directive 2003/89/EC [25] and FALCPA from 2004 [29], as well as legislation in other countries such as Australia and Japan, undoubtedly improve the labeling of allergenic foods. But, whereas in most countries regulation of allergen labeling includes unintended allergen contamination (e.g., carryover during manufacturing), the European allergen law for example is applied exclusively on allergenic ingredients. However, the General Food Law 178/2002/EC [22] imposes general obligations to provide safe food. Article 14 thus states that "1. Food shall not be placed on the market if it is unsafe," and then defines what is considered "unsafe": "2. Food shall be deemed to be unsafe if it is considered to be: injurious to health." The regulation also requires that, in determining safety, attention needs to be paid "to the particular health sensitivities of a specific category of consumers where the food is intended for that category of consumers." Considering also the product safety and product liability

Directives 85/374/EC [30] and 2001/95/EC [31], producers are required to place only safe foods onto the market. Managing cross-contact/cross-contamination issues requires the food producer to undertake some type of risk assessment. For the producer, in order to establish if a risk exists, testing for the potential presence of allergens, which are also used on the same production line or in the same factory, is essential. An assessment of the presence/absence of the allergen is needed, but if present, a level of the allergen present is needed to assess if this might be harmful at these levels for allergic consumers.

There are indications that there are levels of allergens (thresholds) below which an allergen poses only a small risk of causing harm to an allergic consumer [32]. However, commonly accepted trigger levels have yet to be established (with the exception of gluten), and Directive 2003/89/EC [25] gives no threshold or guidance to what constitutes a safe level.

However, in some countries, there have been attempts to set management threshold values. The Swiss authorities—in close cooperation with leading allergologists—defined an action limit of 1 part per 1000 in 2001. This limit represented a compromise between the specific food safety needs of allergic individuals and industrial food production practices at that time. If unavoidable, contaminations of above 1 g/kg (or liter) must be declared as ingredients, whereas contaminations below 1 g/kg do not need to be declared by law [33]. However, 1 mg/g peanut in, for example, a 50-g snack bar, could cause anaphylactic shock and subsequent death.

The Australian Food and Grocery Council (AFGC) published an industry guideline in 2007 as guidance for allergen management [34]. A three-level grid was developed to assist in determining if the residual protein from allergenic substances through unavoidable cross-contact presents such a risk that it requires a precautionary labeling statement. Three different action levels with thresholds for each food allergen were defined, derived from published data on the lowest triggering amounts measurable. The thresholds of the first level (no cross-contact statement required) range from 2 mg/kg for egg, peanuts, sesame, tree nuts, and crustacean to 5 mg/kg for milk and 10 mg/kg for soy as well as 20 mg/kg for fish and gluten.

While global food manufactures are usually well aware of the allergen situation and have appropriate measures in place to control it, small- and medium-sized enterprises are not necessarily in the same position. Here, a lack of knowledge is often combined with a very limited budget, which does not always allow the assessment of the situation at the production site.

2.4. FOOD ALLERGEN TESTING WITHIN A FOOD BUSINESS CONTEXT

2.4.1. The Hazard Analysis Critical Control Point (HACCP) Concept

The underpinning philosophy of modern food safety management techniques rests on quality assurance. In brief, this can be described as optimizing systems to reduce to an acceptable level the probability (risk) of a defective product (hazard) being produced. For the purposes of this discussion, the term defective applies to both food safety and other product attributes. In terms of food safety, the quality management system favored within the EU is HACCP, and there is a legal requirement that all businesses manage food safety in accordance with its principles [35]. HACCP recognizes that food processing alone does not necessarily reduce the risk of a food safety defect occurring. Reliance therefore must be made on supporting (prerequisite) systems [36]. Examples of such systems and relevant to allergen safety extend from those commonly associated to the manufacturing process (e.g., supplier quality assurance and sanitation) to label/wrapper design and, in particular, information provided relating to ingredient composition and required by legislation (discussed previously), as well as simply ensuring that the correct packaging is used for any particular product.

In common with any other food safety hazard, it is necessary to identify any step within the process, which can either significantly increase or reduce the risk of the hazard occurring. Such points are referred to as "critical control points" (CCPs). Codex Alimentarius [36] defines a CCP as *a step at which control can be applied and is essential to prevent or eliminate a food safety hazard or reduce it to an acceptable level.*

A CCP is thus a process step that either brings about a transition from hazardous to safe or, if uncontrolled, has the potential to render the product unsafe. A priori, in order to manage a CCP, it is necessary to be able to continuously monitor the process parameter that affects the safety of the product. A classical example of a CCP often quoted is that of a thermal kill step (e.g., pasteurization or sterilization). In this case, the factor is the temperature–time combination necessary to achieve a particular degree of bacterial kill. If one considers a simple example such as the production of pasteurized milk, the CCP is the pasteurizer. Given the state of the art, the pasteurizer's operating parameters are set on the basis of research demonstrating the various time–temperature combinations necessary to achieve a particular degree of kill in the relevant food matrix (*validation*). Once in operation, the performance of the pasteurizer is monitored by measuring the temperature of the milk leaving the holding tube and the time it is held there (usually by measuring the flow rate of the pump). The efficacy of the management system regarding the operation of the pasteurizer is *verified* by audit and appropriate microbiological analysis of the processed milk. As demonstrated in this example, it is important to note that, almost invariably, monitoring involves the measurement of those process parameters on which the risk reduction process depends, rather than the end point (in this case, the microbiological loading of the milk as it leaves the pasteurizer) itself.

The question then arises: are there steps in the actual food manufacturing process itself which can be managed to assure against adventitious allergen contamination? Food manufacturing processes are almost as diverse as the products made from them. Given the physical and chemical nature of food allergens, it is difficult to conceive of a current manufacturing process that could effect allergen removal or inactivation. Consequently, the CCP concept has limited utility in managing allergen contamination within a food manufacturing environment, and attention should focus more on prerequisite programs. This observation is borne out by consideration of food allergen-related product recalls notified to the United Kingdom Food Standards Agency [37]. Figure 2.1 reflects a summary of an analysis of these recalls (relating to over 150 notified recalls in a 21-month period, as of December 18, 2008.

Consideration of this analysis indicates that the single largest cause of product recalls relates to labeling deficiencies, in that information provided on the wrapper did not reflect the reality of the product. The major source of this problem relates to incomplete ingredient declarations, while the smaller group represents situations where although ingredient declarations were appropriate, the accompanying allergen advice (including precautionary labeling) was not. Deficiencies in food labeling can have a number of potential origins. Based on individual experience, the key contributory factor appears to be human error. This is seen to arise from three different sources:

- inaccurate transfer of information from those responsible for formulation to those with responsibility for wrapper design,
- failing to communicate the information to the consumer by those responsible for generating the wrapper, and
- not transmitting changes in product formulation to those responsible for wrapper design.

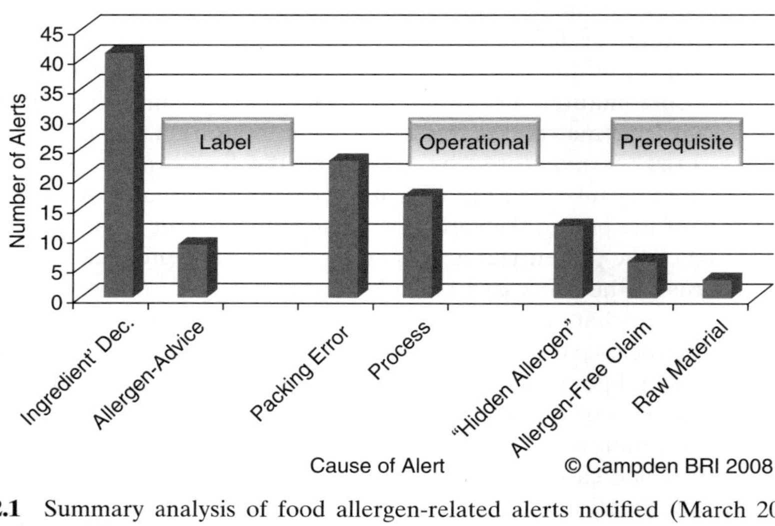

Fig. 2.1 Summary analysis of food allergen-related alerts notified (March 2007 to December 2008).

Errors that occurred at an operational level were the second largest cause of product recalls, that is, when the product was manufactured and packed. The most prevalent operational failure appears to be related to using the wrong packing for the item concerned. Process errors include using additional ingredients not declared on the wrapper, for example, applying a cheese topping to convenience food products where such toppings were not intended.

The smallest category of recalls relates to direct failures in what can be regarded to be the traditional concept of prerequisite programs. These include cross-contact and failures in supplier quality assurance ("hidden allergen" and raw materials; see Fig. 2.1). Possibly the most disturbing set of recalls, both in this category and in the analysis as a whole, are the five recalls associated with products carrying an allergen-free claim. In all of these cases, products which stated on their label that they were free from a particular allergenic food were, on analysis, found to contain the allergen in question. This places the individual consumer allergic to that particular ingredient at greater risk, given his or her greater chance of selecting that particular product.

2.4.2. Food Allergens and Food Allergen Testing in the Context of Food Safety Management

Food allergens provide a fundamental challenge to the food safety management practitioner, in that not only are they inherent constituents of a food but also that their adverse effect is experienced by a proportion of the consumer base. It is therefore unlikely that many food businesses will have a process step that will substantially reduce the allergen content of a particular food. In terms of food safety management, Alldrick [38] proposed that the hazard to be addressed by food businesses could be defined as *the inadvertent consumption of a food allergen by a susceptible individual.*

Food allergen analysis is of particular relevance to verifying the efficiency of prerequisite programs. Failures in this area have already been discussed above. It is worth noting that in the case of recalls relating to products with an "allergen-free" claim, two are related to gluten, two to casein, and one to soya. In terms of the "hidden allergen syndrome," recalls are related to milk, gluten, fish, tree nuts, and peanuts. Most of the recalls were associated with milk, which may reflect the commercial cost and "relative" ease of detection for milk, rather than the frequency of events implicating those allergens. It may also relate to the difficulties factories may encounter in managing dry powder ingredients such as skimmed milk powder, caseinates, and whey protein isolates. The data should therefore be considered under this aspect. A case in point relates to chocolate since, of the 10 recalls under this heading, 6 were related to chocolate-containing confectionery items.

As a case study, chocolate confectionery merits further consideration. Examination of alerts recorded under the Rapid Alert System for Food and Feed (RASFF) indicates that of 40 alerts recorded on the RASFF system for chocolate confectionery from 2005 until mid-February 2009, 26 referred to milk, 5 to peanuts, and 4 to soya followed by 4 to tree nuts (hazelnuts or almonds) and 1 to gluten. In a recent study, Pele et al. [39] examined 254 chocolate confectionery products sourced from 10 member states (not including the United Kingdom) for the presence of hazelnuts and peanuts. None of these products declared either food allergen as an ingredient; however, over 50% of those products, for which there was no precautionary labeling, actually tested positive for hazelnut (23% for peanut).

2.5. CONCLUSION

In many respects, allergens are different from other supply chain hazards. Mycotoxins for example enter the supply chain at a very early stage usually already on the field or during storage of grain products. Allergens on the other hand can enter the supply chain as a consequence of events before, during, or after manufacture of the food. Product contamination can therefore occur due to contaminated ingredients, poor ingredient or finished product storage practices, and cross-contact due to either close proximity to other manufacturing containing allergen-containing foods or poor sanitation practices between production runs of allergen-containing or allergen-free products.

However, cross-contamination is only one of several issues that may make proper labeling of products and therefore consumer choice difficult. For one, legislation in different countries differs significantly. Whereas in the United States eight allergens require labeling (Table 2.2), in Canada, the list of allergens to be labeled and declared is extended to a further two totaling to 10 groups of allergens. In Europe, four more allergens are added, while in Japan, up to 25 are listed, which only partially overlap with other legislatures. The complexity increases with the level of detail; for instance, the list of nuts to be labeled differs between regulations in different countries and regions.

Furthermore, there are no labeling threshold levels for allergens yet (except for sulfite and gluten, neither of them being true allergens). Even though Switzerland has set a mandatory labeling threshold for allergens at 1000 mg/kg, this seems exceedingly high considering that many allergic patients react at doses in the milligram range; for example, for peanut, this level in portion sizes associated with most products could trigger anaphylactic shock and consequently cause death. Thresholds established for the major food allergens should be based on transparent scientific clinical and epidemiological data and reevaluated periodically as new data and tools become available.

The regulatory framework and usage for precautionary ("may contain") labeling in different countries is already complex enough and makes the choice for the allergic consumer very difficult. However, it does not and cannot provide complete protection, and indeed, it must be used with circumspection if it is to retain any value as a risk management tool. Ultimately, this requires the definition of management thresholds, reflective of a tolerable level of risk.

Definition of this level will recognize that total absence of risk is unattainable and will therefore aim to minimize the overall public health risk, based on analysis of all the factors which affect it.

Challenges from the analytical side are hardly less daunting. To determine allergen levels (see Chapter 13), two main techniques are currently being used: enzyme-linked immunosorbent assay (ELISA) and polymerase chain reaction (PCR). The presence of DNA proved by PCR may not necessarily indicate the presence of offending allergenic structures, meaning that the protein that triggers the allergic reaction may not be present while the DNA still is. This poses a major problem in interpreting the results of such tests in terms of risk to the allergic consumer. In terms of labeling, this may be considered prima facie evidence for making an appropriate declaration.

ELISA, currently the most common technique, allows the detection of proteins qualitatively and quantitatively in the low milligram per kilogram range. This technique, however, also has a number of drawbacks as, for example, false-positive results due to cross-reactivity of the antibody used (e.g., antibodies against tropomyosin in crustaceans also detect the closely related protein cockroaches), false negatives especially in processed products, or very different quantitative results. Food matrices can significantly interfere both qualitatively and quantitatively with allergen detection. These facts make it very difficult for food manufacturers or ingredient manufacturers to properly control allergen contamination.

There is an urgent need for a close collaboration between all affected stakeholder groups for which the EU-funded MoniQA project provides a platform. Authorities and food industry need robust analytical tools that have to be validated properly. This will help to make sure that any future allergen regulation is enforceable and ultimately allows the allergic consumer an informed choice.

ACKNOWLEDGMENTS

This work was completed on behalf of the MoniQA WG Food Allergens and funded under the EU FP6 Programme, Topic 5.4.5.1—Quality and safety control strategies for food (Contract No. FOOD-CT-2006-036337). The working group appreciates the contribution of the following external stakeholders: Audrey Dunn Galvin, René Crevel, Sylvia Pfaff, Georg Schäppi, Roger Wood, Andrew Damant, and Samuel Godefroy. Audrey DunnGalvin, Sylvia Pfaff, Rene Crevel, and E. N. Clare Mills acknowledge funding through the EU EuroPrevall project (FOOD-CT-2005-514000). E. N. Clare Mills also acknowledges funding from the Biotechnology and Biological Sciences Research Council through the competitive strategic grant to the Institute of Food Research. Sandra Kerbach and Bert Popping acknowledge the contributions of Eurofins in-kind.

REFERENCES

1. Bush, R.K., Hefle, S.L. (1996). Food allergens. *Critical Reviews in Food Science and Nutrition*, *36*, 119–163.
2. Taylor, S.L., Hefle, S.L., Bindslev-Jensen, C., Bock, S.A., Burks, A.W., Christie, L., Hill, D.J., Host, A., Hourihane, J.O., Lack, G., Metcalfe, D.D., Moneret-Vautrin, D.A., Vadas, P.A., Rance, F., Skrypec, D.J., Trautman, T.A., Malmeheden Yman, I., Zeiger, R.S. (2002). Factors affecting the determination of threshold doses for allergenic foods: how much is too much? *The Journal of Allergy and Clinical Immunology*, *109*, 24–30.
3. Breiteneder, H., Mills, E.N.C. (2005). Molecular properties of food allergens. *The Journal of Allergy and Clinical Immunology*, *115*, 14–23.
4. Johansson, S.G.O., Hourihane, J.O., Bousquet, J., Bruijnzeel-Kooman, C., Drejborg, S., Haahtela, T., Kowlaski, M.L., Mygind, N., Ring, J., van Cauwenberge, P., van Hage-Hemsten, M., Wüthrich, B. (2001). A revised nomenclature for allergy—an EACCI position statement from the EACCI nomenclature task force. *Allergy*, *56*, 813–824.
5. Hischenhuber, C., Crevel, R., Jarry, B., Mäki, M., Moneret-Vautrin, D.A., Romano, A., Troncone, R., Ward, R. (2006). Safe amounts of gluten for patients with wheat allergy or coeliac disease. *Alimentary Pharmacology and Therapeutics*, *23*, 559–575.
6. Bereczky, S., Montgomery, S.M., Troye-Blomberg, M., Rooth, I., Shaw, M.A., Färnert, A. (2004). Elevated anti-malarial IgE in asymptomatic individuals is associated with reduced risk for subsequent clinical malaria. *International Journal of Parasitology*, *34*, 935–942.
7. Sicherer, S.H., Munoz-Furlong, A., Sampson, H.A. (2004). Prevalence of seafood allergy in the United States determined by a random telephone survey. *The Journal of Allergy and Clinical Immunology*, *114*, 159–165.
8. Young, E., Stoneham, M.D., Petruckevitch, A., Barton, J., Rona, R. (1994). A population study of food intolerance. *The Lancet*, *343*, 1127–1130.
9. Sampson, H.A. (2005). Food allergy—accurately identifying clinical reactivity. *Allergy*, *60*, 19–24.
10. Steensma, D.P. (2003). The kiss of death: a severe allergic reaction to a shellfish induced by a good-night kiss. *Mayo Clinic Proceedings*, *78*, 221–222.
11. Hefle, S.L. (2001). Hidden food allergens. *Current Opinion in Allergy and Clinical Immunology*, *1*, 269–271.
12. Codex Alimentarius Commission. (1999). Codex general standard for the labelling of prepackaged foods. *CODEX STAN 1-1985, Revised 1-1991*, 1–11.
13. De Blok, B.M., Vlieg-Boerstra, B.J., Oude Elberink, J.N., Duiverman, E.J., DunnGalvin, A., Hourihane, J.O., Cornelisse-Vermaat, J.R., Frewer, L., Mills, C., Dubois, A.E. (2007). A framework for measuring the social impact of food allergy across Europe: a EuroPrevall state of the art paper. *Allergy*, *62*, 733–737.
14. Miles, S., Fordham, R., Mills, E.N.C., Mugford, M. (2005). A framework for measuring costs to society of IgE-mediated food allergy. *Allergy*, *60*, 996–1003.
15. DunnGalvin, A., de Blok, B.M.J., Dubois, A.E.J., Hourihane, J.O. (2008). Development and validation of the food allergy quality of life—parent adminis-

tered questionnaire (FAQLQ-PF) for food allergic children aged 0-12 years. *Clinical and Experimental Allergy*, *38*, 977–986.

16. Flokstra-de Blok, B.M., DunnGalvin, A., Vlieg-Boerstra, B.J., Oude Elberink, J.N., Duiverman, E.J., Hourihane, J.O., Dubois, A.E. (2008). Development and validation of the self-administered food allergy quality of life questionnaire for adolescents. *The Journal of Allergy and Clinical Immunology*, *122*, 139–144.
17. DunnGalvin, A., Dubois, A.E.J., de Blok, B.M.J., Hourihane, J.O. (2007). Child vs maternal perception of HRQoL in food allergy: developmental trajectories and evolution of risk behaviour. *Allergy*, *62*, 70–166.
18. Sampson, M.A., Munoz-Furlong, A., Sicherer, S.H. (2006). Risk-taking and coping strategies of adolescents and young adults with food allergy. *The Journal of Allergy and Clinical Immunology*, *117*, 1440–1445.
19. Cornelisse, J.R., Voordouw, J., Yiakoumaki, V., Theodoridis, G., Frewer, L.J. (2008). Food-allergic consumers' labelling preferences: a cross-cultural comparison. *European Journal of Public Health*, *18*, 115–120.
20. Hefle, S.L., Furlong, T.J., Niemann, L., Lemon-Mule, H., Sicherer, S., Taylor, S.L. (2007). Consumer attitudes and risks associated with packaged foods having advisory labeling regarding the presence of peanuts. *The Journal of Allergy and Clinical Immunology*, *120*, 171–176.
21. Noimark, L., Gardner, J., Warner, J.O. (2009). Parents' attitudes when purchasing products for children with nut allergy: A UK perspective. *Pediatric Allergy and Immunology*, *20*, 500–505.
22. European Parliament and Council. (2002). Directive 2002/178/EC. *Official Journal of the European Union*, *L60*, 70–80.
23. Food Quality Protection Act. (1996). Public Law 104-170-Aug 3. *110 STAT*, 1489.
24. European Parliament and Council. (2000). Directive 2000/13/EC. *Official Journal of the European Union*, *L109*, 29–42.
25. European Parliament and Council. (2003). Directive 2003/89/EC. *Official Journal of the European Union*, *L308*, 15–18.
26. European Parliament and Council. (2007). Directive 2007/68/EC. *Official Journal of the European Union*, *L310*, 11–14.
27. European Parliament and Council. (2009). Regulation 41/2009/EC. *Official Journal of the European Union*, *L16-3*, 1–3.
28. Codex Alimentarius Commission. (2008). Codex general standard for foods for special dietary use for persons intolerant to gluten. *CODEX STAN 118-1979, Revised 2008*, 1–3.
29. Food Allergen Labeling and Consumer Protection Act. (2004). Public Law 108-282-Aug 2, *18 STAT*, 905.
30. European Parliament and Council. (1985). Directive 85/374/EC. *Official Journal of the European Union*, *L210*, 29.
31. European Parliament and Council. (2001). Directive 2001/95/EC. *Official Journal of the European Union*, *L11*, 4.
32. Crevel, R.W.R., Ballmer-Weber, B.K., Holzhauser, T., Hourihane, J.O., Knulst, A.C., Mackie, A.R., Timmermans, F., Taylor, S.L. (2008). Thresholds for food allergens and their value to different stakeholders. *Allergy*, *63*, 597–609.

33. Lebensmittelkennzeichnungsverordnung. (2005). Verordnung des Eidgenössischen Departements des Inneren über die Kennzeichnung und Anpreisung von Lebensmitteln 817.022.21. *Article 8*, 7–8.

34. The Australian Food and Grocery Council. (2007). Food industry guide to allergen management and labeling. Available at http://www.allergenbureau.net/allergen-guide/afgc-allergen-guide/

35. European Parliament and Council. (2004). Directive 2004/852/EC. *Official Journal of the European Union, L139*. Available at http://eur-lex.europa.eu/LexUriServ/LexUriServ.do?uri=OJ:L:2004:139:0001:0054:EN:PDF

36. Codex Alimentarius. (2003). Recommended international code of practice general principles of food hygiene. *CAC/RCP 1-1969, Revised 4-20031*. Available at http://www.codexalimentarius.net/web/index_en.jsp

37. Food Standards Agency. (2008). Allergy alert. Available at http://www.food.gov.uk/safereating/allergyintol/alerts/

38. Alldrick, A.J. (2006). Managing allergy issues. *The World of Food Ingredients*, October/November, pp. 64–65.

39. Pele, M., Brohée, M., Anklam, E., Van Hengel, A.J. (2007). Peanut and hazelnut traces in cookies and chocolates: relationship between analytical results and declaration of food allergens on product labels. *Food Additives and Contaminants*, *24*, 1334–1344.

3

CRITERIA TO DETERMINE PRIORITY ALLERGENS: TREE NUT ALLERGY REVIEW

Jupiter M. Yeung

3.1. INTRODUCTION

In 2004, the U.S. Congress passed the Food Allergen Labeling and Consumer Protection Act (FALCPA). This act defined the major food allergens, including tree nuts. However, the list for tree nuts, as a food group, does not define in specific terms what is considered a tree nut. The statute only lists three examples of tree nuts: almonds, pecans, and walnuts. In October 2006, the U.S. Food and Drug Administration (FDA) defined tree nuts as the following: "almond, beechnut, Brazil nut, butternut, cashew, chestnut (Chinese, American, European, sequin), chinquapin, coconut, filbert/hazelnut, ginkgo nut, hickory nut, lychee nut, macadamia nut/bush nut, pecan, pine nut/piñon nut, pili nut, pistachio, shea nut, walnut (English, Persian, Black, Japanese, California), heartnut, and butternut" in the latest edition of the *Guidance for Industry: Questions and Answers Regarding Food Allergens, Including the Food Allergen Labeling and Consumer Protection Act of 2004* (Edition 4) [1]. However, there are concerns in the food and beverage industry that the list includes nuts that have minimal health risk, which devalues labeling as a risk management measure and, thus, reduces rather than increases consumer safety. On behalf of the food and beverage industry, the Grocery Manufacturers Association (GMA) examined the allergenic potential of all 19 "tree nuts" defined by the

Allergen Management in the Food Industry, Edited by Joyce I. Boye and Samuel Benrejeb Godefroy

FDA and found little science supporting the labeling of several FDA-listed "tree nuts" as "major food allergens."

3.2. TREE NUT ALLERGY

Among foods causing allergic reactions, tree nuts have attracted considerable attention. Allergies to these foods are common. In the United States, an estimated 1.4% of the overall general population is allergic to peanuts, tree nuts, or both [2, 3], which translates into 4.2 million Americans. Reported allergies to tree nuts include walnut, cashew, Brazil nut, almond, pecan, pistachio, hazelnut, and macadamia nut [2]. Tree nut allergy frequently has an onset in the first few years of life and can result in severe and potentially fatal allergic reactions. Fatalities due to ingestion of tree nuts include almond, Brazil nut, cashew, hazelnut, pecan, pistachio, and walnut. The treatment of tree nut allergies includes avoidance of the specific nuts and educating the tree nut-sensitive individuals and their families of the potential for adverse reactions caused by accidental ingestion. While tree nut allergy is generally thought to be lifelong, about 9% of nut-allergic individuals outgrow their tree nut allergy [4].

3.3. CRITERIA TO DETERMINE A MAJOR FOOD ALLERGEN

We have previously proposed the criteria to determine a major food allergen of public health importance in the context of food labeling [5]. Without scientifically defined criteria, proliferation of allergen lists is likely to occur. Such practices may lead to an unnecessary elimination of enjoyable foods containing important nutrients, compromising food choice. The principal criteria to be used for identifying food allergens that are associated with frequent allergic reactions are prevalence in the population, severity of the reaction, and the threshold levels of the allergen when available. While anaphylaxis is often used as a guideline to assess the severity of reactions, prevalence in the population, a critical indicator of public health concern, is an obligatory criterion for determination of a major food allergen. For example, wheat is a major food allergen and about 1% of the general population has wheat allergy or celiac disease. However, allergic reactions to wheat are generally not severe, and there is no published report of fatality. This is further supported by Hefle et al. [6] who documented over 160 foods with evidence of allergies; there could well be one or more persons in the world who are at risk of developing severe reactions to many of these 160 foods. For instance, one chickpea-induced death was reported by Pumphrey [7]; however, chickpea is not a major food allergen.

On the other hand, the International Life Sciences Institute (ILSI) Europe Food Allergen Task Force proposed either a positive double-blind, placebo-controlled food challenge (DBPCFC) or well-documented anaphylaxis as the main inclusion criteria for definition of food allergens that would require labeling [8], but not all major tree nuts have documented DBPCFC to support their inclusions.

It is important to recognize that, until recently, there was no consensus agreement on the definition of anaphylaxis [9, 10]. Consequently, some of the earlier case reports on anaphylactic reactions may not be classified as such under the existing definition. Anaphylaxis is generally believed to be a severe, potentially fatal, systemic allergic reaction that occurs suddenly after contact with an allergy-causing substance, and is characterized by life-threatening upper airway obstruction, bronchospasm, and hypotension [11, 12]. A universal definition of anaphylaxis was recently proposed as follows: *anaphylaxis is a serious allergic reaction that is rapid in onset and may cause death* [9]. Representatives of this working group were from 16 different international organizations or governmental bodies from North America, Europe, and Australia. For purposes of our assessment, when possible, we tried to use the current definition of anaphylaxis when assessing literature reports of adverse reactions. Given the limited information available in some reports, it was not possible to predict with certainty whether the reported reactions should be classified as "anaphylaxis."

To date, science only supports the inclusion of nine nuts of public health importance in the list of tree nuts because they are associated with prevalence and severity of reactions that warrant their classification as major food allergens. Including tree nuts that have either no evidence of allergic reaction or for which the reactions are lacking in prevalence and severity only creates confusion to allergic consumers and the food and beverage industry. Labeling "tree nuts" that have minimal risk could potentially devalue labeling as a risk management measure and, thus, reduce rather than increases consumer safety.

3.3.1. GMA-Proposed Tree Nut List

The food and beverage industry's objective is to propose a rational tree nut list that can protect public health yet does not unnecessarily limit food choices. The proposed list is based on recent scientific reviews on tree nuts by well-known clinicians and by using the lists prepared by international regulatory agencies, a consumer group, and the food industry. In addition, GMA reviewed the available scientific literature and concluded these nine tree nuts of public health importance meet the criteria for inclusion in a list of major allergens given the prevalence and severity of reactions reported in the literature:

1. almond
2. Brazil nut
3. cashew
4. hazelnut
5. macadamia

6. pecan
7. pine nut
8. pistachio
9. walnut.

These nine tree nuts have also been used by the food industry in its voluntary labeling efforts years before FALCPA was enacted. The list is also reflective of consumer and physician education over the last 10 years. Using this well-established list from North America, Europe, and Australia promotes harmonization of food allergen labeling, which in turn facilitates global trade. All nine tree nuts in the list satisfy the severity and prevalence criteria as major food allergens. Deaths have been documented for these tree nuts, with the exception of macadamia and pine nut, which are included in the list based on the weight of evidence on severity of reactions.

These tree nuts are also found in the list used by the scientific community (Table 3.1), some regulatory agencies (Table 3.2), a consumer group (Table 3.3), and the food industry as indicated below:

A. Medical groups specializing in tree nut allergy (Table 3.1), including University of Johns Hopkins [4], University of California, Davis [13], Monash University, Australia [14] (i.e., same as GMA-proposed list minus pecan and pistachio), and University of Cambridge, UK [15] (i.e., same as GMA-proposed list minus macadamia, pine nut, and pistachio).
B. International regulatory agencies (Table 3.2) such as the Canadian Food Inspection Agency (CFIA) [16], Health Canada [17], European Union (EU) [18], (i.e. same as GMA-proposed list minus pine nut), Food

TABLE 3.1 Comparison of Published Tree Nut Lists—Clinical Classification

	J Allergy Clin Immunol	*Curr Allergy Asthma Rep*	*Clin Exp Allergy*	*Clin Exp Allergy*
Tree Nut	Fleischer et al. [4]	Teuber et al. [13]	de Leon et al. [14]	Clark and Ewan [15]
Almond	✓	✓	✓	✓
Brazil nut	✓	✓	✓	✓
Cashew	✓	✓	✓	✓
Hazelnut	✓	✓	✓	✓
Macadamia	✓	✓	✓	
Pecan	✓	✓		✓
Pine nut	✓	✓	✓	
Pistachio	✓	✓		
Walnut	✓	✓	✓	✓

TABLE 3.2 Comparison of Tree Nut Lists—Regulatory Classification

Tree Nut	CFIA	Health Canada	EU	FSANZ	FDA	Senate Report 108-226[a] (on FALCPA)
Almond	✓	✓	✓	✓	✓	✓
Brazil nut	✓	✓	✓	✓	✓	✓
Cashew	✓	✓	✓	✓	✓	✓
Hazelnut	✓	✓	✓	✓	✓	✓
Macadamia	✓	✓	✓	✓	✓	✓
Pecan	✓	✓	✓	✓	✓	✓
Pine nut	✓	✓		✓	✓	✓
Pistachio	✓	✓	✓	✓	✓	✓
Walnut	✓	✓	✓	✓	✓	✓
Others				Chestnut Hickory nuts	Beechnut Butternut Chestnut Chinquapin Coconut Ginkgo nut Hickory nut Lychee nut Pili nut Shea nut	Chestnut

[a]According to the Committee on Health, Education, Labor, and Pensions (HELP), in Title II of S. 741, the term "tree nuts" refers to a variety of individual nuts, including almonds, Brazil nuts, cashews, chestnuts, filberts/hazelnuts, macadamia nuts, pecans, pine nuts, pistachios, and walnuts (page 6 of the report).

TABLE 3.3 Tree Nut List for Kids—FAAN

Tree Nut	FAAN
Almond	✓
Brazil nut	✓
Cashew	✓
Hazelnut	
Macadamia	
Pecan	✓
Pine nut	✓
Pistachio	✓
Walnut	✓

Standards Australia New Zealand (FSANZ) [19] (i.e., same as GMA-proposed list plus chestnut and hickory nut), and the FDA [1] (i.e., same as GMA-proposed list plus beechnut, butternut, chestnut, chinquapin, coconut, ginkgo nut, hickory nut, lychee nut, pili nut, and shea nut).

C. *The Consumer Group, Food Allergy and Anaphylaxis Network (FAAN).* The February/March 2007 Food Allergy News for Kids (published by FAAN) [20] has a "Let's Learn about Tree Nuts" section (Table 3.3). The tree nuts listed and pictured are almonds, Brazil nuts, cashews, chestnuts, pecans, pine nuts, pistachios, and walnuts. There is no mention of other nuts such as coconut or shea nut.

   Interestingly, FAAN does adopt the FDA list of tree nuts on its label card, "How to Read a Label for a Tree Nut Free Diet" [21]. While FAAN includes coconut on the label card, there is a Q&A on the FAAN website noting that "Coconut, the seed of a drupaceous fruit, has typically not been restricted in the diets of people with tree nut allergy. However, in October of 2006, the FDA began identifying coconut as a tree nut. The available medical literature contains documentation of a small number of allergic reactions to coconut; most occurred in people who were not allergic to other tree nuts. Ask your doctor if you need to avoid coconut" [22].

D. *Industry.* The risk-based GMA tree nut list is the same as the list advocated by the International Tree Nut Council [23]. These nine tree nuts have been used by the food industry in its voluntary labeling efforts for years before FALCPA was enacted.

3.3.1.1. FDA Tree Nut List. The FDA announced their tree nut list in the *Guidance for Industry: Questions and Answers Regarding Food Allergens, Including the Food Allergen Labeling and Consumer Protection Act of 2004* (Edition 4) in October 2006 [1]. The list consists of almond, beechnut, Brazil nut, butternut, cashew, chestnut, chinquapin, coconut, filbert/hazelnut, ginkgo nut, hickory nut, lychee nut, macadamia nut, pecan, pine nut, pili nut, pistachio, shea nut, and walnut. FDA's "tree nuts" that are not consistent with the GMA proposed list are shown in Table 3.4.

3.3.2. Rationale for Selection for the GMA-Proposed Tree Nut List

3.3.2.1. Prevalence and Severity of Reactions to Tree Nuts. Tree nuts are a well-defined cause of food allergy. Allergies to tree nuts, peanuts, or both affect about 1.4% of the general population in the United States. As a food group, tree nuts are the most frequent cause of fatal anaphylactic reactions. Anaphylaxis caused by food allergy may differ clinically from other causes of anaphylaxis. Death is usually caused by respiratory failure, and patients who have asthma are at greatest risk for severe reactions [24]. A recent analysis of 32 fatal food allergic reactions in the United States from 1994 to 1999 showed tree nuts to be responsible for 31% of the deaths [3]. A follow-up study between 2001 and 2006 reported eight fatalities caused by tree nuts out of 31 food fatalities [25]. The greatest number of fatalities still occurred in adolescents and young adults. Similarly, Pumphrey and Gowland [26] reported that 7 out of 48 deaths in the United Kingdom were due to tree nuts between 1999 and 2006. The majority of deaths also occurred between 11 and 30 years of

TABLE 3.4 Comparison of GMA and FDA Tree Nut Lists

Tree Nuts	GMA	FDA
Almond	✓	✓
Brazil nut	✓	✓
Cashew	✓	✓
Hazelnut	✓	✓
Macadamia nut	✓	✓
Pecan	✓	✓
Pine nut	✓	✓
Pistachio	✓	✓
Walnut	✓	✓
Beechnut		✓
Butternut		✓
Chestnut		✓
Chinquapin		✓
Coconut		✓
Ginkgo nut		✓
Hickory nut		✓
Lychee nut		✓
Pili nut		✓
Shea nut		✓

age. A summary of published tree nut fatalities is tabulated in Table 3.5. Fatality represents the extreme outcome of the severe nature of the adverse events. Any tree nut with a reported fatality must be included in the major food allergen list. Tree nut fatalities include almond (three cases), Brazil nut (four cases), cashew (two cases), hazelnut (two cases), pecan (five cases), pistachio (one case), and walnut (sixteen cases). Only pine nuts and macadamia nuts in the GMA list do not have a reported lethal outcome.

3.3.2.2. Specific Tree Nut Review.

3.3.2.2.1. Almond. Allergy to almond was the third most common tree nut allergy reported by Sicherer et al. [27] with 15% of the 1667 self-reported nut-allergic registrants reporting almond allergy. In a later self-reported survey, Sicherer et al. [2] reported that almond remained the third most common tree nut to cause allergy in the United States after walnut and cashew with 32 of 82 respondents reporting allergy. Reactions ranged from hives to life-threatening anaphylaxis to fatality. Similarly, Ewan [28] reported 14 almond-allergic patients among 62 tree nut-allergic patients in 1 year from 1993 to 1994. Clark and Ewan [15] reported that 34 of 1000 peanut- or nut-allergic patients showed their strongest reaction to almond. Due to its frequent occurrence and severity of the reactions including fatalities, inclusion of almond in the tree nut list is warranted.

TABLE 3.5 Summary of Published Tree Nut Fatalities

	J Allergy Clin Immunol	*J Allergy Clin Immunol*	*Curr Opin Allergy Clin Immunol*	*J Allergy Clin Immunol*	*Clin Exp Allergy*	*N Engl J Med*	*Allergy Proc*	*JAMA*
Tree Nut	Bock et al. [25]	Pumphrey and Gowland [26]	Pumphrey [7]	Bock et al. [3]	Pumphrey [78]	Sampson et al. [37]	Boyd [65]	Yunginger et al. [64]
Almond	1		2					
Brazil nut			2	2				
Cashew						2		
Hazelnut	1		1					
Macadamia								
Pecan				2		1	1	1
Pine nut								
Pistachio				1				
Walnut	1		6	3	5	1		
General/mixed/ unidentified	5	9	11	2	10			

Numbers denote cases reported.

3.3.2.2.2. Brazil Nut. Allergy to Brazil nut is a relatively common nut allergy and can be fatal [29]. Senna et al. [30] reported a severe episode of anaphylaxis occurred during a skin prick test of fresh Brazil nut that required epinephrine and intravenous steroids, while others [31–33] documented case reports of anaphylaxis to Brazil nuts. Sicherer et al. [34] documented that 4 of 54 nut-allergic patients reported a reaction to Brazil nut. Similarly, Ewan [28] reported 18 Brazil nut-allergic patients out of 62 peanut and tree nut allergics. Peanuts were the commonest cause of allergy (47) followed by Brazil nut (18), almond (14), and hazelnut (13). Clark and Ewan [15] reported that 162 of 1000 peanut- or nut-allergic patients (16%) showed their strongest reaction to Brazil nut.

Pastorello et al. [35] found that all the Brazil nut-allergic patients had specific immunoglobulin E (IgE) against a 9-kDa allergen, indicating that the allergen underlying clinical reactions to Brazil nut is a 2S albumin. Arshad et al. [36] reported 12 cases of allergy to Brazil nut. Due to its frequent occurrence and severity of the reactions including fatalities [7, 32], inclusion of Brazil nut in the tree nut list is warranted.

3.3.2.2.3. Cashew. Cashew causes severe symptoms and has caused death [37]. Cashew allergy is described as an evolving clinical problem. Davoren and Peake [38] reviewed 213 children with peanut or tree nut allergy over a 42-month period and found that anaphylaxis to cashew nut (74.1%) was more common than to peanut (30.5%), even though 83.1% of those children have peanut allergy and only 12.6% have cashew allergy. Children with cashew allergy are at risk of anaphylaxis. Rance et al. [39, 40] investigated 42 children with cashew allergy. Fifty-six percent had skin symptoms, 25% had respiratory signs, and 17% had digestive signs.

Allergy to cashew was found to be the second most common tree nut allergy [27] with 20% of the 1667 nut-allergic registrants reporting cashew allergy. Similarly, Sicherer et al. [34] reported 11 of 54 nut-allergic patients as cashew allergic. Clark and Ewan [15] reported that 29 of 1000 peanut- or nut-allergic patients showed their strongest reaction to cashew. Moneret-Vautrin et al. [41] reported that 40% of 140 peanut-allergic patients were sensitized to cashew. Tariq et al. [42] reported that 1 of 1218 of a birth cohort (0.08%) was allergic to cashew. Moreover, Rasanen et al. [43] reported a cross-reactivity of pectin with cashew nut extract in one patient, and Ferdman et al. [44] described a pectin anaphylaxis associated with cashew allergy.

Quercia et al. [45] reported that a patient allergic to pistachio developed anaphylaxis after eating cashew. Hourihane et al. [46] also reported 29 patients with cashew nut allergy aged 1–30 years. Fourteen of these reacted to minimal contact without actually eating cashew. Fourteen reported wheeze, and 11 reported collapse or feeling faint. Cashew has recently become widely available in the form of butter spreads. Cashew causes symptoms in formal food challenges and has caused death. Garcia-Menaya et al. [47] described clinical manifestations in three cases of cashew-related anaphylaxis. A case of acute reactions 20 minutes after eating a piece of chocolate candy containing cashew was described by Rasanen et al. [43], and Tariq et al. [42] also reported a generalized urticaria in one child. Due to its frequent occurrence and severity of the reactions including reported fatalities [37], inclusion of cashew nut in the tree nut list is warranted.

3.3.2.2.4. Hazelnut. Hazelnuts are a common cause of food allergy. Children can be sensitized to hazelnut at an early age [34]. Remarkably, a large proportion had never ingested hazelnut to their knowledge, and most allergic reactions to nuts in childhood occur after the first known exposure. As reported by Beyer et al. [48], allergic reactions to hazelnuts range from mild oral allergy syndrome (OAS) caused by cross-reactivity between tree pollen and hazelnut proteins to severe anaphylactic reactions. They reported 14 patients with hazelnut-induced systemic reactions. Clinical symptoms observed by Schocker et al. [49] in 26 patients included OAS in 7 of 26 patients, anaphylaxis (6 of 26), angioedema (6 of 26), urticaria (2 of 26), and food-dependent, exercise-induced anaphylaxis in a single patient.

Pastorello et al. [50] characterized hazelnut allergens by studying the specific IgE reactivity of 65 patients with positive DBPCFC results and 7 patients with severe anaphylaxis to hazelnut. The major allergen of hazelnut is an 18-kDa protein homologous to Bet v 1, and the 9-kDa allergen is presumably a lipid transfer protein (LTP). Other major allergens have molecular weights of 47, 32, and 35 kDa. Hansen et al. [51] found that 5 of 17 patients reacted to roasted hazelnut with OAS, while all 17 reacted to raw hazelnut. Furthermore, the median dose for reaction to raw hazelnut was more than doubled with roasted hazelnut.

From selected subjects with a history of allergic reactions on ingestion of hazelnut, Ortolani et al. [52] found that of 86 patients, 67 (77.9%) had a positive DBPCFC result. Challenge doses were 1.4–20 g of ground hazelnut. Reactions to hazelnut ranged from oral and gastrointestinal symptoms to systemic symptoms.

Wensing et al. [53, 54] attempted to determine the distribution of minimum provoking doses of hazelnut in 31 hazelnut-allergic patients using 1, 3, 10, 30, 100, and 300 mg and 1 g of hazelnut protein (corresponding to 6.4, 19, 64, 190, 640, 1900, and 6400 mg of hazelnut meal, respectively) challenging doses. All reactors showed itching of the mouth as the first symptom; one showed generalized urticaria and lip swelling and another slight lip swelling at 1-mg dose. Some gastrointestinal symptoms, including abdominal pain and nausea, were also reported. The challenge was continued until objective symptoms were observed or subjective symptoms persisted for 1 hour (seven patients). Four patients reacted to the lowest dose. Threshold doses for eliciting subjective reactions varied from a dose of 1 up to 100 mg hazelnut protein (equivalent to 6.4–640 mg of hazelnut meal). Extrapolation of the dose–response curve showed that 50% of the hazelnut-allergic population would suffer from an allergic reaction after ingestion of 6 mg (95% confidence interval [CI], 2–11 mg) of hazelnut protein. Objective symptoms were observed in two patients after 1 and 1000 mg, respectively.

Flinterman et al. [55] studied the clinical relevance of hazelnut sensitization and eliciting doses (EDs) in childhood by DBPCFC. Twenty-eight sensitized children (8 females, 20 males; age 4–16 years) with suspected allergy to hazelnut were recruited. The challenge was composed of nine portions of defatted hazelnut flour in series: 10 μg, 100 μg, 500 μg, 1 mg, 10 mg, 100 mg, 300 mg, 1 g, and 3 g (protein content, 15.5%). The last dose was performed in an open challenge and consisted of 10 hazelnuts (5 g; approximately 635 mg of protein), because this amount could not be masked in edible portions for children. The challenge was discontinued after the occurrence of an objective reaction. Doses as large as 1 mg of hazelnut flour were tolerated by all children. The ED for OAS (≥10 mg; 1.6 mg of protein) and for objective symptoms (≥3.00 mg; 46.5 mg of protein) was observed. While fewer than half of the children with sensitization to hazelnut appeared to be allergic to hazelnut in this study, due to its frequent occurrence and severity of the reactions including reported fatalities [7, 21], inclusion of hazelnut in the tree nut list is warranted.

3.3.2.2.5. Macadamia Nut. Macadamia nut was native to Australia before being brought to Hawaii; hence, it is also called Australian nut, or Queensland nut. There are five reported cases of allergic reactions to macadamia nut even though it is commonly consumed [56–60].

Sutherland et al. [57] first reported a life-threatening anaphylactic reaction to macadamia nut. An 18-year-old female experienced oral itching immediately after eating a slice of flourless orange cake made with macadamia nut meal. Within 5 minutes, this progressed to severe anaphylaxis. Upon arrival at the hospital, the patient was hypotensive, which required immediate treatment of epinephrine, prednisolone, diphenhydramine, and overnight hospitalization. A similar reaction was reported by Pallares [58]. A 36-year-old male developed uvular and posterior tongue angioedema, dysphagia, chest tightness, chest pain, and chest pruritus 5 minutes after eating a chocolate-covered macadamia nut. He was treated with intravenous diphenhydramine, methylprednisolone, and femotidine (a histamine H_2 blocker). Another severe anaphylaxis was reported by Häberle and Hausen [56]; a 38-year-old developed urticaria, shortness of breath, and shock shortly after eating macadamia nuts.

Lerch et al. [59] documented two patients with allergic reactions to macadamia nut. A 42-year-old man developed generalized pruritus, itching of the throat, rhinitis, dyspnea, and dizziness 5 minutes after eating a macadamia nut. Another 34-year-old male repeatedly developed severe OAS after eating macadamia and other tree nuts, including hazelnut, walnut, Brazil nut, and almond. In addition, a 1-year-old boy suffered from OAS after putting a macadamia nut in his mouth [60].

Reactions to macadamia nuts can manifest from OAS to severe anaphylaxis. While there are no reports of death to macadamia nut, the weight of evidence, particularly the reports of severe anaphylaxis, warrant its inclusion in a list of tree nuts.

3.3.2.2.6. Pecan. Allergy to pecan was the fourth most common of the tree nut allergies reported by Sicherer et al. [27] with 9% of the 1667 nut-allergic registrants reporting pecan allergy. Previously, Sicherer et al. [34] reported 13 of 54 nut-allergic patients as pecan allergic. In the United Kingdom, pecan may not be as common a source of tree nut allergy with Clark and Ewan [15] reporting 8 of 1000 patients showing their strongest reaction to pecan based on clinical history. Cross-reactivity between pecan and walnut is known [61]. Malanin et al. [62] and Berrens [63] reported a neo-allergen exclusively present in aged or heated pecan nuts, but not in fresh pecans. They suggested that a Maillard reaction product might be the allergen.

Yunginger et al. [64] was the first to report fatal anaphylaxis from accidental ingestion of pecan, which occurred in a 16-year-old male. Subsequently, four more cases of fatality were reported [3, 37, 65]. Due to its frequency in occurrence and severity of the reactions including several reported fatalities, inclusion of pecan in the tree nut list is warranted.

3.3.2.2.7. Pine Nut. Pine nut is commonly consumed in southwestern United States, Mexico, China, and Mediterranean countries. It is used in the preparation of various dishes, particularly pastries, deserts, salads, and pesto sauce. Despite the wide use of pine nuts, allergy to pine nuts has been reported less frequently than other tree nuts. Reactions, however, were usually severe, and emergency room treatments due to severe anaphylaxis immediately after eating pine nuts have been documented [30, 48, 54, 66–72]. While most reactions occur in adults, young children can be affected. No fatality due to pine nut ingestion has been reported. Pine nut is known to cross-react with almond [69]. Pine nuts are not currently listed in Annex IIIa of the EU directive on labeling of prepackaged foods [18]. Due to the severe nature of the reactions, the weight of evidence, particularly the reports of severe anaphylaxis, inclusion of pine nut in the major tree nut list is warranted.

3.3.2.2.8. Pistachio. While anaphylaxis against pistachio is uncommon, one fatal accident due to pistachio ingestion has been reported [3]. Pistachio nut contains several protein allergens able to trigger type I hypersensitivity reactions. Fernandez et al. [73] reported two cases of pistachio allergy that cross-reacted with cashew, and Jansen et al. [74] reported three patients with adverse food reactions due to pistachio and mango. Using specific IgE determination by CAP-inhibition and leukocyte histamine release, Parra et al. [75] demonstrated the cross-reactivity of pistachio with cashew and other dried fruits belonging to taxonomically unrelated botanical families. Liccardi et al. [76] described two uncommon cases of OAS in a 54-year-old man and a 3-year-old girl, both after eating pistachio nuts. The man had put three whole pistachios with shells in his mouth and cracked them with his teeth before the onset of symptoms. A DBPCFC was performed (0.1–50 mg of protein), for ethical reasons, only in the adult patient. No reaction was observed after ingestion of pistachio protein up to 50 mg in a capsule. Interestingly, they observed a positive intraoral itching and angioedema, which required precautionary antihistamine treatment, only after slight scratching of the oral mucosa at 0.1-mg capsule. Since no reaction was recorded after using a placebo capsule on the scratched mucosa, it was suggested that slight injury of the oral mucosa might enhance the local response. Due to its frequent occurrence and severity of the reactions, inclusion of pistachio in the tree nut list is warranted.

3.3.2.2.9. Walnut. Walnuts rank third in per capita consumption of tree nuts in the United States and can be associated with systemic IgE-mediated reactions in some individuals [77]. Allergy to walnuts was the most common of the allergies to tree nuts reported by Sicherer et al. [27] with 34% of the 1667 nut-allergic registrants reporting walnut allergy. Similarly, Sicherer et al. [34] reported 26 of 54 nut-allergic patients as walnut allergic. Clark and Ewan [15] described 1000 patients allergic to either peanut or at least one tree nut. Thirty study subjects reacted most strongly to walnut. Walnut-allergic patients often have allergies to other tree nuts. Among tree nut fatalities, walnut ranks number one. Sampson et al. [37] reported the first walnut fatal reaction in 1992. Subsequently, 14 more cases were documented [3, 7, 78].

A single case of IgE cross-reaction of walnut to coconut has been reported [79]. Cross-reactivity between walnut and Rosaceae (peach, apple, plum) [80], hazelnut, and Brazil nut [81, 82] are known. Drug-induced (antihypertensive angiotensin converting enzyme [ACE] inhibitor) anaphylaxis to walnut is also known [83]. Due to its frequency in occurrence and severity of the reactions, inclusion of walnut in the tree nut list is warranted.

3.3.2.3. Tree Nuts in FDA's List but Not in the GMA List. Extensive literature review has not revealed any clinical allergy, allergic reactions, or sensitization due to consumption of the seven tree nuts listed below that are included on the FDA list. Due to the lack of incident of allergic reactions to these nuts, inclusion of these nuts in the tree nut list as a major food allergen is not warranted:

1. beechnut
2. butternut
3. chinquapin
4. ginkgo nut
5. hickory nut
6. pili nut
7. shea nut.

3.3.2.3.1. Other Tree Nuts in the FDA's List but Not in the GMA List.

3.3.2.3.1.1. CHESTNUT. Chestnut allergy has been almost exclusively considered in the context of the latex–fruit syndrome. Cross-reactivity in latex allergy has been described with a variety of fruits, including chestnut, avocado, banana, kiwi, and papaya [84–86]. However, clinical reactions to chestnut were rarely reported despite frequent consumption of this food either alone or with preexistent latex allergy. Only two case reports of chestnut allergy, one patient per case, of OAS [87] and anaphylaxis have been documented [88]. The latter case involved a 4-year-old male who had developed OAS plus coughing, wheezing, and dyspnea after eating an acorn. Months later, he experienced similar symptoms after eating a chestnut. No other anaphylaxis was ever reported. Due to low incident of allergic reactions and lack of acute anaphylactic reaction to chestnut, inclusion of chestnut in a list of major allergens is not warranted.

3.3.2.3.1.2. COCONUT. Botanically, a coconut is a simple dry fruit known as a fibrous drupe (not a true nut). Coconut allergy, either alone or with other tree nut allergy, is rare. Cross-reactivity between coconut and walnut or hazelnut is known. There are six clinical reports, for seven patients, mostly from Europe, documented in coconut allergy and coconut-induced allergic reactions [79, 89–93].

Teuber and Peterson [79] first reported two coconut allergic reactions from a 21- and a 50-year-old male. Both sensitive individuals experienced previous reactions to tree nuts, walnut in particular. While symptoms from walnut-induced reactions were clearly described in this report, which included angioedema, nausea, vomiting, and asthma, sometimes with hypotension, reactions from coconut were only described as systemic but "not quite as severe." It is also unknown if hypotension was involved in the reactions to coconut in these two patients. Data from this study demonstrated that the clinical reactivity of these two patients was due to the cross-reactivity of walnut-directed IgE to coconut, but not vice versa. This implies that for a potential serious anaphylaxis to coconut in this study, prior walnut allergy is a prerequisite.

Nguyen et al. [89] described a case report of a 19-year-old male with history of seasonal allergic rhinitis, penicillin allergy, OAS to other foods such as lima beans and various tree nuts (pecan, almond, and walnut), and coconut allergy. His recent coconut reactions involved oral pruritus, generalized hives, and wheezing after ingestion of a coconut-flavored cream topping. In another study, a 3-year-old child experienced abdominal pain, vomiting, and OAS after eating a small piece of fresh coconut [90]. Rosado et al. [91] reported that a 28-year-old man suffered from vomiting and OAS after ingestion of coconut ice cream. This man had a previous reaction with walnut. Couturier et al. [92] reported a suspected coconut reaction of gastrointestinal upset in an 8-month-old baby fed with milk containing coconut. Reactivity was demonstrated by positive reintroduction test and specific IgE test. Stresemann and Scherhorn reported a case of coexisting coconut allergy and inhalation allergy of a stone nut (*Phytelephas*) in a 36-year-old female [93].

While anaphylactic reactions were reported, such reactions were generally mild and nonlife threatening; therefore, they do not fall into the consensus definition of anaphylaxis [15]. Furthermore, the incidents are low. Given the lack of prevalence and severity of reactions, inclusion of coconut in a list of major allergens is not warranted.

3.3.2.3.1.3. LYCHEE NUT. Lychee (also called lichee) is a drupe and is classified as a fruit (not a true nut). There are very few case reports of allergic reactions to lychee in the literature; there have been five reports, each involving only one patient [94–98]. Strangely enough, all five cases were females from 12 to 34 years of age. Clinical symptoms ranged from contact dermatitis to dyspnea. Lychee fruit contains a significant amount of profilin proteins that cross-react with pollens. Due to low incidence of allergic reactions and lack of severe anaphylactic reaction to lychee "nut," inclusion of lychee in the tree nut list as a major food allergen is not warranted.

A review of the scientific literature establishes that there is insufficient data to support the inclusion of *beechnut*, *butternut*, *chinquapin*, *ginkgo nut*, *hickory nut*, *pili nut*, *shea nut*, *chestnut*, *coconut*, and *lychee nut* in a list of major food allergens. Inclusions of "tree nuts" that have either no history of sensitization and elicitation of allergic reactions (beechnut, butternut, chinquapin, ginkgo nut, hickory nut, pili nut, and shea nut) or only a few cases of mild and nonlife-threatening reactions (chestnut, coconut, and lychee nut) contradict the intent of regulatory guidelines such as FALCPA in the United States and leads to an unnecessary elimination of food choices that are enjoyable, nutritious, and convenient to allergic consumers.

3.4. CONCLUSION

Tree nuts are a well-documented cause of food allergy and may constitute a health risk to allergic consumers if not properly managed. Tree nuts, peanuts, or both affect about 1.4% of the general population in the United States. The frequency of nut allergy appears to vary among countries in part because of different dietary habits. Allergic reactions to tree nuts tend to be particularly severe, sometimes life threatening, and fatal reactions shortly following their ingestion have been documented. Nut allergies are managed by avoiding the specific tree nut that may trigger a severe allergic reaction caused by accidental ingestion. Many allergists also recommend avoidance of the entire category of tree nuts due to the potential for both cross-contact during processing and IgE cross-reactivity. Clearly, it is prudent to properly define and label the tree nuts that are "major food allergens" or "priority food allergens" in terms of prevalence and severity of reactions.

GMA proposed a practical and science-based list of tree nuts to inform and protect tree nut-allergic consumers. The list consists of almond, Brazil nut, cashew, hazelnut, macadamia, pecan, pine nut, pistachio, and walnut. The nine tree nuts on this list are consistent with the lists that have been prepared by well-known clinicians, scientific and regulatory authorities, at least one consumer group, and other food industry groups. The underlying data also support the inclusion of these nine nuts of public health importance in a list of tree nuts because they are associated with prevalence and severity of reactions that warrant their classification as "major food allergens" or "priority food allergens." Because there is either no evidence of allergic reactions, or the reactions are lacking in the severity and prevalence needed for classification as a "major food allergen," 10 of the nuts listed in the FDA Q&A guidance *Guidance for Industry: Questions and Answers Regarding Food Allergens, Including the Food Allergen Labeling and Consumer Protection Act of 2004* (Edition 4) [1] should be removed.

The food and beverage industry plans and develops their allergen management strategies based on science and sound regulations, which constitutes a solid foundation for protecting public health. The quality assurance teams in food processing establishments have developed programs to manage the receipt, storage, use, cleaning, and labeling of allergens in the facility. Unfortunately, including additional allergens that are not "major" not only poses a challenge to the general consumer but also complicates the food industry's ability to effectively manage the major allergens in their facilities. While most medical experts advise their patients who have been diagnosed with an allergy to specific tree nuts to avoid all tree nuts, inclusion of a commonly used ingredient such as coconut or its derivatives would unnecessarily restrict food choices. Effective risk communication is an important tool and must be used to give truthful and not misleading labeling information to the public. Including nuts that should not be considered as major food allergens, such as coconuts, that have minimal risk devalues labeling as a risk management measure and, thus, reduces rather than increases consumer safety. Inclusion of rare allergens into allergen management plans will not only be costly, but will also dilute the industry's ability to effectively manage the major allergens in their facilities.

ACKNOWLEDGMENT

The author wishes to thank the GMA Food Allergen Committee, especially the chair and the co-chair, for their insights and guidance.

REFERENCES

1. FDA Guidance for Industry. (2006). Questions and answers regarding food allergens, including the Food Allergen Labeling and Consumer Protection Act of 2004 (Edition 4). Available at http://www.tbt-sps.gov.cn/data/docs/Documents/zsyd/USFDA_Faq_food_Allergens.pdf.
2. Sicherer, S.H., Munoz-Furlong, A., Sampson, H.A. (2003). Prevalence of peanut and tree nut allergy in the United States determined by means of a random digit dial

telephone survey: a 5-year follow-up study. *The Journal of Allergy and Clinical Immunology*, *112*, 1203–1207.

3. Bock, S.A., Munoz-Furlong, A., Sampson, H.A. (2001). Fatalities due to anaphylactic reactions to foods. *The Journal of Allergy and Clinical Immunology*, *107*, 191–193.
4. Fleischer, D.M., Conover-Walker, M.K., Matsui, E.C., Wood, R.A. (2005). The natural history of tree nut allergy. *The Journal of Allergy and Clinical Immunology*, *116*, 1087–1093.
5. Yeung, J.M., Applebaum, R.S., Hildwine, R. (2000). Criteria to determine food allergen priority. *Journal of Food Protection*, *63*, 982–986, and references cited within.
6. Hefle, S.L., Nordlee, J.A., Taylor, S.L. (1996). Allergenic foods. *Critical Reviews in Food Science and Nutrition*, *36*, S69–S89.
7. Pumphrey, R. (2004). Anaphylaxis: can we tell who is at risk of a fatal reaction? *Current Opinion in Allergy and Clinical Immunology*, *4*, 285–290.
8. Bousquet, J., Bjorksten, B., Bruijnzeel-Koomen, C.A., Huggett, A., Ortolani, C., Warner, J.O., Smith, M. (1998). Scientific criteria and the selection of allergenic foods for product labeling. *Allergy*, *53*, 3–21.
9. Sampson, H.A., Muñoz-Furlong, A., Campbell, R.L., Adkinson, N.F. Jr., Bock, S.A., Branum, A., Brown, S.G.A., Camargo, C.A. Jr., Cydulka, R., Galli, S.J., Gidudu, J., Gruchalla, R.S., Harlor, A.D. Jr., Hepner, D.L., Lewis, L.M., Lieberman, P.L., Metcalfe, D.D., O'Connor, R., Muraro, A., Rudman, A., Schmitt, C., Scherrer, D., Simons, F.E.R., Thomas, S., Wood, J.P., Decker, W.W. (2006). Second symposium on the definition and management of anaphylaxis: summary report—second. National Institute of Allergy and Infectious Disease/Food Allergy and Anaphylaxis Network symposium. *The Journal of Allergy and Clinical Immunology*, *117*, 391–397.
10. Wang, J., Sampson, H.A. (2007). Food anaphylaxis. *Clinical and Experimental Allergy*, *37*, 651–660.
11. Pongracic, J.A, Kim, J.S. (2007). Update on epinephrine for the treatment of anaphylaxis. *Current Opinion in Pediatrics*, *19*, 94–98.
12. Brown, S.G. (2006). Anaphylaxis: clinical concepts and research priorities. *Emergency Medicine Australasia*, *18*, 155–169.
13. Teuber, S.S., Comstock, S.S., Sathe, S.K., Roux, K.H. (2003). Tree nut allergy. *Current Allergy and Asthma Reports*, *3*, 54–61.
14. de Leon, M.P., Drew, A.C., Glaspole, I.N., Suphioglu, C., Rolland, J.M., O'Hehir, R.E. (2005). Functional analysis of cross-reactive immunoglobulin E antibodies: peanut-specific immunoglobulin E sensitizes basophils to tree nut allergens. *Clinical and Experimental Allergy*, *35*, 1056–1064.
15. Clark, A.T., Ewan, P.W. (2003). Interpretation of tests for nut allergy in one thousand patients, in relation to allergy or tolerance. *Clinical and Experimental Allergy*, *33*, 1041–1045.
16. Canadian Food Inspection Agency. (2009). Tree nuts—one of the nine most common food allergens. Available at http://www.inspection.gc.ca/english/fssa/labeti/allerg/nutnoie.shtml.
17. Health Canada. (2009). Tree nuts—One of the nine most common food allergens. Available at http://www.hc-sc.gc.ca/fn-an/securit/allerg/fa-aa/allergen_treenut-noix-eng.php.

18. European Union. (2003). Directive 2003/89/EC (2003) as regards indication of the ingredients present in foodstuffs. Available at http://ec.europa.eu/food/food/labellingnutrition/foodlabelling/fl_com2003-89_en.pdf.
19. Food Standards Australia New Zealand (FSANZ). (2006). Food surveillance: Australia New Zealand. Available at http://www.foodstandards.gov.au/scienceandeducation/monitoringandsurveillance/foodsurveillancenewsletter/_tree.
20. The Food Allergy and Anaphylaxis Network (FAAN). (2007). Food allergy news for kids. Available at www.foodallergy.org.
21. The Food Allergy and Anaphylaxis Network (FAAN). (2007). Tree nut card–"how to read a label". Available at http://www.foodallergy.org/allergens/treenut.html.
22. FAAN International Tree Nut Council. (2010). Tree nut allergy. Available at http://www.foodallergy.org/page/tree-nut-allergy.
23. International Tree Nut Council. (2010). Nut information. Available at http://www.nuthealth.org/nut.php3.
24. Keet, C.A., Wood, R.A. (2007). Food allergy and anaphylaxis. *Immunology And Allergy Clinics of North America*, *27*, 193–212.
25. Bock, S.A., Muñoz-Furlong, A., Sampson, H.A. (2007). Further fatalities caused by anaphylactic reactions to food: 2001 to 2006. *The Journal of Allergy and Clinical Immunology*, *119*, 1016–1018.
26. Pumphrey, R.S.H., Gowland, M.H. (2007). Further fatal allergic reactions to food in the United Kingdom, 1999-2006. *The Journal of Allergy and Clinical Immunology*, *119*, 1018–1019.
27. Sicherer, S.H., Furlong, T.J., Munoz-Furlong, A., Burks, A.W., Sampson, H.A. (2001). A voluntary registry for peanut and tree nut allergy: characteristics of the first 5149 registrants. *The Journal of Allergy and Clinical Immunology*, *108*, 128–132.
28. Ewan, P.W. (1996). Clinical study of peanut and nut allergy in 62 consecutive patients: new features and associations. *British Medical Journal*, *312*, 1074–1078.
29. Ridout, S., Matthews, S., Gant, C., Twiselton, R., Dean, T., Arshad, S.H. (2006). The diagnosis of Brazil nut allergy using history, skin prick tests, serum-specific immunoglobulin E and food challenges. *Clinical and Experimental Allergy*, *36*, 226–232.
30. Senna, G., Mistrello, G., Dama, A., Roncarolo, D. (2000). Anaphylaxis to pine nuts and immunological cross-reactivity with pine pollen proteins. *Journal of Investigational Allergology and Clinical Immunology*, *10*, 44–46.
31. Asero, R., Mistrello, G., Roncarolo D., Amato, S., Caldironi, G., Barocci, F., van Ree, R. (2002). Immunological cross-reactivity between lipid transfer proteins from botanically unrelated plant-derived foods: a clinical study. *Allergy*, *57*, 900–906.
32. Borja, J.M., Bartolome, B., Feo, F., Galindo, P.A., Gomez, E. (1999) Anaphylaxis from Brazil nut. *Allergy*, *54*, 1007–1008.
33. Dutau, G., Rance, F., Abbal, M., Juchet, A., Bremont, F. (1998). Brazil nut allergy: a case in an 11-year-old. *Semaine des Hopitaux*, *74*, 910–913.
34. Sicherer, S.H., Burks, A.W., Sampson, H.A. (1998). Clinical features of acute allergic reactions to peanut and tree nuts in children. *Pediatrics*, *102*, 1–6.
35. Pastorello, E.A., Farioli, L., Pravettoni, V., Ispano, M., Conti, A., Ansaloni, R., Rotondo, F., Incorvaia, C., Bengtsson, A., Rivolta, F., Trambaioli, C., Prevedi, M.,

Ortolani, C. (1998). Sensitization to the major allergen of Brazil nut is correlated with the clinical expression of allergy. *The Journal of Allergy and Clinical Immunology*, *102*, 1021–1027.

36. Arshad, S.H., Malmberg, E., Krapf, K., Hide, D.W. (1991). Clinical and immunological characteristics of Brazil nut allergy. *Clinical and Experimental Allergy*, *21*, 373–376.
37. Sampson, H.A., Mendelson, L., Rosen, J.P. (1992). Fatal and near-fatal anaphylactic reactions to food in children and adolescents. *The New England Journal of Medicine*, *327*, 380–384.
38. Davoren, M., Peake, J. (2005). Cashew nut allergy is associated with a high risk of anaphylaxis. *Archives of Disease in Childhood*, *90*, 1084–1085.
39. Rance, F., Bidat, E., Bourrier, T., Sabouraud, D. (2003). Cashew allergy: observations of 42 children without associated peanut allergy. *Allergy*, *58*, 1311–1314.
40. Rance, F., Bidat, E., Bourrier, T., Sabouraud, D. (2003). Cashew nut allergy in children. *The Journal of Allergy and Clinical Immunology*, *111*, S252.
41. Moneret-Vautrin, D.A., Rance, F., Kanny, G., Olsewski, A., Gueant, J.L., Dutau, G., Guerin, L. (1998). Food allergy to peanuts in France—evaluation of 142 observations. *Clinical and Experimental Allergy*, *28*, 1113–1119.
42. Tariq, S.M., Stevens, M., Matthews, S., Ridout, S., Twiselton, R., Hide, D.W. (1996). Cohort study of peanut and tree nut sensitisation by age of 4 years. *British Medical Journal*, *313*, 514–517.
43. Rasanen, L., Makinen-Kiljunen, S., Harvima, R.J. (1998). Pectin and cashew nut allergy: cross-reacting allergens? *Allergy*, *53*, 626–628.
44. Ferdman, R.M., Ong, P.Y., Church, J.A. (2006). Possible association with cashew allergy. *Annals of Allergy, Asthma & Immunology*, *97*, 759–760.
45. Quercia, Q., Rafanelli, S., Marsigli, L., Foschi, F.G., Stefanini, G.F. (1999). Unexpected anaphylaxis to cashew nut. *Allergy*, *54*, 895–897.
46. Hourihane, J.O., Harris, H., Langton-Hewer, S., Kilburn, S.A., Warner, J.O. (2001). Clinical features of cashew allergy. *Allergy*, *56*, 252–253.
47. García-Menaya, J.M., Gonzalo-Garijo, M.A., Moneo, I., Fernández, B., García-González, F., Moreno, F. (2000). A 17-kDa allergen detected in pine nuts. *Allergy*, *55*, 291–293.
48. Beyer, K., Grishina, G., Bardinak, L., Grishin, A., Sampson, H.A. (2002). Identification of an 11S globulin as a major hazelnut food allergen in hazelnut-induced systemic reactions. *The Journal of Allergy and Clinical Immunology*, *110*, 517–523.
49. Schocker, F., Luttkopf, D., Scheurer, S., Petersen, A., Cistero-Bahima, A., Enrique, E., San Miguel-Moncin, M., Akkerdaas, J., van Ree, R., Vieths, S., Becker, W.M. (2004). Recombinant lipid transfer protein Cor a 8 from hazelnut: a new tool for in vitro diagnosis of potentially severe hazelnut allergy. *The Journal of Allergy and Clinical Immunology*, *113*, 141–147.
50. Pastorello, E.A., Vieths, S., Pravettoni, V., Farioli, L., Trambaioli, C., Fortunato, D., Luttkopf, D., Calamari, M., Ansaloni, R., Scibilia, J., Ballmer-Weber, B.K., Poulsen, L.K., Wutrich, B., Hansen, K.S., Robino, A.M., Ortolani, C., Conti, A. (2002). Identification of hazelnut major allergens in sensitive patients with positive double-blind, placebo-controlled food challenge results. *The Journal of Allergy and Clinical Immunology*, *109*, 563–570.

51. Hansen, K.S., Ballmer-Weber, B.K., Luttkopf, D., Skov, P.S., Wuthrich, B., Bindslev-Jensen, C., Vieths, S., Poulsen, L.K. (2003). Roasted hazelnuts—allergenic activity evaluated by double-blind, placebo-controlled food challenge. *Allergy*, *58*, 132–138.

52. Ortolani, C., Ballmer-Weber, B.K., Hansen, K.S., Ispano, M., Wüthrich, B., Bindslev-Jensen, C., Ansaloni, R., Vannucci, L., Pravettoni, V., Scibilia, J., Poulsen, L.K., Pastorello, E.A. (2000). Hazelnut allergy: a double-blind, placebo-controlled food challenge multicenter study. *The Journal of Allergy and Clinical Immunology*, *105*, 577–581.

53. Wensing, M., Koppleman, S.J., Penninks, A.H., Bruijnzeel-Koomen, C.A.F.M., Knulst, A.C. (2001). Hidden hazelnut is a threat to allergic patients. *Allergy*, *56*, 191–192.

54. Wensing, M., Penninks, A.H., Hefle, S.L., Akkerdaas, J.H., van Ree, R., Koppelman, S.J., Bruijnzeel-Koomen, C.A., Knulst, A.C. (2002). The range of minimum provoking doses in hazelnut-allergic patients as determined by double-blind, placebo-controlled food challenges. *Clinical and Experimental Allergy*, *32*, 1757–1762.

55. Flinterman, A.E., Hoekstra, M.O., Meijer, Y, van Ree, R., Akkerdaas, J.H., Bruijnzeel-Koomen, C.A., Knulst, A.C., Pasmans, S.G. (2006). Clinical reactivity to hazelnut in children: association with sensitization to birch pollen or nuts? *The Journal of Allergy and Clinical Immunology*, *118*, 1186–1189.

56. Häberle, M., Hausen, B.M. (2002). Soforttyp-Allergie nach Genuss von Macadamia-Nussen. *Allergo Journal*, *11*, S42.

57. Sutherland, M.F., Czarny, D., O'Hehir, R.E., Suphioglu, C. (1999). Macadamia nut anaphylaxis: demonstration of specific IgE reactivity and partial cross-reactivity with hazelnut. *The Journal of Allergy and Clinical Immunology*, *104*, 889–890.

58. Pallares, D.E. (2000). Allergy to macadamia nut. *Annals of Allergy, Asthma & Immunlogy*, *85*, 385–386.

59. Lerch, M., Egger, C., Bircher, A.J. (2005). Allergic reactions to macadamia nut. *Allergy*, *60*, 130–131.

60. Senti, G., Mallmer-Webber, B.K., Wülrich, B. (2000). Nusse, samen und kerne nut allergogischer sicht. *Schweizerische Medizinische Wochenschrift*, *130*, 1795–1804.

61. Roux, K.H., Teuber, S.S., Sathe, S.K. (2003). Tree nut allergens. *International Archives of Allergy and Immunology*, *131*, 234–244.

62. Malanin, K., Lundberg, M., Johansson, S.G.O. (1995). Anaphylactic reaction caused by neoallergens in heated pecan nut. *Allergy*, *50*, 988–991.

63. Berrens, L. (1996). Neoallergens in heated pecan nut: products of Maillard-type degradation? *Allergy*, *51*, 277–278.

64. Yunginger, J.W., Sweeney, K.G., Sturner, W.Q., Giannandrea, L.A., Teigland, J.D., Bray, M., Benson, P.A., York, J.A., Biedrzycki, L., Squillace, D.L., Helm, R.M. (1988). Fatal food-induced anaphylaxis. *The Journal of the American Medical Association*, *260*, 1450–1452.

65. Boyd, G.K. (1989). Fatal nut anaphylaxis in a 16-year-old male: case report. *Allergy Proceedings*, *10*, 255–257.

66. Ano, M.A., Maselli, J.P., Sanz, M.F., Fernandez-Benitez, M. (2002). Allergy to pine nut. *Allergologia et Immunopathologia*, *30*, 104–108.

67. Attou, D., Faverge, B., Bonein, M., Mecheri, M., Bensekhria, S., Gratecos, L.A. (2000). Pine nut allergy in a child. *Revue Francaise d'Allergologie et d'Immunologie Clinique*, *40*, 190–192.

68. Rubira, N., Botey, J., Eseverri, J.L., Marin, A. (1998). Allergy to pine nuts in children. *Allergie et Immunologie*, *30*, 212–216, and references therein.

69. de las Marinas, D., Vila, L., Sanz, M.L. (1998). Allergy to pine nuts. *Allergy*, *53*, 220–222.

70. Jansen, A., Vermeulen, A., Dieges, P.H., van Toorenenbergen, A.W. (1996). Allergy to pine nuts in a bird fancier. *Allergy*, *51*, 741–744.

71. Roux, N., Hogendijk, S., Hauser, C. (1998). Severe anaphylaxis to pine nuts. *Allergy*, *53*, 213–214.

72. Beyer, A.V., Gall, H., Peter, R.U. (1998). Anaphylaxis to pine nuts. *Allergy*, *53*, 1227–1228.

73. Fernandez, C., Fiandor, A., Martinez-Garate, A., Martinez Quesada, J. (1995). Allergy to pistachio: crossreactivity between pistachio nut and other Anacardiaceae. *Clinical and Experimental Allergy*, *25*, 1254–1259.

74. Jansen, A., de Lijster de Raadt, J., van Toorenenbergen, A.W., van Wijk, R.G. (1992). Allergy to pistachio nuts. *Allergy Proceedings*, *13*, 255–258.

75. Parra, F.M., Cuevas, M., Lezaun, A., Alonso, M.D., Beristain, A.M., Losada, E. (1993). Pistachio nut hypersensitivity: identification of pistachio nut allergens. *Clinical and Experimental Allergy*, *23*, 996–1001.

76. Liccardi, G., Mistrello, G., Noschese, P., Falagiani, P., D'Amato, M., D'Amato, G. (1996). Oral allergy syndrome (OAS) in pollinosis patients after eating pistachio nuts: two cases with two different patterns of onset. *Allergy*, *51*, 919–922.

77. Teuber, S.S., Dandekar, A.M., Peterson, W.R., Sellers, C.L. (1998). Cloning and sequencing of a gene encoding a 2S albumin seed storage protein precursor from English walnut (*Juglans regia*), a major food allergen. *The Journal of Allergy and Clinical Immunology*, *101*, 807–814.

78. Pumphrey, R.S. (2000). Lessons for management of anaphylaxis from a study of fatal reactions. *Clinical and Experimental Allergy*, *30*, 1144–1150.

79. Teuber, S.S., Peterson, W.R. (1999). Systemic allergic reaction to coconut (*Cocos nucifera*) in 2 subjects with hypersensitivity to tree nut and demonstration of cross-reactivity to legumin-like seed storage proteins: new coconut and walnut food allergens. *The Journal of Allergy and Clinical Immunology*, *103*, 1180–1185.

80. Asero, R., Amato, S., Mistrello, G., Roncarolo, D. (2002). Allergy to minor allergens of Brazil nut. *Allergy*, *57*, 1080–1081.

81. Asero, R., Mistrello, G., Roncarolo, D., Amato, S. (2004). Walnut-induced anaphylaxis with cross-reactivity to hazelnut and Brazil nut. *The Journal of Allergy and Clinical Immunology*, *113*, 358–360.

82. Senna, G., Bonadonna, P., Crivellaro, M., Schiappoli, M., Passalacqua, G. (2005) Anaphylaxis due to Brazil nut skin testing in a walnut-allergic subject. *Journal of Investigational Allergology and Clinical Immunology*, *15*, 225–227.

83. Moneret-Vautrin, D.A., Blain, H., Kanny, G., Bloch, Y. (1998). Anaphylaxis to walnuts and pine nuts induced by ACE. *Allergy*, *53*, 1233–1234.

84. Condemi, J.J. (2002). Allergic reactions to natural rubber latex at home, to rubber products, and to cross-reacting foods. *The Journal of Allergy and Clinical Immunology*, *110*, S107–S110.

85. Fernandez de Corres, L., Moneo, I., Munoz, D., Bernaola, G., Fernandez, E., Audicana, M., Urrutia, I. (1993). Sensitization from chestnuts and bananas in patients with urticaria and anaphylasis from contact with latex. *Annals of Allergy*, *70*, 35–39.

86. Tomitaka, A., Matsunaga, K., Akita, H., Suzuki, K., Ueda, H. (2000). Four cases with latex allergy followed by anaphylaxis to chestnut. *Arerugi*, *49*, 327–334 (in Japanese).

87. Antico, A. (1996). Oral allergy syndrome induced by chestnut (*Castanea sativa*). *Annals of Allergy, Asthma & Immunlogy*, *76*, 37–40.

88. Zapatero, L., Baeza, M.L., Sierra, Z., Molero, M.I. (2005). Anaphylaxis by fruits of the Fagaceae family: acorn and chestnut. *Allergy*, *60*, 1542.

89. Nguyen, S.A., More, D.R., Whisman, B.A., Hagan, L.L. (2004). Cross-reactivity between coconut and hazelnut proteins in a patient with coconut anaphylaxis. *Annals of Allergy, Asthma & Immunology*, *92*, 281–284.

90. Tella, R., Gaig, P., Lombardero, M., Paniagua, M.J., Garcia-Ortega, P., Richart, C. (2003). A case of coconut allergy. *Allergy*, *58*, 825–826.

91. Rosado, A., Fernandez-Rivas, M., Gonzalez-Mancebo, E., Leon, F., Campos, C., Tejedor, M.A. (2002). Anaphylaxis to coconut. *Allergy*, *57*, 182–183.

92. Couturier, P., Basset-Stheme, D., Navette, N., Sainte-Laudy, J. (1994). A case of coconut oil allergy in an infant: responsibility of maternalized infant formulas. *Allergie et Immunologie*, *26*, 386–387.

93. Stresemann, E., Scherhorn, K. (2000). Inhalational stone nut dust allergy and coconut allergy. *Allergologie*, *23*, 302–303 (in German).

94. Saraswat, A., Kumar, B. (2005). Anaphylactic reaction to apple, banana and lychee: what is common between botanically disparate plant families? *International Journal of Dermatology*, *44*, 996–998.

95. Niggemann, B., Reibel, S., Hipler C., Wahn, U. (2002). Anaphylactic reaction to lychee in a 12-year-old girl: cross-reactivity to latex? *Pediatric Allergy and Immunology*, *13*, 64–67.

96. Scarpa, S., Brianti, E. (1996). Anaphylactic reaction to *Litchi chinensis*—case report. *Allergy*, *51*, 89.

97. Giannattasio, M., Serafini, M., Guarrera, P., Cannistraci, C., Cristaudo, A., Santucci, B. (1995). Contact urticaria from litchi fruit (*Litchi chinensis* Sonn.). *Contact Dermatitis*, *33*, 67.

98. Fah, J., Wuthrich, B., Vieths, S. (1995). Anaphylactic reaction to lychee fruit: evidence for sensitization to profilin. *Clinical and Experimental Allergy*, *25*, 1018–1023.

4

THE CANADIAN CRITERIA FOR THE ESTABLISHMENT OF NEW PRIORITY FOOD ALLERGENS: EVIDENCE FOR THE INCLUSION OF MUSTARD AND INSUFFICIENT EVIDENCES FOR GARLIC AND ONION AS PRIORITY ALLERGENS IN CANADA

OLGA M. PULIDO, ZOË GILLESPIE, AND
SAMUEL BENREJEB GODEFROY

4.1. INTRODUCTION

As a result of the public comments on its proposed allergen labeling regulations (see http://www.hc-sc.gc.ca/fn-an/label-etiquet/allergen/index-eng.php), Health Canada received a number of comments from the general public, patient groups, health professionals, consumer organizations, and governmental agencies about the need to consider mustard, onion, and garlic as potential food allergens of concern for Canadians.

In this context, Health Canada embarked on the development of criteria in order to determine the scientific validity of including new foods or food ingredients on the list of priority food allergens in Canada.

This chapter presents the Canadian criteria applied to the assessment of scientific information obtained from the available literature, to investigate

Allergen Management in the Food Industry, Edited by Joyce I. Boye and
Samuel Benrejeb Godefroy

whether mustard, garlic, and onion should be considered as priority allergens with regard to the proposed enhanced labeling regulations in Canada. Two systematic reviews were conducted: one for mustard and one for onion and garlic. To ensure a consistent and transparent approach when assessing the potential allergenicity of a food or food ingredient in terms of established criteria, methods for the management and evaluation of available scientific information were developed. Specifically, these methods provide guidance for

- systematic data collection,
- criteria for assessing the strength of evidence,
- organization and tabulation of data, and
- criteria for evaluating the severity of clinical reactions.

The information obtained and evaluated using these methods were assessed using the Canadian criteria for establishing new priority food allergens. The strength-of-evidence approach determines whether the available evidence fulfills the criteria and substantiates the addition of a new food or food ingredient to the list of priority food allergens in Canada. These practices aim to facilitate a consistent scientific approach for declaring new priority food allergens and for the application of the proposed "enhanced food allergen labeling regulations in Canada."

4.2. CANADIAN CRITERIA FOR THE ESTABLISHMENT OF NEW PRIORITY FOOD ALLERGENS

In 1999, the World Health Organization (WHO) convened a food allergens expert panel to provide guidance to the Joint Food and Agriculture Organization of the United Nations (FAO)/WHO Expert Committee on Food Additives (JECFA), which is the committee that advises the Codex Alimentarius Commission[1] on food additives and other chemicals and ingredients in food [1]. The food allergens expert panel was tasked with the establishment of criteria for amending the Codex list of priority allergenic foods. The panel advised that the identification of priority food allergens should be based on the following criteria: the existence of a cause-and-effect relationship, based on positive double-blind, placebo-controlled food challenge (DBPCFC) or unequivocal reports of reactions, including severe symptoms associated with exposure to the food commodity and prevalence data in children and adults, supported by clinical studies relying on DBPCFC studies from the general population of several countries. The expert panel acknowledged that availability of such data for infants, some foods, and in certain regions of the world would represent a challenge. As an alternative, the use of comparative prevalence data in groups of allergy patients from several countries supported by DBPCFC data would be appropriate [1, 2].

National food regulatory agencies have used this guidance to build on the Codex list and develop their own lists of priority foods that should be targeted for mandatory labeling on foods available for sale in the country or region under their oversight. In Canada, Health Canada has the mandate to establish food standards, policies, regulations, and guidelines with an oversight on labeling requirements associated with health, safety, and nutritional quality concerns. Proposed amendments to the *Food and Drug Regulations* in Canada are meant to enhance the labeling of prepackaged food products by requiring the mandatory declaration of the sources of the priority food allergens, gluten, and sulfites (≥10 ppm), when present in a prepackaged food product. In Canada, at the present time, these are almonds; Brazil nuts; cashews; hazelnuts; macadamia nuts; pecans; pine nuts; pistachios; walnuts; peanuts; sesame seeds; wheat including kamut and spelt, triticale, rye, barley, and oats;[2] eggs; milk; soybeans; crustaceans; fish and shellfish; and sulfites present in quantities equal to or in excess of 10 ppm [2]. Current estimates are that food allergies affect as many as 6% of young children and 3–4% of adults [3].

In order to determine the scientific validity of including new foods or food ingredients on the Canadian list of priority food allergens, Health Canada has adopted the criteria for amending the Codex list of priority allergenic foods. In accordance with the JECFA guidelines, these criteria include the following [1]:

1. the existence of a credible cause–effect relationship, based on positive DBPCFCs or unequivocal reports of reactions with typical features of severe allergic or intolerance reactions;
2. reports of severe systemic reactions following exposure to the foodstuff; and
3. assessment of available prevalence data in children and adults, supported by appropriate clinical studies with subjects from the general population of several countries or alternatively available prevalence data from clinical studies with groups of allergy patients from several countries supported as per criterion 1.

In addition to the above listed criteria, the allergenic potency of a food or food ingredient is also considered. In this context, the term allergenic potency is defined as the amount of food or food ingredient required to elicit a reaction in an already sensitized individual [4].

Consideration are also to be given to the potential exposure of Canadians to the food or food ingredient with specific consideration given to whether the current applications of the food or food ingredient is hidden because it is exempt from declaration in the list of ingredients on food packages, as per subsections B.01.009 one and two of the *Canadian Food and Drug Regulations*.[3] The intent of the proposed amendments to the *Food and Drug Regulations* is to enhance food labeling as a *public health tool* for food-allergic consumers, enabling consumers to avoid allergens to which they are susceptible and allowing informed choices of safe food sources. The proposed regulatory amendments would require the declaration of potential hidden sources of priority food allergens in prepackaged food products.

Furthermore, consideration is given to whether it is subject to the proposed definition of a food allergen. In Canada, in the proposed regulatory amendment (1220—Enhanced Labeling for Food Allergens and Gluten Sources and Added Sulfites) (http://www.hc-sc.gc.ca/fn-an/label-etiquet/allergen/project_1220_info-eng.php), the definition of a food allergen emphasizes that the protein portion of the food is responsible for eliciting an allergic reaction. This definition is based on the fact that protein is the portion of the food to which an individual with a food allergy or celiac disease will react [5]. Therefore, protein from any of the defined food allergens, or any modified protein (including any protein fraction), that is derived from any of these foods is considered a health risk to individuals with food allergies. Additional consideration is also given to other relevant factors in the Canadian context, particularly in terms of risk management, for example, allergen cross-reactivity.

The Canadian criteria are applied to the assessment of scientific information obtained from a systematic review of available scientific literature. Methods for the management and evaluation of available scientific information have been developed in order to ensure a consistent and transparent approach when assessing the potential allergenicity of a food or food ingredient.

4.2.1. Systematic Data Collection

An electronic database search was conducted utilizing, but not limited to, current versions of the following databases: Ovid Medline, Ovid Embase, and FSTA Direct. The abstracts for all identified publications directly relevant to the allergenicity of the food or food ingredient being assessed were reviewed and sorted for inclusion or exclusion based on the following criteria:

Include publication if

a. relevant to humans (adults or children),
b. relevant to an allergy via oral exposure through foodstuff, and
c. relevant to the identification and characterization of the specific allergenic proteins.

Exclude publication if

a. experimental study assessing the allergenicity using animal models or *in vitro* methods or
b. relevant to humans but the route of exposure is not via the oral route through foodstuff, for example, occupational exposures (dermal/respiratory).

Excluded references were placed in a separate database file (e.g., occupational allergies) for future use/access if required. References fulfilling the criteria for inclusion were transferred to a current reference software manager and reprints requested. Complete publications were reviewed by one or more investigators and data abstracted using previously agreed upon data abstraction form. Any discrepancy in data abstraction or interpretation was resolved by consensus.

4.2.2. Criteria for Assessing the Strength of Evidence

In order to determine the strength of the evidence provided by the publications in the database, the study methodology was categorized and rated in accordance with the guidelines established by the Joint Task Force on Practice Parameters. This joint task force was comprised of specialists in the field of allergy and immunology. The guidelines to establish the strength of clinical recommendations by rating categories of evidence from clinical studies were supported by three U.S. national allergy and immunology societies: the American College of Allergy, Asthma and Immunology (ACAAI), the American Academy of Allergy, Asthma and Immunology (AAAAI), and the Joint Council of Allergy, Asthma and Immunology (JCAAI). In effect, the criteria used to assess the strength of the evidence contributed by each publication (in descending order from the strongest to weakest evidence) were as follows: meta-analysis of randomized controlled trials, randomized controlled trials, nonrandomized controlled trials, quasi-experimental studies, nonexperimental descriptive studies (comparative, correlation, or case-controlled studies), expert committee reports or opinions, or clinical experience of respected authorities, laboratory-based studies [6, 7].

4.2.3. Organization and Tabulation of Data

All publications fulfilling the selection criteria were reviewed and categorized based on the criteria for assessing the strength of evidence. Publications that could not be categorized by these criteria were referenced in the results under an appropriate heading. Evidence from publications that fulfilled the strength-of-evidence criteria were tabulated under the following categories:

1. pivotal clinical studies (evidence from meta-analysis and randomized and nonrandomized controlled trials),
2. nonpivotal clinical studies (evidence from quasi-experimental studies),
3. other relevant studies (evidence from nonexperimental descriptive studies, comparative/correlation), and
4. case reports (evidence from nonexperimental descriptive studies).

Within these categories, studies were organized chronologically by publication date. The following parameters were tabulated from each publication in categories 1–3: reference (author, country of origin); study methodology (i.e., method of allergy assessment); subjects (adults vs. children, age, sex); clinical history (including premorbid conditions and family history); symptoms and signs (before challenge); symptoms and signs (after challenge with emphasis on the severity of reaction); diagnostic tests (confirm allergenic response); eliciting dose; eliciting allergen (source and type of the allergens, including type of food eaten when reaction was elicited); prevalence if provided; and comments (relevant notes made by the investigator[s]).

The following parameters were tabulated from each publication in category 4: reference (author, country of origin); cases (number of cases, sex, age); clinical history (including premorbid conditions and family history); symptoms and signs (with emphasis on the severity of reaction); diagnostic tests (confirm allergenic response); eliciting dose; eliciting allergen (source and type of allergens, including type of food eaten when reaction was elicited); and comments (relevant notes made by the investigator[s]).

4.2.4. Criteria for Evaluating the Severity of Clinical Reactions

A wide range of symptoms are known to be associated with food allergies [5]. These include

- gastrointestinal symptoms, for example, vomiting, diarrhea, abdominal pain (colic);
- respiratory manifestations, for example, rhinitis, asthma;
- cutaneous manifestation, for example, urticaria, edema, angioedema (AE); and
- anaphylaxis (an acute generalized reaction that may include skin, respiratory, cardiovascular, and gastrointestinal symptoms and may result in death).

There are currently several grading systems for acute systemic hypersensitivity reactions. For consistency in the evaluation of the severity of the symptoms described in different publications, the clinical criteria and grading system of anaphylaxis as outlined in the publication by Brown [8] were applied as follows:

a. Severe reactions include symptoms that are strongly associated with hypotension and hypoxia (life-threatening upper airway obstruction) or neurological compromise: confusion, collapse, loss of consciousness, and incontinence. Preexisting asthma and lung disease are viewed as an increased risk of hypoxia. In children, anaphylaxis is most often caused by a bronchospasm associated with food intake; there is also usually a background of atopy and asthma [9].

b. Moderate reactions include diaphoresis, dizziness, pre-syncope, dyspnea, stridor, wheezing, chest/throat tightness, nausea, vomiting, and abdominal pain.
c. Mild reactions are limited to the skin (urticaria, erythema, and AE). However, when AE affects the face and involves the glottis, it is associated with hypoxia and graded as severe.

Other information that should be considered when assessing the literature on the severity of the reaction includes the clinical severity based on the judgment of the primary care healthcare provider and any description of the use of epinephrine, emergency medical attention, or hospitalization.

The methods developed for the management and evaluation of available information including the systematic data collection, the organization and tabulation of data, the criteria for assessing the strength of evidence, and the criteria for evaluating the severity of clinical reactions help to ensure a consistent and transparent approach when assessing the potential allergenicity of a food or a food ingredient. Specifically, this practice aims to facilitate a consistent scientific approach for considering amendments to the Canadian list of priority food allergens, which are subject to enhanced labeling requirements in Canada.

4.3. MUSTARD: EVIDENCE FOR ITS INCLUSION AS A PRIORITY FOOD ALLERGEN IN CANADA

4.3.1. Background

In response to public inquiries, an assessment of mustard as a potential food allergen of concern in Canada was conducted. In order to determine the validity of including mustard on the list of priority food allergens in Canada, a systematic literature review of the available information on the allergenicity of mustard following the Canadian criteria (Section 4.2) was undertaken.

4.3.2. Methods

4.3.2.1. Systematic Data Collection of Mustard Database. An electronic database search of publications in English, French, or Spanish was conducted as described in section I. These included current versions of the following databases: Ovid Medline, Ovid Embase, and FSTA Direct. The following specific search terms relevant to mustard as food allergen were used: Mustard Plant; Sinapis; mustard*; brassica (alba or juncea or nigra); (allerg* or hypersensi* or intoleran* or anaphyla* or urticaria* or hive*); (food challenge or rechallenge). There was no limitation on the dates of publication, except for those within the databases.

4.3.2.2. Organization and Tabulation of Data. Studies fulfilling the selection criteria for mustard as potential food allergen, as per the criteria described above, were reviewed and assessed based on the strength of evidence. Publications that could not be categorized by these criteria were referenced as described in Section 4.3.3.1, Characterization of mustard. Evidence from publications that fulfilled the strength-of-evidence parameters are tabulated below under the following categories:

- 4.3.3.2 Pivotal clinical studies (Table 4.1: evidence from randomized and nonrandomized controlled trials)
  1. DBPCFC
     a. Randomized
     b. Nonrandomized
  2. Single-blind, placebo-controlled food challenge (SBPCFC)
     a. Nonrandomized
- 4.3.3.3 Nonpivotal clinical studies (Table 4.2: evidence from quasi-experimental studies)
- 4.3.3.4 Other relevant studies (Table 4.3: evidence from nonexperimental descriptive studies [comparative/correlation])
- 4.3.3.5 Case reports (Table 4.4: evidence from nonexperimental descriptive studies)
  1. Canadian reports
  2. International reports

4.3.3. Results

A total of 358 publications were identified through the database search using the terms denoted above. However, based on the inclusion and exclusion criteria to support regulatory recommendations, only 27 publications fulfilled the strength-of-evidence categorization and tabulation criteria as previously indicated (Tables 4.1–4.4). Publications reporting the same data were grouped and counted as one entry. Additional 17 publications were reviewed but not included in the risk analysis. These publications provided information with regard to the general characterization of mustard as a food allergen and were considered relevant to the evaluation.

4.3.3.1. Characterization of Mustard. Mustard is an herbaceous flowering plant (Angiospermae) belonging to the family Brassicaceae (formerly known as Cruciferae), which includes, but is not limited to, cabbage, cauliflower, brussels sprouts, turnips, radishes, broccoli, and fodder (rape) crops [10]. Brassica species (*Brassica napus, Brassica campestris, Brassica juncea*, and *Brassica nigra*) are grown in Canada, Chile, Europe, China, India, and Pakistan, and have been used as a source of industrial and food oil, mustard, and vegetable greens in several parts of the world [11]. Rapeseed (*B. napus*) is also known as turnip rape, colza, ravison, sarson, toria, and canbra. Canola is a cultivar developed from rapeseed and is part of the Brassicaceae family and is related to mustard and other plants in the same family [12].

Mustard seeds are sold whole, ground into powder, or processed further into prepared mustard around the world, including in Canada. Prepared mustard is commonly used as a condiment, and mustard seeds and powder are increasingly being used in cooking and in processed and prepackaged foods as a seasoning or flavoring agent, emulsifier, and water-binding agent for texture control [13]. Mustard sauce or other sauces, which may contain mustard (such as mayonnaise, curry, ketchup sauces, and dips), are commonly used at home as well as in fast food and higher-end restaurants.

The major types of mustard seeds used in cooking and food processing are white (*Sinapis alba* or yellow mustard), brown (*B. juncea* or oriental mustard), and black (*B. nigra* or black mustard). Commercially sold mustard powder is usually a mixture of ground black and white mustard seeds, and prepared mustard sauce is composed of mustard seeds, salt, vinegar, wheat flour, and other spices and additives. White mustard seeds are much larger and considerably less pungent than the brown variety and are the main ingredient in North-American-style mustards. White and brown mustard seeds are blended to make English-style mustards. Brown mustard seeds are the main ingredient in European- and Chinese-style mustards [13].

All three types of mustard seed are available in North America. In fact, Canada is a world leader in the international mustard seed market accounting for about 35% of world production and 50% of global exports [14]. Yellow mustard seeds contain 20–30% protein, 24–35% oil, 6–12% other lipids, and 12–18% carbohydrates [13]. The protein content of wild mustard seeds (*B. campestris*) has been estimated to be ~26% [11]. Mustard protein isolates prepared from oriental mustard (*B. juncea*) seed have protein contents of 45–50% [15]. The high protein content of a variety of mustard and rapeseeds makes them attractive potential sources of food-grade vegetable protein [11, 15].

Mustard seeds also contain irritants that may cause nonimmune reactions mimicking allergic reactions, for example, capsaicin, the irritant ingredient of capsicum, and isothiocyanates. Capsaicin is capable of releasing substance P, which may induce non-immunoglobulin E (IgE)-mediated mast cell degranulation [16]. It is therefore important to base the diagnosis of a mustard allergy on evidence of IgE-mediated response. Early studies investigated whether the adverse reactions associated with the consumption of mustard were attributable to an isothiocyanate sensitivity. Mustard extracts treated with myrosinase, which degrades isothiocynates, did not reduce the cutaneous allergenic potency of the extract. Similar treatment with proteolytic enzymes found that the cutaneous allergenicity was reduced; this demonstrated that the allergic responses to mustard are elicited by protein [17].

The major allergenic proteins in mustard seeds have been identified and characterized. The major allergen of mustard is a 2S albumin, which is a seed storage protein composed of one heavy chain and one light chain (39 and 88 amino acids), linked by two disulfide bridges [18]. This seed storage protein has also been isolated from rapeseed, leguminous plants (peas and soya), walnuts, sesame seeds, and Brazil nuts, and is resistant to thermal degradation [18–21].

The major 2S albumin of yellow mustard is *Sin a 1*, and it is a thermostable protein that is resistant to digestion by trypsin and degradation by other proteolytic enzymes [22–26]. *Sin a 1* is able to interact with membrane lipids [27]. This interaction is postulated to facilitate the uptake of *Sin a 1* at the intestinal barrier, thus increasing the resistance of *Sin a 1* to protease digestion. *Sin a 1* binding to B-cell membranes would allow cross-linking of cell surface proteins, promoting B-cell activation and a subsequent immune response [27].

Characterization of the major allergen found in oriental mustard (*B. juncea*), *Bra j 1*, revealed that *Bra j 1* and *Sin a 1* have a homologous epitope [25, 28, 29]. These findings imply that individuals known to be sensitive to one species of mustard seed are likely to show sensitivity to other species. Furthermore, a marked *in vitro* cross-reactivity between the principal allergen of rapeseed (*Bn III*) and *Sin a1* have been described in the literature [19, 30]. Less is known about cross-reactivity between mustard seed and leafy vegetables in the Brassicaceae species, for example, cauliflower, brussels sprouts, and turnips. While some authors report [31] cross-sensitization and cross-reactivity between mustard and vegetables from the Brassicaceae family, others indicate that it is rare [10]. The proposed reason for the latter observation is that the proteins in Cruciferae leafy vegetables are thought to be more susceptible to digestion and thermal degradation than the allergenic proteins in mustard seeds [32, 33]. Mustard greens known as leaf mustard from *B. juncea* are also used as a leafy vegetable. It is not known whether the protein contained in the leaf mustard is allergenic or if it is more susceptible to degradation as other leafy vegetables in the same family.

The available scientific literature regarding mustard as a potential food allergen focuses on mustard seed and their products, protein characterization, and cross-reactivity; allergy clinical trials; and exposure from potentially hidden sources. More information would be required to address any potential concern with leafy vegetables in the same family.

4.3.3.2. Pivotal Clinical Studies. Three clinical trials are considered pivotal to the strength of evidence to support regulatory recommendations and are tabulated in Table 4.1. Two DBPCFC studies [31, 34] and one SBPCFC study [35, 36] were identified in the literature.

Of the DBPCFC studies, one study design was randomized and conducted mostly with adult subjects [31], and the other study design was nonrandomized conducted mostly with children [34]. In both studies, the number of subjects recruited and who actually participated in the challenges was limited. Out of 38 subjects who were recruited in the Figueroa et al. [31] trial, only 24 subjects participated in the oral challenge. As well, of the 30 subjects who were recruited in the Morisset et al. [34] trial, only 24 of them participated in the oral challenge. The most sensitive subjects who had a history of anaphylaxis were excluded from the oral challenges in both of these trials. Both groups of investigators specified that mustard products used in the trials were free of metabisulfite, which ensured that any reactions observed after the challenge were attributable to a mustard allergy. Both DBPCFC studies masked the strong taste of mustard in other food products.

Figureoa et al. [31] challenged subjects with increasing doses of masked yellow mustard sauce (80, 240, 800, 2400, and 6480 mg) containing *S. alba* mustard seeds (14% w/v) at 15-minute intervals until a clinical reaction was observed or a cumulative dose of 10 g of the mustard sauce was administered. Morisset et al. [34] challenged subjects with increasing doses of masked mustard seasoning (10, 30, 100, 300, and 900 mg) containing 33.6% of *B. juncea* mustard seeds at 20-minute intervals until a cumulative dose of 1340 mg of mustard seasoning was administered or until a clinical reaction was observed.

In the Figueroa et al. [31] trial, 14 out of the 24 challenged subjects (58%) were considered to show a positive reaction specific to mustard. The most frequent symptom observed was oral allergy syndrome (OAS) in 10 subjects (71%). This reaction was considered mild and was characterized by pruritus and mild AE of the lips, tongue, palate, and throat, and was followed by a rapid resolution of symptoms. One subject showed AE and bronchial asthma (BA) after mustard sauce ingestion and another subject reacted with systemic anaphylaxis. In these two cases, the eliciting dose was 156.8 mg of mustard sauce, and the reaction was considered moderate and severe, respectively, according to the criteria of Brown [8]. The lowest dose eliciting a reaction was 44.8 mg of mustard sauce. The mean cumulative reactive dose of mustard sauce was 891.4 ± 855.2 mg, calculated as an equivalent to 124.8 ± 119.7 mg of mustard seeds (*S. alba*).

The results of this study also showed a significant association between mustard hypersensitivity and mugwort pollen sensitization (97% of patients). All patients showed sensitization to at least one other food product within the Brassicaceae family, including cabbage, cauliflower, and broccoli, and about 40% of these sensitized individuals were symptomatic. These manifestations comprised of mild symptoms such as OAS to severe reactions such as anaphylaxis or BA. Cross-reactivity between mustard and other plant-derived foods of the Brassicaceae family was confirmed. Hence, these authors denote

TABLE 4.1 Pivotal Clinical Studies (Mustard)

Author, Country	Study Design Details	SUB	Clinical Hx	Symptoms and Signs[a] (before CH)
1. DBPCFC a. Randomized				
Figueroa et al. [31], Spain	Prospective Questionnaire and CH • Increasing dose: 80, 240, 800, 2400, and 6480 mg w/ a 15-minute interval until symptoms appeared or cumulative max dose of 10-g Mus followed by an open arm up to 25 g Mus 24/38 (83%) CH w/ Mus 14/38 (37%) excluded because of SEV of symptoms or did not agree to enter the food CH	38 SUB Age: 5/38 ≤ 14 y Average: 21.9 ± 8.6 y Sex: 20 F, 18 M	Hx of A in 11% of SUB (exclude from CH) Hx of atopy in 92% of SUB Hx of primary Res Sen in 83% of SUB	A: 11% EIA: 3% OAS: 47% U/AE: 42%
b. Nonrandomized				
Morisset et al. [34], France	Doses 10, 30, 100, 300, and 900 every 20 minutes with a cumulative dose of 1340 mg Mus seasoning amount selected based on routine consumption 24 SUB DBPCFC 6 SUB SBPCFC	30 SUB Ch: 28/30 A: 2/30 Age: 3–20 y Sex: 11 F, 19 M	Hx of Rxn to ingestion of Mus Screened for Mus allergy by SPT and IgE	U, AE, AD, BA, abdominal pain, diarrhea
2. SBPCFC a. Nonrandomized				
Rancé et al. [35, 36], France	SUB selected for by positive Mus SPT Compared SUB to 22 controls without Hx of food allergy Increasing doses: 1, 5, 10, 20, 50, 100, 250, and 500 mg of Mus	36 Ch Age: 10 months to 15 y Average: 5.5 y Sex: 22 M, 14 F	15/30 Ch w/ previous Hx of food allergy Family Hx of atopy 81% 8/15 SUB (53.3%) exhibited Rxn to Mus under the age of 3 years	Of the 54 initial clinical features: AD 52% U/AE 37% BA 9% Laryngeal edema + OAS + C 2%

[a]Symptoms and signs: A, anaphylaxis; AD, atopic dermatitis; AE, angioedema; BA, bronchial asthma; C, conjunctivitis; EIA, exercise-induced anaphylaxis; GI, gastrointestinal; OAS, oral allergy syndrome; SK, skin; U, urticaria.
[b]Severity of reaction: refer to Methods [8].
[c]Diagnostic tests: IgE, serum immunoglobulin E; SPT, skin prick test.
[d]To avoid reactions due to sulfite intolerance.
CH, challenge; Ch, children; DBPCFC, double-blind, placebo-controlled food challenge; F, female; Hx, history; M, male; max, maximum; mod, moderate; Mus, mustard; Res, respiratory; Rxn, reaction; Sen, sensitization; SEV, severity; sev, severe; SBPCFC, single-blind, placebo-controlled food challenge; SUB, subjects; w/, with; y, years old.

Symptoms and Signs[a] SEV of Rxn[b] after CH	Diagnostic Tests[c]	Eliciting Dose	Eliciting Allergen	Prevalence	Comments
Positive Rxn 14/24 (58%) Type of Rxn: OAS 10/14 (71%) SEV: mild AE + BA 1/14 (7%) SEV: mod A 1/14 (7%) in a SUB without previous Hx of A SEV: sev	SPT to a panel of aeroallergens and food extracts IgE to mugwort pollen, Mus, cabbage, broccoli, and so on	Mean cumulative dose (until Rxn appeared or max dose reached): 891.4 ± 855.2 mg of Mus sauce equivalent to 124.8 ± 119.7 mg of Mus Eliciting dose in most severe cases: 156.8 mg of Mus sauce Lowest dose eliciting an Rxn was 44.8 mg of Mus sauce	Mus sauce Metabisulfite free[d]	58% of SUB positive for Mus allergy	Cross-Rxn w/ mugwort pollen: 97% SUB Other food Sen 42% SUB Sen to Brassicaceae 100% SUB Assoc. EIA 2% SUB
Positive Rxn 7/30 (23%) Type of Rxn: SK, for example, pruritus, erythema 5/7 (72%) SEV: mild GI/Res, for example, abdominal pain, diarrhea, sneezing, wheezing 4/7 (57%) SEV: mod	SPT to Mus seed, Mus flour, and metabisulfite-free Mus Mus-specific IgE	Lowest dose-inducing symptoms: 1 Ch 40 mg Mus (0.8 mg of protein) Subject Sen by Mus pollen and rape pollen Another Ch 440 mg Mus	Mus seasoning (*B. juncea* seed) containing 34% Mus seed and 6% Mus protein Metabisulfite free[d]	23% SUB positive for Mus allergy	SEV of certain Rxns argues for an informative labeling; Mus often masked allergen in many manufactured sauces
Mus allergy confirmed in 15/36 (42%) Most common Rxn: U 14/15 (93%) SEV: mild	SPT IgE	1–936 mg of Mus powder Mean cumulative dose: 153 mg of Mus powder	Mus powder	42% SUB positive for Mus allergy Symptoms started ≤3 years of age in 53% of the SUB	67% of SUB were also allergic to other foods—peanuts, eggs, and milk Possible Sen in utero or lactation Mus in baby food

cross-reactions between mustard and taxonomically related foods. Exercise-induced anaphylaxis (EIA) was also associated with 2% of the cases.

In the Morisset et al. [34] trial, 7 out of the 30 challenged subjects (23%) were considered to show a positive reaction specific to mustard. Symptoms of positive reactions included eczema, urticaria, rhinitis, conjunctivitis, abdominal pain, diarrhea, pruritus, sneezing, erythema, and wheezing with a predominance of skin manifestations, followed in frequency by respiratory/gastrointestinal symptoms. Skin manifestations alone are considered mild reactions, whereas gastrointestinal symptoms and respiratory symptoms are considered moderately severe [8]. There were no reports of anaphylaxis or symptoms indicative of hypoxia or hypotension, which are considered severe reactions. Although the cumulative dose aimed during the test was 1340 mg of mustard seasoning, eliciting doses were noted at 440 and 40 mg of mustard seasoning. The dose of 40 mg of mustard seasoning resulted in the subject experiencing rhinitis and urticaria. This dose of mustard seasoning (40 mg) was calculated to be equivalent to 13.5 mg of mustard seeds (*B. juncea*), which is roughly equivalent to 0.8 mg of mustard proteins (*B. juncea* mustard seeds contain 6% mustard protein). This subject was described as being sensitized to mustard pollen and to rape pollen. It was reported that the subject lives in an area of a mustard seasoning factory, which emits an unpleasant smell. In this case, skin reactivity was observed with the two species of mustard (*B. nigra* and *B. juncea*), indicating cross-sensitization.

Another difference between these two DBPCFC studies worth noting is that Morisset et al. [34] found that a positive skin prick test (SPT) and the presence of specific IgE as determined by the radioallergosorbent test (RAST) were not predictive of a positive outcome. In contrast, Figueroa et al. [31] demonstrated a significant relationship between SPT mean wheal diameter (performed with a commercial mustard extract) and challenge outcome, with a cutoff value of 8 mm, a specificity of 90%, and a sensitivity of 50%.

The SBPCFC study by Rancé et al. [35, 36] investigated 36 children (22 males and 14 females) aged 10 months to 15 years (average age was 5.5 years) who had positive mustard SPT and compared these subjects to 22 control subjects without a history of food allergies. Specifications for the mustard seeds used for the SPT were provided (mustard seed powder, including *S. alba* and *B. juncea*, 1:10 w/v, protein concentration 5 mg/mL); however, it was not clear whether the same source of mustard was used for the oral challenge.

Of the 36 challenged subjects, 15 had positive reactions (42%) and 21 were considered not allergic to mustard. Of the subjects with positive reactions to mustard, eight (53%) of the subjects had initially exhibited reactions to mustard under the age of 3 years. Based on this latter observation, the authors suggest that there may be sensitization to mustard through lactation or in utero.

After the challenge, symptoms included urticaria (14 cases), rhinoconjunctivitis (3 cases), AE (1 case), OAS (1 case), and eczema (1 case). These reactions are considered mild [8]. However, the initial clinical features of the subjects included asthma, laryngeal edema with OAS, and rhinoconjunctivitis. These symptoms pose a higher risk for hypoxia and are therefore considered more severe than the predominantly dermatological manifestations observed after the SBPCFC. The SBPCFC cumulative reactive dose varied from 1 to 936 mg of mustard powder. The mean cumulative reactive dose was 153 mg of mustard powder. No reactions to placebo were observed. It is also worth noting that (24/36) 67% of the subjects were also allergic to other foods, including peanuts, eggs, and milk.

4.3.3.3. Nonpivotal Clinical Studies. Six studies that were conducted using an open allergenicity assessment, which included mustard as one of the foodstuffs tested, were identified in the literature. These assessments utilized labial (LFC) or oral food challenges (OFC) and/or a combination of SPT, RAST, and determinations of serum IgE specific to mustard in order to verify an allergic response and quantify the frequency of reactions to certain foodstuff. These studies are tabulated in Table 4.2.

Niinimäki et al. [37] conducted SPT and RAST on 50 subjects with a reported history of reactions to spices and pollen. Mustard was included in a battery of spices tested in a 5% (w/v) test solution. Of the subjects tested, 58% had positive reactions for mustard to either one or both SPT or RAST. Furthermore, there were positive SPT results for three children, ages 1–1.5 years old, who were breast-fed for 11 months and had never orally ingested mustard. These results suggest the possibility of the transfer of mustard allergens through human milk and supports the view expressed by Rancé et al. [35, 36] that there is a possible sensitization to mustard though lactation or in utero.

Rancé et al. [38] conducted a similar study designed to assess the prevalence of allergic reactions to various spices among children with a history of food and pollen allergies. In this study, 83 children were evaluated with SPT and IgE against a variety of specific spices including mustard. Of the 83 subjects, 23 (28%) had SPT and IgE positive results for mustard. Out of the 23 subjects with SPT and IgE positive results, 11 reacted clinically to mustard; they showed symptoms associated with OAS, urticaria, and conjunctivitis. These symptoms are considered mild [8]. The mustard allergy was further confirmed by either LFC or OFC in 7 of the 23 subjects.

Rancé and Dutau [39] examined over 25 food allergens among 142 children with a history of food allergies. Subjects submitted to LFC for various foodstuffs, and when the results of the LFC were negative, SBPCFC were conducted. Twenty-three subjects (16%) had a positive response to mustard, 16 by LFC, and 7 by SBPCFC. Mustard was the third most common food allergen in this study; egg (75%) and peanut (60%) were the most common. Rancé and Dutau [40] also reported a similar prevalence of mustard allergy (12%) among 45 children allergic to three or more different foods. In 2002, Rancé and Dutau reported a mustard allergy prevalence of 7% among children previously identified as having BA by pulmonary function [41]. However, the high prevalence of mustard allergy among children in France, as described by

TABLE 4.2 Nonpivotal Clinical Studies (Mustard)

Author, Country	Study Design Details	SUB	Clinical Hx	Symptoms and Signs[a] (before CH)
Niinimäki et al. [37], Finland	Open study assessing allergy to spices including Mus using SPT and RAST	50 SUB Age: 1–47 y Average: 16.2 y Sex: 25 M, 25 F 3 Ch: Age: 1–1.5 y (never ingested Mus; breast-fed for 11 months)	Hx allergy to spices and birch pollen 64% Hx of atopy 96% Res/OC Sen	Gastric pain R AD 26%
Rancé et al. [38], France	Open study assessing allergy to spices including Mus using SPT and IgE 7/23 SUB had LFC and/or OFC conducted with Mus	83 SUB (Ch) tested for allergy to spices Age: 15 months to 16 y Sex: 50 M, 33 F	Hx of pollen and food allergy	Chronic U or recurrent AE/E
Niinimäki et al. [16], Finland	Open study assessing allergy to spices including Mus using SPT, RAST: IgE at 2-month and 2.9-year intervals	49 SUB Age: 1–51 y Average: 16.5 y Sex: 23 M, 26 F	Hx allergy to spices and birch pollen Hx of atopy	Atopic dermatitis w/ Res symptoms 57% Chronic E
Rancé and Dutau [39], France	Open allergy study assessing over 25 allergens including Mus using SPT, IgE LFC and SBPCFC (when LFC negative)	142 SUB (Ch) Age: 7 months to 15 y Average: 4.5 y Sex: 95 M, 47 F	Hx of food allergy	Multiple presenting symptoms in 66% Ch AD 61% Rash 32% AE 25% BA 24% A 4%
Rancé and Dutau [40], France	Open allergy study assessing foodstuff including Mus using SPT, IgE, and open LFC	45 Ch Age: 3 months to 9 y Average: 2.5 y Sex: 30 M, 15 F	Family Hx of atopy: 78% SUB AD: 93% 44% SUB w/ allergy to ≥3 foods	U 30% AE 26% E 20% BA 10% A 2%
Rancé and Dutau [41], France	Open allergy study assessing foodstuff including Mus documented by DBPCFC in BA patients identified by pulmonary function	163 Ch Age: 2–17 y Average: 7.2 y Sex: 108 M, 54 F	Family Hx atopy 91% Hx of ≥1 food allergies BA for an average of 5.5 y	SK 59% Res 24% GI 12%

[a]Symptoms and signs: A, anaphylaxis; AD, atopic dermatitis; AE, angioedema; BA, bronchial asthma; C, conjunctivitis; E, eczema; GE, generalized eczema; GI, gastrointestinal; OAS, oral allergy syndrome; R, rhinitis; SK, skin; U, urticaria.
[b]Severity of reaction: refer to Methods [8].
[c]Diagnostic tests: IgE, serum immunoglobulin E; RAST, radioallergosorbent test; SPT, skin prick test.
CH, challenge; Ch, children; DBPCFC, double-blind, placebo-controlled food challenge; F, female; Hx, history; LFC, labial food challenge; M, male; Mus, mustard; OC, occupational; OFC, oral food challenge; Res, respiratory; Rxn, reaction; Sen, sensitization; SEV, severity; sev, severe; SBPCFC, single-blind, placebo-controlled food challenge; SUB, subjects; w/, with; y, years old.

Symptoms and Signs[a] SEV of Rxn[b] after CH	Diagnostic Tests[c]	Eliciting Dose	Eliciting Allergen	Prevalence	Comments
Positive Rxn to SPT 29/50 (58%) RAST and SPT correlation good Ch who never ingested Mus had positive SPT Rxn	SPT RAST	0.648 g of Mus dissolved in glycerol and saline to make 5% (w/v) test solution	Commercial powdered Mus	58% SUB positive for Mus by either one or both SPT/RAST	40/50 (80%) SUB positive SPT and RAST to birch pollen 15/50 (30%) SUB positive SPT and RAST to mugwort pollen
Positive SPT for Mus in 23/83 SUB (28%) 6 SUB positive for Mus LFC and 1 SUB positive in OFC Symptoms specific to Mus: OAS/U/C 11/23 (48%) SEV: mild	SPT IgE	Not reported	Commercial extract	39/83 SUB (46%) positive (SPT and IgE) allergy to spices 23/39 SUB (59%) positive allergy to Mus 7/23 SUB (30%) confirmed by LFC or OFC	Pollen allergy existed in 56% of Ch allergic to spices
Positive Rxn to Mus-specific IgE 22/31 (71%) SUB w/ positive SPT	SPT RAST: total IgE Mus-specific IgE	4 mg of powdered spice and 50 μL of saline on the skin	Native Mus (*Sinapis alba* and *Brassica nigra*)	31/49 (63%) SUB positive SPT 22/31 (71%) SUB positive for Mus allergy 22/49 (45%) SUB positive for Mus allergy	Concomitant Rxn to native spices seen in 19/29 SUB who were tested with all spices 38/46 (83%) positive to birch pollen
Positive for Mus allergy 23/142 (16%), 16/23 in LFC, 7/23 in SBPCFC Multiple symptoms per Ch (not specific to Mus) U 74% BA 22% GE 4% A 2% SEV: sev	SPT IgE	1 mg to 5 g for all allergens tested Further details not provided	Extracted from local food	23/202 SUB (11%) positive for Mus allergy	Mus third most common food allergy in study
Positive allergy to Mus 12% SUB	SPT IgE	Mean dose by OFC 900 mg (1 mg to 10 g)	Details not provided	Positive allergy to Mus 12% SUB	None
Asthma induced by food allergens potentially sev A 6%	SPT IgE	Details not provided	Various food extracts	Positive allergy to Mus 7% SUB	Prevalence of asthma induced by food allergen: 10%

Rancé et al., appears to be more frequent in the southwest region rather than in the other regions of France [42]. Niinimäki et al. [16] reported a slightly higher prevalence of positive responses: 22 of 49 (44%) of subjects, with an average age of 16.5 years, and who had a history of allergies to spices and birch pollen, had elevated IgE specific to mustard.

4.3.3.4. Other Relevant Studies. With regard to the literature, three nonexperimental descriptive studies were identified as being relevant to the assessment of the allergenicity of mustard. These studies are tabulated in Table 4.3.

A retrospective analysis conducted by André et al. [43] examined which foods were most frequently associated with anaphylactic reaction over a 9-year period. Mustard was associated with 3% of such reactions. The authors noted an increasing trend in the frequency of sensitization to mustard over time. This observation is in agreement with the opinion expressed in an article by Rancé et al. [38, 44].

In a prospective study, 544 children with a history of food allergies (confirmed by a food challenge) were investigated [45–47]. Of the 544 children, 49 (9%) tested positive for an allergy to mustard via SPT and/or specific IgE. Mustard was the fourth most common allergen identified with this study. In addition, one child within the group reacting positively to mustard was reported as having an anaphylactic reaction. However, the anaphylaxis-eliciting dose was not specified in the report. The cumulative dose used for all allergens tested was from 0.1 to 10g of lyophilized food.

A cohort study conducted by Caballero et al. [33] in 29 subjects, who tested positive for an allergy to mustard by an SPT and mustard-specific IgE, reported anaphylactic reaction in 14 of 29 subjects (48%) with an overall systemic reactions in 19 of 29 subjects (65%). Symptoms ranged from loss of consciousness, dyspnea, AE, generalized urticaria, gastrointestinal symptoms, OAS, conjunctivitis, and rhinitis. The most frequent symptoms were AE (55%) and urticaria (34%) [33]. These symptoms are, for the most part, graded as mild [8]. However, they can be severe in cases of generalized urticaria and rapidly evolving AE, involving the face and neck (including the glottis). Symptoms such as these can present a high risk of airway obstruction and hypoxia.

4.3.3.5. Case Reports. A total of 15 case reports of allergic responses to mustard were identified in the literature. Two case reports documented cases in Canada, and the remaining 13 reports documented international cases including Spain (five reports), Italy (two reports), Sweden (two reports), France (two reports), Germany (one report), and Turkey (one report). These case reports are tabulated in Table 4.4 (Canadian reports and international reports). The reports provide descriptions of the severity of reactions to mustard as well as identifying the sources of mustard exposure.

4.3.3.5.1. Canadian Reports. Yip and Zimmerman [48] reported five cases of a mustard allergy in children (four boys and one girl). Of the five children,

TABLE 4.3 Other Relevant Studies (Mustard)

Author, Country	Study Design Details	SUB	Clinical Hx	Symptoms and Signs[a] (before CH)	Symptoms and Signs[a] SEV of Rxn[b] after CH	Diagnostic Tests[c]	Eliciting Dose	Eliciting Allergen	Prevalence	Comments
André et al. [43], France	Retrospective analysis 9-year period investigating foodstuff most frequently associated with A Rxn	580 SUB 480 Ad, 100 Ch Age: 1–83 y Mean: 30 y Sex: 290 M, 290 F	Hx of adverse Rxn to food	60/580 SUB Hx sev Rxn to food A 52/60 AE 6/60 Bronchospasm 2/60	Not applicable to study design	SPT IgE	Details not provided	Mus	3% of SUB positive for sev Rxn to Mus	An increase in the frequency of Sen (1%) to Mus was noted
Rancé et al. [45–47], France	Prospective prevalence study of food allergy validated by SPT/IgE/LFC	Ch 544 Age: 0–15 Sex: 343 M, 201 F	Hx food allergy confirmed by food challenge Family Hx atopia 71%	AD: 275/544 (51%) U/AE: 165/544 (30%) BA: 47/544 (9%) A: 27/544 (5%)	49/544 (9%) positive Rxn to Mus AD: 21/49 (43%) U/AE: 21/49 (43%) BA: 2/49 (4%) A: 1/49 (sev) (2%) GI: 1/49 (2%) OAS: 1/49 (2%) C: 2/49 (4%)	SPT IgE	0.1–10 g for all allergens tested Dose specific to Mus: not reported	Test substances extracted from local food	49/544 (9%) SUB positive for Mus allergy	Mus fourth most common allergy identified within a population with multiple food allergy
Caballero et al. [33], Spain	Cohort	29 SUB Hx Mus allergy Age: 15–58 y Mean: 27.3 ± 10 y Sex: 10 M, 19 F	Family Hx of atopy 50% SUB allergy to other vegetables 17% allergy to nonvegetables 52% symptoms of pollinosis	Mus allergy confirmed by SPT and IgE	19/29 (65%) SUB SR 14/19 (74%) A AE 16/29 Dyspnea 11/29 U 10/29 OAS 8/29 Most common symptom: U and AE SEV: mild-sev	SPT Total IgE Mus-specific IgE	Details not provided Trace amounts reported to elicit Rxn	Mus extract of *B. nigra* at 1:10 w/v	Not applicable study population 100% positive allergic Rxn to Mus	Mus seed allergens are resistant to digestion and high temperature No challenge Study due to risk of sev Rxn

[a]Symptoms and signs: A, anaphylaxis; AD, atopic dermatitis; AE, angioedema; BA, bronchial asthma; C, conjunctivitis; GI, gastrointestinal; OAS, oral allergy syndrome; SR, systemic reaction; U, urticaria.

[b]Severity of reaction: refer to Methods [8].

[c]Diagnostic tests: IgE, serum immunoglobulin E; SPT, skin prick test.

Ad, adult; CH, challenge; Ch, children; F, female; Hx, history; LFC, labial food challenge; M, male; Mus, mustard; Rxn, reaction; Sen, sensitization; SEV, severity; sev, severe; SUB, subjects; y, years old.

TABLE 4.4 Case Reports (Mustard)

Author, Country	Cases	Clinical Hx	Symptoms and Signs[a] SEV of Rxn[b]	Diagnostic Tests[c]	Eliciting Dose	Eliciting Allergen	Comments
1. Canadian Reports							
Yip and Zimmerman [48], Canada, full publication	5 Ch Sex: 4 M/1 F Ages: Case 1: 2.5 y M Case 2: 2 y M Case 3: 5 y M Case 4: 7 y F Case 5: 3 y M	Case 1: atopic boy w/ Hx of multiple food allergies since 18 months (sesame seed, fish) Case 2: atopic boy w/ Hx of Sen to kiwi, peanut Case 3: boy w/ Hx of BA and multiple food allergies (eggs, sesame, peanuts) Case 4: atopic girl w/ E since infancy; BA; allergy to milk, egg, and peanut Case 5: boy w/ Hx of BA and multiple food allergies	Case 1: three episodes of sev Rxn requiring emergency medical treatment AE and at least one episode of airway obstruction Case 2: three episodes of U immediately after exposure to Mus. No mention of emergency Case 3: at least one episode of Ax reaction to fast food (Mus and sesame) Case 4: U and wheezing following ingestion of Mus. No indication of emergency visit Case 5: vomiting, swelling of the lips. No indication of emergency	SPT positive for several Mus preparations In 4/5 cases, total IgE or Mus-specific IgE not reported	Details not provided	Mus sauce Salad dressing Glazed ham	2/5 sev 2/5 mod 1/5 mild One case due to hidden source of Mus in the glazing of a prepared ham and one case in fast food
Connors et al. [49], Canada, abstract	1 F Age: 50 y	No detailed Hx provided Hx of A-type Rxn to Mus	Symptoms of A after ingestion of Mus	SPT	Details not provided	Fast food Mus sauce	Very limited info provided

2. International Reports							
Panconesi et al. [88], Italy	1 M Age: "young"	Hospitalization twice in 1 year; acute giant U with edema of the glottis after eating pizza	Skin test to Mus antigen extracted from black and white Mus caused intense wheal reaction followed by shock and glottic edema	SPT IgE: RAST	Small amount of Mus contaminating the pizza	Mus not used in preparation of the pizza: reaction to cross-contamination with Mus	
Meding [89], Sweden	1 F Age: 40 y	Atopy Rxn to Mus		SPT black and white Mus, rapeseed, and other Cruciferae			Negative SPT for allyl isothiocyanate Positive SPT for Mus, rapeseeds
Widström and Johansson [90], Sweden	1 F Age: 25 y	Hx of allergy to egg and fish in childhood that cleared	Acute Rxn to Mus/mayo U and AE on face and neck	IgE: RAST positive for white and black Mus and rapeseed	One episode of possible cross-contamination of fast food	Fast food and mayonnaise	
Vidal et al. [91], Spain	2 F Age: Case 1: 47 y Case 2: 15 y	Case 1: Hx pollen allergy Case 2: Hx of A and pollen allergy	Case 1: sev U and facial AE w/ GI and Res symptoms Case 2: U, facial and throat edema, and chest tightness	Positive SPT, total IgE, specific Mus IgE		Mayonnaise and Mus on sandwich and in a salad	
Monreal et al. [92], Spain	1 M and 1 F Age: Case 1: 17 y M Case 2: 14 y F	Case 1: family Hx of atopy, personal Hx of IgE-dependent BA, allergy to molds and grasses, reocurrent U/AE since age 5 Case 2: family Hx of atopy, personal Hx of BA service	Case 1: Rxn to Mus edema lips/tongue, dysphagia, upper Res symptoms—required emergency service Case 2: 1 hour after physical exercise and ingestion of Mus: edema tongue, lips face, U, upper Res distress—required emergency service	Case 1: SPT w/ Mus, aeroallergens positive, total IgE and specific IgE by RAST positive Case 2: SPT for Mus positive, total IgE and specific IgE by RAST positive	Both reported as small amount	Both Mus sauce	Compared SEV of Rxn to Mus allergy w/ penicillin No oral challenge due to SEV of Rxn

TABLE 4.4 Continued

Author, Country	Cases	Clinical Hx	Symptoms and Signs[a] SEV of Rxn[b]	Diagnostic Tests[c]	Eliciting Dose	Eliciting Allergen	Comments
Malet et al. [50], Spain	2 M Age: Case 1: 31 y Case 2: 32 y	Case 1: hypersensitivity to pollen, seasonal R Case 2: no reported Hx of other allergies, contact hypersensitivity to Mus since age 10	Case 1: AE of face, U, dyspnea, after oral and nasal exposure to Mus Case 2: U, AE, pruritus, BA after accidental ingestion of Mus, requiring hospital treatment	Case 1: SPT, total IgE high, specific IgE by RAST positive to Mus seed Case 2: SPT, total IgE high, specific IgE by RAST to Mus seed	At least one case of accidental contact to Mus or even the smell of Mus elicited Rxn	Mus sauce or seed	Minimal quantities elicited a dermo-Res Rxn sev enough in one case to require hospitalization
Jorro et al. [93], Spain	2 M and 1 F Age: Case 1: 43 y M Case 2: 17 y F Case 3: 19 y M	Case 1: Hx of sev Rxn to Mus and AE to shellfish Case 2: Hx of BA, R, U Case 3: Hx of R	All three cases very sev A requiring urgent hospital service Cases 1 and 2: pruritus, edema of face/tongue, upper Res distress Case 3: upper Res distress, U	All cases positive SPT and IgE to Mus		Cases 1 and 3: Mus sauce Case 2: Mus in salad	No oral challenge to Mus due to sev of Rxn Oral challenge to other Cruciferae negative
Valero et al. [94], Spain	3 M and 2 F Age: Case 1: 34 y F Case 2: 31 y M Case 3: 25 y M Case 4: 52 y F Case 5: 33 y M	Four cases w/ Hx of aeroallergy Three cases w/ Hx of food allergies Two cases w/ Hx of Rxn when present in areas where Mus was manipulated SPT on 86 patients with Hx of atopy	U, AE, R, dyspnea, bronchial spasm All cases w/ SK symptoms 4/5 cases w/ Res symptoms All requiring hospital emergency service	All five cases positive to Mus by SPT and IgE 18.4% of the 86 SUB tested by SPT were positive to Mus		Mus sauce	The two cases described earlier in Malet et al. [50] appear to be included in this case report
Kanny et al. [52], France	1 F Age: 38 y	Hx allergy to Mus Hx of exposure to Mus powder in pharmacy where she worked	Oral pruritus, facial edema, U, dyspnea, fainting 20 minutes after eating chicken dip; sev reaction requiring emergency medical attention	Positive SPT/IgE: RAST	Estimated concentration of Mus in the dip: 0.15 mg/100 mg	Chicken dip Hidden source of Mus allergen	Mus is the most allergenic of the commonly used spices; it is able to induce sev A in very small amounts

Frémont et al. [95], France	1 F Age: 22 months	Family Hx atopy Hx of celiac disease and allergy to milk at age 11 months	Failure to thrive Diarrhea Vomiting Coughing fits AE	SPT/IgE positive for Mus, although the culprit food in this case was not Mus		Milk proteins in flour	Emphasis on hidden allergens Case demonstrates Sen to Mus in young baby
Asero et al. [30], Italy	1 M Age: 54 y	Hx of A Rxn after ingestion of sunflower seed in bread		SPT positive to sunflower seeds and Mus IgE-specific positive Cross-reactivity of Mus and sunflower seeds		SPT positive to sunflower seeds and Mus IgE-specific positive Cross-reactivity of Mus and sunflower seeds	Objective of the report was to demonstrate the cross-reactivity of sunflower seeds and Mus
Lingelbach et al. [96], Germany	1 F Age: 40 y	Hx of food and pollen allergies 11 years Hx of EIA	At least two episodes of EIA requiring emergency medical assistance	SPT, specific IgE (RAST)		Fast food	EIA associated to Mus in food Patient rejected oral challenge
Aygencel et al. [97], Turkey	1 F Age: 53 y		AE of tongue and oropharynx, requiring hospital Emergency assistance			Wild Mus: raw leaves of *Brassica* (*Sinapis*) *arvensis* in salad	

[a]Symptoms and signs: A, anaphylaxis; AE, angioedema; BA, bronchial asthma; E, eczema; EIA, exercise-induced anaphylaxis; GI, gastrointestinal; R, rhinitis; SK, skin; U, urticaria.
[b]Severity of reaction: refer to Methods [8].
[c]Diagnostic tests: IgE, serum immunoglobulin E; RAST, radioallergosorbent test; SPT, skin prick test.
Ch, children; F, female; Hx, history; M, male; mod, moderate; Mus, mustard; Res, respiratory; Rxn, reaction; Sen, sensitization; SEV, severity; sev, severe; SUB, subjects; w/, with; y, years old.

three were 3 years of age or less and the other two were 5–7 years old. All of the children had a reported history of multiple food allergies, and three of the five children were atopic.

Reactions to mustard included AE, airway obstruction, urticaria, wheezing, vomiting immediately after exposure, and swelling of the lips. Two cases had at least one episode of airway obstruction or anaphylaxis requiring emergency hospital attention. These symptoms are considered moderate to severe [8]. In at least one case, the allergen was hidden within the glazing of a prepared ham. The sensitivity to mustard was supported by positive SPT in all cases. However, based on the clinical history of the subjects, none of the cases were orally challenged with mustard due to the high risk of a severe reaction.

The other Canadian report concerned a single case. A 50-year-old woman had a history of anaphylactic-type reactions after exposure to mustard. This clinical history was supported by a positive SPT for mustard. Further details about this case were not available [49].

These two Canadian reports provide the only North American data with regard to allergic reactions to mustard. Neither report included testing of IgE specific to mustard or oral challenge.

4.3.3.5.2. International Reports. For most international cases, the information provided was limited because none of the case reports included oral challenges (Table 4.4). However, the case reports provided valuable information regarding the severity of reactions, which were induced by the ingestion of small amounts of mustard. Reactions ranged from acute anaphylaxis to generalized skin manifestation, with the majority of cases reporting severe to moderate acute reaction to mustard. Of the 13 international case reports, describing 22 individual cases of allergic reactions to mustard, 15 individuals had anaphylactic-type reactions that required emergency medical intervention. Reactions occurred after ingestion of small amounts of mustard: some as a result of apparent cross-contamination in fast food, and, in one case, a reaction was apparently elicited by the smell of mustard [50]. All cases tested positive for mustard allergy via SPT and/or IgE. This suggested that the dose may only need to be minute to elicit a severe reaction to mustard in food.

4.3.4. Discussion

An overall assessment of the available literature suggests that a strong scientifically based database exists to assess the potential allergenicity of mustard in food. However, the following limitations of the systematic literature review were taken into consideration when determining the scientific validity of including mustard on the Canadian list of priority food allergens.

There were a small number of DBPCFC and OFC studies identified in the literature (Table 4.1). There are likely several reasons for this. (1) There is an agreement among researchers in the field that the difficulty of masking the strong taste of mustard limits the attempts to perform DBPCFC studies. (2)

Many subjects in mustard allergy studies have previous clinical history indicating a high risk of severe systemic reactions and therefore the use of oral challenges to confirm that mustard elicited the allergic reaction was considered an unethical health risk. (3) A large number of publications identified in the initial database search were excluded from our assessment because the studies addressed dermal, respiratory, and/or occupational exposures. These data were not considered to be directly pertinent to the issue of food allergenicity; however, it is recognized that this information is important for those in the clinical field to assess the possibility of occupational or environmental disorders, particularly in areas where mustard is cultivated and/or processed. It is also important to note that these other routes of exposure can be the source sensitization to mustard proteins; the sensitized individuals may thereafter react to oral exposure. Hence, when collecting clinical information on mustard allergy, all potential routes of sensitization need to be explored.

Looking at the publications that addressed oral exposure to mustard, there were wide inconsistencies in the reporting of the data, including the amount and level of detail of information provided, and the description and interpretation of clinical symptoms. Data relevant to the assessment of mustard as food allergen were often contained within publications with more general study objectives of food allergies, for example, allergy to spices, and not specific to mustard allergy. This fact made the identification of relevant information more challenging and may have led to the exclusion of available information on mustard. Because some investigators have several publications of the same subject in more than one language, data were sometimes included in more than one publication and/or the same data were published in different languages. Hence, when feasible, duplications of data were identified and eliminated before assessment and tabulation.

The overall analysis of the available scientific information fulfills the Canadian criteria for the introduction of mustard in the list of priority allergens in Canada:

The first criterion of the Canadian adopted JECFA recommendations stipulates the existence of a credible cause–effect relationship, based on positive DBPCFC studies or unequivocal reports of reactions with typical features of severe allergic or intolerance reactions. The existing database includes two eligible DBPCFC studies: one with a randomized study design conducted with adult subjects [31] and one with a nonrandomized study design conducted with children [34]. A positive nonrandomized SBPCFC conducted with children [35, 36] was also considered pivotal in the evaluation of the mustard cause–effect relationship because the category of evidence from these studies is considered strong as per the strength-of-evidence criteria established by the ACAAI [6]. In addition to these studies, supporting studies were categorized by the strength of the evidence provided by the study designs and evaluated accordingly (Tables 4.1–4.4).

There is evidence within the database to support the conclusion that the amount of mustard required to elicit a reaction may be very small; however, there is insufficient information to estimate a dose threshold [51]. The mean cumulative dose–response in a pivotal study with mostly adults was 124.8 ± 119.7 mg of mustard seed (*S. alba*) contained in yellow mustard sauce [31], calculated as equivalent of ~32.2 mg of protein. Yellow mustard seeds contain ~20–30 mg of protein [13]. The lowest dose–response reported in the same study was 44.8 mg of mustard seed, calculated as equivalent to ~11.2 mg of protein. In children, the lowest dose–response reported by Morisset et al. [34] was 40 mg of mustard seasoning (*B. juncea* seed), calculated as equivalent to 13.5 mg of mustard seed and 0.8 mg of protein. The mustard seasoning used in the study was considered to contain 6% of protein. Furthermore, most of the case reports describe that the food eliciting the reaction was mustard sauce or mustard hidden in other sauces such as chicken dip, salad dressing, and fast food, or due to cross-contamination. Only one case report [52] estimated the concentration of mustard in the dip responsible for causing the reaction as 0.15 mg/100 mg. Other case reports only indicated that the amount of mustard associated with the allergic response was small or present in trace amounts, which included reports of cross-contamination of fast food, for example, of mustard hidden in the glaze of a ham, and, in one case, the smell of mustard [50].

The prevalence of positive mustard allergies among the challenged subjects in the pivotal clinical studies were as follows: Morisset et al. [34] reported a confirmation that 23% of the children who had a previous reported history of mustard reactions exhibited mustard-specific IgE reactions after an oral challenge with mustard seasoning. This rate is lower than the results of Rancé et al. [35, 36] who reported that 42% of previously sensitized children had an IgE-confirmed reaction to an oral challenge with mustard seed extract and the 58% reported by Figueroa et al. [31] in atopic adults challenged with mustard sauce.

In the assembled evidence base, the vast majority of the study populations were atopic or have a family history of atopy. In the Figueroa et al. [31] trial, 92% of the subjects had a history of atopy, and in the Rancé et al. [35, 36] trial, 81% of the children had a family history of atopy. The term atopy describes the genetic predisposition to become IgE sensitized to allergens commonly occurring in the environment and to which everyone is exposed but the majority do not produce a prolonged IgE antibody response. Thus, atopy is a clinical definition of an IgE antibody high responder [53, 54]. The use of atopic study populations in clinical trials is not considered a misrepresentation of the risk of the allergenic potential of mustard because this sensitive segment of the general population represents the majority of those individuals who are susceptible to food allergies. The capacity of mustard to elicit an IgE-mediated response is valid whether the individual is atopic or not.

Subjects with a clinical history of anaphylaxis were excluded from participating in the OFCs. In the Figueroa et al. [31] trial, 11% of the subjects were excluded based on their clinical history of severe systemic reactions. Nonetheless, anaphylaxis was reported in the challenge studies. One out of 14 subjects without a history of anaphylaxis had an anaphylactic reaction after being challenged with mustard sauce [31]. Anaphylaxis in association with exercise (EIA) was also noted. Other clinical presentations included OAS, urticaria, and AE, which are considered mild reactions [8]. However, when these reactions are complicated by laryngeal edema, BA, or respiratory symptoms, particularly in individuals with a previous history of asthma, these reactions are considered moderate to severe.

The second criterion of the Canadian adopted JECFA recommendations calls for reports of severe systemic reactions following the exposure to the foodstuff. In the DBPCFC studies, the observed frequency of severe reactions and anaphylaxis was higher in adults than in children [31, 34]. Anaphylactic reactions are reported in 2% of children [39, 45–47] and in up to 48% of adults with a confirmed mustard allergy [33]. In a Canadian case report, two of the five children described had severe reactions to the ingestion of mustard, which required emergency medical intervention. These cases were not confirmed by an oral challenge with mustard due to the high risk of another severe reaction. Of the 13 international case reports describing 22 individual cases of allergic reactions to mustard, 15 individuals had anaphylactic-type reactions that required emergency medical intervention. Other severe reactions described in case reports included laryngeal edema, generalized urticaria, and BA.

The third and last criterion of the Canadian adopted JECFA recommendations requires the assessment of all available Canadian prevalence data in children and adults, supported by appropriate clinical studies or alternatively available data from other countries. Data on the prevalence of mustard allergy are not available for Canada or for many other regions of the world. However, mustard allergy has been reported as the third/fourth most common food allergy among children in some regions of France [39–41, 45–47] and is probably the most common allergy to spices [16, 37, 38]. Furthermore, mustard is affirmed on the most recent list of 14 allergens to be declared on labels (updated in 2007) by the European Commission [55]. As well, mustard is recognized as an allergen by the International Union of Immunological Societies [56].

Other notable information from the evidence base includes the reported occurrence of mustard allergy symptoms beginning in subjects under the age of 3 years. In one trial, this age group represented up to 53% of the subjects [35, 36], and in another study [37], positive SPT results were reported for three children, ages 1–1.5 years old, who were breast-fed for 11 months and had never orally ingested mustard. Several hypotheses are discussed in the literature as to a possible explanation for the early onset of mustard allergies. Suggestions include sensitization in utero, during lactation, or the presence of mustard in baby foods. This issue will require further investigation.

Furthermore, there is also supporting evidence of cross-sensitization between mustard, pollen, and other aeroallergens. Figueroa et al. [31] reported 83% of adult cases of mustard allergy had primary respiratory sensitization (mustard dust exposure or cross-reactivity to aeroallergens). It is apparent that sensitization and reactions to mustard can be elicited via cross-sensitization with other aeroallergens or through contact or inhalation of mustard dust in areas where mustard is cultivated. This is important for Canadian consumers because Canada is a major producer of mustard seeds.

In addition to the Canadian adopted JECFA recommendations [1] for the introduction of food to the Codex priority list of food allergens by individual countries, the proposed amendments to the Canadian *Food and Drug Regulations* proposed in July 2008 define food allergens with an emphasis on the protein portion of the food being responsible for eliciting allergic reactions. The major allergenic proteins in mustard have been identified and characterized. Sin a 1 is the seed storage protein in yellow mustard associated with allergic reactions. It is resistant to degradation by heat and digestive enzymes and interacts with membrane lipids [23, 25, 27]. These characteristics suggest that the allergenic proteins in mustard can remain intact throughout food processing and digestion, which would elicit an allergic reaction in a susceptible individual. Furthermore, the seed storage proteins found in mustard have also been isolated from Brazil nuts, walnuts, and sesame seeds, which are foods currently defined as priority food allergens in Canada [18, 20, 21]. Similar structural features of Sin a 1 with proteins characterized in other types of mustard indicate that individuals who are sensitive to one species of mustard are likely to show sensitivity to other species.

Prepared mustard is commonly used as a condiment. Mustard seeds and powder are becoming increasingly used in cooking. They are also used in processed and prepackaged foods as a seasoning or flavoring agent, emulsifier, and water-binding agent for texture control [13]. In Canada, the current Canadian *Food and Drug Regulations* exempt components of certain ingredients, preparations, and mixtures from declaration in the list of ingredients on food packages. Mustard falls under this exception; hence prepacked foods for sale in Canada can contain undeclared sources of mustard.

Since food-allergic consumers must rely on information provided on food labels in order to avoid foods that contain the ingredients to which they are likely to react, Health Canada has proposed regulatory amendments to identify potentially hidden sources of food allergens. Once the proposed regulatory amendments are adopted, mustard is to be added to the list of priority allergens, and Health Canada would require the declaration of mustard on the label of prepackaged food products, either in the list of ingredients or in a statement beginning with the words "Contains:" when mustard protein is present in the prepackaged food product.

4.3.5. Conclusions on the Evidence for the Inclusion of Mustard as a Priority Food Allergen in Canada

An assessment of the assembled evidence base on mustard allergenicity provides international data supporting a credible cause–effect relationship; reports of severe systemic reactions including anaphylaxis following exposure to very small amounts of mustard within foodstuff; evidence that mustard allergy is common in some regions of Europe and has been affirmed on the European Commission's list of priority allergens; Canadian case reports supporting the occurrence of mustard food allergies in children and adults in Canada; evidence that all three types of mustard seed are available in Canada and mustard is used in cooking and in processed and prepacked foods; results of characterization studies indicating that the allergenic proteins in mustard are resistant to degradation by heat, thus likely to withstand food processing; they are also resistant to degradation by digestive enzymes; information that mustard is used in food processes that can result in "hidden" sources of food allergen; and evidence that individuals known to be sensitive to allergenic proteins from one type of mustard are likely sensitive to other types. Additional factors that make mustard allergy relevant to the Canadian scenario include the potential cross-reactivity between mustard and rapeseed and the facts that Canada is a major producer of both these crops and sensitization to mustard can be acquired through dermal and respiratory exposure.

This scientific evidence provides a sufficient strength of evidence to fulfill the Canadian criteria required to add mustard to the Canadian list of priority allergens.

4.3.6. Recommendations

Based on the conclusions of this assessment, it has been recommended that mustard be added to the Canadian list of priority food allergens and that the proposed amendments to the *Food and Drug Regulations* relating to the labeling of food allergens apply accordingly.

4.4. GARLIC AND ONIONS: INSUFFICIENT EVIDENCE TO BE INCLUDED ON THE LIST OF PRIORITY FOOD ALLERGENS IN CANADA

4.4.1. Background

In response to public inquiries, the scientific validity of including garlic and/or onion on the Canadian list of food allergens was assessed following the Canadian criteria described above. Using information obtained from a systematic review of available literature regarding the potential allergenicity of garlic and/or onion, an assessment was conducted following the guidelines established to identify a new priority allergen.

4.4.2. Methods

4.4.2.1. Systematic Data Collection on Onion and Garlic as Potential Food Allergens. An electronic database search of publications in English, French, or Spanish was conducted utilizing the same databases as for the mustard assessment. The following specific search strategies relevant to onion and garlic as potential food allergen were used: Allium; garlic* or onion* or chive* or shallot* or leek* or allium or alliaceae or allisa* or allibessen or alloton or kwai or kyolic or salicap* or sanhelios or sapec or xund or carisano or alliin* or allicin; exp Hypersensitivity; exp Urticaria; allerg* or hypersensi* or intoleran* or anaphyla* or urticaria* or hive*; sensiti* or toleran*; food challenge or re-challenge; exp Bulbous Vegetable; Garlic Extract; Garlic Oil; Onion Extract. No date limitations were set except those of the database searched.

4.4.2.2. Organization and Tabulation of Onion and Garlic Data. Studies fulfilling the selection criteria were reviewed and assessed based on the strength of evidence. Publications which could not be categorized by these criteria are referenced under Section 4.4.3.1, Characterization of garlic and onion. Evidence from publications that fulfilled the strength-of-evidence parameters are tabulated under the following categories:

4.4.3.2 Clinical studies (Table 4.5)—*evidence from quasi-experimental studies*

4.4.3.3 Other relevant studies (Table 4.6)—*evidence from nonexperimental descriptive studies (comparative/correlation)*

4.4.3.4 Case reports (Table 4.7)—*evidence from nonexperimental descriptive studies*

4.4.3. Results

A total of 411 publications were identified through the database search using the terms denoted above. However, based on the inclusion and exclusion criteria, only 20 fulfilled the strength-of-evidence categorization and tabulation criteria as previously described (Tables 4.5–4.7). Additional 16 publications provided information with regard to the characterization of garlic and onion and were considered relevant to the evaluation. Thus, a total of 36 publications from the scientific literature were considered relevant to the assessment of garlic and/or onion as food allergens.

4.4.3.1. Characterization of Garlic and Onion. Garlic and onion are herbaceous perennial flowering plants belonging to the family Alliaceae (formerly classified under the lily family [Liliaceae]). Within the Alliaceae family the genus *Allium* includes vegetables with bulbs or corms [57]. A bulb is an underground bud with thick, fleshy scales, and a corm is a short, solid, vertical underground stem with papery thin leaves [58]. Garlic (*Allium sativum* L.), onions (*Allium cepa*), shallots (*Allium oschaninii*), leeks (*Allium porrum*), and chives (*Allium schoenoprasum*) are classified under this genus [57].

Both garlic and onions have been used as food or medicine for thousands of years [59]. They are rich in volatile sulfur-containing compounds that are responsible for their pungent flavor and odor as well as for many of their health-promoting effects. Due to the pungent flavor of these vegetables, they are typically used as a seasoning or condiment in food preparations. In addition to the use of fresh garlic and onion in cooking, many commercial preparations are available for both garlic and onion, including, but not limited to, pickled, dehydrated, powdered, oil, or oil macerates.

Garlic is one of the most investigated medicinal plants. Between 1960 and 2007, more than 3000 research papers were published on the chemistry and biological effects of garlic and garlic preparations. These studies mainly focus on the cardiovascular, antimicrobial, and anticancer effects of garlic [60]. Similar claims have been made for onions; however, there is less detailed information available in the literature.

The amount of chemical constituents in garlic varies substantially based on where and how it is grown, as well as how the end product is prepared and stored [61]. Intact garlic cloves contain only a few active compounds, the main chemical constituent being the amino acid alliin, an alkyl derivative of cysteine alkyl sulfoxide. The content of alliin may vary from 0.2% to 2% fresh weight of garlic. Crushing, chewing, or cutting (or exposing dehydrated, pulverized garlic to water) of garlic cloves release the enzyme alliinase that rapidly lyses the cytosolic cysteine sulfoxide to form sulfenic acid, which immediately condenses to form diallylthiosulfinate (allicin). The formation of allicin occurs within 0.2–0.5 minutes at room temperature. Allicin represents 70–80% of the total thiosulfinates and is the least stable of the thiosulfinates formed. The thiosulfinates released from crushed garlic are reactive molecules and undergo a number of transformations, depending on the temperature, pH, and solvent conditions [60].

Similar to garlic, the enzyme alliinase is released when the cells in onion are broken through crushing, chewing, or cutting. The enzyme rapidly breaks down amino acid sulfoxides and generates sulfenic acids and subsequently allicin. As described for garlic, allicin can generate numerous transformation products depending on the environmental conditions to which it is exposed and the unstable sulfenic acids can further spontaneously rearrange into a volatile gas called syn-propanethial-S-oxide, which is an eye irritant [62].

The majority of identified research publications on onion and garlic adverse effects focused on characterizing the irritant and contact sensitization properties rather than investigating possible oral allergenic affects of garlic and/or onion. Low-molecular-weight proteins, which are usually responsible for allergic contact dermatitis reactions, have been detected in garlic extracts via patch tests and dermal sensitization experiments in guinea pigs and identified as diallyl disulfide, allylpropyl disulfide, and allicin [63, 64]. However, an irritant type of reaction could not be excluded for allicin in these studies. Eliciting agents of systemic allergic reactions are typically proteins of a higher molecular weight [63]. Most known food allergens have a molecular weight between 10 and 70 kDa [65].

Immunoblotting analyses conducted with serum from individuals in case reports have identified IgE-binding protein bands with molecular masses of approximately 12 and 40–50 kDa in both garlic and onion. Using the serum of a woman who experienced EIA after the consumption of young unripe garlic, a 12-kDa band was identified on an IgE immunoblot for extracts of young garlic, garlic, onion, leek, hazelnut, and mugwort pollen [66]. Enrique et al. [67]

also reported a single 12-kDa band using onion extract and the serum of a woman who developed urticaria after consuming raw onion. Asero et al. [68] found bands with molecular masses of 10, 20, and 40 kDa using garlic extract and the serum of a woman who had urticaria after the ingestion and contact with raw and cooked garlic and also reported that the serum of a man who experienced urticaria/AE after consuming raw onion had IgE reactivity to 15- and 43-kDa bands in an onion extract [69]. Further characterization of these proteins has not been described in the literature [58, 70].

In contrast, Kao et al. [71] studied the antigenicity, allergenicity, and IgE-binding cross-reactivity of a 56-kDa protein purified from the sera of 15 subjects with reported garlic allergies. The isolated protein was identified as alliin lyase. Skin tests confirmed that alliin lyase elicited IgE-mediated hypersensitive responses in subjects with garlic allergies. IgE cross-reactivity and IgE-inhibition analyses demonstrated that garlic alliin lyase showed high cross-reactivity with alliin lyase from other *Allium* species such as leek, shallot, and onion. Furthermore, carbohydrate epitopes were found to contribute to the binding of IgG and IgE to alliin lyase [71]. The analysis of the homology between the alliin lyase of garlic and onion by Nock and Mazelis [72] indicated that the enzymes have very little homology in structure; similar sized subunits and the carbohydrate moieties are quite different in each case. These results may explain why reports of cross-reactivity among different Alliaceae vegetables vary greatly within the literature. It is also suggested in the literature that the varying reports of cross-reactivity may be due to the level of cross-reactivity varying among individuals [73, 74]. However, the effect of the carbohydrate moieties in garlic and onion on allergenicity and cross-reactivity requires further investigation because carbohydrate epitopes can bind human IgE from allergic subjects and have a role in cross-reactivity between allergens from unrelated sources [75, 76].

None of the studies that identified IgE-reactive proteins in garlic and/or onion conducted enzymatic or heat stability tests on the proteins. However, available evidence from case studies indicates that the allergenic proteins are susceptible to digestive enzymes and heat. Two subjects who had positive SPT and serum IgE results for onion did not experience allergic reactions after consuming onion powder [77], and several cases reported a history of allergic reactions after consuming raw garlic and onion but a tolerance to cooked garlic and onion.

4.4.3.2. Clinical Studies. Pivotal clinical studies that include evidence from meta-analysis of randomized controlled trials, individual randomized controlled trials, or nonrandomized controlled trials were not available for the allergenicity assessment of either garlic or onion. However, six studies were identified in the literature that were conducted using an open allergenicity assessment that included garlic (two studies), onion (one study), or both garlic and onion (three studies) as part of the foodstuffs tested. These assessments utilized LFC or OFC and/or a combination of SPT, RAST, and determinations of serum IgE specific to garlic and/or onion, in order to verify an allergic response and quantify the prevalence of reactions to certain foodstuff. These studies are presented in Table 4.5.

Rancé and Dutau [39] examined over 25 food allergens among 142 children with a history of food allergies. Subjects were submitted to LFC for various foodstuffs, and when the results of the LFC were negative, SBPCFC was conducted. One subject (1%) had a positive LFC response to garlic. Egg (75%), peanut (60%), and mustard (16%) were the most common food allergens in this study. In 2002, Rancé and Dutau also reported a garlic allergy prevalence of 1% among 163 children previously identified as having BA by pulmonary function [41].

Valdivieso et al. [77] performed SPT for onion allergies among 106 subjects randomly selected from an allergy clinic. Of the 106 subjects, eight (8%) had positive SPT results for onion and four of those eight subjects experienced clinical symptoms such as rhinoconjunctivitis, dyspnea, eczema, and BA. These four subjects participated in bronchial, nasal, and/or oral provocation tests. All subjects showed positive reactions to heated and nonheated onion extracts. Two subjects experienced intense rhinoconjunctivitis and one subject had chest tightness and wheezing and dyspnea after exposure to the smell of onions. Only two subjects participated in the double-blind oral provocation with 2 g of onion powder; however, neither experienced allergic responses. In the author's opinion, these results suggest that onion is a respiratory allergen.

An open study examining the frequency of reported food-induced symptoms, food allergy, or food intolerance in 169 subjects monosensitized to grass pollen, reported that the number of subjects with food intolerances was higher than that of subjects with food allergies [78]. A positive food allergy was defined by evidence of IgE sensitization to food and the demonstration of OAS, gastrointestinal symptoms, and urticaria–AE within 2 hours of the ingestion of small quantities of food. Out of the 169 subjects, 19 (11%) were considered to have food allergies and 46 (27%) were considered to have food intolerances. Positive serum-specific IgE results for garlic and onion were reported in 58% (7/12) and 33% (3/9) of subjects tested, respectively. Oral challenge confirmed that one subject had a garlic allergy and one subject had an onion allergy (1/169; 0.6%). Garlic intolerance was exhibited in six subjects (4%) and onion intolerance in four subjects (2%). Subjects who were challenged with garlic and/or onion mainly experienced urticaria and gastrointestinal symptoms. No respiratory symptoms were observed among the subjects [78]. These allergic reactions are considered mild to moderate in severity [8].

A retrospective study was conducted using 26 food allergens and 14 subjects who had a history of food-dependent EIA [79, 80]. Of the 14 subjects, 10 (71%) were positive for onion and 9 (64%) for garlic via SPT and/or RAST IgE. One subject participated in a food–exercise challenge after consuming garlic. However, no reaction was observed. Onion was not tested in a food–exercise challenge. Authors noted the clinical history data for the subjects was not as helpful in the interpretation of positive SPT/RAST results for garlic, onion,

TABLE 4.5 Clinical Studies (Grouped by Relevance to Garlic, Onion, and Both Garlic and Onion and Organized Chronologically by Publication Date within Groups)

Author, Country	Study Design Details	SUB	Clinical Hx	Symptoms and Signs[a] (before CH)
Garlic				
Rancé and Dutau [39], France	Open study assessing over 25 allergens including garlic using SPT, prick + prick test IgE LFC and SBPCFC (when LFC negative)	142 SUB (Ch) Age: 7 m–15 y Average: 4.6 y Sex: 95 M, 47 F	Hx of food allergy	Multiple presenting symptoms in 66% Ch AD 61% Rash 32% AE 25% BA 24% A 4%
Rancé and Dutau [41], France	Open allergy study assessing foodstuff including garlic documented by DBPCFC in BA SUB identified by pulmonary function	163 SUB (Ch) Age: 2–17 y Average: 7.2 y Sex: 109 M, 54 F	Family Hx atopy 91% Hx of ≥1 food allergies BA for an average of 5.5 y	SK 59% Res 24% GI 12%
Valdivieso et al. [77], Spain	SPT for onion performed on 106 SUB randomly chosen from allergy clinic; positive responders participated in one or more of the following: bronchial or nasal provocation, double-blind oral provocation, IgE (RAST)	106 SUB randomly chosen from allergy clinic 8 SUB positive SPT Rxn (8%) 4/8 SUB has clinical symptoms to onion exposure Age: 31–45 y Sex: 1 M, 3 F	Hx of food and pollen allergies	Pollen or onion induced: RC (4/4) D (2/4) E (1/4) BA (2/4)
Onion				
Valdivieso et al. [77], Spain	SPT for onion performed on 106 SUB randomly chosen from allergy clinic; positive responders participated in one or more of the following: bronchial or nasal provocation, double-blind oral provocation, IgE (RAST)	106 SUB randomly chosen from allergy clinic 8 SUB positive SPT Rxn (8%) 4/8 SUB has clinical symptoms to onion exposure Age: 31–45 y Sex: 1 M, 3 F	Hx of food and pollen allergies	Pollen or onion induced: RC (4/4) D (2/4) E (1/4) BA (2/4)
Garlic and Onion				
Boccafogli et al. [78], Italy	Open study assessing frequency of food-induced symptoms or allergy in SUB monosensitized to grass pollen SPT, specific IgE, and oral challenge (0.45-mg food administered w/ 10-fold increase in 30-minute intervals until cumulative dose of 5 g if no Rxn observed administration up to 50 g in 30-minute intervals	169 SUB monosensitize to grass pollen Age: 9–54 y Mean: 28 y Sex: 84 M, 85 F Control: 50 SUB monosensitize to *Dermatophagoides* Age: 10–52 y Mean: 30.5 y Sex: 25 M, 25 F	Hx of BA and/or R during grass pollen season or household dust Positive SPT, specific IgE, and nasal challenge for aeroallergens	Pollen group had increased frequency of adverse food Rxn 81/169 (48%) compared with control 3/50 (6%) Symptoms: U 28/169 (17%) GI 28/169 (17%) Res 13/169 (8%)

Symptoms and Signs[a] SEV of Rxn[b] after CH	Diagnostic Tests[c]	Eliciting Dose	Eliciting Allergen	Prevalence	Comments
Positive Rxn to garlic 1/142 in LFC	SPT, prick + prick test, IgE	1 mg to 5 g for all allergens tested Further details were not provided	Commercial extract and extract from fresh foods	1/142 (1%) positive for garlic allergy LFC	No comments relevant to garlic
Asthma induced by food allergens potentially sev 6% of SUB reported A Rxn	SPT IgE	Details not provided	Various food extracts Further details not provided	Positive allergy to garlic 1% SUB	Prevalence of asthma induced by food allergens: 10%
2 SUB had intense RC and 1 SUB had chest tightness, wheezing, and D after exposure to smell of onions Oral provocation in 2 SUB did not elicit an Rxn	SPT 106 SUB Bronchial provocation 2 SUB Nasal provocation 1 SUB Double-blind oral provocation 2 SUB	SPT—516 μg protein/mL Bronchial and nasal—1:1 v/v onion extract Oral—2-g onion powder	Fresh onion, onion powder	8 SUB (8%) positive SPT 3/8 SUB positive IgE onion 4/8 SUB pollen-specific IgE 2/4 SUB garlic-specific IgE	4 SUB had immediate positive SPT to fresh onion, garlic, fried onion and leek All showed positive reaction to heated and nonheated onion extract Results suggest Res allergy
2 SUB had intense RC and 1 SUB had chest tightness, wheezing and D after exposure to smell of onions Oral provocation in 2 SUB did not elicit an Rxn	SPT 106 SUB Bronchial provocation 2 SUB Nasal provocation 1 SUB Double-blind oral provocation 2 SUB	SPT—516 μg protein/mL Bronchial and nasal—1:1 v/v onion extract Oral—2-g onion powder	Fresh onion, onion powder	8 SUB (8%) positive SPT 3/8 SUB positive IgE onion 4/8 SUB pollen-specific IgE 2/4 SUB garlic-specific IgE	4 SUB had immediate positive SPT to fresh onion, garlic, fried onion and leek All showed positive Rxn to heated and nonheated onion extract Results suggest Res allergy
Garlic and onion mainly associated w/ U and GI symptoms 19/169 SUB (11%) considered to have food allergies 1/19 garlic 1/19 onion 46/169 SUB (27%) considered to have food intolerance 6/46 garlic 4/46 onion	SPT Specific IgE Oral challenge	Details not provided	Garlic and onion No further information provided	19/169 SUB (11%) considered to have food allergies 1/169 garlic (0.5%) 1/169 onion (0.5%) (IgE-sensitized, OAS, GI, and/ or U/AE within 2 hours after ingestion of small quantity)	Number of SUB with food intolerance higher than that of SUB w/ food allergy Cross-reactivity between pollen allergens and food allergens may explain food allergy association but not higher incidence of food intolerance, increased intestinal permeability to macromolecules hypothesized

TABLE 4.5 Continued

Author, Country	Study Design Details	SUB	Clinical Hx	Symptoms and Signs[a] (before CH)
Romano et al. [79, 80], Italy	Retrospective food-dependent EIA study in SUB with Hx of EIA within 2 hours after meal 26 food allergens tested by SPT, IgE assay 8/14 SUB participated in food–exercise challenge	14 SUB w/ Hx of food EIA Age: 16–27 y Mean: 20.5 ± 3.6 Sex: 10 M, 4 F	Family Hx of allergy 50% Hx pollen allergy 29%	Hx of A
Asero et al. [81], Italy	Open allergy study assessing vegetable foods including garlic and onions to identify which vegetables are safe for lipid transfer protein (LTP) allergic SUB to consume	49 SUB (monosensitized to LTP) Age: mean 29 y Sex: 21 M, 28 F Controls: 24 sensitized to birch pollen 18 sensitized to profilin 16 sensitized to LTP and birch pollen	Hx of Rx to vegetables including garlic and onions	Hx of OAS, U, AE, BA, A No SUB reported Hx of allergy to garlic 2/49 (4%) reported Hx of allergy to onion

[a]Symptoms and signs: A, anaphylaxis; AD, atopic dermatitis; AE, angioedema; BA, bronchial asthma; D, dyspnea; E, eczema; EIA, exercise-induced anaphylaxis; GI, gastrointestinal; OAS, oral allergy syndrome; R, rhinitis; RC, rhinoconjunctivitis; SK, skin; U, urticaria.
[b]Severity of reaction: refer to Methods [8].
[c]Diagnostic tests: IgE, serum immunoglobulin E; RAST, radioallergosorbent test; SPT, skin prick test.
CH, challenge; Ch, children; DBPCFC, double-blind, placebo-controlled food challenge; F, female; Hx, history; LFC, labial food challenge; M, male; Res, respiratory; Rxn, reaction; SEV, severity; sev, severe; SBPCFC, single-blind, placebo-controlled food challenge; SUB, subjects; w/, with; y, years old.

parsley, and basil because subjects generally did not recall if these ingredients were part of the meal that caused EIA.

Asero et al. [81] conducted an open allergy study assessing vegetables including garlic and onion in order to identify which vegetables are safe for consumption by individuals with a lipid transfer protein (LTP) allergy. LTP is the main food-related allergen identified in southern Europe [67]. Forty-nine subjects were selected who were monosensitized to LTP and control groups of subjects sensitized to pollen (24 subjects), profilin (18 subjects), and LTP and pollen (16 subjects) were also selected. A history of reactions after consuming garlic or onion was reported in zero and two subjects, respectively. The

Symptoms and Signs[a] SEV of Rxn[b] after CH	Diagnostic Tests[c]	Eliciting Dose	Eliciting Allergen	Prevalence	Comments
14/14 allergic to two or more food allergens 10/14 positive for onion and 9/14 positive for garlic by SPT/IgE-specific levels 1 SUB participated in food–exercise challenge after consuming garlic—no Rxn observed	SPT RAST (IgE)	Not provided	Onion or garlic no further details provided	10/14 (71%) positive for onion 9/14 (64%) positive for garlic No Rxn observed after garlic exercise challenge; onion not tested	Hx data less helpful interpreting positive SPT/RAST for garlic, onion, parsley, and basil because patients could not recall if they were part of the EIA meal
8/36 (22%) SUB positive SPT to garlic 16/41 (39%) SUB positive SPT to onion 4/13 (31%) controls positive SPT to garlic 6/14 (43%) controls positive SPT to onion Results not considered statistically significant	SPT	Not provided	Commercial vegetable food extracts No further details provided	8/36 (22%) SUB positive SPT to garlic 16/41 (39%) SUB positive SPT to onion 4/13 (31%) controls positive SPT to garlic 6/14 (43%) controls positive SPT to onion Results not considered statistically significant	Specificity of SPT poor, many patients showed positive SPT to foods that they reported eating without any problems, including garlic and onions

positive SPT results for garlic and onion were not statistically significant between the monosensitized LTP subjects and the controls. Out of 36 LTP subjects, 22% (8/36) and 31% (4/13) of controls had a positive SPT to garlic, and 39% (16/41) of LTP subjects and 43% (6/14) of controls had a positive SPT to onion. The specificity of the SPT was considered poor by the authors.

4.4.3.3. Other Relevant Studies. Three nonexperimental, descriptive studies were identified in the literature search as being relevant to the assessment of the allergenicity of onion (one study) and both garlic and onion (two studies). These studies are tabulated in Table 4.6.

TABLE 4.6 Other Relevant Studies (Grouped by Relevance to Garlic, Onion, and Both Garlic and Onion and Organized Chronologically by Publication Date within Groups)

Author, Country	Study Design Details	SUB	Clinical Hx	Symptoms and Signs[a] (before CH)
Onion				
Castanón et al. [82], Mexico	Retrospective analysis of 1419 SUB w/ food allergies to determine frequency of hypersensitivity to foods	1419 SUB (Ch) Age: 12 months to 18 y Average: 12.8 y Sex: details not provided	Hx of food allergies	Most allergy affect age groups: 4–7 y (49%) 1–3 y (24%) 12–17 y (14%)
Garlic and Onion				
André et al. [43], France	Retrospective analysis of 9-year period investigating foodstuff most frequently associated with A Rxn	580 SUB 480 Ad 100 Ch Age: 1–83 y Mean: 30 y Sex: 290 M, 290 F	Hx of adverse Rxn to food	60/580 (10%) SUB Hx sev Rxn to food A 52/60 (87%) AE 6/60 (10%) Bronchospasm 2/60 (3%)
Moneret-Vautrin et al. [83], country not cited	Retrospective analysis of a food allergy database (589 SPT) and one case report of onion allergy (case 1) 1 SUB (case 2) DBPCFC for garlic 1 SUB (case 3) SBPCFC for garlic 1 SUB (case 4) LFC for garlic	Food allergy database 589 SUB: 402 Ch Age: <15 y 187 Ad Age: not reported Sex: not reported	Hx of food allergy	Case 1 w/ systemic mastocytosis and A induced by celery, carrot, onion, hazelnut Case 2 w/ U and AE Case 3 w/ R, BA, and U from fruits and vegetables Case 4 w/ U and A from garlic and onion soup

[a]Symptoms and signs: A, anaphylaxis; AE, angioedema; BA, bronchial asthma; R, rhinitis; U, urticaria.
[b]Severity of reaction: refer to Methods [8].
[c]Diagnostic tests: IgE, serum immunoglobulin E; SPT, skin prick test.
Ad, adult; CH, challenge; Ch, children; DBPCFC, double-blind, placebo-controlled food challenge; F, female; Hx, history; LFC, labial food challenge; M, male; mod, moderate; Rxn, reaction; SEV, severity; sev, severe; SBPCFC, single-blind, placebo-controlled food challenge; SUB, subjects; w/, with; y, years old.

Symptoms and Signs[a] SEV of Rxn[b] after CH	Diagnostic Tests[c]	Eliciting Dose	Eliciting Allergen	Prevalence	Comments
Not applicable to study design	SPT	Details not provided	Onion Further details not provided	4% SUB positive SPT to onion	Onion identified in group of foods responsible for 58% of allergic Rxn (fish, milk, seafood, soy, beans, orange, onion, tomato, chicken, nut, lettuce, and strawberry)
Not applicable to study design	SPT IgE	Details not provided	Garlic and onion	Garlic and onion were not among the top 19 foods most frequently associated with A Rxn (celery 18% to chamomile 1%)	Evolution of sensitization over 1984–1992 indicate an overall increasing trend for garlic and onions Increased consumption and more attentive clinical examinations are credited for this evolution
Case 1: no challenge reported Case 2: U after 45 minutes SEV: mild Case 3: laryngeal pruritus, rhinorrhea, cough SEV: mild-mod Case 4: positive LFC result	SPT, DBPCFC, SBPCFC, LFC	Case 2: 1000-mg garlic Case 3: 500-mg garlic	No details provided about garlic used in challenge reports	20/265 positive SPT for garlic (8%) 7/263 positive SPT for onion (3%)	In this series, two spices responsible for half cases observed: garlic and fennel seed

Castanón et al. [82] conducted a retrospective analysis of 1419 subjects with food allergies in order to determine the frequency of hypersensitivity to specific foods. Positive SPT results for onion occurred among 4% (56 subjects) of the study group. Garlic was not tested in this study.

A retrospective analysis conducted by André et al. [43] examined which foods were most frequently associated with an anaphylactic reaction over a 9-year period in France. Neither garlic nor onions were among the 19 foods most frequently associated with anaphylactic reactions in 580 subjects. The authors noted an increasing trend in the frequency of sensitization to garlic and onions over time; however, increased consumption and more attentive clinical examinations were credited for this evolution.

Moneret-Vautrin et al. [83] conducted a retrospective analysis of a food allergy database that contained the SPT results of 589 subjects. Of these, four cases were identified as relevant to garlic or onion food allergy (three garlic and one onion). The garlic cases were challenged either by DBPCFC or SBPCFC or an LFC. All three garlic cases were positive. In one case, after the ingestion of 1000 mg of garlic, urticaria was observed. This reaction is considered to be mild in severity [8]. In another case, the ingestion of 500 mg of garlic leads to laryngeal pruritus, rhinorrhea, and coughing. This reaction was considered moderate in severity [8]. The last case reported positive LFC results but no further information was provided. The case of the onion allergy was not challenged because the subject had a history of systemic mastocytosis and anaphylaxis after the consumption of celery, carrot, hazelnut, and onion. This publication also reports SPT positive responses for garlic in 20 (10 adults and 10 children) (8%) subjects out of 265 tests and 7 (five adults and two children) (3%) subjects out of 263 tests for onion.

4.4.3.4. Case Reports. A total of 12 case reports of allergic responses to garlic (six cases), onions (five cases) or both garlic and onion (one case) were identified in the literature. These case reports are tabulated in Table 4.7 and provide descriptions of the severity of reactions to garlic or onion as well as identifying the sources of garlic or onion exposure.

For most cases, the information provided was limited as only three of the case reports included oral challenges and only one reported the eliciting dose. However, the case reports provided valuable information regarding the severity of reactions and condition under which the reaction was elicited. Reactions ranged from acute anaphylaxis to generalized skin manifestation. Of the 12 case reports, seven (three garlic, three onion, one both) individuals reported anaphylactic-type reactions, and four (one garlic, two onion, one both) required emergency medical interventions. The case reports provided evidence that the allergenic proteins in garlic and onions are susceptible to enzymatic digestion and/or heat. Several cases reported a history of allergic reactions after consuming raw garlic and onion but a tolerance to cooked garlic and onion. An oral challenge with cooked garlic provided negative results [74]. Four of the five cases reported allergies to raw onion; however,

TABLE 4.7 Case Reports (Grouped by Relevance to Garlic and Onion and Organized Chronologically by Publication Date within Groups)

Author, Country	Cases	Clinical Hx	Symptoms and Signs[a] SEV of Rxn[b]	Diagnostic Tests[c]	Eliciting Dose	Eliciting Allergen	Comments
Garlic							
Burden et al. [98], United Kingdom	1 M Age: 58 y	Sev SK Rxn on the hands after taking garlic extract	Double-blind oral provocation w/ garlic tablets once daily for a week resulted to D of the hands; no Rxn to placebo SEV: mild	Oral challenge	Not provided	Garlic extract and garlic tablet	Lack of systemic Rxn reported when garlic is ingested by SUB w/ contact allergy may either be dose related or due to the relevant antigens being heat liable and destroyed by cooking
Asero et al. [68], Italy	1 F Age: 35 y	Hx of systemic U/AE associated with both raw and cooked garlic No Hx of pollen allergy or to other common food allergens	U/AE but no GI symptoms reported after ingestion of garlic SEV: mild	SPT	Not provided	Commercial garlic extract and fresh garlic	In a series conducted by this author w/ over 300 SUB w/ both pollen allergy and OAS, no SUB reported garlic intolerance, suggesting lability of cross-reactive antigen No GI symptoms of this case suggest garlic allergens destroyed by enzymatic digestion
Pérez-Pimiento et al. [66], Spain	1 F Age: 23 y	Hx of allergy to pollen and dried fruit and food-dependent EIA	An A Rxn after eating young garlic; emergency medical treatment was required SEV: sev	SPT Prick + prick IgE	Not provided	Young garlic (unripe garlic)	SUB tolerated garlic cloves and other Liliaceae but was sensitized to pollen and foods distantly related to garlic No oral challenge (sev Rxn)
Pires et al. [74], Portugal	1 M Age: 16 months	Hx of milk and egg white allergy Hx of tolerance of cooked garlic and onion	Oral challenge with raw garlic positive generalized U SEV: mild Oral challenge with cooked garlic negative	SPT Prick + prick	Not provided	Commercial garlic extract and fresh garlic	Suggest allergic fraction that this case is sensitized to is denatured by heat

TABLE 4.7 Continued

Author, Country	Cases	Clinical Hx	Symptoms and Signs[a] SEV of Rxn[b]	Diagnostic Tests[c]	Eliciting Dose	Eliciting Allergen	Comments
Yin and Li [99], China	Case 1: 1 M Age: 42 y Case 2: 1 F Age: 16 y	Both cases had Hx of hay fever	Case 1: an A Rxn after eating young garlic (U, AE, throat swelling, D, syncope, and hypertension) Case 2: an A Rxn while walking for 10 minutes after eating young garlic (EIA)	SPT IgE for various foods	Not provided	Young garlic	Limited details provided in the abstract Challenge study refused by subjects
Onion							
Hicks and Tanner [85], United States	1 M Age: 27 y	Hx of food-dependent EIA requiring emergency medical attention	EIA (lightheadedness, generalized pruritus, numbness in jaw, facial swelling) Played basketball after eating SEV: sev	SPT positive for corn, soybeans, peanuts, walnuts, cabbage, and onion Self-conducted oral challenge with chicken, peanut, potatoes, and onion negative without exercise challenge	Not provided	Several foods including onion	Challenge to cooked onion not performed
Arena et al. [86], Spain	1 F Age: 44 y	No Hx of allergies to aeroallergens or other foods	An A Rxn after eating raw/lightly cooked onion SEV: sev	Onion-specific IgE Positive for raw onion, negative for cooked onion	Not provided	Raw onion	SUB sensitized to thermolabile antigenic fraction of onion Cooked onion tolerated
Asero et al. [69], Spain	1 M Age: 45 y	Hx of OAS and gastric pain after ingesting peaches	Sev systemic U/AE Rxn after eating raw onion SEV: mod Tolerate cooked onion	SPT Onion-specific IgE Total IgE positive	Not provided	Raw onion	No reaction to cooked onion

Pérez-Calderón et al. [84], Portugal	1 F Age: 26 y	Hx of RC and BA due to pollen allergies	Several episodes of A-type Rxn after eating onion and exercising (EIA), at least one episode requiring emergency medical treatment SEV: sev	SPT Prick + prick Total IgE positive Food challenge at rest negative Food challenge w/cooked onion positive after exercise 10 minutes, 2 hours after eating	30-g cooked onion	Raw and cooked onion	Observed cross-reactivity w/ other members of the Lilaceae family (garlic, asparagus, or leek)
Enrique et al. [67], Spain	1 F Age: 19 y	Hx of RC and U to mugwart and oral pruritus after eating raw onion and peach	U after eating raw onion SEV: mild	SPT Onion-specific IgE positive	Not provided	Raw onion	Cooked onion extract not tested
Garlic and Onion							
Belleau and Blaiss [100], United States	1 M Age: 5 y	Hx of perennial allergic R, atopic dermatitis, mild intermittent asthma, egg and peanut allergy	An A Rxn requiring medical intervention after consuming steak seasoned with "Kraft Greek Vinaigrette Dressing"	SPT positive for peanuts, Kraft dressing, whole egg, garlic RAST positive for garlic, onion, and various foods	Not provided	Garlic and onion in salad dressing	Challenge study considered inappropriate based on SEV of Rxn

[a]Symptoms and signs: A, anaphylaxis; AE, angioedema; BA, bronchial asthma; D, dyspnea; EIA, exercise-induced anaphylaxis; GI, gastrointestinal; OAS, oral allergy syndrome; R, rhinitis; RC, rhinoconjunctivitis; SK, skin; U, urticaria.
[b]Severity of reaction: refer to Methods [8].
[c]Diagnostic tests: IgE, serum immunoglobulin E; RAST, radioallergosorbent test; SPT, skin prick test.
F, female; FC, food challenge; Hx, history; M, male; mod, moderate; Rxn, reaction; SEV, severity; sev, severe; SUB, subjects; w/, with; y, years old.

an EIA food challenge had positive results after consuming cooked onion and exercising but negative results without exercise. This reaction to cooked onion was elicited after the consumption of 30g of onion and running for 10 minutes [84].

4.4.4. Discussion

An overall assessment of the scientific literature provides very limited information on onion and garlic as potential food allergens. The evidence identified through this review suggests that cooked onion and garlic are less likely to induce an allergic response than when the product is uncooked [84]. This information is relevant since the food labeling regulations and the list of priority allergens addresses prepacked food and specific consideration is given as to whether the food or food ingredient may become hidden.

The following limitations of the systematic literature review were taken into consideration when determining the scientific validity of including garlic and/or onion on the Canadian list of priority food allergens.

A large number of publications identified in the initial database search were excluded of our assessment because the studies were not relevant to food allergy, but to other aspects such as potential benefits or to allergies elicited through other routes of exposure such as dermal, respiratory, and/or occupational. These data were excluded because they were not considered to be pertinent to the issue of food allergy; however, it is recognized that this information is important for those in the clinical field in assessing the possibility of occupational or environmental disorders, particularly in areas where garlic and/or onion is grown and processed.

While there were very few studies that looked specifically at onion and garlic as potential food allergens, relevant data were sometimes contained within publications with other objective such as food allergy in general. This fact made the identification of relevant information more challenging and may have led to the exclusion of available information on garlic and/or onion.

Based on the identified limitations of the systematic literature review, the strength of evidence is considered insufficient to fulfill the Canadian criteria for the introduction of food to the priority list of food allergens.

The first criterion of the Canadian adopted JECFA recommendations stipulates the existence of a credible cause–effect relationship, based on positive DBPCFC studies or unequivocal reports of reactions with typical features of severe allergenic or intolerance reactions. In the absence of DBPCFC studies, supporting studies were evaluated as per the strength of evidence provided by the study designs [6] (Tables 4.5–4.7).

The positive results obtained from diagnostic tests (SPT/IgE), which were utilized in the majority of the supportive studies, were not considered sufficient to substantiate a cause–effect relationship because evidence within the database suggests that SPT and IgE diagnostic tests may not provide an accurate reflection of the potential for allergic reactions after the consumption of garlic and/or onion. This available evidence is not sufficient to support the determination of a credible cause–effect relationship for the allergenicity of garlic and/or onion as food.

Valdivieso et al. [77] reported a prevalence (8%) of onion allergies among 106 adults randomly selected for SPTs from a food allergy clinic. However, only four of these eight subjects experienced clinical symptoms associated with onion exposure, and only three of these eight subjects had positive onion-specific IgE results. Furthermore, an oral provocation with onion was conducted with two of these subjects, and both had negative results [77]. The standard tests utilized in this study (SPT and IgE) did not provide an accurate indication of allergic reactions following the consumption of onion. The Boccafogli et al. [78] study further showed inconsistent results between positive IgE and oral challenge results. In this study, 58% (7/12 cases) and 33% (3/9 cases) had positive serum-specific IgE for garlic and onion, respectively, but only one case in each of the groups had a confirmed allergy after an oral challenge. Asero et al. [81] also reported that the results of SPT were not reliable for garlic and onion as many of the positive SPT subjects in their study consumed garlic and onion without experiencing any symptoms. Furthermore, studies that only reported the prevalence of positive SPT responses to garlic and/or onion among study populations did not consider the irritant properties of the sulfur-containing compounds within garlic and onions. Without further confirmation of a food allergy, these irritant effects cannot be discounted when interpreting the results. Without evidence from a DBPCFC study, the small number of positive oral challenges reported in the supportive studies and the uncertainty in the results due to the inconsistencies observed between the reported positive diagnostic results, and the few reports of allergenic or intolerance reactions after oral challenges, the strength of evidence in the current database is not considered sufficient to establish a credible cause–effect relationship for the oral allergenicity of garlic and/or onion.

Moreover, this review did not identify DBPCFC studies designed specifically to assess garlic and/or onion as potential food allergens. The studies available were not designed specifically to assess the allergenicity of garlic and/or onion through oral exposure, and consequently, qualitative and quantitative details about the garlic or onion preparations were not provided. Furthermore, study designs assessing numerous foodstuffs did not report whether appropriate controls were employed in order to differentiate between allergic responses and the irritant effects associated with the sulfur-containing compounds within garlic and onions.

The second criterion of the Canadian adopted JECFA recommendations calls for reports of severe systemic reactions following the exposure to foodstuff. Anaphylactic reactions associated with the consumption of garlic and/or onions have been reported in specific cases, although the prevalence of severe anaphylaxis-type reactions reported to be associated with the ingestion of garlic and/or onion is considered low. A retrospective analysis reported the incidence frequency of the top 19 foodstuffs involved in anaphylactic reactions over a 9-year period in France (1984–1992) [43]. Garlic and/or onion were not among the top 19 foods associated with anaphylaxis. The foods associated with anaphylactic reactions ranged from 1% to 18% incidence frequency in sensitive individuals. Of the 12 specific cases reported, seven (three garlic, three onion, one both) individuals reported anaphylactic-type reactions and four (one garlic, two onion, one both) required emergency medical interventions. In some cases, subjects who experienced anaphylactic reactions to a particular form of garlic and/or onion on the other hand exhibited tolerance to garlic and/or onion in a different form. The case that reported an anaphylactic reaction after the consumption of young unripe garlic also reported that the subject had a tolerance to the consumption of garlic cloves and onion [66]. Two cases that reported EIA associated with the consumption of onion reported that no allergic reaction was observed after consuming onion and refraining from exercise [84, 85], and the other case reported reactions only associated with consumption of raw onion and a tolerance for cooked onion [86]. The tolerance for cooked garlic and/or onion was also reported in several other cases that exhibited less severe reactions than with cooked preparations of these vegetables. It is suggested that the lability of the antigens in garlic and onions to heat and digestive enzymes are the reasons for these observed tolerances and may explain the previously noted low prevalence of anaphylactic reactions associated with the ingestion of garlic and/or onion. In an open study examining the frequency of food-induced symptoms among 169 subjects monosensitized to grass pollen, the consumption of garlic or onion was mainly associated with urticaria and gastrointestinal symptoms [78]. These symptoms are considered mild to moderate reactions [8].

The third, and last, criterion of the Canadian adopted JECFA recommendations requires the assessment of all available Canadian prevalence data in children and adults, supported by appropriate clinical studies or alternatively available data from other countries supported as per the first criterion. Currently, prevalence data are not available for Canada or other regions of the world.

Although some scientific evidence indicates that some individuals experience a severe reaction to the consumption of garlic and/or onion, the prevalence of food allergies to garlic and/or onion in children and adults remains unknown, and there are insufficient clinical data to establish a credible cause–effect relationship for the oral allergenicity of garlic and/or onion. Therefore, the overall strength of evidence to fulfill the Canadian criteria required to add new allergens to the list of priority allergens is considered inadequate at this time.

Despite the limitations of the current database, in accordance with the proposed amendments to the *Food and Drug Regulations* (1220—Enhanced Labeling for Food Allergen and Gluten Sources and Added Sulfite) (http://www.hc-sc.gc.ca/fn-an/label-etiquet/allergen/project_1220_info-eng.php), this review gave consideration to the likelihood of allergic reactions occurring as a result of nondeclared sources of garlic and/or onion in prepackaged foods. Based on available evidence, the potential for severe allergic reactions after the consumption of hidden sources of garlic and/or onion within prepackaged products is considered unlikely. The prevalence of severe allergic reactions after the ingestion of garlic and/or onion are considered low compared with other food allergens based on the results of a retrospective study conducted over a 9-year period. Furthermore, the majority of reported allergic reactions occur after the consumption of raw garlic and/or onion. Although the allergenic proteins in garlic and onion that elicit systemic allergic reactions have yet to be fully identified and characterized, reports of tolerances to cooked garlic and/or onion indicate that the antigens are labile to heat and/or digestive processes. Based on this evidence, the likelihood of consuming raw garlic and/or onion present in prepackaged foods is considered minimal, as it can reasonably be assumed that the vast majority of prepackaged foods undergo heating either as part of the manufacturing process or during the preparation of the foodstuff for consumption. Also of note, there were no reports of exposure to garlic and/or onion through hidden sources or cross-contamination of foodstuff. There is insufficient information in the database to estimate a dose threshold for garlic or onion, as only three case reports provided the eliciting amount of garlic or onion ingested before an allergic reaction was observed. However, the lowest eliciting dose in the database was reported in a case who experienced laryngeal pruritus, rhinorrhea, and coughing after ingesting 0.5 g of garlic (30 mg of protein[4]) [83]. The current limited data indicates that moderate amounts[5] of garlic and/or onion may be required to elicit allergic reactions in sensitized individuals; therefore, the potential risk of allergic reactions being elicited after the consumption of garlic and/or onion used as an undeclared spice or seasoning is considered minimal.

4.4.5. Conclusions on the Lack of Evidence for the Inclusion of Garlic and Onion as Priority Food Allergens in Canada

An assessment of the assembled evidence base for garlic and/or onion does not provide sufficient evidence to fulfill the Canadian criteria required to add new allergens to the list of priority allergens. Although there are few scientific publications that suggest that some individuals may experience severe reactions to the consumption of garlic and/or onion, particularly if uncooked, the prevalence of food allergies to garlic and/or onion in children and adults remains unknown, and there are insufficient clinical data to establish a credible cause–effect relationship for the oral allergenicity of garlic and/or onion. Furthermore, based on the information within the current database, the potential for severe allergic reactions as a result of hidden sources of garlic and/or onion in prepackaged foods is considered minimal.

In conclusion, at this time, the overall strength of evidence is considered inadequate to support the declaration of garlic and/or onion as priority food allergens, and therefore the *Food and Drug Regulations* (1220—Enhanced Labeling for Food Allergen and Gluten Sources and Added Sulfite) is not applicable to the use of garlic and/or onion in foodstuff.

4.4.6. Recommendations

Based on the conclusions of this report, it has been recommended that garlic and/or onions not be included on the Canadian list of priority food allergens at this time.

4.5. SUMMARY

Using the Canadian criteria established to consider the addition of new foods/ingredients to the list of priority allergenic foods in Canada, we conducted two systematic literature reviews of the available information on mustard, garlic, and onions as potential food allergens in order to determine whether there is sufficient scientific evidence to justify including mustard and/or garlic and/or onion on the list of defined priority food allergens in Canada.

For mustard, the following evidence was available to substantiate the addition of mustard to the list of food allergens in Canada:

- There are Canadian case reports documenting the occurrence of mustard food allergies in children and adults in Canada.
- A credible cause–effect relationship between oral exposure to mustard and allergic reactions is supported by positive DBPCFC studies designed to assess mustard as a food allergen.
- Reports describe severe systemic reactions, including anaphylaxis following exposure to very small amounts of mustard within foodstuff.
- Mustard is affirmed on the most recent list of 14 allergens to be declared on labels (updated in 2007) by the Commission of the European Communities [55], and mustard is recognized as an allergen by the International Union of Immunological Societies [56].
- All three types of mustard seed are available in Canada, and mustard is used in cooking and in processed and prepacked foods.
- Results from characterization studies of allergenic proteins indicate that proteins in mustard are resistant to degradation by heat and digestive enzymes, which makes these proteins more likely to withstand food processing.
- The thermostable allergenic proteins in mustard have the potential to be hidden within certain ingredients, preparations, and mixtures in processed and prepackaged foods.
- Individuals known to be sensitive to allergenic proteins from one type of mustard seed are likely sensitive to other types.
- Additional factors that make mustard allergy relevant to the Canadian scenario include the potential cross-reactivity between mustard and rapeseed and the facts that Canada is a major producer of both of these crops and sensitization to mustard can be acquired through dermal and respiratory exposure.

This scientific evidence provided a sufficient strength of evidence to fulfill the Canadian criteria for the addition of a new food to the list of priority allergens in Canada. In satisfying the criteria, mustard was recommended for addition to the Canadian list of food allergens; the Food Allergen Labeling Regulations, which require enhanced labeling for priority allergens in prepackaged foods, will be applied accordingly in order to ensure that mustard-allergic consumers in Canada are duly protected.

For garlic and onion, the assessment of the assembled evidence base does not provide sufficient evidence to fulfill the Canadian criteria required to add new allergens to the list of priority allergens. Although there is scientific evidence that suggests that some individuals experience severe reactions to the consumption of garlic and/or onion, particularly if uncooked, the prevalence of food allergies to garlic and/or onion in children and adults remains unknown, and there are insufficient clinical data to establish a credible cause–effect relationship for the oral allergenicity of garlic and/or onion. Furthermore, based on information within the current database, the potential for severe allergic reactions as a result of hidden sources of garlic and/or onion in prepackaged foods is considered minimal.

Therefore, at this time, it is recommended that garlic and onions not be included on the Canadian list of priority food allergens and that the proposed amendments to the *Food and Drug Regulations* (1220—Enhanced Labeling for Food Allergen and Gluten Sources and Added Sulfite) would not be applicable to the use of garlic and/or onion in foodstuff.

NOTES

1. The Codex Alimentarius Commission was created in 1963 by FAO and WHO to develop food standards, guidelines, and related texts such as codes of practice under the Joint FAO/WHO Food Standards Programme. The main purposes of this program are to protect the health of the consumers and ensure fair trade practices in the food trade, and to promote coordination of all food standards work undertaken by international governmental and nongovernmental organizations.
2. Oats are included in the list of cereal grains capable of inducing adverse effects in persons with celiac disease. However, a recent review conducted by Health Canada indicates that most individuals with celiac disease can tolerate moderate amounts of oats free from contamination with other sources of gluten [87].
3. Subsections B.01.009 one and two of the Food and Drug Regulations in Canada specifically exempt components of certain ingredients, preparations, and mixtures from declaration in the list of ingredients, such as spices and seasonings. As a result, prepackaged food products may contain undeclared (hidden) sources of food allergens.

4. Assuming the case consumed raw garlic, nutrition data indicate that 136 g of raw garlic contains 9 g of protein (http://www.nutritiondata.com).
5. Published lowest observable adverse effect levels (LOAELs) for food allergens: egg (0.13–1.0 mg of protein [P]), peanut (0.25–10 mg of P), milk (0.36–3.6 mg of P), tree nuts (0.02–7.5 mg of P), soy (88–522 mg of P), and fish (1–100 mg of P).

ACKNOWLEDGMENTS

The authors give special thanks to Dr. Sheila Dubois and her team for contributing on the development of the systematic review methodology used in this work, as well as for her expert opinion in peer reviewing this manuscript. The authors acknowledge the contribution of Mr. Frank Geraghty and Mr. Michael Abbott for reading of the manuscript. Special thanks to Ms. Jani Lalonde for her assistance in the management of the references.

REFERENCES

1. Taylor, S.L. (1999). Prospects for the future: emerging problems—food allergens. *Conference on International Food Trade Beyond 2000: Science-Based Decisions, Harmonization, Equivalence and Mutual Recognition*. Melbourne, Australia: Codex Alimentarius Commission (CAC), October 11–15. Available at http://www.fao.org/docrep/meeting/X2670e.htm
2. Godefroy, S.B., Popping, B. (2009). International regulatory environment for food allergen labelling. In: *Molecular Biological and Immunological Techniques and Applications for Food Chemists*, Popping, B., Diaz-Amigo, C., Hoenicke, K. (Eds.). Hoboken, NJ: John Wiley & Sons, pp. 267–292.
3. Rona, R.J., Keil, T., Summers, C., Gislason, D., Zuidmeer, L., Sodergren, E., Sigurdardottir, S.T., Lindner, T., Goldhahn, K., Dahlstrom, J., McBride, D., Madsen, C. (2007). The prevalence of food allergy: a meta-analysis. *The Journal of Allergy and Clinical Immunology*, *120*, 638–646.
4. Björkstén, B., Crevel, R., Hischenhuber, C., Lovik, M., Samuels, F., Strobel, S., Taylor, S.L., Wal, J.M., Ward, R. (2008). Criteria for identifying allergenic foods of public health importance. *Regulatory Texicology and Pharmacology*, *51*, 42–52.
5. See Chapter 1 of this book.
6. Chapman, J.A., Bernstein, I.L., Lee, R.E., Oppenheimer, J., Nicklas, R.A., Portnoy, J.M., Sicherer, S.H., Schuller, D.E., Spector, S.L., Khan, D., Lang, D., Simon, R.A., Tilles, S.A., Blessing-Moore, J., Wallace, D., Teuber, S.S. (2006). Food allergy: a practice parameter. *Annals of Allergy, Asthma & Immunology*, *96*, s1–s68.
7. Bernstein, I.L., Li, J.T., Bernstein, D.I., Hamilton, R., Spector, S.L., Tan, R., Sicherer, S., Golden, D.B., Khan, D.A., Nicklas, R.A., Portnoy, J.M., Blessing-Moore, J., Cox, L., Lang, D.M., Oppenheimer, J., Randolph, C.C., Schuller, D.E., Tilles, S.A.,

Wallace, D.V., Levetin, E., Weber, R. (2008). Allergy diagnostic testing: an updated practice parameter. *Annals of Allergy, Asthma & Immunology*, *100*, s1–s148.

8. Brown, S.G. (2004). Clinical features and severity grading of anaphylaxis. *The Journal of Allergy and Clinical Immunology*, *114*, 371–376.
9. Brown, S.G., Mullins, R.J., Gold, M.S. (2006). Anaphylaxis: diagnosis and management. *The Medical Journal of Australia*, *185*, 283–289.
10. Rancé, F. (2003). Mustard allergy as a new food allergy. *Allergy*, *58*, 287–288.
11. Alanís-Guzmán, G., Wesche-Ebeling, P., Maiti, R. (1995). Chemical, nutritional and functional characterization of proteins extracted from wild mustard (*Brassica campestris*, Brassicaceae) seeds from Nuevo Leon, Mexico. *Economic Botany*, *49*, 260–268.
12. Canadian Grain Commission. (2009). Canadian rapeseed. Available at http://grainscanada.gc.ca/rapeseed-colza/rcm-mrc-eng.htm#description
13. Agriculture and Agri-Food Canada (AAFC). (2007). Mustard seed. Available at http://www4.agr.gc.ca/AAFC-AAC/display-afficher.do?id=1175116081724&lang=eng#stat
14. Agriculture and Agri-Food Canada (AAFC). (2008). Canada's agriculture, food and beverage industry. Available at http://www.ats.agr.gc.ca/supply/3311_e.htm
15. Marnoch, R., Diosady, L.L. (2006). Production of mustard protein isolates from oriental mustard seed (*Brassica juncea* L.). *Journal of the American Oil Chemists' Society*, *83*, 65–69.
16. Niinimäki, A., Hannuksela, M., Mäkinen-Kiljunen, S. (1995). Skin prick tests and in vitro immunoassays with native spices and spice extracts. *Annals of Allergy, Asthma & Immunology*, *75*, 280–286.
17. Leanizbarrutia, I., Muñoz, D., Fernández de Corrès, L. (1987). Cutaneous allergy to mustard. *Contact Dermatitis*, *17*, 262–263.
18. Menéndez-Arias, L., Moneo, I., Domínguez, J., Rodríguez, R. (1988). Primary structure of the major allergen of yellow mustard (*Sinapis alba* L.) seed, Sin a I. *European Journal of Biochemistry/FEBS*, *177*, 159–166.
19. Bartolomé, B., Méndez, J.D., Armentia, A., Vallverdú, A., Palacios, R. (1997). Allergens from Brazil nut: immunochemical characterization. *Allergologia et Immunopathologia*, *25*, 135–144.
20. Bush, R.K., Hefle, S.L. (1996). Food allergens. *CRC Critical Reviews in Food Science and Nutrition*, *36*, s119–s163.
21. Teuber, S.S., Dandekar, A.M., Peterson, W.R., Sellers, C.L. (1998). Cloning and sequencing of a gene encoding a 2S albumin seed storage protein precursor from English walnut (*Juglans regia*), a major food allergen. *The Journal of Allergy and Clinical Immunology*, *101*, 807–814.
22. Domínguez, J., Cuevas, M., Ureña, V., Muñoz, T., Moneo, I. (1990). Purification and characterization of an allergen of mustard seed. *Annals of Allergy*, *64*, 352–357.
23. González De La Peña, M.A., Monsalve, R.I., Batanero, E., Villalba, M., Rodríguez, R. (1996). Expression in *Escherichia coli* of Sin a 1, the major allergen from mustard. *European Journal of Biochemistry/FEBS*, *237*, 827–832.
24. Menéndez-Arias, L., Domínguez, J., Moneo, I., Rodríguez, R. (1990). Epitope mapping of the major allergen from yellow mustard seeds, Sin a I. *Molecular Immunology*, *27*, 143–150.

25. Monsalve, R.I., Gonzalez de la Peña, M.A., Menendez-Arias, L., Lopez-Otin, C., Villalba, M., Rodriguez, R. (1993). Characterization of a new oriental-mustard (*Brassica juncea*) allergen, Bra j IE: detection of an allergenic epitope. *The Biochemical Journal*, *293*, 625–632.
26. Palomares, O., Cuesta-Herranz, J., Rodríguez, R., Villalba, M. (2005). A recombinant precursor of the mustard allergen Sin a 1 retains the biochemical and immunological features of the heterodimeric native protein. *International Archives of Allergy and Immunology*, *137*, 18–26.
27. Oñaderra, M., Monsalve, R.I., Mancheño, J.M., Villalba, M., Martinez del Pozo, A., Gavilanes, J.G., Rodriguez, R. (1994). Food mustard allergen interaction with phospholipid vesicles. *European Journal of Biochemistry*, *225*, 609–615.
28. González de la Peña, M.A., Menéndez-Arias, L., Monsalve, R.I., Rodríguez, R. (1991). Isolation and characterization of a major allergen from oriental mustard seeds, BrajI. *International Archives of Allergy and Applied Immunology*, *96*, 263–270.
29. Palomares, O., Vereda, A., Cuesta-Herranz, J., Villalba, M., Rodríguez, R. (2007). Cloning, sequencing, and recombinant production of Sin a 2, an allergenic 11S globulin from yellow mustard seeds. *The Journal of Allergy and Clinical Immunology*, *119*, 1189–1196.
30. Asero, R., Mistrello, G., Roncarolo, D., Amato, S. (2002). Allergenic similarities of 2S albumins. *Allergy*, *57*, 62–63.
31. Figueroa, J., Blanco, C., Dumpiérrez, A.G., Almeida, L., Ortega, N., Castillo, R., Navarro, L., Pérez, E., Gallego, M.D., Carrillo, T. (2005). Mustard allergy confirmed by double-blind placebo-controlled food challenges: clinical features and cross-reactivity with mugwort pollen and plant-derived foods. *Allergy*, *60*, 48–55.
32. Astwood, J.D., Leach, J.N., Fuchs, R.L. (1996). Stability of food allergens to digestion in vitro. *Nature Biotechnology*, *14*, 1269–1273.
33. Caballero, T., San-Martín, M.S., Padial, M.A., Contreras, J., Cabañas, R., Barranco, P., López-Serrano, M.C. (2002). Clinical characteristics of patients with mustard hypersensitivity. *Annals of Allergy, Asthma & Immunology*, *89*, 166–171.
34. Morisset, M., Moneret-Vautrin, D.A., Maadi, F., Frémont, S., Guénard, L., Croizier, A., Kanny, G. (2003). Prospective study of mustard allergy: first study with double-blind placebo-controlled food challenge trials (24 cases). *Allergy*, *58*, 295–299.
35. Rancé, F., Dutau, G., Abbal, M. (2000). Mustard allergy in children. *Allergy*, *55*, 496–500.
36. Rancé, F., Abbal, M., Dutau, G. (2001). Mustard allergy in children. *Pediatric Pulmonology*, Suppl. 23, 44–45.
37. Niinimäki, A., Björkstén, F., Puukka, M., Tolonen, K., Hannuksela, M. (1989). Spice allergy: results of skin prick tests and RAST with spice extracts. *Allergy*, *44*, 60–65.
38. Rancé, F., Juchet-Gibon, A., Fejji, S., Bremont, F., Dutau, G. (1994). Sensibilisation aux épices chez l'enfant. *Revue Française d'Allergologie et d'Immunologie Clinique*, *34*, 475–479.
39. Rancé, F., Dutau, G. (1997). Labial food challenge in children with food allergy. *Pediatric Allergy and Immunology*, *8*, 41–44.

40. Rancé, F., Dutau, G. (1998). Multiple allergy syndrome. *Revue Française d'Allergologie et d'Immunologie Clinique*, *38*, 267–274.
41. Rancé, F., Dutau, G. (2002). Asthma and food allergy: report of 163 pediatric cases. *Archives of Pediatrics*, *9*, s402–s407.
42. Moneret-Vautrin, D.A., Kanny, G., Morrisset, M. (2006). *Les Allergies Alimentaires de l'Enfant et de l'Adulte*. Paris: Masson, pp. 180.
43. André, F., André, C., Colin, L., Cacaraci, F., Cavagna, S. (1994). Role of new allergens and of allergens consumption in the increased incidence of food sensitizations in France. *Toxicology*, *93*, 77–83.
44. Rancé, F. (2000). Les allergies alimentaires actuelles de l'enfant. [Food allergies nowadays of the child]. *Allergie et Immunologie*, *32*, 366–376.
45. Rancé, F., Kanny, G., Dutau, G., Moneret-Vautrin, D.A. (1998). Clinical features in food allergy. *Revue Française d'Allergologie et d'Immunologie Clinique*, *38*, 900–905.
46. Rancé, F., Kanny, G., Dutau, G., Moneret-Vautrin, D.A. (1999). Food hypersensitivity in children: clinical aspects and distribution of allergens. *Pediatric Allergy and Immunology*, *10*, 33–38.
47. Rancé, F., Kanny, G., Dutau, G., Moneret-Vautrin, D.A. (1999). Food allergens in children. *Archives of Pediatrics*, *6* (Suppl. 1), 61s–66s.
48. Yip, L.Y., Zimmerman, B. (1999). Mustard allergy: uncommon allergy with a common spice. *Canadian Journal of Allergy and Clinical Immunology*, *4*, 76–78.
49. Connors, L.A., Yang, W.H., Lacuesta, G.A. (2006). Case reports of seed anaphylaxis: mustard, flax and sunflower seed. *The Journal of Allergy and Clinical immunology*, *117*, s52.
50. Malet, A., Valero, A., Lluch, M., Bescos, M., Amat, P., Serra, E. (1993). Hypersensitivity to mustard seed. *Allergy*, *48*, 62–63.
51. Taylor, S.L., Hefle, S.L., Bindslev-Jensen, C., Atkins, F.M., Andre, C., Bruijnzeel-Koomen, C., Burks, A.W., Bush, R.K., Ebisawa, M., Eigenmann, P.A., Host, A., Hourihane, J.O., Isolauri, E., Hill, D.J., Knulst, A., Lack, G., Sampson, H.A., Moneret-Vautrin, D.A., Rance, F., Vadas, P.A., Yunginger, J.W., Zeiger, R.S., Salminen, J.W., Madsen, C., Abbott, P. (2004). A consensus protocol for the determination of the threshold doses for allergenic foods: how much is too much? *Journal of the British Society for Allergy and Clinical Immunology*, *34*, 689–695.
52. Kanny, G., Fremont, S., Talhouarne, G., Nicolas, J.P., Moneret-Vautrin, D.A. (1995). Anaphylaxis to mustard as a masked allergen in "chicken dips". *Annals of Allergy, Asthma & Immunology*, *75*, 340–342.
53. Johansson, S.G., Hourihane, J.O., Bousquet, J., Bruijnzeel-Koomen, C., Dreborg, S., Haahtela, T., Kowalski, M.L., Mygind, N., Ring, J., van Cauwenberge, P., van Hage-Hamsten, M., Wüthrich, B. (2001). A revised nomenclature for allergy. An EAACI position statement from the EAACI nomenclature task force. *Allergy*, *56*, 813–824.
54. Johansson, S.G., Bieber, T., Dahl, R., Friedmann, P.S., Lanier, B.Q., Lockey, R.F., Motala, C., Ortega Martell, J.A., Platts-Mills, T.A., Ring, J., Thien, F., Van Cauwenberge, P., Williams, H.C. (2004). Revised nomenclature for allergy for global use: report of the Nomenclature Review Committee of the World Allergy

Organization, October 2003. *The Journal of Allergy and Clinical Immunology*, *113*, 832–836.

55. Commission of the European Communities (CEC). (2007). Directives. *Official Journal of the European Union, L310/11–L310/14*. http://eur-lex.europa.eu/JOIndex.do.
56. International Union of Immunological Societies (IUIS). (2009). Allergen nomenclature. Available at http://www.allergen.org/Allergen.aspx
57. Fay, M.F., Chase, M.W. (1996). Resurrection of Themidaceae for the Brodiaea alliance, and recircumscription of Alliaceae, Amaryllidaceae and Agapanthoideae. *International Association for Plant Taxonomy (IAPT)*, *45*, 441–451.
58. McGovern, T.W., LaWarre, S. (2001). Botanical briefs: garlic—*Allium sativum*. *Cutis*, *67*, 193–194.
59. Zohary, D., Hopf, M. (2000). Cereals. In *Domestication of Plants in the Old World*. New York: Oxford University Press, pp. 16–88.
60. Singh, V.K., Singh, D.K. (2008). Pharmacological effects of garlic (*Allium sativum* L.). *Annual Review of Biomedical Sciences*, *10*, 6–26.
61. Morbidoni, L., Moore Arterburn, J., Young, V., Mullins, D., Mulrow, C., Lawrence, V. (2001). Garlic: its history and adverse effects. *Journal of Herbal Pharmacotherapy*, *1*, 63–83.
62. Scott, T. (1999). What is the chemical process that causes my eyes to tear when I peel an onion? Available at http://www.scientificamerican.com/article.cfm?id=what-is-the-chemical-proc
63. Bleumink, E., Doeglas, H.M.G., Klokke, A.H., Nater, J.P. (1972). Allergic contact dermatitis to garlic. *British Journal of Dermatology*, *87*, 6–9.
64. Papageorgiou, C., Corbet, J.P., Menezes-Brandao, F., Pecegueiro, M., Benezra, C. (1983). Allergic contact dermatitis to garlic (*Allium sativum* L.). Identification of the allergens: the role of mono-, di-, and trisulfides present in garlic. A comparative study in man and animal (guinea pig). *Archives of Dermatological Research*, *275*, 229–234.
65. Taylor, S.L., Lehrer, S.B. (1996). Principles and characteristics of food allergens. *CRC Critical Reviews in Food Science and Nutrition*, *36*, s91–s118.
66. Pérez-Pimiento, A.J., Moneo, I., Santaolalla, M., de Paz, S., Fernández-Parra, B., Domínguez-Lázaro, A.R. (1999). Anaphylactic reaction to young garlic. *Allergy*, *54*, 626–629.
67. Enrique, E., Malek, T., De Mateo, J.A., Castelló, J.V., Lombardero, M., Barber, D., Salcedo, G. (2007). Involvement of lipid transfer protein in onion allergy. *Annals of Allergy, Asthma & Immunology*, *98*, 202.
68. Asero, R., Mistrello, G., Roncarolo, D., Antoniotti, P.L., Falagiani, P. (1998). A case of garlic allergy. *Journal of Allergy and Clinical Immunology*, *101*, 427–428.
69. Asero, R., Mistrello, G., Roncarolo, D., Amato, S. (2001). A case of onion allergy. *Journal of Allergy and Clinical Immunology*, *108*, 309–310.
70. Borrelli, F., Capasso, R., Izzo, A.A. (2007). Garlic (*Allium sativum* L.): adverse effects and drug interactions in humans. *Molecular Nutrition and Food Research*, *51*, 1386–1397.
71. Kao, S.H., Hsu, C.H., Su, S.N., Hor, W.T., Chang, W.H., Chow, L.P. (2004). Identification and immunologic characterization of an allergen, alliin lyase, from

garlic (*Allium sativum*). *Journal of Allergy and Clinical Immunology*, *113*, 161–168.

72. Nock, L.P., Mazelis, M. (1989). Lack of homology between the alliin lyases of garlic and onion. *Phytochemistry*, *28*, 729–731.
73. Sanchez-Hernandez, M.C., Hernandez, M., Delgado, J., Guardia, P., Monteseirin, J., Bartolomé, B., Palacios, R., Martinez, J., Conde, J. (2000). Allergenic cross-reactivity in the Liliaceae family. *Allergy*, *55*, 297–299.
74. Pires, G., Pargana, E., Loureiro, V., Almeida, M.M., Pinto, J.R. (2002). Allergy to garlic. *Allergy*, *57*, 957–958.
75. Hiemori, M., Bando, N., Ogawa, T., Shimada, H., Tsuji, H., Yamanishi, R., Terao, J. (2000). Occurrence of IgE antibody-recognizing N-linked glycan moiety of a soybean allergen, Gly m Bd 28K. *International Archives of Allergy and Immunology*, *122*, 238–245.
76. Iacovacci, P., Pini, C., Afferni, C., Barletta, B., Tinghino, R., Schininà, E., Federico, R., Mari, A., Di Felice, G. (2001). A monoclonal antibody specific for a carbohydrate epitope recognizes an IgE-binding determinant shared by taxonomically unrelated allergenic pollens. *Journal of the British Society for Allergy and Clinical Immunology*, *31*, 458–465.
77. Valdivieso, R., Subiza, J., Varela-Losada, S., Subiza, J.L., Narganes, M.J., Martinez-Cocera, C., Cabrera, M. (1994). Bronchial asthma, rhinoconjunctivitis, and contact dermatitis caused by onion. *The Journal of Allergy and Clinical Immunology*, *94*, 928–930.
78. Boccafogli, A., Vicentini, L., Camerani, A., Cogliati, P., D'Ambrosi, A., Scolozzi, R. (1994). Adverse food reactions in patients with grass pollen allergic respiratory disease. *Annals of Allergy*, *73*, 301–308.
79. Romano, A., Di Fonso, M., Giuffreda, F., Quaratino, D., Papa, G., Palmieri, V., Zeppilli, P., Venuti, A. (1995). Diagnostic work-up for food-dependent, exercise-induced anaphylaxis. *Allergy*, *50*, 817–824.
80. Romano, A., Di Fonso, M., Giuffreda, F., Quaratino, D., Papini, S., Palmieri, V., Venuti, A., Zeppilli, P. (1998). Food-dependent excercise-induced anaphylaxis: diagnostic value of in vitro and in vivo allergologic tests. *International Journal of Immunopathology and Pharmacology*, *11*, 35–40.
81. Asero, R., Mistrello, G., Roncarolo, D., Amato, S. (2007). Detection of some safe plant-derived foods for LTP-allergic patients. *International Archives of Allergy and Immunology*, *144*, 57–63.
82. Castanón, L.A., López, J.P., del Rio Navarro, B.T., Rosas Vargas, M.A., Ortiz, L.L., Monge, J.J.L. (2002). Hipersensibilidad por prueba cutanea a alimentos en pacientes alergicos en el Hospital Infantil de Mexico Federico Gomez. *Revista Alergia Mexico*, *49*, 74–79.
83. Moneret-Vautrin, D.A., Morisset, M., Lemerdy, P., Croizier, A., Kanny, G. (2002). Food allergy and IgE sensitization caused by spices: CICBAA data (based on 589 cases of food allergy). *Allergie et Immunologie*, *34*, 135–140.
84. Pérez-Calderón, R., Gonzalo-Garijo, M.A., Fernández de Soria, R. (2002). Exercise-induced anaphylaxis to onion. *Allergy*, *57*, 752–753.
85. Hicks, C.D., Tanner, S.B. (1998). A case of food-dependent exercise-induced anaphylaxis to onion. *Annals of Allergy, Asthma & Immunology*, *80*, 86.

86. Arena, A., Cislaghi, C., Falagiani, P. (2000). Anaphylactic reaction to the ingestion of raw onion. A case report. *Allergologia et Immunopathologia*, *28*, 287–289.
87. Pulido, O.M., Gillespie, Z., Zarkadas, M., Dubois, S., Vavasour, E., Rashid, M., Switzer, C., Godefroy, S.B. (2009). Introduction of oats in the diet of individuals with celiac disease: a systematic review. In: *Advances in Food and Nutrition Research*, Taylor, S.L. (Ed.). Burlington, MA: Academic Press, pp. 235–285.
88. Panconesi, E., Sertoli, A., Fabbri, P., Giorgini, S., Spallanzani, P. (1980). Anaphylactic shock from mustard after ingestion of pizza. *Contact Dermatitis*, *6*, 294–295.
89. Meding, B. (1985). Immediate hypersensitivity to mustard and rape. *Contact Dermatitis*, *13*, 121–122.
90. Widström, L., Johansson, S.G. (1986). IgE-mediated anaphylaxis to mustard. *Acta Dermato-Venereologica*, *66*, 70–71.
91. Vidal, C., Díaz, C., Sáez, A., Rodriguez, M., Iglesias, A. (1991). Anaphylaxis to mustard. *Postgraduate Medical Journal*, *67*, 404.
92. Monreal, P., Botey, J., Pena, M., Marin, A., Eseverri, J.L. (1992). Mustard allergy. Two anaphylactic reactions to ingestion of mustard sauce. *Annals of Allergy*, *69*, 317–320.
93. Jorro, G., Morales, C., Brasó, J.V., Peláez, A. (1995). Mustard allergy: three cases of systemic reaction to ingestion of mustard sauce. *Journal of Investigational Allergology and Clinical Immunology*, *5*, 54–56.
94. Valero, A.L., Amat, P., Bescos, M., Lluch, M., Serra, E., Malet, A. (1995). Mustard seed allergy: report of five cases. *Revista Espanola de Alergologia e Inmunologia Clinica*, *10*, 193–198.
95. Frémont, S., Kanny, G., Bieber, S., Nicolas, J.P., Moneret-Vautrin, D.A. (1996). Identification of a masked allergen, alpha-lactalbumin, in baby-food cereal flour guaranteed free of cow's milk protein. *Allergy*, *51*, 749–754.
96. Lingelbach, A., Rakoski, J., Ring, J. (2003). Exercise-induced anaphylaxis to cabbage and mustard. *Allergy and Clinical Immunology International*, *15*, 181–183.
97. Aygencel, G., Akıncı, E., Keles, A., Demircan, A. (2007). Tongue and uvular edema caused by *Brassica arvensis*. *Marmara Medical Journal*, *20*, 179–181.
98. Burden, A.D., Wilkinson, S.M., Beck, M.H., Chalmers, R.J. (1994). Garlic-induced systemic contact dermatitis. *Contact Dermatitis*, *30*, 299–300.
99. Yin, J., Li, H. (2007). Anaphylaxis caused by younger garlic: two cases report in China. *Journal of Allergy and Clinical Immunology*, *119*, s34.
100. Belleau, J., Blaiss, M.S. (2002). Anaphylaxis to garlic and onion. *Journal of Allergy and Clinical Immunology*, *109*, s337.

PART II

GENERAL PRINCIPLES FOR ALLERGEN MANAGEMENT AND CONTROL

5

ALLERGEN MANAGEMENT AND CONTROL AS PART OF AGRICULTURAL PRACTICES

Vernon D. Burrows

5.1. ALLERGEN MANAGEMENT AND CONTROL AS PART OF AGRICULTURAL PRACTICES

For many individuals, the task of acquiring their daily food requirements is analogous to them walking through an allergen minefield. With every decision they make, they are literally a step away from challenging their immune system, which may result in their body responding with either a minor or major allergic reaction. These individuals have to try to determine if with every food choice they make, they are consuming a dreaded allergen. In many instances, they know from reading and from experience which commodities or processed foods are unsafe so they can avoid them, but, in other cases, they are unsure if an allergen is present in processed foods. They have to decide if they should take a chance that the food is safe or whether the risk is too great for them to consume. These individuals are in constant fear of making a poor choice. In many situations, they do not have enough accurate information to make a sound decision so they usually avoid the food. Their diet, thus, becomes more restrictive than need be based on fear rather than on knowledge.

Systems have to be developed to aid such patients who have been identified as allergic to specific plants and animal species so that they are confident that they can avoid the consumption of troublesome allergens. Since there are a huge number of allergens in nature, and because individuals in many cultures

Allergen Management in the Food Industry, Edited by Joyce I. Boye and Samuel Benrejeb Godefroy

and regions of the world show different degrees of sensitivity to both common and uncommon allergens, it is important that the content of any product that is ingested be known with as much accuracy as possible. For example, it is a relatively simple matter for sensitive patients to identify foods like shrimp or peanuts, but the problem is more difficult when pieces of these, or any other troublesome species, cannot be identified and are present in foods because they impart a desirable flavor or they impart important functional properties. Accurate labeling of packaged foods or any ingested product such as pharmaceuticals is helpful or essential, but often manufacturers and chefs are not aware of the problems they create by not informing themselves that adding certain ingredients to their products can be a real problem for clients suffering from severe allergies. Some allergies, such as an allergy to peanuts, are well-known in many societies so manufacturers are aware that they must clearly label their products and thoroughly clean their equipment between runs if peanuts were included in the manufacture of a previous product. Both patients and manufacturers of foods must educate themselves if they want to avoid sickness or troublesome legal cases. The patient has to read labels and ask questions, whereas manufacturers have to become knowledgeable about the common allergens and consult medical experts about the advisability of including certain ingredients into foods. They also have to avoid cross-contamination of ingredients in their processing plants or, in the case of chefs, in their kitchens. Price should not be the only criterion in choosing ingredients. For example, it may be cheaper to use wheat starch, which contains traces of the protein gluten, to provide bulk in the manufacture of pharmaceutical pills, but wheat protein is a problem for celiac disease (CD) patients on a gluten-free diet. For CD patients, why not use rice starch, which is fine grained and safe for CD patients. Rice starch may carry some rice protein as an impurity, but it is safe for CD patients to consume. Food processors are not necessarily to blame for using impure commodities because pure commodities are commonly not available. Systems have to be devised both in the manufacture of pure agricultural species, in the movement of foods to processors, in the formulation of processed foods, in the cleanliness of the processing facilities, and in the accurate identification of ingredients for sensitive clients. The goal of the agricultural and health sectors is to make certain that safe, pure commodities are available to sensitive clients even if only in niche markets. This is a very big subject, and the author will deal with allergen management and control as part of agricultural practices.

All commodities used for food have to be grown somewhere or captured from natural habitats (wild berries, fruits, nuts, bulbs, mushrooms, ferns, etc.), and protocols have to be devised, in concert with regulatory systems, to ensure the safety of the food for the broad range of the human population. This applies to food products produced in any country, whether the product is produced domestically or imported from other countries. Unfortunately, standards for purity differ internationally, and contaminants such as weed seeds and other crop species also differ from the common contaminants in the importing country. Importers and regulatory authorities have to be knowledgeable and vigilant on what allergens will likely be present in imported foods. They should also be aware of common procedures and chemical pesticides used by foreigners to successfully grow and process foods. Often the pesticides applied in foreign countries to control diseases and weeds are peculiar to their region, and the pesticides are not registered for use in the areas where the food is consumed. Often these pesticides have not been tested for their ability to generate allergenic reactions in countries where the food is consumed. Authorities have to be aware of the potential danger of these impurities (chemicals or foreign seeds) that may cause allergenic reactions. The protein allergens present in the seeds of wild species have often not been studied in great detail, so it is prudent to only permit the sale and use of weed-free crops but this does not solve the problem of food crops carrying residual pesticides. To be aware of all the problems that may arise that will pose problems to people suffering different allergies is next to impossible. Many allergies have not been traced back to impure crops. Alternatively, if a food commodity has been found, through extensive scientific investigation, to be safe for allergic individuals, all efforts should be directed toward making it available to consumers in a pure form that is properly identified.

Once the safety of a particular commodity has been judged to be adequate for most people, a dedicated agricultural production, harvesting, cleaning, crop inspection, storage, transport, processing, and food inspection and labeling system has to be devised and followed to deliver this safe food to the marketplace. A study of the process will quickly reveal that the number of opportunities for contamination is enormous. The difficulty is that not everyone in the food chain is informed about the consequences of sloppy work nor do they realize how sensitive some individuals are to low concentrations of allergens. The quest for insisting on purity is usually related to price of the commodity. If allergic individuals understand why the growing and processing of pure foods is more costly than impure foods, they have to determine whether the benefits they would derive from consuming pure food can be justified with their budget.

Since approximately 60% of the carbohydrate and 50% of the protein in the human diet is derived from cereal grains, it is important to enumerate the principles involved in the production of pure cereals. Breeders, seed growers, and regulators have worked on this problem for many years in many countries, developing new varieties to meet the quality standards and specifications of both the food and feed markets. All "breeder lines," which are bulked to produce a new variety, are each true breeding and pure for those traits that the breeder has selected them for during the segregating generations following the making of a new hybrid. The lines are uniform for important agronomic and quality traits such as grain yield per hectare, threshing efficiency, height, uniform maturity, resistance to lodging, winter hardiness, seed weight and shape, bushel or test weight, protein content, quality of the protein, milling yield, reaction to prevalent races of important disease organisms, plus several other traits that are particular to the crop species. Various genetic protocols have been developed and followed by geneticists and breeders to develop new varieties, but when a new variety is registered and released, it is inspected by regulatory agencies, and if it passes inspection, it is considered to be pure. After release, seed of the new variety has to be increased many times by specialized growers before a commercial seed is sold to ordinary farmers to grow. In Canada, pedigree seed is inspected throughout its passage through the various classes of seed from breeder, select, foundation, and registered to certified seed. Bags of seed in these various classes carry the variety name and an official tag identifying the class. When a certified seed is grown by farmers, the seed then becomes commercial seed and is not permitted to carry the variety name. It is out of the pedigree system and carries the highest levels of contamination. Each time a variety is grown, the risk of contamination with foreign seeds occurs, so the tolerance for impurities increases at each stage of the multiplication system. This system was designed for seed growers and ordinary farmers and not for manufactures of nonallergenic foods. However, later in this chapter, the Professional Advisory Board (PAB) of the Canadian Celiac Association (CCA) has used a part of the pedigree system to produce safe oats for CD patients. They used this system rather than devise a new system for oats to be used to produce pure seed for CD patients because all the necessary safeguards were in place in the official pedigree system including an inspection system. It is important for the reader to know of some of the precautions that should, or must, be taken to prevent contamination of pure seed so that it and the products made from it can be safely consumed by individuals suffering from specific allergies.

5.2. PRECAUTIONARY PROCEDURES TO AVOID CONTAMINATION

5.2.1. Dedicated Systems

All of the participants in the system must be committed to make available safe foods to allergic individuals by using dedicated procedures and dedicated equipment to produce, isolate, process, label, market, and distribute safe food to the marketplace. Because the size of the market is relatively small and the costs of producing very pure seed are higher than producing impure commercial seed, the number of growers and processors required is relatively small. If only a few seed companies and growers are willing to become involved, it may be an incentive to purchase equipment to be used exclusively for producing and processing pure seed of one crop kind.

5.2.2. Begin with Pure Seed

Begin the production system with pure pedigreed seed carrying official identification tags signifying that the seed has been inspected by an official agency both in the field and again later after the grain is harvested and cleaned. After the grain passes official inspection, it is sealed and the sealed bag is labeled with an official tag signed by the breeder. This should be followed by a dedicated protocol that will keep its progeny pure each time it is grown to meet the desired specifications of the intended market. The seed has to be multiplied several times to obtain enough seed for processing into product. In a dedicated system, the producer must know in detail what market he or she wishes to serve and be committed to, taking all the precautions in the assignment of land and machinery that are necessary to deliver a safe product. In Canada, this operation is carried out by select seed growers who are members of the Canadian Seed Growers Association (CSGA) because they have been trained previously during a 2-year probationary period to produce pure pedigreed seed.

5.2.3. Choice of Land to Grow Pure Seed

The importance of proper crop rotations cannot be overemphasized in the growing of pure seed. Many cereal grains will produce a sparse density of volunteer plants on the same land that was used the previous season to grow the same, or a different, cereal. Never grow cereal grains on the same land in two or more consecutive years if the object is to produce pure seed. This is true for both spring- and winter-type cereals. It is preferable to grow a pure cereal on land that was used to grow crops like soybeans, lentils, flax, canola, or corn the previous 1 or 2 years. This will give many opportunities for volunteer cereal seeds in the soil to germinate and be eliminated from the soil by tillage or in northern countries by freezing winter conditions.

5.2.4. Purchasing Fertilizer

There is a danger when purchasing bulk fertilizer that the rail cars and trucks carrying fertilizer are not thoroughly cleaned prior to loading and delivery. These vehicles may contain traces of other crops or weed seeds resident in crevices in the floor of the vehicle, and these seeds become mixed with the fertilizer and are delivered to the farmer. These contaminants largely go undetected at the time of delivery, and the seeds are spread with the fertilizer on the land to contaminate the new crop. One cannot solve this problem completely, but buying fertilizer in sealed bags can minimize contamination.

5.2.5. Preparing Clean Equipment

Many pieces of agricultural equipment are required, at various times, to grow and process pure-seeded crops. Great care and time must be taken by producers to thoroughly clean all equipment before it is used especially if growers use the same equipment to grow and clean other crops on the same farm. It would be helpful if a complete set of the equipment could be reserved for growing the pure grain each year. This is not always practical unless the specialized market for very pure seed is very large, and the grower earns the reputation as a reliable supplier of this market. The type of equipment referred to here involves fertilizer spreaders; tillage equipment to prepare a proper seedbed; soil packers to firm the soil after planting; sprayers to control insect pests and weeds; swathers to cut the grain and arrange the swathed grain in rows to dry; harvesting combines to thresh the grain; trucks to transport it; storage bins to hold the grain; seed cleaners to remove weed seeds, chaff, and straw; magnets to remove pieces of metal and loose nuts and bolts; destoners to remove small stones that may have been picked up from soil by a combine set to pick up lodged crop; gravity tables to classify grain; and bagging facilities to hold the grain and protect it from becoming contaminated after tagging. Some small pieces of equipment are easier to clean than large machines, but the most difficult equipment to clean are the combine harvester, the seed cleaning units including the destoner, gravity tables, augers, and the various sized sieves that are used to separate large (corn) and small crop seeds (canola, mustard, and flax) and many different weed seeds. Anyone who has worked with harvesting and cleaning equipment will testify that to be certain that a piece of equipment is free of contamination is more optimistic than realistic. It is amazing how many "nooks and crannies" in a combine will hide contamination because a combine was designed to only thresh grain but not to keep it pure. As a breeder, I had to insist that my field technician dismantle many of the working parts of the combine harvester before use and to employ forced air, and at times forced streams of water, to clean the combine before putting the main components back together. We would spend 1 day to clean the combine, and the next day, we would use it to harvest a breeder plot that took us about 2 hours to complete. This is why it is wise for anyone producing pure seed to have dedicated equipment available that they can use each year to process the same specific commodity. Dedicated equipment requires much less cleaning.

5.2.6. Official Crop Inspection by Regulatory Authorities

Official inspection by field and seed inspectors is an essential step in producing pure seed. There are specific regulations that must be followed dealing with the identity and pedigree status of the crop being grown, the cropping history of the land, and the identity of previous crops. The CSGA supplies written information to pedigree seed growers stating "isolation" distances that must be followed between specific crops to ensure crop purity. Failure to follow these directions means the crop will not pass official inspection. Isolation distances are required to prevent physical seed mixing caused by strong winds and cross pollination. Following harvest and cleaning, a laboratory analysis of properly sampled harvested seed must be performed to establish its purity.

Once the grain has been cleaned and has passed official inspection, it should be placed in new "tote bags," sealed with official tags and transported to a dedicated processing facility. Many different processing facilities may be involved depending on crop identity, and they range from facilities to make flour, pasta, cake flours, and bread to malting facilities, beverage manufacturers, to the suppliers of breakfast cereals such as rolled oats, steel cut oats, oat flour, or deep-fried groats. The processor may want to sample each bag with an official grain probe before it enters his or her facility to be confident that the seed is pure. The processor does not want to contaminate his or her facility with impure seed. Up to this point, all inspections have been accomplished using visual methods, but once the grain has been made into product, the processor may have to establish purity by relying on chemical or official immunological tests. In the case of ensuring pure oats for CD patients, an R5 enzyme-linked immunosorbent assay (ELISA) test is employed.

5.2.7. Dedicated Food Processing Facilities

It is advisable that any company wishing to sell a pure food should operate a dedicated processing facility and follow strict rules to avoid contamination. Many large food companies do not wish to develop niche markets for allergic individuals because of the smaller estimated size of the market and the costs involved in building separate facilities. They are also apprehensive about lawsuits arising should one of their products cause harm to a client.

5.2.8. Hazard Analysis and Critical Control Point (HACCP)

All food processing companies must have HACCP certification to ensure that the facility can process, and the owners can legally sell, food products. This means that regulatory officials will inspect the facility on a regular basis for cleanliness.

5.2.9. Equipment and Additives

In the manufacture of different products, attention must be devoted to thorough cleaning of equipment between runs, and caution must be exercised when adding noncereal products such as flavorings and spices into the product. Additives such as adjuncts, starches, proteins, fats or oils, or extruded products should be tested to be certain that they do not cause allergic reactions to clients.

5.2.10. Third-Party Inspection of Processing Plant

Third-party inspectors should be hired to ensure the safety of manufactured products even though this adds to the cost of production.

5.2.11. Marketing

5.2.11.1. Labeling, Identity, and Purity. All products sold in the market place should be accurately labeled to inform the consumer of the identity and purity of all the ingredients in the product. The level of purity has to be defined for each food product. Some individuals react violently to an allergen, such as those suffering a nut allergy, whereas others are able to tolerate trace amounts of a different allergen without exhibiting a violent adverse reaction. In either case, the purity of the food product must be stated clearly on the label for the benefit of the client.

5.2.11.2. Packaging. Pure foods should be sold in sealed packages so that there is no contamination in the health food store or supermarket. Some health food stores sell grain such as rolled oats in open bins, and it is easy for patrons to put scoops, and maybe even product, back into the wrong bin after use.

5.3. NEW DEVELOPMENTS IN ALLERGEN MANAGEMENT FOR SPECIFIC GRAINS (E.G., OATS)

In response to the requests delivered to the CCA by its membership, the association embarked on a program to develop a strategy to make pure oats (*Avena sativa* L.) available to patients suffering from CD. CD patients internationally were, and are, advised to not consume commercial oats because they may be contaminated with trace amounts of various species of wheat (*Triticum* species), barley (*Hordeum vulgare* L.), rye (*Secale cereale* L.), and triticale (wheat × rye hybrids) seeds, which contain the protein gluten. The CD patient cannot digest gluten completely into its constituent amino acids, which leaves a 33-mer peptide [1, 2] that is absorbed by the small intestinal wall where it initiates an autoimmune reaction leading to an inactivation of the villi located on the inner intestinal surface. These villi normally function by absorbing digested food ingredients. This immune reaction leads to many health conditions in the patient that can only be remedied by the patient adopting a gluten-free diet for the remainder of his or her life. Fortunately, the 33-mer protein does not arise after the digestion of oat, rice, or corn proteins, so most CD patients can tolerate these grains. Contamination of oats with gluten-containing seeds can occur at any point in the production, processing, or marketing chain.

Several modern studies have concluded that "pure oats" are considered safe for patients suffering from CD [3], and this is why the CCA embarked on a program to develop a strategy and a procedure for producers and processors to make "pure oats" available to CD patients. Rice and corn are usually not contaminated with wheat, barley, or rye gluten because rice farmers do not commonly grow wheat, barley, or rye, and if they grow corn, small oat seeds can be easily separated from large corn kernels before processing into food.

Many adult patients are diagnosed with CD as adults, and they miss their oatmeal porridge, cookies, and many of the other food products that use oat as an ingredient in foods: foods such as meat loaf, oat/rice side dishes, yogurt, soups, drinks, custards, pilaf, snack foods, and fish and poultry stuffings to name a few products. Many patients said they would dearly love to resume eating pure oats if they were available.

This quest for returning oats into the CD diet has been given greater importance because studies in the past 15–20 years have revealed that oat is rapidly becoming recognized as both a medical as well as a very good nutritional crop. Oat groats contain approximately twice as much protein as rice, and their amino acid profile is very similar to that of rice grains. The addition of oat groats to rice improves the protein content of oat/rice side dishes in addition to improving flavor. The fat content of oat groats is approximately equal to corn but four times higher than rice and its fatty acid composition is 80% unsaturated (oleic and linoleic). Its soluble dietary fiber (β-glucan) content is 6–7%, and numerous studies have shown that it helps hypercholesterolemic patients lower their low-density lipoprotein (LDL) cholesterol [4], and this has been verified [5] in a 10-year study to evaluate the Food and Drug Administration's decision to permit food processors to claim a medical benefit from eating oats. The starch content of oat groats consists of both amylose and amylopectin, which is lower than rice or wheat and may contribute to its known lower glycemic index. Oats also contain a good supply of B vitamins and minerals such as Ca and P, and oats have good quality bran. Oats also contain many unique phytochemicals called avenanthramides [6, 7], which have both antioxidant and anti-inflammatory properties, and it was shown [8] that some avenanthramides had an inhibitory effect on the expression of proinflammatory cytokines in coronary heart disease. Stix [9] reviewed the pertinent medical literature and has stated that inflammation underlies a broad range of human diseases including heart disease, stroke, diabetes, Alzheimer's disease, depression and schizophrenia, cancer, and Crohn's disease. In the case of cancer, Stix reported that anti-inflammatory drugs do not kill cancer cells but they inhibit tumors from spreading to various parts of the body. Stix stated that "anti-inflammatory drugs may join traditional chemotherapies which could keep solid tumours or premalignancies localized to one place." Since oats contain some unique anti-inflammatory phenolic compounds, these compounds and oats may find a greater role in the health of all consumers including CD patients. It may only be necessary to consume pure oat variety groats with elevated levels of avenanthramides to combat serious diseases in the future.

As stated above, it is impossible at present, to guarantee 100% purity of oats for CD patients. Although it may be desirable to have such a high level of purity, it is almost impossible to achieve 100% purity for most patients. Alternatively, medical researchers and government regulators internationally have tried to arrive at a safe level of gluten in prepared foods for CD patients. The level now adopted by Codex Alimentarius (adopted in 1979, amended in 1983, revised in 2008) as safe is 20 ppm [10]. Operationally, the protocol and position statement published by the CCA [11] supports the lower level of 5 ppm in pure oat products. The strategy and procedures that were formulated and adopted by the PAB of CCA to achieve this goal are now described.

To achieve any level of purity that is considered safe, it is necessary to adopt a production and inspection procedure that can be monitored by government inspectors so that CD patients can be assured that the labeled product is safe for them to consume. The PAB decided to take advantage of the seed purity and inspection system [12] that is already in place in Canada to service farmers and the seed industry (seed regulations of the Canadian Seeds Act). In the pedigree system, the amount of impurities permitted in the various classes of seed (breeder, foundation, registered, and certified) increases progressively from the purest breeder seed to the most contaminated certified seed.

Breeder seed comes from the breeder of the variety and consists of a bulk of breeder lines that are selected, for phenotypic and quality uniformity such as plant height, days to maturity, kernel seed size, kernel color and test weight; reaction to important races of disease; and protein, oil, and β-glucan composition. A 2-kg reference sample of breeder seed is kept by the Canadian Food Inspection Agency (CFIA) for legal purposes should anyone challenge, in the future, the identity of seed of the variety they have purchased.

Trained seed growers belonging to the CSGA who have completed a 2-year probationary period as qualified select seed growers are allowed to grow allotments of breeder seed. They are required to remove any impurities from the crop, and they arrange for the crop to be inspected by CFIA inspectors to make certain that the plants and seed are eligible for select seed status. Select seed is then grown on a much larger area to produce foundation seed. The crop is again inspected in the field by CFIA inspectors, and properly sampled seed is sent to government seed laboratories to check for purity and germination percentage. To obtain the foundation #1 rating, the seed shall only contain from 0 to 1 wheat, barley, rye, or triticale seed per kilogram of oat groats. If the crop meets or exceeds this level of purity, the CFIA issues the grower with documentation to verify the grains status, and the seed is made secure. The gluten content per kilogram with this degree of contamination can be calculated to be less than 5 ppm, which is much lower than the 20 ppm required by Codex Alimentarius.

The CCA has adopted the classification of foundation #1 as the purity classification to recommend to CD patients for their consumption. Eating pedigree seed at this level of seed purity is costly because stopping seed multiplication at the foundation #1 level prevents the seed grower from growing repeated multiplications of seed to end up at the certified seed level, which is the class of seed commonly sold to ordinary farmers. It is also the class of pedigreed seed with the highest level of contamination, which cannot be tolerated by CD patients. Many CD patients are willing to pay the price to be able to eat foundation #1 seed.

Once the seed is sent to a dedicated food processor, it may be steel cut, flaked, or ground into flour. The level of purity of the final processed food product must be verified because contamination with gluten-containing grains can occur during delivery or within the confines of the processing facility. In the field or seed laboratory, purity is determined by visual means by highly trained inspectors, but once the grains are cut, flaked, or ground into flour, purity is monitored using an R5 ELISA test. This test is able to detect a gluten concentration of 5 ppm or higher but cannot detect lower levels of gluten. This low level is safer than is required, but the CCA has adopted this standard because it is possible that the Codex may wish to adopt a lower standard in the future.

Products made from oats as described above have to be labeled so that CD patients can quickly identify oat products that are safe to consume. Terms such as "gluten free" or "only oats" are probably not true because if the oats are grown and processed as described, they will likely contain some gluten but the amount will be less than 20 ppm and likely below 5 ppm. To avoid the problem of having to list on the package the actual gluten content, the CCA prefers to use their copyrighted logo called "PAVENA," which can be affixed to the package to inform the CD patient that it is safe for them to consume. PAVENA is a composite word that stands for pure Avena, and Avena is the generic Latin name for oats. CD patients will have to inform themselves what PAVENA means in terms of how the oats were grown, inspected, and processed for their safety.

Experience has shown that some individuals react adversely to even pure oats. Their difficulty may be caused by an extreme sensitivity to the small amount of gluten in the oat or to other proteins not related to the gluten fraction in the kernel. CD patients are not accustomed to eating substantial amounts of soluble dietary fiber in oats or oat bran, but patients should initially consume modest amounts of PAVENA until their system adjusts to the increase in fiber. One could speculate ad infinitum on what causes a few patients to react adversely to oats, and the causal agents might reside in herbicides or pesticides that were used to grow the oat crop or to toxins liberated by even low infection of the seed caused by fungi present on or in the oats. Some very sensitive patients are advised to stop consuming PAVENA oats and obtain their important dietary fiber from other food sources. At present, two food companies in Canada, Cream Hill Estates and Farm Pure Seeds, are selling pure seed or products to CD patients in Canada and the United States.

As stated earlier in this chapter, ideally one would like to formulate a strategy that would produce less expensive and preferably even purer oats for CD patients. The author is in the process of developing such a new system at Ottawa, which is based upon changing the color of oat hulls to dark brown through breeding so that these dark-colored grains can be efficiently separated from all gluten-containing seeds using modern optical, high speed, seed sorting equipment after harvest. The hulls of covered seeded oats grown by farmers today usually are either white, yellow, or light tan. The strategy is to first remove all grains that contaminate the brown hulled oats, including gluten-containing seeds, and then to dehull the pure brown oats in a separate operation to produce pure groats in a dedicated facility. By following this procedure, growers will be able to bypass field inspections and the pedigree system and may be able to even use commercial seed providing the bulk of the oat seed is dark brown. If the seed is heavily contaminated with wheat, barley, or rye, the brown oats may have to be put through the seed sorting machines several times to obtain the desired level of purity.

Once pure oat groats are obtained, they should be assayed for gluten content using the R5 ELISA test and stored in clean tote bags ready for transporting to a certified HACCP food processing facility. Successful seed lots should carry the PAVENA trademark label. All agricultural commodities and products that may contain even traces of wheat, barley, rye, or triticale should be carefully monitored to ensure purity and safety for patients suffering from CD. Purity is the watch word for those servicing the celiac community.

REFERENCES

1. Shan, L., Molberg, O., Parrot, I., Hausch, F., Filiz, F., Gray, G.M., Sollid, L., Khosla, C. (2002). Structural basis for gluten intolerance in celiac sprue. *Science*, *297*, 2275–2279.
2. Wieser, H., Koehler, P. (2008). The biochemical basis of celiac disease. *Cereal Chemistry*, *85*, 1–13.
3. Janatuinen, E.K., Kemppainen, T.A., Julkunen, R.J.K., Kosma, V-M., Mäki, M., Heikkinen, M., Uusitupa, M.I.J. (2002). No harm from five year ingestion of oats in celiac disease. *Gut*, *50*, 332–335.
4. Anderson, J.W., Chen, W.J.L. (1986). Cholesterol lowering properties of oat products. In: *Oats—Chemistry and Technology*, Webster, F.H. (Ed.). St. Paul, MN: American Association of Cereal Chemists, pp. 309–333.
5. Andon, M.B., Anderson, J.W. (2008). The oatmeal-cholesterol connection: 10 years later. *American Journal of Lifestyle Medicine*, *2*, 51–57.
6. Collins, F.W. (1989). Oat phenolics: avenanthramides, novel substituted N-cinnamoylanthranilate alkaloids from oat groats and hulls. *Journal of Agricultural and Food Chemistry*, *37*, 60–66.
7. Collins, F.W., McLachlan, D.C., Blackwell, B.A. (1991). Oat phenolics: avenalumic acids, a new group of bound phenolic acids from oat groats and hulls. *Cereal Chemistry*, *68*, 184–189.
8. Guo, W., Wise, M.L., Collins, F.W., Meydani, M. (2008). Avenanthramides, polyphenols from oats, inhibit IL-β-induced NF-κB activation in endothelial cells. *Free Radical Biology and Medicine*, *44*, 415–429.
9. Stix, G.A. (2007). A malignant flame. *Scientific American*, *297*, 60–67.
10. Codex Alimentarius Commission. (2008). Draft revised standard for gluten-free foods at September 7, ALINORM 08/31/26. In: *Report of the 31st session of the Codex Committee on Nutrition and Foods for Special Dietary Uses*, June 30–July 5, Joint FAO/WHO Food Standards Programme, Geneva, Switzerland, pp. 6–8.
11. Rashid, M., Butzner, D., Burrows, V., Zarkadas, M., Case, S., Molloy, M., Warren, R., Pulido, O., Switzer, C. (2007). Consumption of pure oats by individuals with celiac disease: a position statement by the Canadian Celiac Association. *Canadian Journal of Gastroenterology*, *21*, 649–651.
12. Department of Justice. (1985). Canada Seeds Act (RS, 1985, c S-8). Act current to April 9, 2010. Available at http://laws.justice.gc.ca/en/S-8/text.html

6

PRINCIPLES AND PRACTICES FOR ALLERGEN MANAGEMENT AND CONTROL IN PROCESSING

WARREN E. STONE AND JUPITER M. YEUNG

6.1. INTRODUCTION

Managing allergens during food processing is of vast importance in today's food industry. Food allergies affect an estimated 6–8% of children under the age of 3, and 1–2% of adults in industrialized countries. While most allergic reactions are mild in nature, some of these sensitized individuals can develop serious or life-threatening allergic reactions when exposed to allergenic proteins. Since there is currently no remedy for food allergies, the only successful method to manage one's allergy is to avoid foods containing the causative allergens.

The overall objective of an allergen management program can be summarized as avoiding the production and subsequent distribution of products containing undeclared allergens into the food chain. This can be further defined as ensuring that any product destined for consumption and containing allergens is properly labeled. While each operation will have its own unique situations, five categories are typically involved in successful allergen management: administrative and management functions, controls to minimize the potential for cross-contact, management of work-in-process (WIP) and rework, effective sanitation practices, and label control programs. Each of these categories can contain many subcomponents depending on the individual operation. In all cases, the level of control should be commensurate with the potential risks presented.

Allergen Management in the Food Industry, Edited by Joyce I. Boye and Samuel Benrejeb Godefroy

6.2. MANAGEMENT RESPONSIBILITIES

Management has the responsibility for establishing appropriate policies and procedures and providing the necessary resources for an infrastructure that will facilitate allergen management such that there are no undeclared allergens in its products. Management must ensure that each facility that stores, manufacturers, and distributes products has an active and effective allergen management plan. While a plan may contain common components between different locations, effective allergen management plans, like Hazard Analysis and Critical Control Point (HACCP) plans, should be location and product specific. In most instances, the plan will also be line specific as well.

6.2.1. The Allergen Management Team

One of the first steps in developing an allergen management program is the formation of an allergen management team. An effective team is often a multidisciplinary unit with members from quality assurance, manufacturing, engineering, warehousing, and regulatory affairs. In addition, since they play a key role in allergen management, it is important to have representation from the line personnel on the team. While some individuals on the team may have extensive knowledge of allergens or procedures for allergen management, this is not a prerequisite for membership. On the other hand, some, particularly smaller, companies may not have ample personnel with such knowledge and, therefore, may need to hire consultants or obtain assistance from other outside experts. Outside assistance can also be found from trade associations, government organizations, universities, and online resources. Assembling this highly visible team with authority and responsibility to set plant-wide allergen policy represents another form of communicating management's commitment to the allergen management program.

Responsibility for developing procedures in accordance with the company's allergen management policies and objectives will lie with the allergen management team. Since the allergen management plan will be facility, product, and line specific, the team members must have extensive knowledge concerning the establishment's structures, equipment, operational procedures, personnel practices, ingredients, products, sanitation programs, labels, and label control practices. They should also have current information and knowledge concerning allergens, best practices for allergen management, and regulations pertaining to allergens and allergen labeling. In keeping with the objectives that products containing allergens are labeled correctly and no products contain undeclared allergens, the allergen management team will want to conduct a thorough allergen risk assessment for each product produced, stored, and/or distributed at a given facility.

6.2.2. Risk Assessment

Risk analysis is a key component of a successful allergen management plan. It is the backbone on which all future decisions and plans are based. An allergen management plan will not be effective in preventing undeclared allergens reaching the marketplace, regardless of how well it is followed, if the risk assessment is not conducted properly and the risks that warrant control are not identified. Risk assessment for the development of comprehensive allergen management plans involves two stages: risk identification and risk evaluation. The following are general procedures the allergen management team can follow to identify and evaluate the risk of undeclared allergens being present in products made at their facility. It is often helpful to perform the first three tasks before attempting to undertake the remaining ones:

- identify allergen-containing materials associated with each product and process;
- identify the potential presence of unlabeled allergens in incoming materials;
- map the traffic patterns of all ingredients, products, and allergen-containing materials through the process facility;
- identify specific procedures or areas where cross-contact may occur, including handling of rework and WIP materials;
- identify potential labeling/packaging problems; and
- assess the likelihood and probability that any of the above could lead to an undeclared allergen in any finished product.

It is vitally important for the allergen management team to document their evaluations of each risk identified. These records can be retained for reference when there is a need to review and/or update the allergen management plan. Detailed information on risk assessment and risk management is provided elsewhere in this book.

6.2.3. Supplier Relations

An often overlooked facet of allergen management is the assessment and evaluation of practices employed by one's suppliers. Food processors and their suppliers should work together to assure that ingredients and packaging materials are labeled properly and that they contain no undeclared allergens.

Prior to approving suppliers, customers should assess whether the supplier has procedures in place that can be used to highlight the presence of allergens and to minimize the potential for the inadvertent presence of allergens in the materials that they provide. In conducting an assessment, customers have a variety of methods at their disposal. Supplier inspection surveys, facility audits, product testing, and evaluation and review of product specification compliance are but a few of the approaches that could be employed. Whatever tactics are utilized, the assessment should be designed to provide a level of knowledge that programs are, or are not, present, effective, and operating at a level that will ensure that no ingredients contain undeclared allergens and those allergens in the ingredients are properly declared. Elements of successful programs will be discussed in greater detail later in this chapter.

Customers and suppliers should agree on all controls necessary for the proper handling throughout the food chain of allergen-containing ingredients. This may be accomplished by the use of a written ingredient specification that delineates, among other items, receiving, packaging, labeling, transportation, and storage requirements for ingredients containing allergens. To maximize supplier customer communications, the supplier and customer should each sign or initial the specification. Often, customers may want to reference a given specification on their purchase orders.

In addition to developing pertinent purchasing specifications, customers often require a letter of guarantee indicating that materials received from the supplier have been manufactured and/or handled in a manner that assures that the specifications were met and that the materials comply with applicable laws and regulations including those related to the declaration of allergens. In certain instances, the customer may require that a certificate of analysis (COA) accompany each lot of material from the supplier. The COA is used to indicate that a particular lot has been tested and meets the specification. For example, if the customer is aware that its supplier of cornmeal also manufactures peanut meal in the same facility, they may require a COA indicating that the cornmeal does not contain detectable peanut residues.

While requirements will depend on the material involved, information on specifications, site-specific procedures, and customers included on a COA could include [1]

- full description of the commodity,
- lot numbers for products in the shipment,
- date of production,
- date product was shipped,
- quantity of product covered by the COA (e.g., 40 cases at 70 lb each),
- laboratory conducting the testing (supplier's lab or outside laboratory),
- results of analyses,
- methods of analysis (name of test kit used, method number from the Association of Official Analytical Chemists [now known as AOAC International], etc.), and
- descriptions of sampling plans used to generate results contained on the COA.

6.2.4. Training

All employees involved in the production of foods in a factory that handles allergenic ingredients should complete a training session on food allergens. The sessions can be part of new employee orientation, as well as ongoing training that can be repeated as often as necessary. Consideration should be given to the fact that employees have varying backgrounds, knowledge, abilities, and language skills. For these reasons, the information presented should be relevant, applicable, and brief. Information that could be covered for most employees includes

- what are food allergies and allergens,
- health consequences of unintended exposure of allergens for food-allergic consumers,
- issues involving cross-contact,
- labeling and mislabeling issues.
- recall statistics showing the percentage involving allergen labeling issues, and
- allergen management strategies.

Members of the allergen management team can attend training with content more oriented toward regulatory issues, current research on allergens, risk evaluation, and allergen testing.

6.3. PROGRAMS DESIGNED TO PREVENT CROSS-CONTACT

6.3.1. Good Manufacturing Practices (GMPs) and Standard Operating Procedures (SOPs)

Prevention of cross-contact often starts with compliance to basic and fundamental requirements for the production of safe and unadulterated food. Referred to as "good manufacturing practices" or "good hygienic practices," these GMP/GHP programs provide the basic environmental and operating conditions necessary for effective control of common hazards in the food industry [2]. Various sections of these regulations deal with issues that indirectly impact food safety. Some of these sections are more relevant to the issue of allergen management than others. For example, the sanitary design and maintenance of facilities and equipment as well as employee practices can have a direct bearing on the effectiveness of an allergen management program. Without programs such as the GMP/GHPs, specific allergen management measures may be ineffective in assuring the production of safe foods. Furthermore, the GMP/GHPs simplify the development and maintenance of an allergen management program.

Key components of an allergen management program may have to be developed and documented depending on the individual requirements of a given facility. Detailed SOPs can be developed for measures such as control of ingredients, storage and handling of ingredients, packaging material and finished goods, scheduling, and control of rework and WIP materials. The corresponding monitoring procedures can also be developed in the form of an SOP. It is important to verify that these allergen management and monitoring procedures are effective and are being implemented. Therefore, written procedures should be developed for conducting such verification activities as well.

6.3.2. Receipt of Ingredients

Receiving personnel can have specialized training in procedures for proper receipt, handling, and storage of allergen-containing ingredients. Employees may be trained to understand color coding or other marking procedures such as colored placards or distinct shrink-wrapping that are used for allergen control. These procedures should be documented in an SOP.

Prior to breaking seals or otherwise opening a delivery vehicle, receiving personnel can verify that goods are being received from an "approved supplier." Approved supplier status should ensure that the supplier's allergen management procedures have been reviewed and found to be acceptable or consistent with one's own allergen management strategies. The customer can then inspect all ingredients and the vehicle's cargo space prior to acceptance. The inspection can determine if there are allergen-containing materials on the load or if the packaging integrity of allergen-containing ingredients has been compromised creating an opportunity for cross-contact. The vehicle that delivered the ingredients can be inspected after unloading as well to see if any allergen-containing foods are present that could have been hidden when the cargo area was full.

Some allergen-containing ingredients are received in tankers or railcars and unloaded into bulk storage systems. Bulk storage silos or tank farms need stringent controls to ensure that nonallergen components cannot be mixed with allergen-containing ones. Such controls can include tamper-evident seals on delivery vehicle entry and discharge points, verification of prior adequate washout, prior load information, and clean transfer areas and equipment. Access to ports and pipelines used in the receipt of bulk ingredients should be restricted by using locks and/or seals. Where multiple receiving ports are available, distinct and proper labeling of each port or pipeline connection is also recommended.

6.3.3. Storage and Transfer of Ingredients, WIP, Rework, and Finished Goods

Movement and storage of allergen-containing ingredients, WIP, rework, and finished goods should be done in a manner that minimizes the opportunity for cross-contact. The design of storage systems and transfer protocols will depend on the nature of cross-contact risks, the characteristics of the allergens involved, and individual operations in a facility. Retrieval of the wrong material and subsequent usage in a product can have serious implications, the least of which is a product recall. As noted earlier, controls to eliminate hazards should be commensurate with the risks presented.

One of the best ways to prevent cross-contact between allergen-containing materials and other items during warehousing is to use segregated storage areas. These areas are reserved for the storage of allergens, and ideally, specific allergens will have designated locations. While this is an ideal situation, the practicality of employing it is often limited. Many facilities do not have available storage space to implement a program such as this. Also, component materials frequently can contain more than one allergen. Many baked goods, for example, can include dairy, wheat, soy, eggs, and possibly, tree nuts.

A common alternative to segregated storage is to use distinct identifiers for allergen-containing materials. Colored shrink-wrap or colored placards are frequently used to designate allergenic compounds. The use of colored shrink-wrap has the added benefit that it can enhance the integrity of a given package. In operations that use only a few allergens, different colors of shrink-wrap or different colored placards could designate specific allergens (e.g., blue for dairy and red for tree nuts). Distinct pallets or unique totes or bins can also be designated for allergen use only.

Whatever method of demarcation and/or identification is selected, including the use of segregated storage, allergen-containing materials need to be well marked when placed in storage. Somewhere on the container, pallet, or bin, there should be information indicating the name of the material, WIP, or rework. Other information could include the date of receipt or production, an inventory control number, and the name of the allergen present in the material. Computerized bar code generation and scanning technology could also be used to generate and record necessary information for allergen management as well. In the case of palletized items where the units will not be used in pallet quantities, care should be taken to place the necessary information in a location that will not be lost as the pallet is used. Some companies choose to place this information on the lower containers located on a pallet, or on the pallet sideboards. If information is placed directly on pallet sideboards, additional effort should be employed to ensure that the information is removed from the pallet once the pallet becomes empty. Failure to do so could result with two sets of information on a given pallet. This is the type of confusion and miscommunication that can lead to the retrieval and subsequent use in production of the wrong ingredient, WIP, or rework. An event such as this often leads to a major product recall.

It is important to use secure storage containers for the warehousing of allergen-containing ingredients or foods. This is especially prudent when storing rework or WIP materials that can often find themselves stored in makeshift enclosures (see Section 6.3.5, Management of rework and WIP). Storage containers should provide two important properties; they should preclude the opportunity for the contents to be contaminated and for the contents to contaminate other items. When possible, containers that hold allergen-containing materials should be movable on their own (e.g., totes on wheels) without the use of forklifts, pallet jacks, and other equipment. If this cannot be facilitated, care must be taken to avoid the spread of allergens to other parts of the factory through these transport and lifting devices.

Handling and transport of allergen-containing ingredients, WIP, and rework should also be performed in a manner that minimizes any potential for cross-contact. The allergen management team should evaluate traffic patterns and product flows for all items containing allergens and adopt policies and procedures that offer the optimum protection against cross-contact. Written protocols in the form of SOPs and adequate training of appropriate personnel will go a long way toward accomplishing these objectives.

6.3.4. Processing System Design

Processing systems should be designed to minimize cross-contact opportunities, prevent the accumulation of food residues, and facilitate cleaning and inspection operations. It is relatively easy to incorporate proper design features to minimize the potential for cross-contact of allergens into the overall plant layout and process design when building a new facility. On the other hand, when working in an existing facility, which is typically the case in the industry, dealing with allergen cross-contact issues is characteristically more taxing. The challenge is to adequately manage allergens in the facility with minimal disruption to existing operations and processes.

The use of dedicated processing systems is considered one of the most effective methods for dealing with allergen cross-contact. The dedication of processing systems can take different forms. In one application of this principle, an entire section of a facility is set aside for the manufacture of a particular allergen-containing product. None of the processing lines in that section has any cross-connections to other lines in the plant. At other times, only a particular processing line and adjacent equipment are set up for a particular allergen-containing product. Lines manufacturing other products that do not contain allergens may be close by, but there is no physical connection or shared equipment between those lines.

While a company may not be able to isolate entire processing lines and pieces of equipment for the production of products containing a specific allergen, the dedication of utensils, tools, storage areas, smocks or uniforms, and even personnel is often possible. Such procedural control measures significantly reduce the risk of allergen cross-contact and can be implemented without major capital investment. The manner in which each of these measures is used will be based on the risks of allergen cross-contact that exist in the specific processing situation.

When the use of dedicated systems is not feasible, adequate control may be achieved by physically separating unit operations handling allergen-containing materials from the rest of the plant. The degree of separation is a function of procedural measures, such as controlling the movement of materials and personnel throughout the facility, and physical containment of allergen-containing products in tanks, pipes, and enclosed processing rooms. The operation may or may not be separated from the rest of the facility by walls or other enclosures. Adequate control can be achieved through diligent adherence to the procedures and proper supervision.

Production scheduling can be a powerful tool to minimize the risk of cross-contact associated with producing allergen-containing products on shared equipment. It can often be utilized to offset limitations inherent in facility design. Control is attained entirely by procedural means, which are relatively easy to implement. However, production scheduling is also more vulnerable to human error than line dedication and physical separation. Oftentimes, scheduling is not used solely to reduce risk; rather, it is considered an adjunct to other measures. Scheduling can enhance allergen management through various means including setting the order of production, reducing frequent changing from one product to another by running lines for longer hours, running allergen-containing product lines when other lines are idle, and manufacturing allergenic product last in a series of product runs.

6.3.5. Management of Rework and WIP

The term "rework" refers to finished or partially finished products that are later reincorporated into the manufacturing process. For example, ice cream in bulk containers left over from the last production run and reintroduced into the processing line to manufacture another ice cream product is considered rework. Likewise, if a finished product is added back to the same product line because of a packaging defect, the add-back material is considered rework. WIP items are partially finished products that are in between different production stages. For instance, cheese that is shredded and transferred to a cooler to await incorporation into a pizza is considered WIP. Bread that is baked and then stored overnight before packaging is another example of WIP material.

The generation of WIP products is often inherent to a given processing operation but the same cannot be said of rework. Rework that contains an allergen is inherently risky to handle. Consequently, any efforts by a firm to reduce or eliminate rework can significantly reduce the allergen management risks involved. While not feasible for all operations, some businesses have restricted the use of rework material containing allergens to the particular production run that generated the rework or to use in "like" or same product.

Whenever possible, rework or WIP products that contain allergens should be used immediately during a production run. When that is not possible and these products have to be stored until further use, care must be taken to prevent any cross-contact of these materials with other products in the storage area. These products should be clearly identified to minimize the possibility of their use on the wrong processing line. Appropriate inventory and traceability records need to be kept in order to document the control of these materials during storage and to be able to appropriately trace allergen-containing materials. For further information, see the Section 6.3.6, Documentation and record keeping, located elsewhere in this section.

6.3.5.1. Storage of Rework and WIP. There is one significant difference between regular ingredients and rework/WIP materials that can affect the ability to control incidental cross-contact. Ingredients are usually received in their original packages designed to withstand the normal wear and tear of distribution without loss of container integrity (e.g., cans, drums, bins, bags, cases). The sturdiness of the packages also minimizes the potential for their contents to be released. In other words, the basic integrity of the packaging materials, combined with palletization features such as stretch wrap, greatly lowers the chance for cross-contact of allergen-containing materials with other products in the storage area. Rework and WIP materials, on the other hand, are usually stored in temporary containers such as totes, pails, and carts. Moreover, the types of covers or closures for these containers can range from nonexistent to flimsy (cardboard covers, foil wrap, shrink-wrap) to secure (fastened lids). These temporary and "homemade" containers have greater potential to break or otherwise spill allergen-containing ingredients into their surroundings or on to other non-allergen-containing products.

Dedicated containers, closures, totes, and so on, should be used to store rework and WIP products. If that is not feasible, the containers and lids should be thoroughly washed before being reused using a procedure that has been validated to facilitate the removal of allergenic residues (see Section 6.3.10, Validation and verification of "allergen-clean" sanitation standard operating procedure [SSOPs], later in this chapter). Disposable plastic inserts or liners can also be used in the containers. All containers should be clearly marked to indicate the contents and the presence of allergens in a product. In operations that deal with few allergenic ingredients, a color scheme could be used to designate the presence of a specific allergen. In facilities that handle numerous allergens, color schemes may be used in addition to other identification systems. For further information, see Section 6.3.3, Storage and transfer of ingredients, WIP, rework, and finished goods.

In addition to being properly identified during storage, in order to avoid accidental use of an allergen-containing material in a non-allergen-containing product, rework and WIP products need to be accurately inventoried during storage. Entering the identification information from WIP and rework containers into an inventory control system could facilitate tracking and reconciliation of usage if the need should arise.

6.3.5.2. Usage of Rework and WIP Materials. For product labeling considerations, as well as food safety concerns, allergen-containing rework should only be added back to products that contain the same allergen. This practice is referred to as "like-into-like" or "like product." For example, chocolate ice cream with peanuts may be added to chocolate fudge ice cream that contains peanuts, almonds, and pistachios (provided that all of the ingredients in the chocolate ice cream are also in the chocolate fudge ice cream) but not the reverse. On the other hand, hummus containing walnuts cannot be added to hummus that does not contain walnuts.

Reentry of rework and WIP products into the process stream should be tightly controlled in order to minimize the potential for faulty product mixing. The level of control is a function of various factors, such as the staging and transfer of these materials, the reentry points in the process, and the equipment used for add-back. One approach could be to assemble all allergen-containing items for a specific batch in a dedicated staging area. Ideally, non-allergen-containing ingredients should be excluded from this area. The transfer of rework and WIP materials from the staging area to the processing line should be accomplished without cross-contact with other ingredients or products. Tracking the allergen-containing materials through the entire process also needs to be documented, including reconciling rework at the end of the day or shift [3].

Without properly documented and executed reentry procedures, the necessary level of control for handling allergen-containing rework and WIP potentially cannot be achieved. These procedures could utilize work instructions on staging, transferring, and safely adding back allergen-containing materials, including standard methods of dispersion and prevention of spillage. They also could include instructions on how to handle reentry machinery including preoperative inspections and cleanup. Record keeping information is also important. Spillage may require specific instructions for collection and disposal.

6.3.6. Documentation and Record Keeping

As with any other formal facility program, the management of allergens relies heavily on employee conformance with the intent and specifics of the program. One method to assure conformance is by developing and using written procedures and/or SOPs that provide enough detail and information for employees to follow. Such procedures should be easy to understand, applicable to the operation, and current. SOPs should be developed for each and every facet of the allergen manual program. When dealing with the potential of undeclared allergens, the consequences of not following prescribed procedures are too severe to leave execution of the plan dependent on verbal instruction.

Monitoring records are perhaps the most important daily operations records since they document the proper implementation of the allergen management measures, or lack thereof. The obvious purpose for monitoring is to generate data, either through measurements or observations that indicate that all procedures were followed according to the allergen management plan and that they were effective. When monitoring activities yield unacceptable results, a corrective action designed to control or remedy an out-of-compliance situation should ensue. Monitoring records should provide enough information for the person performing the monitoring and a reviewer to determine if a given allergen management procedure was conducted properly or if corrective actions are needed. Corrective actions should also be documented with significant detail.

A secondary use of monitoring data is the analysis of trends, which may provide a mechanism for detecting potential problems before they become acute. This would allow a processor to initiate process improvements before a particular situation becomes out of control.

Verification is a concept discussed in Section 6.3.7, Sanitation practices. However, verification activities, and records documenting these activities, should also be part of the overall allergen management plan. Verification includes those activities, other than monitoring, that are designed to ensure that the allergen management plan is being implemented properly. Since verification encompasses overall compliance with the allergen management plan, it can include activities ancillary to the actual plan such as verification of GMP/GHP programs or other indirectly related practices. Common verification activities include, but are not limited to, calibration, record review, and independent checks.

Processors may want to consider a documentation system to track the allergen-containing ingredients as well as rework/WIP materials from receipt though final usage. While specific systems often differ from plant to plant, certain basic records can be necessary to keep track of these materials such as batch sheets that identify allergen-containing items, designate which product/formula the materials will be added to, the processing line, production date, and the corresponding batch number. Batch or formula sheets could specifically identify those ingredients that are allergenic to alert operators that they are dealing with allergens and should use special handling procedures. This is especially useful when the allergen may not be easily identified (e.g., "albumin" instead of "eggs"). Also, at a rework/WIP reentry point, information on staged containers could be reconciled with a pre-authorized production batch sheet. Notation of this reconciliation step could be entered onto the batch sheet, along with the operator's initials and the time of the activity. Practice or "mock" tracing exercises can be performed routinely to verify the system is functioning as expected.

Allergen management depends, to a large degree, on maintaining sanitary conditions in a plant that go beyond basic plant sanitation or GMP/GHP compliance. Therefore, detailed procedures on how to achieve these conditions should be developed and their compliance verified and documented. SSOPs are specific written procedures that can be the key in providing food contact surfaces that are free of allergenic residues and cross-contact from the environment by allergenic substances that have escaped proper cleaning. For greater detail, see Section 6.3.7, Sanitation practices. Nevertheless, documented SSOPs designed to deliver allergen-clean equipment should be part of an overall allergen management program. Monitoring and verification of these SSOPs can also be part of the program. This could include procedures and records to verify the concentration of a detergent used in a procedure, or similar step.

Finally, records should be kept in support of the original risk assessment. Notes on why a given risk is part of the allergen plan when others are not could be valuable when updating a plan or explaining it to auditors or government agents. Any support documentation such as letters from suppliers, scientific articles, or regulatory agency guidance can also be kept to support decision-making activities associated with the risk assessment or control measures. Notes should also be kept to support the adequacy of given sampling frequencies where appropriate.

6.3.7. Sanitation Practices

A company's sanitation program plays a vital role in the management of allergens. In the context of allergen management, the main purpose of sanitation is the removal of product residues through the use of proper cleaning techniques. Removal of microorganisms plays a secondary role, as they are the target of the sanitizing step that typically follows cleaning.

For sanitation measures to be effective, equipment and plant structures should employ accepted standards of sanitary design and installation. Machinery should be cleanable and accessible and the cleaning step should result in the removal of food soils and chemical residues. Documented and validated SSOPs should be carried out by a trained, competent, and well-staffed sanitation crew. The crew should utilize sanitation systems and tools that are appropriate for the job and should have ample time allotted to perform an adequate cleaning job. Sanitation efficacies can be independently verified, as appropriate.

Most food operations rely on the use of water and detergents and the proper application of cleaning equipment to keep their facilities clean. Wet cleaning, as it is commonly referred to, is preferred where feasible since allergenic proteins tend to be soluble in hot soapy water. Alkaline detergents and, in some cases, acid detergents are useful in removing proteins. When detergents are employed, their concentration should be checked with sufficient frequency to ensure they are at the desired strength and the results are documented on appropriate forms. Clean-in-place (CIP) systems are highly beneficial where feasible because they can be engineered to be very consistent once validated.

Low-moisture foods rely on low water activity for their microbiological stability. The equipment soils left behind on equipment by these foods do not have enough available moisture to support microbial growth. However, if water is introduced during wet cleaning, microorganisms may start to grow in product residues. Therefore, for cleaning dry food lines, activities such as scraping, vacuuming, brushing, or wiping are usually employed. Wet cleaning, if done at all, is restricted to disassembled equipment parts, and these equipment parts are often cleaned in a separate location. Also, these parts should be completely dried before reassembly. Compressed air can be effective in removing soil from hard-to-reach areas, and drying damp equipment, but its practicality is limited because it can disperse dust and other debris, and blow allergens from one area to another. Therefore, it should only be used with discretion. Residues of high-fat products, such as nut butter or milk chocolate, can be scraped from equipment surfaces but the results emanating from such a method can be inconsistent due to variability in tools and strength of the employee performing the SSOP. Consequently, any procedure employing such a technique should be properly validated and routinely verified (see Section 6.3.10, Validation and verification of "allergen-clean" SSOPs). While scraping could be effective for high-fat products, this method would be ineffective in removing flour dust in a flour mill, where vacuums are essential.

Vacuum cleaning is one of the most effective means of removing dry particles from surfaces without spreading dust. This method also does not require a secondary means for collecting the debris. Portable vacuum systems are usually used in smaller operations. Because of their mobility, care has to be taken to prevent these units from contributing to the spread of allergens in a plant. Central vacuum systems may be encountered in larger operations. If the system hoses are portable, care must be taken to ensure that the hoses do not contribute to the spread of allergens.

"Push-through" is another approach used in dry cleaning operations, often with highly viscous products. Push-through involves pushing the product through a process system until all allergenic residues have been eliminated from the system. Push-through can be accomplished with the subsequent product or with some inert ingredient such as salt or flour. This method can be especially useful in locations such as piping systems where product contact surfaces are not entirely accessible. Push-through should not be used without proper validation, which will be discussed in subsequent portions of this chapter.

6.3.8. Sanitary Design and Construction

By incorporating sanitation features into the design and construction of buildings and equipment, cleanup efforts can be maximized. This is important because many allergen management tools rely on a certain level of accessibility and cleanability of food contact and nonfood contact surfaces to be effective. The primary goal of sanitary equipment design is to construct equipment that remains as clean as possible during operation and that is easily accessible and cleanable. Since most pieces of equipment contain food contact surfaces, cleanability issues have a direct impact on the ability to remove allergen containing material from such surfaces.

Specific design features often vary from equipment to equipment and depend on other factors such as what type of product is run or whether wet, dry, or push-through cleaning is used. The following list presents some of the most desirable features of easily cleanable equipment [4, 5]:

- All surfaces and areas should be readily accessible and should be designed for quick dismantling with simple or no tools.
- All food contact surfaces should be smooth, nonporous, and nonabsorbing, and withstand cleaning, sanitizing solutions, and hot water.
- To ensure proper removal of product, there should be no interior ledges, recesses, or unfinished welds. All permanent joints should be butted and continuously welded.

- All corners should be round or coved.
- Structural components should be round or tubular to avoid accumulation of debris and to facilitate cleaning.
- Mixing blades should be welded to the drive shaft or be of one piece construction. Shaft and blades should be removable.
- Pipes, fittings, pumps, and so on, should be readily dismantled for cleaning and inspection or cleanable by CIP procedures. All threads should be located outside.
- Liquid handling equipment or equipment that requires wet cleaning should be constructed for easy draining.
- All product zones should be covered. The covers should be easily removable for cleaning.

For further information on sanitary design of equipment, see the references at the end of this chapter.

Poorly designed or constructed food processing plants are often plagued by inherent sanitation problems. Poor plant layout and lack of sanitary facility design features may result in a situation where sanitary conditions cannot be maintained in the factory, thus compromising the successful management of allergens. Although building requirements often depend on specifics of the operation, certain sanitary design features are universally applicable to successful allergen management programs [5]:

- There should be sufficient space for the various plant operations.
- Overhead piping and ductwork in processing areas should be minimized or avoided entirely, if possible.
- Floors should be constructed to withstand cleaning, sanitizing solutions, steam and hot water, acids, sugars, fats, and heavy traffic.
- The juncture between floors and walls should be coved to avoid duct collection and provide for ease of cleaning.
- Walls should be smooth, flat, and resistant to wear and corrosion.
- Ceilings with exposed beams and other structural components should be designed to eliminate crevices and ledges.

6.3.9. Factory Airflow

All air handling systems consist of two principal components: the exhaust component provides for the removal of air from the building, and the makeup component introduces outside air into the plant. The interaction of these two components affects parameters such as air pressure differential and airflow through the plant. Inadequate sanitation of the hardware components of an air handling system, negative air pressure in areas where food is exposed, and lack of control over makeup air can contribute to the spread of allergen-containing material in a factory.

In order to minimize the potential for the accumulation and spread of allergens in the air handling system of a processing facility, the following basic sanitary design and construction features can be used:

- All incoming air can be filtered. The filter size can fluctuate depending on the air quality required. High-efficiency particulate air (HEPA) filters may maximize air quality.
- All components of an air handling system should be on a preventive maintenance and sanitation schedule.
- Makeup air units and air ducts can have access for cleaning and inspection.
- Intake and exhaust stacks should *not* be located in close proximity to each other.
- Ceiling cooling units and fans should be cleaned and sanitized on a regular basis to prevent the accumulation of food residues, some of which could contain allergenic components.

6.3.10. Validation and Verification of "Allergen-Clean" SSOPs

Management of the risks posed by allergens relies heavily on the maintenance of a high level of sanitation in the production facility. In order to achieve such sanitary goals certain conditions, in addition to employment of sanitary design principles in the facility and its equipment, should be met. Cleaning and changeover procedures need to be proven effective; the results of the procedures should be validated to ensure that residues of target allergens will be removed when an SSOP is followed correctly. Additionally, the proper implementation of these validated procedures should be monitored and verified.

To explore the efficacy of a given SSOP, it will be helpful to look at the concept of "allergen clean." Allergen clean usually means that food contact surfaces and areas around the processing line are visibly clean, that is, free of visible residues. Whenever an allergen-free product is run after an allergen-containing item, or when products that contain different allergens are produced in sequence, special sanitation measures should be taken between production runs. These sanitation measures need to be adequate to render the processing equipment allergen clean.

Validated cleaning procedures are an integral part of an effective food allergen control program. It is important, however, to distinguish between validation and verification, particularly for the day-to-day operation of an allergen control program. Validation entails collecting and evaluating technical information to determine that a process, when properly implemented, will effectively control a hazard or produce a desired end point. Within an allergen control program, a cleaning procedure would be considered validated when the procedure is shown to adequately remove food allergenic residues from food contact surfaces below the established critical limits. A validated process needs to be documented to assure consistent implementation.

The validation of sanitation practices for shared equipment is very important to the overall allergen management program. Once visually clean can be achieved on a reliable basis, additional validation should be provided to demonstrate that allergenic proteins are removed when SSOPs are followed. Allergen-specific test kits can provide additional validation data to indicate effectiveness of SSOPs designed to remove allergenic residues. Since regulatory thresholds for allergens are not established, the processing facilities can establish their risk-based critical limits or use the detection limit of the test, for their critical control points. Commercial enzyme-linked immunosorbent assay (ELISA) test kits, both quantitative and qualitative, are available from several companies and can be used to validate sanitation practices for peanut, egg, milk, wheat, soy, crustacean shellfish, some tree nuts, sesame seeds, buckwheat, lupine, and mustard. When attempting to validate SSOPs that remove multiple allergenic residues (e.g., milk, soy, and eggs), processors could target their validation strategies to detect the most allergenic allergen (e.g., eggs in this case).

One protocol that can be used to validate SSOP effectiveness for removing allergenic residues is to analyze the first product produced after a changeover for the presence of the previous allergens present in the product produced prior to the changeover. Test kits may be used to determine if detectable residues exist in the first product manufactured after changeover. When using CIP systems, similar examination of the final rinse water for allergenic residues can also serve to validate an SSOP.

An alternative validation protocol is to use a qualitative test kit, such as dipsticks, to assess whether equipment surfaces are free of allergenic residues after application of an SSOP. However, like microbial surface swabs, it is impossible to predict how much contamination might exist in the finished product based on a positive swab test of a food contact surface. When using this technique, it is advisable to swab areas that are especially difficult to clean. "Dead spots" in the processing systems (valves, joints, corners, etc.) present cleaning challenges and should be included in the sampling plan. A positive swab test would indicate more cleaning is needed until a negative result can be obtained.

Validation is particularly important in the case of push-through situations. In some cases involving push-through of thick, highly viscous liquids, allergen residues have been found hours after changeover [6]. In this case, only dedicated processing equipment can realistically provide prevention from cross-contact. In other cases, however, push-through has been effective and test kits are one of the best methods to determine the volume of push-through needed to prevent allergen cross-contact.

After successful initial and subsequent negative test results, the efficacy of the SSOP can be considered validated. However, validation tests should be repeated occasionally to verify continued successful performance. Any changes in an SSOP or the detection of the allergen of concern can require reassessment and possibly revalidation. Changes in product formulation, preparation procedures, or processing operations may also trigger a need for reassessment of the SSOP.

While validation is designed to check the adequacy of the SSOP, verification includes those activities, other than monitoring, that determine a validated procedure is being implemented properly and the system is operating according to plan. Many methods can be used to verify the allergen-clean status of a processing system. Visual inspection is a commonly used verification method whereby accessible equipment surfaces are checked for the presence of visible soil. Less accessible equipment parts, such as pumps, valves, and pipes may have to be taken apart or left unassembled after cleaning so they can be visually inspected. Adenosine triphosphate (ATP) bioluminescence technology can be used for verification of previously validated cleanup procedures. The use of ATP technology is often paired with visual inspections of processing systems. Verification can also include periodic evaluation using allergen test kits. By periodically testing equipment surfaces, final CIP rinse water, or the first product off a freshly cleaned line, the efficiency of the SSOP can be verified.

The outcome of the verification activities will determine whether or not a system is considered allergen clean. All verification activities and results should be recorded on appropriate forms or checklists, and signed off by authorized personnel. This determination, in turn, may be used by authorized personnel for a positive release of the system for the next production run. Any devices if used for further verification, such as test kits and ATP meters, should be appropriately calibrated, and a calibration record should be maintained as well.

6.3.11. Label Control Programs

Proper allergen management includes safeguards to ensure that each container has an accurate label that declares all allergens that are present in the product, and meets all applicable laws and regulations. Improper declaration of allergens through unsuitable packaging is a leading cause of food recalls in the United States, Canada, Australia, and New Zealand [7–9]. Two important aspects of label management include controls for label and packaging design and for inventory of these items.

Label and packaging design controls can often be overlooked, especially in a copacking relationship. But it is important to remember that even in a copack relationship, both parties have equal legal responsibility for proper labeling: the party that supplied the packaging and the party that used the packaging and shipped mislabeled products to the marketplace.

Common design controls include having orders in writing for artwork and labeling copy such as label drafts, sketches, or black and white layouts. The labeling copy should be reviewed by someone with knowledge of regulatory requirements, including declaration of allergens, to ensure compliant designs. Allergen declarations must use commonly understood terms in consumer-friendly language (e.g., milk, not whey or casein). Label and package proofs should always be approved in writing. Processors may want to consider identity coding of labels and packaging through the use of numbers or color sequences when designing labels and packaging. Labels and preprinted packaging should be received and stored in a manner that prevents comingling of the packaging.

Proper control and usage of labels and preprinted packaging is also of fundamental importance to ensure that allergens are properly declared on all packages. During the labeling and packaging process, only the packaging currently being processed should be allowed onto the factory floor. Return of unused packaging to inventory should be done so that there is no comingling of labels and preprinted packaging. Prudent processors may have procedures to check labels and preprinted packages prior to and during production and labeling operations.

6.4. ALLERGEN CONTROL AND PRODUCT DEVELOPMENT

6.4.1. Product Formulation

Whenever possible, product developers may want to avoid including any allergenic ingredient that does not have a functional effect in a product. By using allergen-containing ingredients only when they are essential, companies will have fewer allergenic materials in their factories and lessen the opportunity for cross-contact with other products. For example, using corn flour instead of wheat flour, vegetable shortening instead of butter, or a highly refined oil instead of a cold-pressed one all reduce the presence of allergen-containing ingredients in a factory.

Even when a particular allergenic ingredient may be crucial to product character, proper investigation can be made to select the ingredient with the least potential for cross-contact. For example, peanut butter would be far less likely to spread allergenic dust than crushed peanuts or peanut meal. Conversely, peanut butter might represent a greater sanitation challenge than the other two ingredients.

6.4.2. Reformulating Products

Reformulation of a product to include allergen-containing ingredients may lead to contamination of other lines or products in the absence of proper controls. When modifying formulas or extending new varieties of existing products, companies should consider whether an allergen-containing ingredient is critical to the product character. While there may be business reasons for choosing such an ingredient, deliberate formula modifications that introduce allergens should be taken with caution. The addition of a new allergenic ingredient to a plant's inventory will require, at minimum, reassessment of allergen management practices.

Reformulating existing products to include new allergenic ingredients not previously part of the formula requires clear communication with the consumer. Allergen-sensitive consumers, who may have been consuming the product for some time, need to be informed of the new potential hazard. This could be in the form of a label banner or highlight indicating something like "Now with peanuts!"

Businesses may also want to consider discontinuing, reformulating, or contract-packing minor lines that are the sole users of allergenic ingredients in a factory's product line. Even the presence of one allergen calls for a plant to have an allergen management program.

6.4.3. Factory Trials

Factory trials of products containing allergens require strict measures to avoid allergen cross-contact with existing products. This is especially true if the factory does not already handle allergens or a particular allergen. Information on the potential presence of a new allergenic ingredient should be made available to those factory personnel responsible for allergen management plans before any new allergenic ingredients are delivered to the factory. To deal with the new ingredient, the factory may have to coordinate activities between receiving, storage, production, inventory control, quality, and sanitation departments.

6.4.4. Consumer Testing

Product development laboratories create numerous variations of proposed items before culminating in a decision for final merchandise. Often, these concept products are affixed with laboratory-generated labels stating simply "New Product #1" and "New Product #2." These test or concept items will be sampled by corporate buyers, taste panels, and other individuals.

Samples of new products leaving the product development kitchen, even for modest "sales presentations," should be clearly marked as containing allergenic ingredients. Labeling on containers should contain common names of allergens such as "milk" or "eggs." Such terms should be used in place of, or in addition to, less common names such as "casein" or "albumin."

6.5. CONCLUSION

Food allergies affect millions of consumers worldwide, and some evidence suggests that this number could be growing. Presently, a cure for food allergies does not exist; the only successful prevention for allergic consumers is to avoid foods containing the offending protein compounds.

Some may say that helping consumers avoid foods to which they are allergic involves two facets of food manufacturing: proper labeling of foods and avoiding potential cross-contact of allergenic foods with nonallergenic foods. This statement is far too one-dimensional. As simple as it sounds, experienced food professionals realize that this goal entails an extremely complex template of well-designed and effective protocols operating under a supportive and resourceful management structure. To achieve these two objectives and thus prevent consumption of undeclared allergens by sensitive consumers, it is critical that the sourcing, storage, use, and shipment of allergen-containing foods be actively managed and controlled. Of no less importance is the attention to scientific detail required in certain allergen management disciplines such as management commitment, product development, sanitation practices, and risk assessment. Only when all of these aspects are operating in tandem, and at the utmost efficiency, can truly effective control be achieved that protects the food supply for all consumers.

REFERENCES

1. Grocery Manufacturers Association. (2008). *Food Supply Chain Handbook*. Available at http://www.gmabrands.com/publications/GMA_SupplyChain2.pdf
2. FDA. (2009). Good manufacturing practices. *Current Good Manufacturing Practice in Manufacturing, Packing or Holding Human Food, 21: Food and Drugs*, CFR Part 110. Available at http://ecfr.gpoaccess.gov/cgi/t/text/text-idx?c=ecfr&sid=7df51e8fdea81ea4e71d091cde43af0a&rgn=div5&view=text&node=21:2.0.1.1.10&idno=21
3. Stone, W.E., Jantschke, M., Stevenson, K.E., Parkinson, N.G. (2009). Basics of food allergen management. In: *Managing Allergens in Food Processing Establishments*, Stone, W., Stevenson, K. (Eds.). Washington, DC: Grocery Manufacturers Association, pp. 11–24.
4. Imholte, T.J., Imholte-Tauscher, T. (1999). *Engineering for Food Safety and Sanitation—A Guide to the Sanitary Design of Food Plants and Food Plant Equipment*, 2nd edition. Medfield, MA: Technical Institute of Food Safety.
5. Katsuyama, A.M. (1993). *Principles of Food Processing Sanitation*, 2nd edition. Washington, DC: The Food Processors Institute.
6. Taylor, S.L., Helfe, S.L. (2005). Allergen control. *Food Technology*, *59*, 40–43, 75.
7. Food Standards Australia New Zealand. (2008). Food recall statistics. Available at http://www.foodstandards.gov.au/foodmatters/foodrecalls/foodrecallstats.cfm
8. Schmit, J. (2007). Recalls often cite faulty ingredient lists. *USA Today*. Available at http://www.usatoday.com/money/industries/food/2007-12-26-food-label-recalls_N.htm
9. Vierk, K., Falci, K., Wolyniak, C., Klontz, K. (2002). Recalls of foods containing undeclared allergens reported to the US Food and Drug Administration, fiscal year 1999. *The Journal of Allergy and Clinical Immunology*, *109*, 1022–1026.

7

ALLERGEN MANAGEMENT AND CONTROL IN THE FOODSERVICE INDUSTRY

M. HAZEL GOWLAND

7.1. BIOGRAPHY

Hazel Gowland has been allergic to nuts and peanuts since 1960 and has survived some severe reactions since. Since 1994, she has worked for and with the U.K. Anaphylaxis Campaign (www.anaphylaxis.org.uk) to protect those at risk from severe food allergies through assessment, management, and communication of allergy risks throughout the food supply chain. She investigates allergy-related deaths and "near misses" and works at policy level to improve understanding of risk practice and behavior, both among allergic people and food business operators (www.allergyaction.org). She also develops and delivers training courses and materials for a wide range of clients including manufacturers, retailers, caterers, food enforcement inspectors, auditors, trainers, and school and care personnel (www.allergytraining.com).

7.2. INTRODUCTION

> Eating out with an allergy is really simple. You just need to know what went into the food deliberately and what else might have gotten in along the way. (Food allergic consumer)

People seeking to avoid a particular food or foods encounter numerous barriers. Eating out poses an even higher risk because of the complexities of food

Allergen Management in the Food Industry, Edited by Joyce I. Boye and Samuel Benrejeb Godefroy

> production in catering establishments, lack of knowledge among catering staff, food enforcement officers and allergic consumers alike, and the fact that allergic consumers do not have the benefit of an ingredients list to guide them. [1]

Of course, both these assertions are quite true. At the simplest level, a person at risk from food allergy only needs to be protected from the particular allergens to which he or she is allergic in the particular food or drink he or she is about to consume. On the other hand, ensuring that information about ingredients and possible allergen contaminants is accurate and can be communicated effectively to the consumer at risk is far from simple.

This chapter examines the challenges to consumers avoiding particular foods to protect their health and to food business operators at all levels who are engaged in supplying their food. We examine current and recent eating patterns and the trend to have more of our food prepared by strangers. We look at the culture of hospitality and the romance of eating for pleasure against the apparent increase in consumer demand to avoid a particular ingredient(s) for a variety of reasons including the potential for life-threatening risk. We study the legal aspects of preparing, protecting, describing, and selling food and drink, and the shortcomings on both sides of the dining "contract" among both consumers and food business operators. We examine a range of "real-life" cases and the lessons learned from a wide range of scenarios in recent decades. Finally, we look at recent initiatives to address food allergy risks including models for best practice and training and how such risks may be addressed in the future.

7.3. BACKGROUND

The following are the author's comments from a personal perspective:

> I have had a severe allergy to nuts and peanuts since 1960 when I was 14 months old, and have survived a number of severe reactions over the years. My parents had called me Hazel and throughout my childhood, I was the only person anybody knew with a nut allergy. As a child, nearly all the food I was offered had been made by close family members from simple ingredients. Both my grandmothers had excellent food skills, not least because they had had to feed their families throughout the 1930s and the rationing during and after World War II. We ate foods as they were in season and the traditional dishes of northern Europe and England in particular. Although my parents knew that I was at risk from nuts and peanuts, these were still around in our home and places where we went. I knew what they looked and smelled like, the types of dishes in which they would appear, and what it felt like to eat them by mistake. I had occasional minor "near misses" and regularly left parties and celebrations wheezy, blotchy, and puffy from low-level exposure to nuts and peanuts.

Thankfully, most reactions were controlled quickly, and I rarely needed hospital treatment [2].

At the same time, our eating habits were beginning to change. As a child, most of the bread I ate had been made at home. Holidays were always self-catering, so the food we ate was always similar to what we had at home. As one of four children, eating in restaurants was a rarity. I can remember my first meal in a Chinese restaurant when I was probably about 11 and all the novel attractions—the table settings, tea with petals in elegant bowls, hot towels, and then the familiar mouth tingle and consequent panic as I reacted to "something" and had to abandon the meal and take medication. As I grew up, eating food prepared by others was becoming more frequent.

My own first job was in a department store café in 1974, preparing and serving snacks and drinks. I began to pick up the basics of commercial food safety and also realized that my potentially life-threatening allergy was entirely unrecognized in the early hazard analysis and risk controls of that era. In due course, I studied languages and worked abroad in a children's holiday camp in France where we were serving three meals a day to over 200 children from a wide range of backgrounds. Thirty years ago, the "special dietary needs" were minimal. There might have been a few celiacs, but nobody else had a nut or any other food allergy. I then worked with some highly trained and very competent French chefs in a German resort hotel. If asked, they might have given you a short textbook lesson in the risks associated with scombrotoxins in fish, but nobody was aware of food allergies.

So what has changed?

7.4. FOOD PREPARED BY STRANGERS

Eating out … throws into sharp relief narrow concerns with food as merely a means of subsistence, for eating out seems to be expanding as a form of entertainment and a means to display taste, status, and distinction. Also significant is the willingness of people to swap their private domestic food provisioning arrangements for commercial or communal alternatives [3].

Spending on eating out overtakes meals at home. (Headline—UK Office of National Statistics 2006 [4])

Consider what people eat today. How much of our food was prepared from scratch at home using basic ingredients? How much was made in a factory for easy preparation and service, perhaps at home or perhaps via some kind of food business. The distinctions are ever fuzzier.

For the purposes of this chapter, food or drink is included if it is sold for immediate consumption. This includes anything not sold in factory-sealed packaging, that is, which is sold non-prepacked or "loose." In the European Union (EU) and elsewhere, such food has a defined legal status, particularly regarding its "labeling" or description. This will include open bakery and patisserie, non-factory-packed sandwiches, counter-served delicatessen, for example, cut cheese, cold meats, salads, and similar foods selected and packed by the customer using self-service.

In addition, the chapter will examine all aspects of catering including restaurants; takeaways and home delivery; sandwich and snack bars; hotdog, bagel, and ice cream stands; pick "n" mix sweet counters; stalls; room service; events catering; and all aspects of institutional dining—child care, nurseries and schools, prisons, workplaces, and care for the elderly.

Back in the late 1970s when I was picking up basic foodservice skills, the European law on providing food information was updated to help consumers find out more about foods on sale [5]. At that time, the legislators focused on the labeling of prepacked foods and did not require key information such as ingredients for non-prepacked foods, *provided that the consumer still receives sufficient information* [5].

Of course at that time, not only were those preparing food in the home usually competent cooks and able to work from basic ingredients, but those preparing and serving food to the public also had a wide range of food skills and knowledge.

The following is the author's comment:

> The chefs with whom I worked in the French holiday camp would prepare everything from scratch including hastily made and beautifully decorated birthday cakes to order. Nowadays, it might be more practical and economic to pick up a cake at the local supermarket or cash and carry, or perhaps have a stock of commercial cakes ready in the freezer. We were not as reliant on food prepared by others.

People eating out would probably encounter somebody who had had a close involvement in the preparation of the food. Local (often family-run) businesses meant that the food supply chain and, therefore, the food information chain were as short as possible. While some restaurateurs were keen to ensure that the dining out experience was romantic and full of mystique, it was still possible to discuss every aspect of the food on sale with the person who had prepared and served it. In addition, as nobody else seemed to have any food allergies in the late 1970s, it did not really matter too much that food information regulation was a bit vague.

7.5. THE ADVENT OF DIETARY NEEDS—SPECIAL OR NOT

While few people in the late 1970s seemed to have food allergies, customers in general seemed to be a lot less particular than they are nowadays. There were not many vegetarians, and few people knew what a vegan was. Quite a lot of food was probably organic, but nobody realized it at the time or asked for it specially. In the same way, much more food was local and served in season but nobody asked much about that either. Fewer members of the U.K. population were following religious or cultural dietary rules, and the tendency to obesity with related diabetes and heart disease from diets high in saturated fats, sugar, and salt was not yet common.

Conversations with staff working in a range of food businesses in 2008 indicated that they were aware of customers who avoid different foods. Personal preference and moral or ethical reasons as well as religious dietary laws may be cited as reasons for avoiding particular ingredients or dishes. At a recent meeting with 20 young U.K. catering college students (who also had concurrent foodservice work placements), all of them had served a customer who had asked to avoid a particular food (unpublished interviews with the author) [6]. Most of these requests were for health reasons including allergies, intolerances, and celiac disease, and one student had witnessed a severely allergic customer having a reaction from food served where he was working.

Although accurate data are hard to come by, the U.K. government Food Standards Agency quotes prevalence figures for immunoglobulin E (IgE)-mediated food allergy of 1–2% among adults and 5–8% among children, for example, in Guidance on Allergen Management and Consumer information FSA/1064/0606 P7 2006 [7]—see also House of Lords Allergy Report 2007 P31 4.12–4.14 [8]. The patient/support group Coeliac UK quotes prevalence data for celiac disease in the United Kingdom of at least 1 in 100 [9].

One of the practical problems is the need for detailed communication of the precise reason behind any avoidance by the consumer to the food business operator. The word "allergy" is widely used, not always by those at risk from IgE-mediated allergies, to the extent that some food business staff have difficulty in associating the declared dietary need with the appropriate risk management precautions. This will be discussed further in some of the case studies.

Figure 7.1 provides a simple analysis of the reasons why a consumer might seek to avoid a particular food or foods [10]. On the left-hand side marked *personal* are food choices made for personal reasons. These may merge with choices made under the heading *moral/ethical* such as ecological or political considerations or animal welfare. The next heading *religious* provide a summary of some of the more common religious or cultural diets. All the boxes on the right-hand side under the heading *health* cover food choices for health protection and the prevention of short- and long-term symptoms. The box with the thicker border indicates food allergen avoidance for those with IgE-mediated allergies.

Any food business operator preparing or selling food to or for the public needs to be aware of these different subgroups and to be ready to respond to their individual needs. The only way that this can happen in practice is for the consumer at risk to be very clear about what exactly he or she needs to avoid, and for the food business operator to be ready, willing, and able to respond.

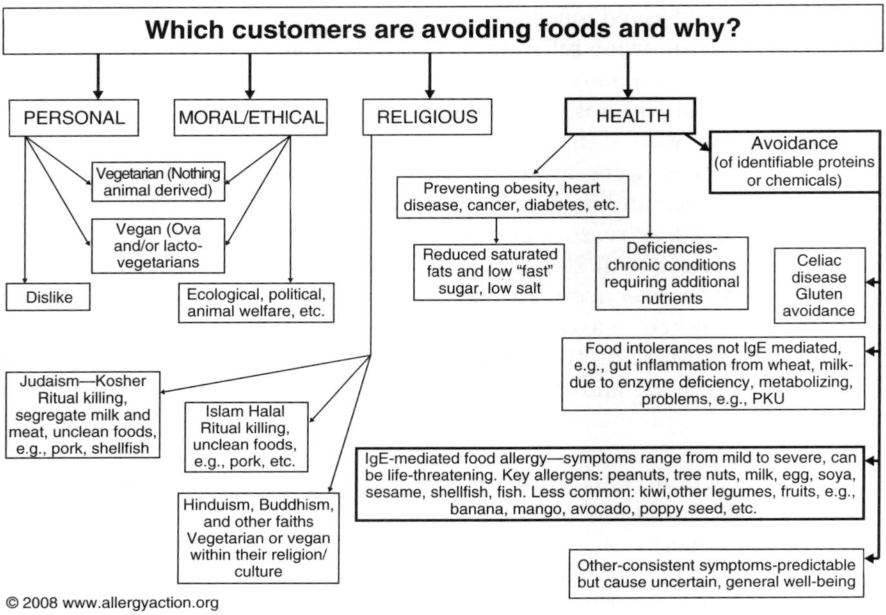

Fig. 7.1 Special dietary needs—from www.allergyaction.org.

Fig. 7.2 Menu leaflet—Chinese takeaway.

7.6. SIGNS OF THE TIMES

Figure 7.2 is taken from a menu leaflet delivered to homes on behalf of a local Chinese restaurant and takeaway in 2005. Along the foot of the menu it says, *kindly inform us of any food allergies prior to ordering your meal.* Although

Fig. 7.3 "I'm a veggie."

Fig. 7.4 "I'm special."

the text is in smaller print than the main menu, it is printed on a red background and might encourage somebody with an allergy to mention it when ordering.

Figures 7.3 and 7.4 are cards that were arranged at place settings at a Christmas dinner and dance party in a London hotel in 2007. Those guests with food allergies and other food avoidance needs had given advanced notice in writing of their dietary requirements. They sat at the places indicated by the

Fig. 7.5 Countertop notice.

cards, and staff members were able to identify them and make sure they received the food that had been prepared specially for them. Although the event was noisy and guests were sometimes in semidarkness or with disco lighting, the serving staff had been well trained and were well informed and helpful.

Many menus and notices displayed in catering and food retail operations nowadays encourage customers to talk to staff and to discuss their dietary needs.

Figure 7.5 is a countertop notice from a department store café taken in 2006. The notice on the left offers information about ingredients on request, and the one on the right focuses particularly on nuts.

> Product information: details of the ingredients used in all our menu items are available on request. Please see any of our staff for further information. Nut allergies: some of our products contain *nuts*. If you require further information please ask to speak to any member of our foodservices team.

In the United Kingdom, allergies to peanuts and tree nuts are the most common food allergies in adults and older children (6541 of the 7900 Anaphylaxis Campaign members were avoiding nuts and/or peanuts in July 2008—from the membership database).

In some respects, this has made it easier to raise awareness of food allergies among caterers and others supplying food to the public, but it can also mislead them into thinking that other allergens are not important. Nuts and peanuts can be used as a model for hazard analysis and training for risks from other allergens, but it is vital that food business operators are able to identify *any* ingredient used deliberately or which may have gotten into another product or dish along the way.

7.7. WHICH ALLERGENS?

While nuts and peanuts are believed to be the most common food allergens among adults and older children in the United Kingdom, other food allergen

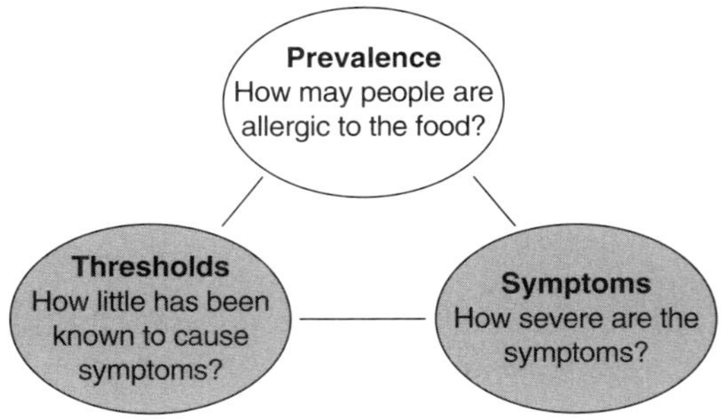

Fig. 7.6 The allergy risk triangle.

ingredients also cause a wide range of symptoms and need to be taken into account [8]. In addition to prioritizing the potential risk between different consumers with different dietary requirements, food business staff may also try to prioritize the potential risk from different food ingredients that they use. Such prioritization is an inexact science and involves consideration of the allergy risk triangle in Figure 7.6.

Many models for optimizing food allergen controls in foodservice originate in food production, particularly in the manufacture of prepacked foods. Since November 2005 in Europe, ingredients information on prepacked foods has been more closely regulated to enable consumers at risk from allergies and intolerances to identify key allergens used deliberately as ingredients.

These are

- cereals containing gluten
- crustaceans
- egg
- fish
- peanuts
- milk
- nuts—including almond, hazelnut, walnut, cashew, pecan nut, Brazil nut, pistachio nut, and Queensland or macadamia nut
- soya
- sesame
- celery
- mustard
- sulfur dioxide and sulfites (not food proteins but require labeling at levels above 10 mg/kg to protect people whose asthma is made worse by them).

Further information about the relevant legislation and amendments to regulations may be found in the Appendices to the FSA Guidance on Allergen Management and Consumer Information 2006 [7].

This list was drawn up with guidance from the European Food Safety Authority's expert panel of clinicians and scientists who examine each food allergen considering the criteria in Figure 7.6 and provide risk assessment advice to the European Commission who implements labeling and food safety law.

In addition to improving the availability of key ingredient information for the majority of food-allergic people who need to avoid the above foods, this list also began to help food manufacturers and some foodservice businesses in prioritizing their routine food allergen risk assessment activities. Typically, this would involve collecting information about the potential for cross-contamination by other allergens on the list, and communicating such data through the allergen management system. In cases where such contamination cannot be controlled, the use of "may contain" labeling has become widespread on packaging.

Additional facets to the allergy risk triangle might include the age and potential vulnerability of the consumers, as well as known consumption patterns. Certainly, manufacturers of baby food and products for children have been first to assess the possibility of the presence of milk and egg in products, which do not contain them as ingredients for example.

7.8. WHAT DO WE KNOW ABOUT THOSE AT RISK?

For the purposes of this paper, we will focus on the needs of consumers at risk from IgE-mediated allergies, while at the same time acknowledging that much foodservice allergy training and many of the allergen controls will be of value to a wider range of consumers, including those at risk from intolerances and celiac disease as well as those making other dietary choices.

It is recognized that consumers at risk from food allergy, and particularly those who have been advised by a doctor, or believe that they have a potentially fatal food allergy may find eating out stressful or at least life limiting [2]. For a minority, and particularly for those with potentially severe allergies to a range of foods, eating food prepared by strangers may be impossible or at least very rare. Others may have identified particular ingredients or food practices in certain types of food business, which they would consider too risky to contemplate.

The following is the author's comment:

> In my own case, I choose not to eat in Indian, Chinese, or Thai restaurants, or to eat bread or patisserie from local, artisan, or in-store bakeries. I only eat pre-packed bread from recognized plant bakery manufacturers who do not use nuts or peanuts on-site. There are countries where communicating my allergy risk, local food safety practices, and the use of nuts/peanuts in everyday food would all prove difficult, if not impossible for my potential degree of risk.

In recent research interviews with food-allergic teenagers, Dr. Michael Gallagher at the University of Edinburgh asked them to describe their food allergy. Words used included *severe*, *deadly*, *annoying*, and *hassle* (personal correspondence [unpublished] 2008) [11].

The Food Standards Agency report on Qualitative Research into the Information Needs of Teenagers with Food Allergy and Intolerance in 2005 [12] includes an overview of many aspects of growing up and leaving home with a severe allergy, intolerance, or celiac disease. "Eating out is just too much hassle—we don't do it anymore" (P35) [12] or "we always go self-catering and take food with us" (P39) [12] reflects commonly held views of families contacting the Anaphylaxis Campaign helpline.

As teenagers and young adults start to develop their own social lives and the pressure is on to eat out, share food at other people's houses, travel, holiday, and then live with other young people away from home, those at risk from food allergies in particular can find life very challenging. For starters, it can be more expensive to live with a food allergy. The Food Standards Agency study "May contain labelling"—the consumer's perspective 2002 [13] indicated that it can cost 11% more just to find everyday prepacked products without a "may contain nuts" warning. Celiacs have to pay a premium for everyday substitute items to avoid gluten. Those avoiding milk, egg, wheat and/or a range of foods can have extremely restricted diets and food can be very expensive. For many of these people, eating away from the home is a rare and very limited, if not a nail-biting, experience.

One of the most significant problems for food-allergic people is not wishing to be different or stand out from the crowd. Unless someone is very sure of his or her diagnosis (and many people are not, particularly in the United Kingdom where there are few specialist clinics [14]—over half of those who died from food allergy had had no professional allergy advice) and has a confident and assertive personality, he or she is unlikely to want to make a fuss in many social situations involving food.

As previously indicated [2, 15], teenagers do not want to be different even if it means putting their life at risk. Reports received by the Anaphylaxis Campaign include a nut-allergic young male student collapsing in a shop doorway after sending his friends on ahead because he did not want to be a nuisance, and both male and female young professionals leaving restaurant meals with clients midreaction and taking taxis home to fetch emergency medication without explaining to the fellow diner(s) what the problem was. There can be a tendency to want to hide away, go to the bathroom, and not be witnessed having allergic symptoms. More than one young adult has rushed from a social event to find his or her injectible adrenaline, which is frequently left in his or her student room or bag somewhere else.

"Risk-taking behaviors in teenagers … are generally attributed to a reduced appreciation of potential dangers and a belief that consequences can be controlled" [16]. Pumphrey and Gowland [14] described the death of a 16-year-old who ate a potentially nut-contaminated chocolate because she had her injectible adrenaline with her. It did not save her life.

Additional factors include alcohol and drugs, either of which may impact upon the alertness of the allergic subject (and his or her friends) to identify and respond to symptoms in time, and practical logistics to get help, for example, not being able to describe exactly where you are or call emergency services.

In the same way, inexperienced atopic people may not recognize the added "allergic load" of a flare-up of eczema, asthma, and/or hay fever, as well as the possible impact of exertion, stress, hormonal fluctuations, or a recent infection.

Another factor with improved allergy awareness is that younger allergic people may have been so vigilantly protected from any kind of exposure to some food allergens (e.g., nuts and peanuts) which have now become taboo in homes, child care, and school environments that they are unfamiliar with their use, appearance, smell, or taste. A recent fatal reaction in the United Kingdom involved a 16-year-old boy who knew he had a peanut allergy but who ate a whole chocolate bar containing peanuts without realizing. As described by Grimshaw et al. [17], while the fatty chocolate food matrix may have disguised some of the peanutty taste and prevented telltale early symptoms in the mouth, some warning signals might have been expected before the whole bar was eaten.

The following is the author's comment:

> I am thankful now that my low-level childhood reactions to nuts and peanuts equipped me to recognize what I needed to avoid.

The greatest risk factor, which is hardest to plan for, is not actually expecting any kind of allergic symptoms at all. The Anaphylaxis Campaign has regular contact with families of those who have died from food allergies. Many parents of young adults who have died from anaphylaxis say that their family doctor advised them when their children were younger that the food allergy was "only mild." Pumphrey and Gowland [14] suggested that over half of those recorded to have died from food allergy in the United Kingdom between 1999 and 2006 had only had previous mild reactions and may not have recognized the possibility of a potentially fatal risk. Even among members of the Anaphylaxis Campaign [18], it was "notable that about a third of reactions happen after exposure to unidentified allergens or to foods not previously identified as a risk."

Roberts [19] recognized that identifying the relevant allergen is essential to prevent unnecessary avoidance of foods, which may lead to poor development in children. In older children, teenagers, young adults, and their friends and families, not being able to identify which food triggered the symptoms is extremely stressful. Food allergen avoidance is tough enough when you can identify the food that caused severe symptoms, but a nightmare when you cannot. Finely tuned coping strategies and appropriate levels of confidence and competence are essential.

7.9. EFFECTIVE AVOIDANCE

As outlined at the beginning of this chapter, all food allergy symptoms can be prevented provided the person at risk does not come into contact with the particular food allergen(s) to which they are allergic. In practice, such avoidance can be implemented by the allergic person himself or herself and/or by the food supplier. Effective avoidance depends on either or both parties being competent, informed, willing, and able to identify and avoid the relevant food allergen(s) in anything they might consume/supply.

As discussed above, the foodservice sector has changed beyond recognition in the last 30 years. It can no longer be assumed that dialogue is possible between a customer and a food competent supplier about allergens used as ingredients or handled in the food preparation and service environment.

Figure 7.7—the foodservice spectrum of allergy risk—illustrates two typical scenarios in the food supply chain where it is easier for allergic people to find out what went into a product or dish and whether it is suitable for them. The box on the left (A) describes a scenario where the consumers at risk may discuss food on sale with a professional food handler who can match their needs with his/ or her product knowledge and, if necessary, prepare something especially with confidence and competence. The box on the right (C) describes a scenario at the other end of the food supply chain, where the food has been produced and contained in highly controlled factory conditions and the labeling is available, clear, and accurate.

The middle box (B) describes a scenario in which food allergen avoidance is more likely to fail. The staff providing food information may not have first-hand knowledge of what went into the product or dish, and may not have factory ingredients or "may contain" information available. They may not have cultural knowledge of likely ingredients, particularly in ethnic or traditional recipes. In addition, in their wish to be helpful, they may provide information that is what they think the customer wants to hear, but which may be far from accurate.

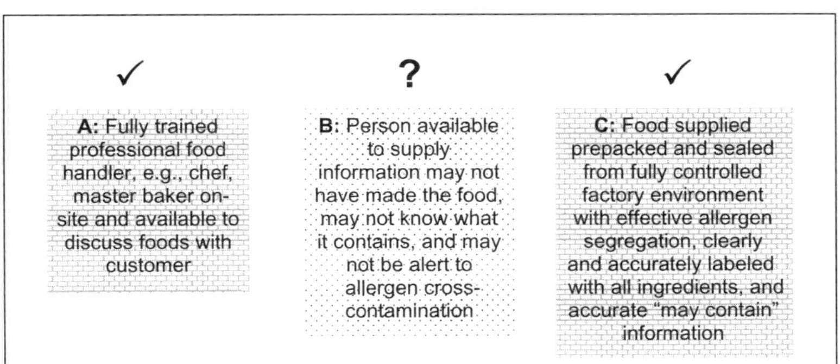

Fig. 7.7 The foodservice spectrum of allergy risk.

7.9.1. Avoidance Failure—Fatal Cases

Like most safety practice, effective food allergen avoidance strategies are often learned from previous failures. Dr. Richard Pumphrey, consultant immunologist continues to study both fatal cases and "near miss" allergic reactions, working closely with the Anaphylaxis Campaign. He has supplied the following data from fatal cases [14, 20].

Figures 7.8 and 7.9 indicate the increase in fatal reactions in people aged over 18 eating out in restaurants, on holiday, and at festive events (e.g., a wedding or party at a friend's house). Particular risks are associated with barbecues where ingredients in marinades and sauces are often hard to identify and with buffets because there is rarely any ingredients information. In addition, at events where the food is brought in by outside caterers or fellow

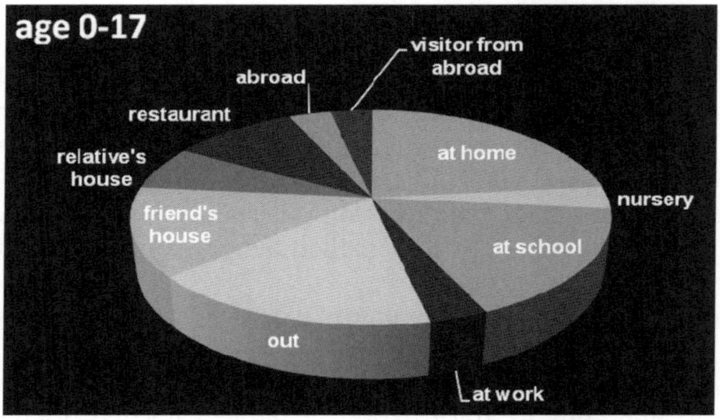

Fig. 7.8 Fatal allergic reactions to food in the United Kingdom (1992–2007). The graph shows where the reaction was triggered for people aged 0–17.

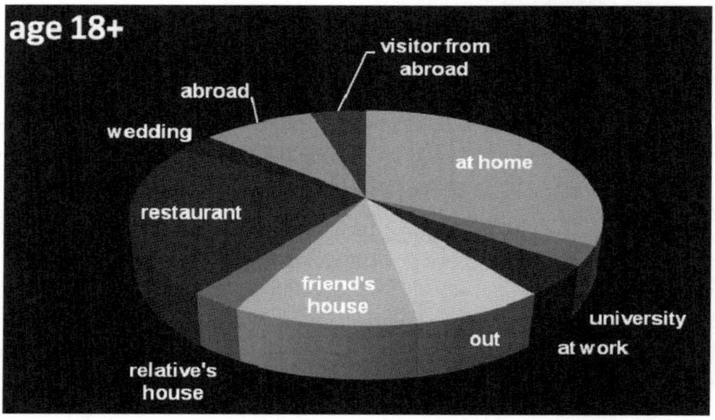

Fig. 7.9 Fatal allergic reactions to food in the United Kingdom (1992–2007). The graph shows where the reaction was triggered for people aged 18+.

guests, there may be no written information available and those who prepared the food may have left the event. Further allergy hazards may be introduced by guests themselves who swap serving utensils and introduce allergen cross-contamination.

Figures 7.10 and 7.11 show Dr. Pumphrey's analysis of the usual allergen avoidance practiced by those who died from food allergy in the United Kingdom in 1992–2000 and indicates that those practicing no or low allergen avoidance were more likely to die from food eaten in restaurants. Dr. Pumphrey has also examined the nature and presentation of the allergen-containing food consumed (Figs. 7.12 and 7.13). More adults than children ate the food whole or served by caterers whereas more children suffered fatal reactions to pre-packed and home-prepared foods than adults.

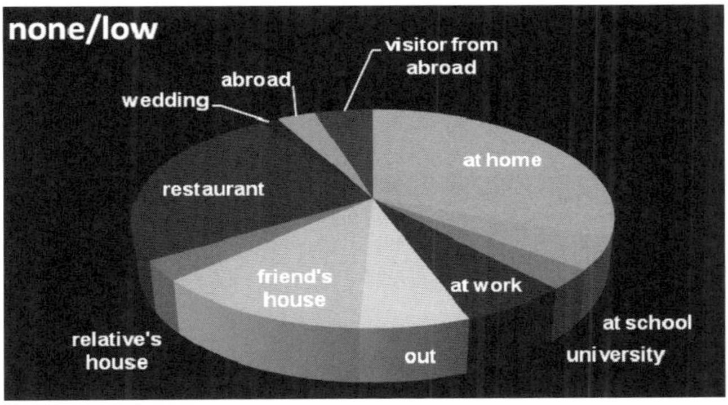

Fig. 7.10 Fatal allergic reactions to food in the United Kingdom (1992–2007). The graph shows situations in which care taken over allergen avoidance was none or low.

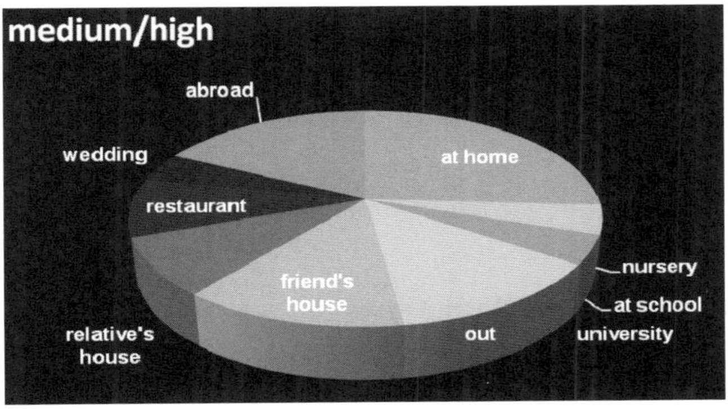

Fig. 7.11 Fatal allergic reactions to food in the United Kingdom (1992–2007). The graph shows situations in which care taken over allergen avoidance was medium or high.

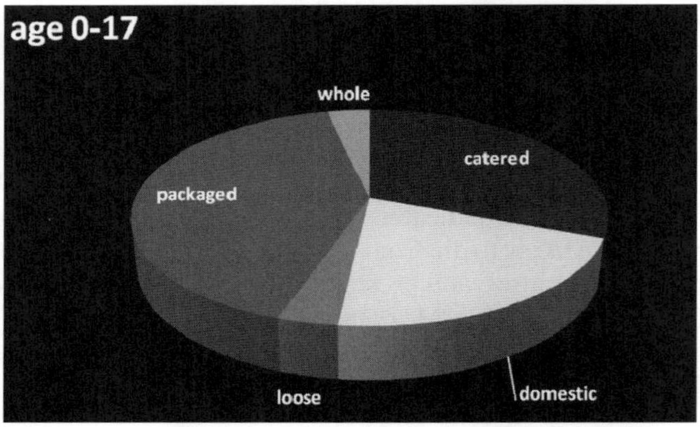

Fig. 7.12 Fatal allergic reactions to food in the United Kingdom (1992–2007). The graph shows how the food was prepared for children.

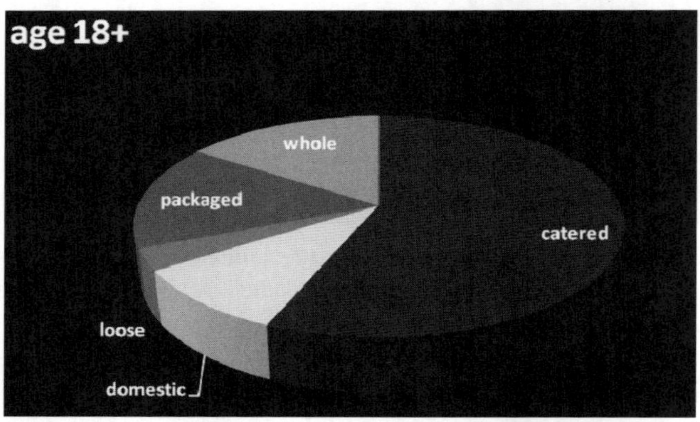

Fig. 7.13 Fatal allergic reactions to food in the United Kingdom (1992–2007). The graph shows how the food was prepared for adults.

7.9.2. Avoidance Failure—"Near Misses"

In addition to fatal food reactions, the Anaphylaxis Campaign collects and analyzes data on reactions requiring emergency medication and sometimes hospitalization. Studies such as Uguz et al. [18] and Derby et al. [21], as well as daily feedback from the helpline, continue to contribute to better understanding of food allergen hazards and potential risks.

7.10. UNRECOGNIZED INGREDIENTS

One of the most common causes of severe reactions is the consumption of an ingredient that was used deliberately in the recipe by the food manufacturer or caterer but not recognized by the food-allergic consumer and/or the person serving the food.

There is a single catering practice, which is widely but not universally practiced in the United Kingdom, that has been responsible for at least 10 deaths and many more "near misses" in peanut-allergic individuals in recent years. The addition of peanut flour as a thickener in curry sauces is rarely identified as a possible allergen hazard by restaurants and takeaway staff. Research undertaken by local government food safety and food standards officers has indicated that 8 out of 20 curry meals sold as peanut free actually contained peanut protein [22–24].

This is a particular problem in student populations in the United Kingdom for whom "going for a curry" often follows a good night out. The Anaphylaxis Campaign and food control officers have undertaken a number of projects in areas frequented by students to raise awareness of food allergy risks among consumers and businesses alike. Similar data from Canada looking at accidental exposure to peanut in children indicated 12 out of 35 reactions in which the food eaten contained peanut as a deliberate ingredient [25].

7.10.1. Case Files: Unrecognized Ingredients

7.10.1.1. Bought-In Desserts in Chain Restaurants. The supplier had run out of the usual stock and sent a substitute product that contained nuts. Service staff did not manage the product information and gave the nut-allergic diner the usual ingredients list and not the one for the substitute product, causing at least two severe reactions requiring hospitalization and time off work.

7.10.1.1.1. Who Made This? Although restaurant recipes are increasingly standardized for budgetary and stock management reasons, a number of reactions have been recorded because individual chefs using the same recipe to make the same dish occasionally add a "secret ingredient" without which in their view the dish would be incomplete. Food-allergic consumers tend to rely on continuity and consistency eating the same product or dish from the same food supplier and have been caught out by this practice.

7.10.1.1.2. Bread Rolls on Side Plates. At least one U.K. fatality and another severe reaction requiring life-saving treatment involved these. In both cases, the bread roll contained nuts used deliberately but unrecognized by the diners or staff. Bread rolls are often served and may be eaten before the main menu choices are made and dialogue about allergen avoidance even begins.

7.10.1.1.3. Weasel Words. Peanut- and nut-containing products and dishes with misleading or unknown names such as coronation chicken and satay sauce have been responsible for fatal reactions in young adults. Pesto ingredients may include nuts, seeds, and cheese; curries may include ghee from milk; and bakery and patisserie items may include praline (hazelnut paste), marzipan (from almonds), and hidden egg in glazes. There are many more examples, and every cuisine has its range of dishes and accompaniments that may be made from unexpected allergens. Food allergen management policies should begin with the identification of key allergens used deliberately in all products and dishes.

7.10.1.1.4. Lost in Translation. One of the recurrent risks associated with eating out and failed allergen avoidance is the problem of communication between consumers and food business staff who do not share a language and/or dining culture. One nut-allergic young woman was misled by an ingredients list in English, which included the word Kajo. This turned out to be cashew to which she suffered a very severe reaction and hospitalization.

7.10.1.1.5. Suspend Your Disbelief. Even with extreme vigilance for allergen avoidance, there have been cases where a superconfident member of the service staff was able to convince an allergic person that a particular product or dish would be OK for them. One example was a waitress in a family restaurant who believed that the ice cream in a dessert dispensing machine was totally milk free. Sadly, it was not, and the child suffered severe symptoms and poor health for some weeks.

7.10.2. Case Files: Cross-Contamination from Allergen Traces

7.10.2.1. Conveyor Belts at Point of Sale. A woman suffered a severe reaction to shellfish. Her husband had shopped at the supermarket and asked the person working on the till to wipe up a wet patch caused by leaking shellfish on the conveyor belt. Once home, he unpacked the shopping on the kitchen worktop. His wife later prepared food on the worktop and unknowingly contaminated her food with shellfish. In the same way, milk and egg leakage on belts in supermarkets and in home delivery vans has caused hidden cross-contamination and triggered allergic reactions.

7.10.2.2. Sandwich Bars. Sandwich production whether in a factory, in a catering setting, or in the home can involve a wide range of food allergens, many of which (e.g., mustard, celery, soya, egg, milk, sesame) are often unrecognized. While keeping fillings in a display fridge and making up sandwiches to order on the counter above may be considered adequate for the control of microbiological risks, repeated use of the same chopping boards, knives, and other serving utensils will introduce the possibility of allergen cross-contamination. Breads may leave residues of nuts or seeds, which may not be visible.

7.10.2.3. Bar Snacks. One fatal case involved a man with a severe peanut allergy. He moved a bowl of peanuts along a bar out of the way. Traces must have remained on his hands and transferred to his glass or directly to his mouth. We also received a report of a bride who nearly died on her wedding day. She had a peanut allergy and ate some crisps from a bowl at the wedding reception. Unknown to her, the bowl had previously contained peanuts. Once they were gone, bar staff topped up the empty bowl with crisps on top of the peanut traces.

7.10.2.4. Service. In cases where food is served and recognized by the allergic diner as likely to contain the allergen they need to avoid, staff may be asked to remove the dish and bring something else. One Canadian bride who died on her honeymoon was served lobster that she sent back as she had a shellfish allergy. Further investigation indicated that the steak requested instead had been put on the same plate. Allergic people frequently report wafers being removed from ice creams, gravy scraped off meat, and garnishes being taken off before the dish is re-served.

7.10.2.5. Airborne Contamination. As a general rule, even the most sensitive food-allergic people can be in a room containing the food allergen they need to avoid provided they take precautions to prevent surface cross-contamination, for example, from shaking hands with other diners. There are, however, some exceptions. One peanut-allergic woman started to suffer breathing difficulties when covering cloths were removed from a wedding buffet and particles wafted around the room. Fish-allergic individuals most frequently report symptoms from passing fish stalls or cooking fish.

7.10.2.6. Gloves. There are two allergy problems from the use of gloves in food preparation and service. If the gloves are made of latex rubber, they may well sensitize and/or trigger symptoms in latex-allergic staff and customers. In addition, they may transfer latex particles into the food, which may trigger allergic symptoms. It is also recognized that food handlers who wear gloves may develop a false sense of security about their cleanliness and might not change them as often as they would usually wash their hands.

7.11. EFFECTIVE ALLERGEN AVOIDANCE IN DIFFERENT ENVIRONMENTS

As a general rule, much food allergen management best practice may be learned from foodservice professionals working in nurseries, schools, colleges, workplaces, and other institutions where they serve the same people every day and where staff are motivated to implement practical allergen controls, and committed to regular updates and improvements to reduce risks.

7.11.1. Child Care and Nurseries

Managing babies and small children with food allergies involves a number of additional challenges for the following reasons:

- The child may not have a history of food allergy, and the first allergic reaction may take place in the nursery.
- Small children are dynamic. They can develop new food allergies and grow out of older ones by the day. This can cause confusion and make preparing up-to-date management plans a particular challenge.
- Small children can have a lot of close contact with one another. Even if particular food allergens such as nuts and peanuts are banned from the nursery environment, other children may have them on their hands and faces.

A 5-month-old baby boy died in a nursery in the United Kingdom in 2002 having been given a breakfast cereal that contained clearly labeled but perhaps unexpected milk and in spite of clear parental instructions.

7.11.2. Schools

Whether or not schools are in a position to implement bans on particular allergens (e.g., nuts and peanuts) as described by Yu et al. [25], many examples of committed allergen controls are to be found in schools, both in the provision of food and drink by the school meals service and supervising the consumption of food brought from home, and also as part of the main curriculum and in peripheral activities such as breakfast, after school, and activity clubs. Many catering teams work exhaustively to read packets and to source ingredients, which enable those at risk to have the same or something very similar to the food other children are enjoying. In this way, the educational aspects of social eating, particularly for primary aged children, are recognized and supported.

Ideally, the control of particular food allergens in the school environment should be integrated into tailored management plans for identified individual children and students whose allergies are fully diagnosed and for whom emergency management and treatment are ready if needed.

In 2005 in Canada, Sabrina's law was passed followed by the tragic death of 13-year-old Sabrina Shannon. She reacted to milk contamination in French fries from the school canteen. The law aims to protect schoolchildren at risk from allergies and its implementation continues. CBC News in December 2007 [26] reported that "the school board must make it easy to identify students with allergies; parents must provide correct documents to the school; and the education ministry provides additional EpiPens (adrenaline injectors) to schools depending on the number of students who register with allergies."

In the Anaphylaxis Canada newsletter for August 2008, their Youth Advisory Panel discussed how to get through a school day with a potentially severe food allergy. Although students recognized that management plans and good general cleaning to remove allergens in the school environment seemed to be effective, they were less keen to be singled out to sit at segregated tables. Muñoz-Furlong [27] also discussed best practice management for children and suggested that such segregation policies should be adapted as the children get older. One example seen in the United Kingdom is to segregate the allergen rather than the food-allergic children. In one primary school with a number of children allergic to nuts and peanuts, there was a "nut table." If children had nuts or peanuts in their packed lunch, they would use this table and then put their rubbish in the nearby bin and wash their hands under staff supervision.

7.11.3. Residential Care

Although there is a perception that children and younger adults are most likely to be food allergic, chefs and catering managers do occasionally contact the U.K. Anaphylaxis Campaign for advice on allergen avoidance for older people, and particularly those who may become more vulnerable and less aware of allergy risks in later years, and also those whose tendency to allergic symptoms is affected by other medication.

7.11.4. Prisons

Managing food-allergic people in prisons is a particular challenge because many of them are young adults who may be particularly vulnerable to severe reactions. The U.S. website mlive.com [28] reports the death of an inmate from eating a peanut butter sandwich in April 2008. It is reported that he did not inform the jail authorities of any allergies when asked on his arrival there.

7.11.5. Workplace Dining

If food-allergic diners are confident in allergen avoidance and able to notify catering personnel of their needs, then much can be done to protect them. Like food-allergic children in schools, they may be recognized by staff who may be able to provide additional relevant information and advice about dishes and products on sale.

7.11.6. Airlines

There are a number of issues about eating on airlines, not least because we are short of data about whether airborne and/or surface food allergen contamination may trigger symptoms in susceptible people, and if so, how long for. In addition, the practicalities of providing in-flight food and snacks are complex. Since most airliners do not have trained chefs on board, option A in Figure 7.7 is not possible. Most airline food is a combination of options B and

C in Figure 7.7. Some may have come in sealed and labeled packaging from highly controlled factory environments where some food allergens were deliberately and effectively eliminated. However, airlines have to source their in-flight meals and snacks from a wide range of destinations, and this may not always be possible. Traceability is a challenge, and food brought on board by other passengers may also introduce allergen risks. A number of reports of food allergy deaths on airlines have been received by the Anaphylaxis Campaign including a case where the passenger's wish for a vegetarian meal led her to eat nuts to which she knew she was allergic.

7.11.7. Holidays and Travel

Food-allergic people traveling and on holiday may encounter some particular risks, which can be classified as follows:

- The food is different. Information about what is in it may not be available. Part of the holiday experience is trying different foods.
- People on holiday may be relaxed and may drink more alcohol than usual. This may affect their allergen avoidance and early recognition of symptoms.
- Food suppliers at the holiday destination may not be familiar with food allergen avoidance or ready to provide information to consumers.
- Consumers at risk and food suppliers may not share the same language.

The Anaphylaxis Campaign has received a report of a groom with a fish allergy who died on his honeymoon in Thailand, apparently after some confusion over food ingredients.

7.11.8. Celebrations

Festive occasions are often the background for failed allergen avoidance. One case involved an egg-allergic father of young children who died at a Sikh wedding. The meal was served in the temple itself where according to the (Guru Da Langar) dietary regime, meat, fish, eggs, and alcohol were not allowed, and the man had expected his meal to be egg free.

A young boy suffered a severe reaction to unexpected nuts at his brother's Bar Mitzvah. Although extensive advance precautions had been taken by the family, on the day of the party, the boy's mother was misinformed by a waitress about ingredients in a dish, and the boy spent the night in hospital.

7.11.9. Menus and Descriptions

In restaurants, takeaways and in situations where food is sold without sealed factory packaging and labeling, the information available is critical in effective allergen avoidance. A promising young student died following a university dinner after eating *strawberry shortbread on a red fruit sauce*. Although she knew she had a nut allergy, unknown to her the shortbread contained ground almonds, which were invisible and unexpected. Advice to caterers and others describing food is that key allergens used as ingredients should always be mentioned in the description of the product or dish.

Another fatal case in the United Kingdom involved a peanut-allergic 13-year-old girl who routinely ate chips with curry sauce from her local takeaway. Staff usually put some peanut butter in the curry sauce but would leave it out for her. On the day she died, her friend had picked up the chips from the takeaway and staff did not know that they were for the allergic girl.

As a general rule, nuts and sometimes peanuts are a relatively expensive ingredient and therefore merit a particular mention on the menu or product description. In more upmarket restaurants, however, there have been reactions to foods that contained even more expensive ingredients where the use of nuts or unrefined nut oils was not mentioned.

7.11.10. Room Service

The practicalities of room service, particularly for a food-allergic diner eating alone, can be very challenging. Usually, room service food is ordered via the hotel telephone. The diner will not see any of the dishes on offer, so any visual clues will not be available (although some hotel menus do now have photos of the meals, which can be useful).

The following is the author's comment:

> In my own case, I ordered a meal in French in a French-speaking hotel, and inquired about every component. I chose the simplest food on offer and was very firm about needing to avoid nuts. The soup was served with a bread roll, which seemed to contain whole grains or seeds. When I phoned back to check, I was told that it contained walnuts. For me, that was a close call. It was also a lesson to the hotel who might otherwise have found themselves managing an allergic emergency or worse.

7.12. FOOD ALLERGY IN CONTEXT

When examining allergen management and control in the foodservice industry, the fundamental principle still applies: "Eating out with an allergy is really simple. You just need to know what went into the food deliberately and what else might have gotten in along the way."

It is increasingly recognized by consumers, food business operators, and those responsible for advising on food safety and food description that allergen avoidance depends on good communication across languages and cultures. MacAuslan [29] described the barriers to food safety best practice from poor literacy and language skills, which may be overcome by accessible and effective training.

Market research undertaken by Mintel in 2003 [30] indicated that although young adults are generally unconcerned about food safety issues, *almost one in three 15- to 24-year-olds would like more information about food allergies, reflecting complications among consumers with severe food allergies (e.g., to nuts), which mostly occur in this age group*.

The case histories and scenarios outlined above suggest that there are still many challenges to effective food allergen avoidance, but also that there is a growing commitment to getting things right. One key factor is the growth in prevalence of food allergies, which is reported as being particularly high among food handlers and their families—perhaps because they handle many different foods every day—which may lead to sensitization and further exposure causing symptoms. Commonly cited allergies to shellfish and nuts in particular mean that allergy awareness among chefs and their colleagues is heightened. High-profile chefs are able to raise awareness of food allergies and give professional confidence and skills in effective allergen management to colleagues.

Another issue, which is often raised, is that of access. Although most food-allergic people would be reluctant to class their food allergy as a disability, it may well restrict their lifestyle and cause them stress and inconvenience. As restaurants install wheelchair access ramps and other support for disabled people, some food-allergic individuals are keen to ensure that their needs are met too.

7.12.1. Best Practice: What Can Be Learned from the Food Manufacturing Sector?

Much best practice for allergen controls in foodservice originates in food manufacturing where automation, bulk production, physical controls, product labeling, and extensive third-party auditing of all food safety and food composition practices have led to intelligent risk management and improved allergen avoidance. In the United Kingdom, industry guidance is available from the Food Standards Agency as a management document *Guidance on Allergen Management and Consumer Information* [7] and a convenient picture leaflet for smaller businesses *Allergy—what to consider when labelling food* [31]. The Australian Food and Grocery Council *Guide to Allergen Management and Labelling* 2007 [32] is a similar food industry document. Other national and international standards and schemes also provide guidance on the elimination of different food allergens from food production and the reduction of risks from mislabeling.

The principles of all such schemes are as follows:

- *Eliminate*—if you don't need to use a food allergen, then do not introduce it into your production environment.
- *Control*—if you do use a food allergen, make sure that it is recognized as a food allergen and that you know exactly where it is and in what form.

- *Communicate*—make sure that everybody knows exactly which allergens are used as ingredients and which may be around as possible contaminants, and can pass on that information accurately to those responsible for the next stage in the supply chain. Product information and temporary or retail labeling must be correct and clear.

7.12.2. Food Allergy in the Food Safety Context

Food safety professionals, auditors, technologists, and labeling experts continue to discuss the place of food allergen controls in food safety and food information management. The following model (Fig. 7.14) has been useful in recent years to explain how food allergen controls and other avoidance needs may be considered in the wider context of standards (composition and description) and safety [10].

Consumers who need to avoid particular foods to protect their health need to know both what went into the food as an ingredient (and which would be covered under food standards labeling and description including written and oral information), as well as whether the food had been protected from risks such as allergen contaminants during preparation and service.

Some advocates of "classic" food safety principles consider that only physical, chemical, and microbiological risks should be addressed by food safety management systems. Although control models that integrate these traditional risks with allergen controls are widely used in food manufacture, they are not always considered suitable in foodservice. The current U.K. government food safety initiative "scores on the doors," for example, which aims to

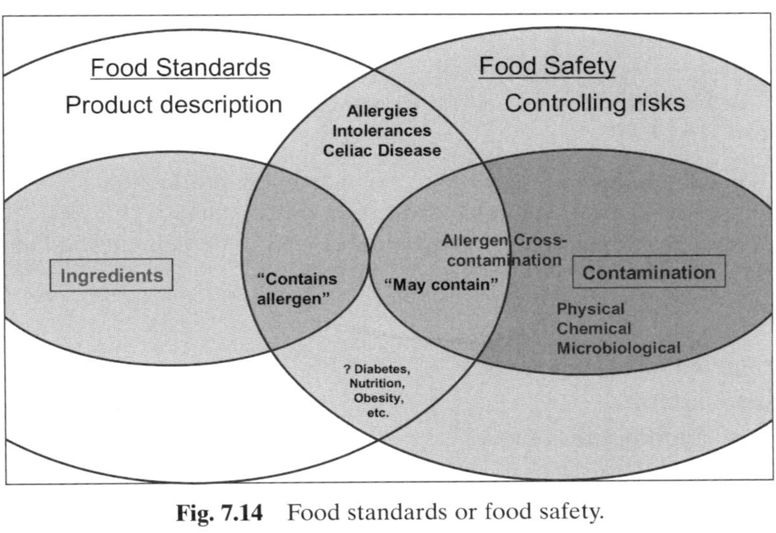

Fig. 7.14 Food standards or food safety.

improve food safety in restaurants and takeaways, will not include the assessment of food allergen controls in its scoring system. A top star score may not indicate any kind of competence in assessing, controlling, or communicating food allergy risks.

7.12.3. Anybody or Somebody?

One of the perennial debates about food allergy in the food safety context is whether food safety systems should be required to protect the public in general or a particular subsection of the population. Food-allergic individuals have occasionally been advised that as they are in the minority, their needs cannot be covered by "classic" food safety controls. On the other hand, it has also been pointed out that microbiological hazards, which are covered by such "classic" controls, do not represent a risk to all consumers. Children, pregnant women, people whose immune systems are compromised, and the elderly are more likely to be affected if such hazards are not controlled. Since people avoiding foods to protect their health are a growing minority and since the consequences of failed allergen avoidance are potentially fatal, the integration of allergen controls in food safety management plans is vital.

7.12.4. How Much Is Too Much?

Excluding allergen controls from "classic" food safety management is also justified in some quarters by the absence of knowledge about thresholds for different food allergens. As described above, food allergic symptoms are unpredictable and may be affected by a wide range of additional factors. Once again, comparison with other "classic" food safety risks indicates that agreed acceptable levels for microbiological risks (for example) may not be appropriate for all consumers.

7.12.5. To Hazard Analysis Critical Control Point (HACCP) or Not to HACCP?

The HACCP method of assessing and managing production risks (which is widely used in food manufacturing and other industry sectors) and its simpler hazard analysis relative can be developed to implement food allergen controls.

The process followed is to

- examine the process steps,
- assess hazards,
- devise appropriate controls,
- implement controls,
- monitor controls,

- review the process, and
- implement guidance and training to support the system.

Manufacturers and some foodservice businesses have integrated allergen controls into this process. Risk assessment may indicate that some of the controls already in place to address physical, chemical, and microbiological risks are also effective for allergens. This additional benefit is then merged into the hazard analysis or HACCP process. In other cases, a separate column for allergen management is added on the right-hand side of the food safety management plan, which ensures that allergen risks are controlled in parallel with other risks. It is then possible to develop assessment methods for the effectiveness of each process step.

7.13. WHAT WORKS IN FOODSERVICE?

There are four effective allergen controls in foodservice [10]:

1. The allergic person can eat a standard menu item.
2. The allergic person can eat a standard menu item with some things left off, for example, sauce, wafer biscuit.
3. Staff can prepare something especially in an area free from the allergen.
4. Staff suggest to the allergic person that precautions currently in place might not be adequate to protect them from contact with the allergen they need to avoid.

If any of the controls above is implemented correctly, then allergen avoidance will be effective.

7.13.1. People, Equipment, and Procedures

Like all good safety systems, these controls will depend on the people, equipment, and procedures available at the time.

7.13.1.1. People. All staff will need to be trained to an appropriate level for their activity. Those responsible for managing ingredients, recipes, and menus; communicating with customers; and controlling product knowledge and information will need to be particularly competent and confident in how to implement the controls.

7.13.1.2. Equipment. One key factor in many foodservice businesses is the available space to store, decant, prepare, cook, and serve foods from which particular allergens need to be eliminated. Sometimes, segregation by space and/or by time, for example, preparing the allergic person's food first before other foods are used, or last after a thorough clean down is the answer. It is also vital that hands, work surfaces, overalls, utensils, cloths, and other kitchen equipment are thoroughly washed and cleaned.

7.13.1.3. Procedures. The procedures in place for managing food information and for segregating foods for physical, chemical, microbiological, and food allergen reasons need to be understood and integrated into everyday practice. This may involve training, color coding, signage, written policies, and regular assessments of staff knowledge. As for every safety regime, it is essential to work through some "what ifs" (i.e., what if a key staff member is absent, what if we run out of a particular ingredient) to assess worst case scenarios and be prepared for contingencies.

7.13.2. Verification and Validation

Whether the HACCP or simpler hazard analysis process is employed, it is essential to undertake regular audits and reviews of staff knowledge and understanding of the controls in place. Food safety management plans usually include a section for "lessons learned," which serve to improve controls for the future.

7.13.3. Setting Management Targets

To be effective, any management system needs to be practical and achievable. Any target set will need to be realistic and ideally measurable. While some food manufacturers may be able to verify their factory controls and validate cleaning processes (e.g., by sending product and swab samples for laboratory testing), this is rarely practical in foodservice.

Instead, it is worth setting targets based on what is easier to assess and what is already in place:

To date, there is limited peer-reviewed data about the effectiveness of cleaning for allergen controls in catering situations [33]; but this research, together with history and experience, indicates that even the most highly sensitive food-allergic people are usually protected by good hand washing, best practice commercial dishwashing, and common sense work methods for not spreading allergens around inadvertently. The U.K. laboratory Reading Scientific Services Limited (RSSL) and the Anaphylaxis Campaign undertook a preliminary study of allergen cleaning in 2006, which indicated the following:

- Peanut and hazelnut proteins are highly tenacious even following rigorous chemical/mechanical treatments.
- Milk proteins are slightly easier to remove.
- Automatic washing is generally better than manual bowl washing for nut proteins.
- Wooden and used chopping boards are extremely difficult to get clean.

- High levels of contamination are picked up and transferred by sponges and cloths used.
- Detergents are mildly better than hot water alone at removing allergens bound in high fat matrices.
- Nut and milk proteins are well removed from hands by soap.

Both the Perry research and feedback from the RSSL study can be used to support best practice washing and cleaning in foodservice.

Foodservice personnel will be more willing and able to address food allergy risks once they understand that much of the effort and many of the controls they are already implementing, for example, regular hand washing, good cleaning, segregation of particular foods, and product identification and labeling will also be effective as allergen controls. This applies to the foodservice sector as well as in manufacturing.

Ahuja and Sicherer [34] reminded us, however, in their recent study that there is no room for complacency among food personnel or consumers at risk. Twenty-four percent of foodservice staff questioned in their study believed that a small amount of the allergen would be safe, and 35% believed that fryer heat would destroy allergens. This is one particular example where controls in place for other food safety reasons will not be effective for allergens (e.g., cooking in a hot wok).

7.14. FOOD ALLERGY AND THE LAW

As discussed previously, it is increasingly common for food allergen controls to be addressed and supervised by those responsible for both food safety and food labeling controls, both in-house and externally by local and national authorities. Although there has been reluctance by some government bodies to direct food control authorities to address food allergen risks in their everyday activities [15, 35, 36], recent developments have been more encouraging [37].

It is awkward that effective food allergen avoidance is dependent on a number of different legal contexts:

- food safety law—relating to protecting foods on sale from contamination and degradation to protect health;
- food standards law—ensuring that the consumer is sold exactly what he or she expects and that information available is accurate and up to date;
- traceability law—relating to being able to trace foods up and down the food supply chain in case a recall is necessary;
- health and safety law—both at work and for the general public who may be deliberately or inadvertently exposed to food allergens (not just in foods, but also in personal care and other products);

- general product safety law—*catchall* legislation to protect consumers from any risks in anything they may buy, for example, by providing information about potential hazards of use; and
- civil law—the civil responsibility of one person or business not to cause harm to another person or business. There have been civil claims following allergic reactions to foods.

7.14.1. The Principle of Due Diligence

This principle is widely recognized in the U.K. food sector. If a food supplier is prosecuted under the food law, then evidence of the management system and controls effective at the time of the offense may be used in mitigation to reduce any penalty. In practice, this means that it is worthwhile for a business to implement and maintain good food safety management systems including those for allergen control. This may also lead to a reduced risk of legal action being taken against the food business operator, which may, in turn, lead to reduced premiums for business liability insurance.

7.14.2. Current and Recent Food Control Authority Initiatives

Early initiatives to support food businesses in the control of food allergens and the adoption of best practice included the Canadian Allergy Awareness Programs [38], food allergy training delivered on behalf of various U.K. local authorities by Allergy Action [10], and workshops developed for Rochdale Metropolitan Borough by environmental health professional Jackie Hall in 2004 [39]. More recently, in 2006–2007, the U.K. Food Standards Agency commissioned and coordinated a program of food allergen online training and workshops for over 1000 food control officers, which was developed and delivered by Allergy Action and partner company, Hygiene Audit Systems [40]. The same partnership also delivered a national conference and seven courses for the Co-operation and Working Together (CAWT) and Safefood organizations on the island of Ireland in 2007. The allergy training DVD training pack stage 1 [41], which was produced by the same team, is in regular use by food control teams throughout the United Kingdom and beyond.

In 2006–2007, the U.K. Food Standards Agency also developed and distributed guidance on the provision of allergen information for non-prepacked foods [42], together with a pictorial guidance leaflet [41] and a laminated kitchen poster [43] to raise staff awareness.

In addition to proactive awareness-raising activities and training, some authorities (e.g., Kent County Council [44], Association of Public Analysts [45]) have also been involved in taking food samples both formally and informally for analysis and communicating the results of such projects in order to alert both consumers and food business operators to the possibility of allergen risks [22, 45, 46].

In the UK in 2009, a market trader was convicted of selling "unsafe food"—in this case a prepacked box of chocolates, which was not labeled in English (i.e., a language easily understood)—and therefore represented a risk to food allergic people. In 2010, a UK takeaway business owner was convicted of selling "unsafe food" which contained peanuts to a customer with a declared peanut allergy. Both prosecutions were taken under EU Food Safety regulation 178/2002 EC (Article 14) and represent legal milestones in using food safety law (and not just food labeling/composition law) to protect food allergic people [47].

7.15. BEST PRACTICE GUIDANCE

While the guidance above is freely available online and other national guidance is also available for use in foodservice, the following notes may be helpful in developing food allergen management plans and the assessment, management, and control of food allergen risks.

7.15.1. Identifying Allergenic Contaminants in Incoming Materials

- Supplier questionnaires and audits—approved supplier lists.
- Detailed information required about both ingredients and possible allergen contaminants in foods bought in.
- Detailed delivery checks to ensure that what is delivered matches exactly what is ordered. Any discrepancies are embargoed until further checks have taken place.
- Contingency shopping—what to do when the usual ingredient runs out?
- Foreign language labeling and specifications, checking poorly translated ingredients for sense.

7.15.2. Dealing with Bulk Unlabeled Ingredients

- Managing information to prevent confusion
- Controlled decanting into smaller containers, keeping foodstuffs covered, not redecanting into dirty/wrong containers
- Managing spillage—prioritizing cleaning for allergen spillage
- Careful management of labeling of part-prepared foods
- Ensuring that utensils such as scoops, scales, and tin openers are washed, rinsed, and dried thoroughly before reuse

7.15.3. Preventing Cross-Contamination during Food Preparation

- What kind of segregation is possible/practical?
- Can you segregate by time, for example, make something before other foods are used or after a clean down?
- Can you segregate by space—work elsewhere?

- Be alert to cross-contamination from dirty cloths and overalls, and poorly washed hands and utensils.
- Service and maintain dishwashing and pot washing equipment and ensure that the appropriate cleaning products and temperatures are used.
- Keep lids on all foods when not in use.

7.15.4. Food Handling and Displays in Supermarkets

7.15.4.1. Fresh Produce—Vegetables, Fruit, and Nuts

- Need to be recognized at point of selection and by staff at the till.
- Be alert to spillage and nuts being cracked on the floor.
- Small children may pick up nuts, seeds, and other allergens and even put them in their mouths.

7.15.4.2. Stickers

- Be careful not to put reduced price stickers over key information, for example, ingredients or "may contain" warnings.

7.15.4.3. Bakery and Patisserie

- If served by staff, they should have good product knowledge and have information about ingredients and allergens in the preparation environment to hand.
- If served by customers, they need to match exactly which product is which.
- Tongs and other utensils may need to be tied down to prevent reuse for another product.
- Stickers for self-pricing should carry a generic "may contain" warning message to alert other people who may consume the food later that this food was sold in an open bakery environment.

7.15.4.4. Selling Fish and Shellfish

- In addition to the problems of transferring leakage from wet fish and shellfish to other surfaces, it is important that the staff serving them are aware of the possibility of cross-contamination between fish and shellfish.
- Once again, the staff need to know their products and be ready to provide customer information.

7.15.4.5. Delicatessen Counters

- Ready-to-eat pies and pastries—staff should be aware of the possibility of wheat, gluten, soya, milk, and egg contamination (for example) to other simpler foods, for example, cold meats.

- Cold meats and ready-to-eat salamis, sausage, and so on—these can sometimes contain unexpected ingredients that are not easily recognizable—for example, mortadella, which contains small pieces of pistachio nut and other sausages, which contain walnuts, mustard, and a range of other nuts and seeds.
- Prepared salads—for example, coleslaw, potato salad—if served by staff, then they will need to be able to find out all ingredients and prevent cross-contamination, for example, by using clean/different serving utensils. If served by customers, notices should raise awareness of cross-contamination risks from using the wrong utensil. Once again, stickers should carry a generic warning that the food came from an open environment with a range of other food allergens.
- Cheese—although customers may think that cheese is a simple food, it can also contain other allergens such as lysozyme, which is derived from egg.

7.15.4.6. Pick 'n' Mix

- Once again, self-service of open foods carries particular risks for those with food allergies. Utensils may be mixed up. Labels may not have been changed when the containers were refilled, and containers may not have been washed between different products.
- This is particularly relevant for unwrapped dried fruit, nuts, sweets, and chocolates.
- However, it is also important to be aware that the foils and wrappers on confectionery may change. A young girl died after eating a Brazil nut chocolate in error because the color of the foil had been changed by the manufacturer.

7.15.5. Basic Training Requirements for Staff

- Food allergy training should be developed to match training needs identified in the food allergy management system.
- All staff need some allergy training in line with their activities and responsibilities.
- Staff training records should be monitored and maintained.
- The first principle of allergy training is that if you are asked for information that you do not have, you must *never guess*.
- Team leaders and supervisors should be supported to undertake formal and informal training for their staff.
- They also need to assess staff competence in the workplace through routine checks and discussions with colleagues.

- They should be provided with information and relevant materials to support this learning.
- Materials should be accessible and visual where practical to overcome any language or learning difficulties.
- Any allergy-related incident should be recorded and discussed.
- Workplace displays and signage can support best practice and improve awareness.
- Managing an allergic emergency—in particular, early recognition of allergic symptoms and the need to ensure that paramedics are brought as quickly as possible to the person at risk. Having clear written directions by the phone may bring help more quickly.

7.15.6. Key Messages Used in Allergy Training Courses [41]

- Know your ingredients and recognize allergens including nuts, peanuts, and seeds.
- Invite dialogue and be ready to provide information to customers who ask.
- Control allergen cross-contamination where possible.
- Tell the truth about a food. Never guess.
- Have an emergency protocol—thinking ahead can save a life.

7.16. CONCLUSION

Those at risk from IgE-mediated food allergies need evidence-based clinical diagnosis and guidance on which foods they should avoid and how. They may also need advice on how to recognize their own allergy risk status and emergency management.

Allergen management and control in the foodservice industry can be implemented effectively for increasing numbers of people at risk from unpredictable IgE-mediated food allergies. This will also be beneficial for others who need to avoid foods including those with intolerances and celiac disease.

- The starting point for any allergen management system is good food information—recipes, ingredients, and written product information are correct and up to date.
- Caterers and other food businesses who already undertake hazard analysis and have controls in place for other risks (e.g., hand washing, cleaning, segregation between ingredients, or part-finished foods) should examine these closely and identify those that are also effective allergen controls. Their food safety management plan should be extended to recognize the double benefit of carrying out these controls effectively.

- All staff should be aware of their role in managing allergens and controlling cross-contamination. They should have regular work-related update training to ensure that they really understand the risks and appropriate controls.

ACKNOWLEDGMENTS

The author acknowledges Keith Gowland; colleagues at the Anaphylaxis Campaign and, in particular, Moira Austin; Dr. Richard Pumphrey, consultant immunologist; Dr. Lisa Ackerley, expert in food safety and allergy risks; and Kate Grimshaw, food allergy research dietitian and expert in food allergen avoidance.

REFERENCES

1. Anaphylaxis Campaign UK. (2007). Memorandum of Evidence to the UK House of Lords Science and Technology Committee Investigating Allergy. Minutes of evidence ordered to be reported July 24th, 2007. http://www.publications.parliament.uk/pa/ld200607/ldselect/ldsctech/166/7013108.htm.
2. Gowland, M.H. (2002). Food allergen avoidance—risk assessment for life. *The Proceedings of the Nutrition Society*, *61*, 39–43.
3. Warde, A., Martens, L. (2000). *Eating Out: Social Differentiation, Consumption and Pleasure*. Cambridge, UK: Cambridge University Press.
4. UK Office of National Statistics. (2006). Spending on eating out overtakes meals at home. *News Release: United Kingdom Input-Output Analyses*, August 18, pp. 1–8.
5. Council Directive 79/112/EEC of 18 December 1978 on the approximation of the laws of the Member States relating to the labelling, presentation and advertising of foodstuffs, Articles 3, 4 and 12. (1979). *Official Journal of the European Communities*, *L33*, 1–14.
6. M.H. Gowland facilitating discussion with catering students at Brixham Community College, Devon, UK. (2007). Unpublished interviews.
7. Food Standards Agency. (2005). Draft of guidance on allergen management and consumer information. Available at http://www.reading.ac.uk/foodlaw/pdf/uk-05052-allergens.pdf.
8. The Science and Technology Committee. (2007). *Sixth Report of Session 2006-07 on Allergy*, Vol. I. London: Authority of House of Lords, p. 31. Available at http://www.publications.parliament.uk/pa/ld200607/ldselect/ldsctech/166/16607.htm#a26.
9. Coeliac UK Website. Myths about coeliac disease. Available at http://www.coeliac.org.uk/coeliac-disease/myths-about-coeliac-disease.
10. Gowland, M.H. (2008). Slide presentation Allergy Action—Advice. Available at www.allergyaction.org.
11. Gowland, M.H. (2008). Personal correspondence with Dr. Michael Gallagher.

12. Food Standards Agency. (2005). Qualitative research into the information needs of teenagers with food allergy and intolerance. *FSA/0991/0905*. Available at http://www.food.gov.uk/multimedia/pdfs/foodinfoteens.pdf.

13. Food Standards Agency. (2002). 'May contain' labelling—the consumer's perspective. *FSA/0582/0502*. Available at http://www.food.gov.uk/multimedia/pdfs/maycontainreport.pdf.

14. Pumphrey, R.S.H., Gowland, M.H. (2007). Further fatal allergic reactions to food in the United Kingdom, 1999-2006. *The Journal of Allergy and Clinical Immunology*, *119*, 1018–1019.

15. Gowland, M.H. (2001). Food avoidance—the patient's viewpoint. *Allergy*, *56*, 117–120.

16. Sampson, H., Muñoz-Furlong, A., Sicherer, S. (2006). Risk-taking and coping strategies of adolescents and young adults with food allergy. *The Journal of Allergy and Clinical Immunology*, *117*, 1440–1445.

17. Grimshaw, K.E.C., King, R.M., Nordlee, J.A., Hefle, S.L., Warner, J.O., Hourihane, J.O. (2003). Presentation of allergen in different food preparations affect the nature of the allergic reaction—a case series. *Clinical and Experimental Allergy*, *33*, 1581–1585.

18. Uguz, A., Lack, G., Pumphrey, R., Ewan, P., Warner, J., Dick, J., Briggs, D., Clarke, S., Reading, D., Hourihane, J. (2005). Allergic reactions in the community: a questionnaire study of member of the anaphylaxis campaign. *Clinical and Experimental Allergy*, *35*, 746–750.

19. Roberts, G. (2007). Anaphylaxis to foods. *Paediatric Allergy and Immunology*, *18*, 543–548.

20. Pumphrey, R.S.H. (2000). Lessons for management of anaphylaxis from a study of fatal reactions. *Clinical and Experimental Allergy*, *30*, 1144–1150.

21. Derby, C.J., Gowland, M.H., Hourihane, J.O. (2005). Sesame allergy in Britain: a questionnaire survey of members of the anaphylaxis campaign. *Paediatric Allergy and Immunology*, *16*, 171–175.

22. BBC News. (2004). Allergy warning about takeaways. August 6. Available at http://news.bbc.co.uk/2/hi/uk_news/england/lancashire/3540268.stm.

23. Leitch, I., Walker, M., Davey, R. (2005). Food allergy: gambling your life on a take-away meal. *International Journal of Environmental Health Research*, *15*, 79–87.

24. Mail Online. (2007). Tikka masala killed teenage girl. December 6. Available at http://www.dailymail.co.uk/news/article-500129/Tikka-masala-killed-teenage-girl.html.

25. Yu, J.W., Kagan, R., Verreault, N., Nicolas, N., Joseph, L., St Pierre, Y., Clarke, A. (2006). Accidental ingestions in children with peanut allergy. *The Journal of Allergy and Clinical Immunology*, *118*, 466–472.

26. Bourgon, L. (2007). In a nutshell: what to do about food allergies in schools. *CBC News*. December 13. Available at http://www.cbc.ca/news/background/foodsafety/allergies-school.html.

27. Muñoz-Furlong, A. (2003). Daily coping strategies for patients and their families. *Pediatrics*, *111*, 1654–1661.

28. Reens, N., McClellan, D. (2008). Inmate never mentioned deadly peanut allergy, Montcalm jail officials say. *The Grand Rapid Press*, April 22. Available at http://blog.mlive.com/grpress/2008/04/inmate_never_mentioned_deadly.html.
29. MacAuslan, E.M.R. (2004). Overcoming the barriers. *Environmental Health Journal*, *3*, 16–18.
30. White, A. (2003). *Market Research Intelligence Report on Food Intolerance and Allergies*. London: Mintel International Group Ltd, Chapter 3, Section 1.
31. Food Standards Agency. (2007). Allergy—what to consider when labelling food. *FSA/1105/1206*. Available at http://www.food.gov.uk/multimedia/pdfs/allergyleaflet.pdf.
32. The Australian Food and Grocery Council. (2007). Food industry guide to allergen management and labeling. Available at http://www.allergenbureau.net/allergen-guide/afgc-allergen-guide/.
33. Perry, T.T., Conover-Walker, M.K., Pomès, A., Chapman, M.D., Wood, R.A. (2004). Distribution of peanut allergen in the environment. *The Journal of Allergy and Clinical Immunology*, *113*, 973–976.
34. Ahuja, S., Sicherer, S. (2007). Food-allergy management from the perspective of restaurant and food establishment personnel. *Annals of Asthma and Immunology*, *98*, 344–348.
35. Leitch, I., Blair, I., McDowell, D. (2001). The role of environmental health officers in the protection of allergic consumers. *International Journal of Environmental Health Research*, *11*, 51–61.
36. Rothstein, H. (2005). Escaping the regulatory net: why regulatory reform can fail consumers. *Law and Policy*, *27*, 520–548.
37. UK Parliamentary Business. (2007). House of Lords Science and Technology—sixth report. Available at http://www.publications.parliament.uk/pa/ld200607/ldselect/ldsctech/166/16602.htm.
38. de la Rue Browne, S. (2002). Allergy awareness alert. *Environmental Health International*, *4*, 5–7. Available at http://www.ifeh.org/magazine/ifeh-magazine-2002_v4_n2.pdf.
39. Hall, J. (2004). Determination of the views of consumers with food allergies and an assessment of the awareness of food allergies in small-medium enterprises. A scientific short paper, University of Central Lancashire. Available at http://www.rochdale.gov.uk/PDF/2007-07-24-food-allergy-scientific-paper.pdf.
40. Hygiene Audit Systems. (2009). Available at www.hygieneauditsystems.com.
41. Gowland, M.H. Training organization providing food allergy courses and resources. www.allergytraining.com.
42. Food Standards Agency. (2008). Provision of allergen information for non pre-packed foods. *FSA/1226/0108*. Available at http://www.food.gov.uk/multimedia/pdfs/loosefoodsguidance.pdf.
43. Food Standards Agency. (2007). Pictorial guidance leaflet (FSA/1201/0108). Available at http://www.food.gov.uk/multimedia/pdfs/publication/loosefoodsleaflet1109.pdf.
44. Kent County Council. (2008). Analytical report—sweet and flour confectionary—nut allergy. Available at http://www.tradingstandards.gov.uk/kent/Analyst%20reports/sweetflourconfec.htm.

45. Association of Public Analysts. (2003). Sampling and analysis for food allergens. Available at http://www.publicanalyst.com.

46. Food Standards Agency. (2007). Think allergy. Laminated kitchen poster (FSA/1215/0108). Available at http://www.food.gov.uk/multimedia/pdfs/publication/thinkallergy.pdf.

47. Article 14 (2002). Regulation (EC) No. 178/2002 of the European Parliament and of the Council of 28 January 2002 laying down the general principles and requirements of food law, establishing the European Food Safety Authority and laying down procedures in matters of food safety. Official Journal of the European Communities, L 31, 10.

PART III

PROCESSING FOODS FREE FROM SPECIFIC ALLERGENS

8

PROCESSING FOODS FREE FROM DAIRY PROTEINS

JOYCE I. BOYE, SAHUL H. RAJAMOHAMED, AND MICHEL BRITTEN

8.1. INTRODUCTION

Food allergies have increased at a remarkable rate in the industrialized world and are estimated to affect 4–8% of children and about 3–4% of the adult population [1–5]. Eight foods (the "big eight") (i.e., cow's milk, egg, peanut, soy, tree nuts, crustaceans, fish, and wheat) are classified as priority allergenic foods because they represent the most common foods to which consumers are sensitized. Cow's milk allergy (CMA) poses a particularly serious challenge because cow's milk is consumed by a large percentage of the population, is used in the preparation of many different food products, and is a major allergen for infants and children. There are currently no cures for food allergy; thus, CMA patients must be able to identify foods that contain milk proteins and avoid them in their diets. Milk proteins have excellent nutritional and functional properties, and since they are the culprit components in dairy products, their avoidance is particularly bothersome. The chapter will highlight some of the major dairy foods and ingredients commonly used by the food industry, and the major and minor allergenic proteins in milk, and will attempt to provide a list of ingredients and foods that could be used as alternatives to dairy in the formulation of foods labeled as "free from" dairy or "nondairy." Due to the widespread use of dairy and dairy ingredients in food formulation, the chapter will begin with an extensive overview of dairy products and ingredients currently available on the market, and a summary of their properties and some of the processes used for their production.

Allergen Management in the Food Industry, Edited by Joyce I. Boye and Samuel Benrejeb Godefroy

8.2. OVERVIEW OF DAIRY INGREDIENTS, THEIR PROCESSING, AND CHARACTERISTICS

Consumers appreciate dairy foods for their nutritional value and sensory properties. A wide variety of fluid milks, creams, butters, cheeses, and yogurts are available on the market. Dairy products are also used as ingredients in the preparation of cooked meals, spreads, dips, desserts, and sauces. Development of large-scale separation processes supported the emergence of a strong dairy ingredient industry. Today, milk fractions are isolated from milk and used to improve the nutritional or sensory properties of formulated foods. Apart from whole dairy foods, the main classes of dairy ingredients used in food formulation are milk protein concentrates made of casein (CN) and whey protein fractions. These are described in greater detail below.

8.2.1. Dairy Foods

All mammalian species produce milk to feed their young. Since ancient times, the milk from cow, goat, sheep, camel, buffalo, and others has served as a source of nutrition for humans. In many countries today, however, cow's milk has become the major source of dairy foods. Whole milk from cows contains about 3% protein, 3.5–4% fat, and about 5% carbohydrate. Dairy foods are usually considered as final products directly targeted to the consumer. However, a significant portion of dairy products are also used as ingredients in many different processed foods and may be added to food formulations to improve nutritional value, taste, color, or mouthfeel.

Cream, cream cheese, and butter are examples of high-fat products obtained from milk. They provide texture, creaminess, and typical dairy notes to processed foods. Milk fat is characterized by a sharp melting profile. At 5°C, the solid fat fraction is about 60%, which decreases to almost 0% at mouth temperature [6]. Fat crystals interact with each other and contribute to the firmness of foods at low temperature. During mastication, rapid melting of fat alters the rheological character of food, which changes from elastic to viscous. Fat melting absorbs thermal energy and induces a coolness perception when fat-rich dairy products are consumed. Additionally, aeration of creams with more than 30% fat results in the formation of soft foam, stabilized by fat crystals aggregated at the surface of tiny air bubbles [7]. These products are used as whipped cream toppings, and they provide a rich texture and taste to different types of desserts. Milk fat is also a reservoir of characteristic aroma compounds, which are slowly released during food mastication.

Natural, processed, and powdered cheeses are popular ingredients in food formulations. Use of cheese as pizza topping is a widespread application. Shredded cheese melts during cooking to form a homogeneous and elastic coating at the surface of the pizza. The melting properties depend on cheese moisture, mineral balance, and degree of proteolysis. Residual sugars present in cheese facilitate the Maillard reaction and the development of a light brown color. Furthermore, melted cheese releases enough oil to produce a shiny surface and prevent excessive dehydration during cooking. Additionally, cheese powder provides a cheesy flavor to processed foods and is often used as a dry coating ingredient for snacks and bakery products or in the preparation of dry sauce mixes.

Yogurt, a fermented milk product, is characterized by a typical acid taste and a complex aroma profile. Lactic acid produced by the bacterial cultures induces the aggregation of CN micelles and the formation of a soft gel [8, 9]. Yogurt is also available in powdered form and can be used to provide texture and taste to salad dressings and dips and is also a popular ingredient in the formulation of sweet coating for nuts, dry fruits, and nutritional bars.

8.2.2. Milk Powders, Coprecipitates, and Milk Protein Concentrates

Milk is highly perishable, and surplus milk is spray dried to extend its shelf life. Large-scale processes have been developed to produce milk protein concentrates with different compositions and characteristics. Milk and milk protein concentrate powders are easy to use and can be integrated in a large number of food formulations to improve nutritional and sensory qualities.

Whole milk and skim milk powders are produced by evaporation, concentration, and spray drying. These powders are characterized by a high concentration of lactose (more that 50% in skim milk powder). Protein precipitation or membrane separation processes can be applied to milk prior to dehydration to reduce lactose and increase protein content. CN micelles can be disrupted by acidification to form a precipitate at their isoelectric point (pH 4.6) [10]. Since native serum proteins remain soluble at low pH, heat treatment can be applied with the addition of calcium prior to acidification to facilitate the denaturation of serum proteins and their coprecipitation with CNs at low pH. Lactose and minerals are then washed out to obtain a coprecipitate containing 85–90% protein, <5% lactose, and 3–7% ash [11].

Membrane filtration is another technique frequently used to produce total milk protein concentrates [12]. Ultrafiltration membranes with appropriate pore size allow the selective removal of lactose and soluble minerals and retention of milk proteins. Unlike the coprecipitation process, membrane filtration preserves the native state of milk proteins. Milk protein concentrates or isolates processed this way, with protein concentrations ranging from 50% to 90%, are available to food processors. In some instances, larger pore size membranes (microfiltration) can also be used to remove a portion of soluble protein [13]. Concentrates obtained by microfiltration, therefore, contain higher concentration of micellar CN. These concentrates are often called "native phophocaseinate."

Milk powders and protein concentrates have a bland flavor and do not alter the flavor profile of processed foods. They form colloidal dispersions in aqueous media, with strong light scattering power. CN, the major milk protein, is heat stable [10] and makes milk protein ingredients well adapted for cooking applications. At neutral pH, the solubility of low-heat skim milk powders and protein concentrates is higher than 95% [13]. In coprecipitates, serum protein denaturation is responsible for poor solubility. However, increasing the pH and using calcium chelating salts, such as phosphates and citrate, can restore solubility. Milk protein ingredients interact with water, affecting the viscosity or the firmness of final food products. The amphiphilic nature of native milk proteins provides them with excellent emulsifying and foaming properties [10]. They adsorb at oil/water or air/water interfaces and prevent emulsion coalescence or foam collapse.

Essential amino acids represent ~40% of total amino acids in milk proteins [14]. Digestive enzymes are very well adapted to milk proteins, and 95% of ingested protein is absorbed during intestinal transit [15]. The nutritional value of milk protein is, therefore, much higher than that of vegetable proteins [16, 17], but they contain low concentrations of sulfur-containing amino acids, which slightly reduce their nutritional value. CN micelle is a natural carrier for calcium and phosphorus. In milk, two-thirds of total calcium is associated with CN micelles as colloidal calcium phosphate [18]. Milk powders and milk protein concentrates are, therefore, excellent sources of calcium. However, for milk protein coprecipitation, the acidification treatment dissolves colloidal calcium phosphate, which is often lost in the process.

Large amounts of milk powders and milk protein concentrates are used by the dairy industry to standardize the protein content of cheese milk, yogurt, and dairy beverages. These ingredients are also used to impart whiteness and opalescence to foods and drinks. They stabilize fat dispersions and interact with gums and starches to reduce water mobility, increase viscosity, and reduce syneresis in sauces and dressings. They also reduce water loss during cooking of bakery products. Total milk protein ingredients also improve the nutritional value of processed foods and are extensively used in the formulation of nutritional bars and drinks.

8.2.3. CN Ingredients

CN and serum proteins represent respectively 80% and 20% of total milk protein [10, 19]. The characteristics of these two classes of protein are very different and justify their separation. CN can be extracted from milk in various forms to meet the requirements of food processors.

CN is coagulated from nonheated milk by acidification or enzymatic treatment. Fermentation or direct acid addition is used to reduce milk pH. Electrodialysis, ion exchange resins, and carbon dioxide injection are other available processes to acidify milk [20]. During acidification, calcium and phosphorus originally bound to CN micelles dissociate, and CN precipitation occurs at pH < 5.0 [21]. The precipitate, called acid CN, is then separated from whey, which contains serum proteins. After washing and pressing, acid CN is dried on a fluid bed, attrition, or ring dryer. Acid CN is hardly dispersible in water, and it is common practice to convert it to caseinates prior to spray drying. CN conversion is obtained by adding a strong alkali to bring the pH up to 6.7 and restore solubility. For that purpose, sodium hydroxide is used to produce sodium caseinates. Other hydroxides can be used to produce potassium, ammonium, or calcium caseinates. Milk-clotting enzymes are also used to coagulate CN without acidification. Rennet is also commonly used for that purpose to produce the so-called rennet CN. Unlike acid CN, rennet CN retains the calcium and phosphorus originally bound to the micelles. The CN family is formed by four major fractions: α_{s1}-, α_{s2}-, β-, and κ-CNs [10, 19]. Efforts have been made in the past to develop processes to separate these fractions, but only the β-CN fraction is available on a commercial basis. Enzymatic hydrolysis is also used to produce peptides from CNs. Crude CN hydrolysates are available and can be further purified to develop high-value peptide fractions for specific applications.

CN and caseinates obtained in optimal conditions have a bland flavor, but the use of low-quality milk or excessive temperatures or pH during production can result in major flavor defects. Bitter flavors are often associated to CN hydrolysates. Acid and rennet CNs are not soluble in water, but sodium caseinate is soluble and produces viscous and translucent dispersions. The structure of dissociated CNs is highly disordered and strongly interacts with water. Calcium induces the association of CNs and calcium caseinate exists as small and cohesive aggregates [22], resulting in opaque and low-viscosity dispersions. Soluble caseinates (sodium, potassium) are resistant to heat treatments [23] and have excellent interfacial properties. Furthermore, CN is rich in essential amino acids and contains phosphoserine residues, which interact with calcium and might facilitate its absorption in the small intestine [24]. Specific biological activities have also been associated with peptides produced during CN digestion in the human body. Opioid, antihypertensive, antibacterial, and immunomodulating activities are associated with peptides from β- and α_{s1}-CNs [25].

CN and caseinates are used in the preparation of dairy product analogs. Rennet CN is heated with water and chelating salts to emulsify vegetable oil in the preparation of cheese analogs, while sodium caseinate is used to stabilize fat in coffee creamers or whipped cream analogs. CN ingredients are used in bakery products to increase water-holding capacity. It is also used to improve the amino acids profile of breakfast cereals, cookies, pasta, and bread. Wheat protein is deficient in lysine, and adding 4% CN to wheat flour increases the lysine content by 60% [26]. Sodium caseinate is widely used in processed meat industry. It stabilizes fat to produce fine emulsions and improves heat gelation of meat proteins [27]. Sodium caseinate reduces fat and water losses during cooking [27]. Additionally, CN ingredients find applications in the formulation of dietary products. Hyperproteic food products, containing CN ingredients, are used to control obesity, and patients with gastrointestinal disorders are sometimes fed with CN hydrolysates [28, 29]. CN phosphopeptides may also be added to calcium supplements to increase intestinal absorption [24], and a CN peptide extract is commercialized to reduce anxiety [30].

8.2.4. Whey Protein Ingredients

Whey is the by-product of cheese and CN industries. Industrial processes were first developed to recover serum proteins from whey in order to reduce its negative impact on the environment. Today, high-value ingredients are produced from whey and used in the food processing industry.

Thermal precipitation is the first industrial process used to isolate whey proteins. Whey is heated at 90°C for more than 10 minutes, followed by pH adjustment to values between 4 and 5. The protein precipitate is then washed, recovered by centrifugation, and dried. The fully denatured whey protein ingredient obtained by this process is called lactalbumin [10]. Ultrafiltration is an alternative technology to concentrate whey proteins in their native state. Ingredients with protein concentrations ranging from 35% to 85% are produced by ultrafiltration. In order to produce whey protein isolate (WPI) (>90% protein), residual lipids must be removed from whey by microfiltration treatment. Techniques such as ion exchange chromatography are also used at the industrial scale for the production of high-purity WPIs.

Different processes have been developed to isolate individual proteins from whey. Selective precipitation, large pore ultrafiltration, and ion exchange chromatography are used to isolate the major proteins β-lactoglobulin (β-Lg), α-lactalbumin (α-la), and glycomacropeptide from cheese whey [10]. Specialized techniques such as cation exchange chromatography are used for the purification of minor proteins such as lactoferrin (LF) and lactoperoxidase [31]. Additionally, enzymatic whey protein hydrolysates are produced from whey or the individual protein isolates for use in food applications.

Whey protein ingredients, with low-fat content, have a neutral flavor. Lipids are sensitive to oxidation, which could induce off-flavor development in whey protein ingredients with significant fat content. Lactalbumin, made from fully denatured whey proteins, is hardly soluble in aqueous systems and shows poor emulsifying, foaming, and gelling properties. Whey protein concentrates and isolates are soluble in water and produce low-viscosity dispersions at any pH. The globular structure of whey proteins is sensitive to heat treatment, which induces protein unfolding and aggregation. Depending on pH, ionic strength, and protein concentration, a gel, a colloidal dispersion, or a precipitate can form after heating. Whey proteins have excellent foaming properties [32]. They adsorb and spread out at the air/water interface to form stable protein films. Heating whey proteins induces gelation [33] and further increases the stability of whey protein foams [33]. Compared with CNs, whey protein contains more than twice the amount of sulfur-containing amino acids and is an excellent source of branched amino acids [34]. Whey protein ingredients have been shown to increase intracellular glutathione concentration (a powerful antioxidant) in human cells [35]. Specific biological activities have been associated with individual whey proteins and their derived peptides [25].

The dairy industry uses whey protein ingredients to reduce formulation costs. Adding heat-treated whey protein concentrates to cheese milk significantly increases the yields. Specially designed whey protein concentrates are used to improve the texture of ice cream and reduce the syneresis in yogurts. Whey protein concentrate (35% protein) is used as skim milk replacer in food formulations. The heat-induced gelation properties of whey proteins are most appreciated in bakery and meat product applications. Aggregated whey protein is commercially available to replace fat and provide creaminess in low-fat foods (e.g., Simplesse). Egg white substitutes are also produced from WPIs for foaming applications [36, 37]. In recent times, sport nutrition has created an expanding market for whey protein ingredients. As indicated above, whey protein contains branched amino acids (valine, leucine, isoleucine), which have been shown to support muscle development [38]. Additionally, whey proteins and derived peptides are extensively used in protein-rich diets and nutritional supplements. Due to the excellent nutritional properties and functionality, dairy foods and ingredients are used in probably every food category as is evident from the information provided above.

8.3. MILK PROTEIN ALLERGENS

CMA is an immunoglobulin E (IgE)-mediated reaction to proteins in cow's milk, which may induce cutaneous (atopic dermatitis, urticaria, angioedema), respiratory (rhinitis, asthma, cough) and gastrointestinal (vomiting, diarrhea, colic, gastroesophageal reflux) reactions, and, in some extreme cases, even systemic anaphylaxis [39]. Allergy to cow's milk is one of the most common challenges especially in early childhood with an incidence of 1–3% during the first year [40, 41]. While the factors contributing to CMA are not completely understood, it has been suggested that factors such as low stomach pepsin activity at birth, an immature stomach acid generating mechanism, and malfunction of pancreatic and intestinal enzymes may contribute to the stability of cow's milk proteins by limiting their gastric proteolysis, thereby, exposing neonates to intact protein and a higher risk of allergic responses [42]. Other factors such as genetic predisposition, early exposure to foreign food protein (time, dose, and frequency), method of food preparation, and gastrointestinal permeability have been cited as potential contributing factors to the development of food allergy. CMA is normally outgrown in the first 3 years of life [41, 43]; however, some allergic children remain allergic for much longer periods, which can persist into adult years [44]. The mechanisms responsible for the development of tolerance remain unclear.

Cow's milk consists of several proteins that are considered to be antigenic and capable of inducing adverse immune responses. Among them β-Lg and CNs have been classified as major allergens, while α-la, bovine serum albumin (BSA), LF, and immunoglobulins (Igs) are known as minor allergens [39]. The structure and properties of these proteins are described in greater detail below.

8.3.1. Major Cow's Milk Allergens

8.3.1.1. CN. CN accounts for about 80% of the total protein content in milk. The principal CN protein fractions are α_{s1}-CN (32%), α_{s2}-CN (10%), β-CN (28%), and κ-CN (10%) [19, 45]. These proteins are coded by different genes carried on the same chromosome and have little primary structure homology [46]. The CNs, however, have some common features. They are phosphorylated proteins, have a loose tertiary texture, and have highly hydrated structure due to the absence of disulfide bonds with high numbers of proline residues and the presence of hydrophobic amino acids comprising nearly 50% of the protein [19]. Due to their structure, CNs are not significantly affected by severe heat treatments but are very susceptible to all proteinases and exopeptidases [19]. After ingestion, most of the potentially antigenic or allergenic structures are assumed to be modified under acidic gastric conditions by the action of digestive enzymes and uptake through the intestinal mucosa [47]. Six major and three minor IgE-binding epitopes, as well as eight major and one minor immunoglobulin G (IgG)-binding regions, were identified for β-CN, while two major and two minor IgG-binding epitopes were found for κ-CN using the sera of 15 milk-allergic children [48]. Jarvinen et al. [49] reported five IgE-binding epitopes (two on α_{s1}-CN: AA123-132 and AA 69-78; one on α_{s2}-CN: AA171-180; and two on κ-CN: AA155-164 and AA13-22) recognized by patients with persistent CMA. Milk-allergic children with persistent symptoms were also found to have significantly higher levels of specific IgE antibodies to linear epitopes from α_{s1}-CN (AA69-78) and β-CN than children who had achieved tolerance [50].

8.3.1.2. β-Lg. β-Lg is an 18.3-kDa protein accounting for 50% of the total protein in the whey fraction of milk. At physiological pH, it exists as a 36-kDa dimeric protein. Each subunit consists of a 162-residue polypeptide with two disulfide bridges and one free cysteine, which are responsible for its physicochemical properties as well as interaction with CN during heat treatment [51]. There are seven genetic variants [52]. The two main isoforms of β-Lg are the genetic variants A and B, which differ in two point mutations on residues 64 and 118; aspartic acid and valine in β-Lg A; and glycine and alanine in β-Lg [53]. β-Lg is relatively stable to denaturation and is resistant to proteolytic hydrolysis by chymotrypsin and pepsin, but susceptible to trypsin hydrolysis [54]. Seven different IgE binding epitopes and six IgG-binding regions have been identified that cause allergic reactions to humans [55, 56]. Among these, the antigenicity of sequence 124–134 has been suggested, and this sequence may also be responsible for the IgE cross-reactivity between α-la and β-Lg [57]. β-Lg peptides as short as 12–14 amino acid residues account for a significant part of the allergenicity of the whole molecule in some patients [56].

8.3.2. Minor Cow's Milk Allergens

8.3.2.1. α-la. α-la is a monomeric globular protein with a molecular weight of 14.2 kDa. It represents 25% of whey proteins and is comprised of 123 amino acid residues and four disulfide bridges [19, 58]. Several studies have confirmed the presence of linear epitopes in α-la that are able to bind human IgE antibodies [55, 59]. Amino acid residues 1–16, 13–26, 47–58, and 93–102 of α-la recognized IgE antibodies from 11 persistent CMA patients' sera (4–18 years) by immunoblot analysis [55].

8.3.2.2. BSA. BSA has a molecular weight of 66.4 kDa and accounts for around 5% of total whey proteins. It is organized in three homologous domains and consists of nine loops connected by 17 disulfide bonds [60]. Most of the disulfide bonds are protected in the core of the protein and are therefore not easily accessible and quite stable to denaturing conditions [61]. Antigenic epitopes on BSA recognizing human IgE antibodies and antibodies from mice have been identified [62]. Sera from four BSA-allergic children recognized an IgE-binding epitope from the –COOH terminal region (524–598 amino acid residues; 524–542 critical amino acid sequence) by immunoblot, while murine IgG antibodies recognized amino acid residues from the NH_2 terminal region (126–144 amino acid residues).

8.3.2.3. LF. LF is a milk-specific iron-binding protein with a molecular weight of 76.1 kDa representing less than 1% of total whey protein. LF consists of a single polypeptide chain folded into two globular lobes, each of them having high-affinity iron-binding sites, which are connected by a 3-turn helix, and it also contains five potential glycosylation sites [63]. Wal [46] reported that 41 of 92 milk-allergic patients had detectable levels of IgE antibodies to bovine LF and concluded that LF was an important milk allergen. However, to date, no studies have been conducted to identify LF IgE epitopes or T-cell epitopes.

8.3.2.4. IgG. The IgG fraction accounts for about 1% of total milk protein and 6% of whey proteins. IgG has the characteristic "Y-shaped" Ig structure composed of four polypeptide chains linked through intra- and intermolecular disulfide bonds. The monomers consist of heavy and light chains each of these composed of V and C domains. The V domains of H and L chains converge to form the antigen-binding site, while the C regions characterize the isotype of the Ig in cow's milk: IgG, IgA, IgM [64]. The existence of three IgG classes in cattle (IgG1, IgG2, and IgG3) has been reported [65]. Data on the potential allergenicity of bovine Igs are very limited. However, some studies have proposed IgG as another milk allergen due to the observation that IgE antibodies from CMA patients specifically binds bovine IgG [66, 67]. There are, unfortunately, no reports in the literature on either B- or T-cell epitopes of bovine IgG.

8.4. EFFECTS OF PROCESSING ON DAIRY ALLERGENS

Food processing can alter the allergenic properties of proteins by hiding, destroying, or disclosing allergenic epitopes through conformational changes in protein structure, or by changing the digestibility of proteins [68–70]. IgE-binding epitopes are either conformational or linear in nature. Processing can have a disruptive effect on the native protein structure of an allergen, which can result in disruption of the IgE binding epitopes and consequently modify the ability of an allergen to elicit a reaction. On the other hand, food processing could also unveil or create new epitopes, sometimes termed neo-epitopes. This may occur through protein unfolding, which unravels inner portions of the protein structure, not generally available for antibody binding or the covalent modification of a protein by sugars or other food components [71, 72]. Clinical investigations of the allergenic potential of proteins must, therefore, take into consideration both the potential decrease as well as the possible increase of antibody-binding capacity, which can result from protein unfolding during processing. Several chemical and technological approaches have been considered to produce non- or hypoallergenic foods by removal, destruction, proteolytic modification, or masking of epitopes. Examples of these techniques include heat treatments, enzymatic digestion, fermentation, ultrafiltration, microparticulation, high pressure (HP), irradiation, and genetic engineering, some of which are reviewed in greater detail below.

8.4.1. Heating

Milk proteins differ markedly in their resistance to heat (i.e., α-CN is the most heat stable, BSA is the most labile, and β-Lg is relatively heat stable). Pasteurization, preheating, and sterilization are commonly used methods for processing of milk by the dairy industry. Heat treatment of milk at 90°C for 15 minutes decreased the immunoreactivity of α-la and β-Lg to 12.72% and 18.74%, respectively, as compared with raw milk [73]. Earlier studies conducted by Duranti et al. [74] also showed that the immunoreactivity of α-la decreased to 8% in sterilized milk compared with raw milk. Kohno et al. [75], however, analyzed specific IgE binding to native and denatured α-CN in children with CMA and reported no differences in binding activity. In contrast Vila et al. [50] found differences in IgE binding to denatured and native α-CN, and electrophoretic studies showed differences in the structure of the modified proteins.

Immunoreactivity of native and denaturated β-Lg has also been studied, and the results showed that cleavage of the intrachain disulfide bonds within the β-Lg molecule and consequently the loss of the conformation of the molecule had little or no effect on its immunoreactivity, suggesting that linear epitopes are mainly implicated [76, 77]. Treatment of the protein at 90°C for 15 minutes, on the other hand, significantly reduced IgE-binding capacity. However, severe heat treatment (60 minutes at 90°C) did not eliminate all IgE-binding activity [78]. One of the major IgE-binding epitopes on β-Lg is reportedly situated in the region containing the free SH group in Cys 121 and could be exposed after moderate heat treatment and oxidized at high temperature, which could then be identified by IgE antibodies [55]. In contrast to these findings, Kleber et al. [79] reported that heat-induced denaturation (90°C) increased the allergenicity of β-Lg, by exposing formerly hidden antigenic sites. They further reported that heating above 90°C decreased the antigenic response of β-Lg, owing to the loss of the conformational epitopes and to the masking of the sequential epitopes in the course of aggregation and thermally induced protein modification (e.g., the formation of glycosides by Maillard reaction) [80, 81]. The results are, therefore, quite conflicting but may be explained by the extent of aggregation or unfolding of the proteins upon heat treatment.

Up to 90% of β-Lg has been shown to be aggregated in processed milk [82]. Heating of milk may also favor the binding of β-Lg with CN micelles and coaggregation with α-la, which can extensively modify surface availability of epitopes [83]. Walter et al. [84] found that pasteurization caused aggregation of β-Lg and α-la inhibiting uptake by intestinal epithelial cells *in vitro* and *in vivo*. Furthermore, they observed that aggregation redirected uptake to Peyer's patches, which promoted significantly higher Th2-associated antibody and cytokine production in mice than their native counterparts. A higher Th2-associated antibody and cytokine production is correlated with higher allergenic response. Interestingly, they reported that only the soluble forms of β-Lg and α-la elicited anaphylaxis when the allergens were administered orally. Aggregated β-Lg and α-la as well as CN required systemic administration to induce anaphylaxis. This may explain some of the discrepancies in the results obtained and why some allergic individuals tolerate cooked milk much better than raw and pasteurized milk [85–87]. Alteration of the structure of cow's milk proteins during heat treatments can be affected by the temperature, duration, intrinsic characteristics, and physicochemical conditions of the environment. From the reported studies, it may be concluded that heat treatment alone may not be an effective approach for eliminating the allergenicity of cow's milk proteins. Duranti et al. [74] suggested that heating and lactic acid fermentation could be more effective at reducing the immunoreactivity of some milk proteins.

8.4.2. Glycation

Glycation is a process of conjugating proteins with sugars through the Maillard reaction; this alters the functional properties of the proteins by changing their charge, cross-linking, and/or conformation [88]. The Maillard reaction is a ubiquitous reaction of condensation of a reducing sugar with amino groups of proteins, which occurs naturally during heating and storage of food products [89]. Glycation could potentially reduce the allergenicity of proteins since the Maillard reaction can result in the masking of allergenic epitopes.

Chevalier et al. [90] investigated the association behavior and conformation changes of β-Lg after glycation with sugars at 60°C for 3 days. β-Lg heated with sugars showed larger structural modifications depending on the sugar used and the degree of glycation. In general, conjugation of a protein with a larger polysaccharide is likely to be more effective in reducing antigenicity compared with conjugation of a protein with lower-molecular-weight (MW) compounds since larger polysaccharides can cover the epitopes of allergens more effectively than lower-MW compounds [91]. Reduced immunogenicity of β-Lg-carboxymethyl dextran (CMD) conjugates observed by Kobayashi et al. [92] was explained by such a shielding of the epitopes in β-Lg by CMD, which allowed the epitopes to escape from recognition by the immune system. Structural analyses by intrinsic fluorescence, circular dichroism (CD) spectra, and enzyme-linked immunosorbent assay (ELISA) with monoclonal antibodies indicated that the surface of β-Lg in each conjugate was covered by CMD without great disruption of native conformation. Hattori et al. [93] also observed reduced immunoreactivity of β-Lg after conjugation with acid oligosaccharide (ALGO) and phosphoryl oligosaccharides (POs). Conjugation with ALGO and POs substantially enhanced the thermal stability of β-Lg; however, anti-β-Lg antibody response was markedly reduced after immunization with both conjugates in BALB/c, C57BL/6, and C3H/He mice. Peyron et al. [70] reported that with nonspecific interactions with β-Lg, pectin could reduce the accessibility of cleavage sites and/or epitope sequences. The presence of both high-methoxy pectin and low-methoxy pectin enhanced aggregate formation, which apparently influenced digestibility and immunogenicity. Similar responses have been reported for α-la. Enomoto et al. [94] glycated α-la with maltopentose (MP) through the Maillard reaction (MP-α-la) and subsequently phosphorylated it by dry heating. The anti-α-la antibody response decreased significantly by glycation and subsequent phosphorylation. The suppressive effect of α-la on the production of proinflammatory cytokines such as IL-6 after stimulation with lipopolysaccharide was significantly enhanced by glycation with MP and was further enhanced by phosphorylation after glycation. These studies are interesting, and further *in vivo* studies will be useful to determine if glycation actually reduces allergic reactions in humans.

8.4.3. Enzymatic Hydrolysis

Nutritionally, hydrolysis can yield a variety of new peptides that could have many physiological benefits for humans [95]. Among milk proteins, α- and β-CN, β-Lg, and α-la are very sensitive to trypsin, while Igs and BSA are very resistant [96]. Genetic variants of β-Lg may also differ in their rate of hydrolysis with trypsin (β-Lg A>β-Lg B) [97]. Enzymatic hydrolysis yields hydrolysates with low levels of free amino acids and a major fraction consisting mostly of short peptides [98]. The technique has been exploited in the development of different hydrolyzed milk products and depending on the degree of hydrolysis, allergenic epitopes could be destroyed rendering the product hypoallergenic. As the degree of hydrolysis affects the extent to which the allergenicity of milk proteins can be reduced, several factors need to be considered in the processing of such hypoallergenized milk products. The purity of the protein source, pretreatment conditions, enzyme specificity, processing conditions (e.g., pH, temperature, ionic strength), and degree of hydrolysis, method of enzyme inactivation (i.e., heat treatment, acidification, membrane filtration), as well as posthydrolysis treatments (i.e., use of adsorbents for free amino acids removal, membrane separation, etc.), are some of the important parameters that need to be carefully considered in order to optimize the preparation of hypoallergenic formulae [99, 100]. Depending on the degree of protein hydrolysis, products obtained may be classified into extensive or high degree and partially or low-degree hydrolyzed formulae. Residual allergenicity of these products will vary.

The effects of *in vitro* proteolysis on the allergenicity of major whey proteins with different proteases were evaluated by Asselin et al. [101]. Results showed that hydrolysis of β-Lg using trypsin alone or in combination with chymotrypsin or pepsin with chymotrypsin significantly decreased the allergenicity of β-Lg, but it did not abolish allergenicity. Similarly, Pahud et al. [102] reported that oral administration of trypsin-hydrolyzed whey protein reduced anaphylactic reactions in guinea pigs compared with milk and untreated whey protein. The antigenicity and allergenicity of hydrolyzed milk proteins will depend on the MW of the remaining peptides, and it may be difficult to determine the limit under which the remaining peptides are small enough to guarantee hypoantigenicity and hypoallergenicity. Businco et al. [103] reported that hydrolysate formulae containing only peptides with a MW of <6 kDa rarely elicit anaphylactic reactions. Peptides in some hydrolyzed milk formulae have MWs of <3 kDa [29]. While these peptides may be big enough to contain two epitopes, they rarely are able to cross-link IgE antibodies. van Beresteijin et al. [104] suggested that the minimal molecular mass for peptides to provoke IgE reactions falls between 3 and 5 kDa. They further indicated that residual antigenicity of milk proteins is based not only on the MW of the residual peptides, but also on the molecular structure, which will vary as a function of the type of proteases used to produce the hydrolysate.

Thus, although enzymatic hydrolysis may eliminate allergenic epitopes, peptides generated may still react with IgE antibodies due to residual antigenicity. Haddad et al. [105] reported that partial digestion of bovine milk with pepsin or pepsin and trypsin resulted in peptides from β-Lg that still bound to IgE antibodies in the sera of patients with CMA. Görtler et al. [106] studied the antigenicity (ability to bind human IgG and IgE antibodies) of five different hypoallergenic infant formulae using immunoblotting and radioallergosorbent test (RAST). They found that all tested formulae still contained antigens with MWs ranging from 3 to 67 kDa although the antigen content of the hydrolyzed formulae was considerably lower when compared with nonhydrolyzed infant formulae and a mixture of the major cow's milk proteins. Reduced IgE-binding capacity of hydrolyzed formulae has also been observed by other workers [56, 107, 108]. In the study conducted by Görtler et al. [106], 5 out of 12 cow's milk-allergic children possessed IgE antibodies that reacted with the hydrolyzed formulae, which raises questions about their safety for CMA patients. Ragno et al. [109] also noticed allergic reactions to an extensively hydrolyzed whey preparation in up to 13% of a CMA population and to an extensively hydrolyzed CN preparation for 10% of these individuals. It is of no surprise, therefore, that Svenning et al. [110] remarked that the use of single endo and exo proteolytic enzyme of animal, plant, or microbial origin has shown no promise in minimizing the anitigenicity of whey proteins.

The combination of enzymatic hydrolysis with a preceding heat treatment may, however, markedly enhance tryptic and peptic hydrolysis of milk proteins and further reduce their antigenicity [111, 112]. Enzymatic hydrolysis in combination with ultrafiltration has also been reported as an efficient method to reduce the antigen content of hypoallergenic infant formulae. *In vitro* and *in vivo* studies with animals have shown a reduction in sensitizing capacity and anaphylaxis in orally sensitized animals, when treated with such ultrafiltered hydrolysates [106, 113]. It should be mentioned, however, that the process does not eliminate antigenicity. Moreover, as mentioned earlier, proteolytic digestion irrespective of other peripheral treatments could potentially reveal new epitopes that could enhance antigenicity [114]. As a result, the American Academy of Pediatrics recommends that only extensively hydrolyzed milk proteins (<1 kDa) or amino acid formulations be used for infants with CMA [115].

8.4.4. Fermentation

Jedrychowski and Wroblewska [116] reported that sterilization of milk followed by lactic acid fermentation reduced the immunoreactivity of cow's milk proteins. Fermentation of sterilized cow's milk was done using mesophilic (288 cultures) and thermophilic lactic acid bacteria (161 cultures). A 99% decrease in antigenicity of α-la and β-Lg was reported for both mesophilic (*Leuconostoc mesenteroides* ssp. *cremoris*, *Lactococcus lactis* ssp. *lactis* (biovar. *diacetilactis*) and thermophilic (*Lactobacillus delbrueckii* ssp. *bulgaricus*, *Lactobacillus helveticus*, *Streptococcus salivarius* spp. *thermophilus*, *Lactobacillus casei*, *L. delbrueckii* spp. *lactis*, and *Lactobacillus acidophilus*) lactic acid bacteria fermented milk as analyzed by ELISA using rabbit antibodies. Similarly, Maier et al. [117] also noted a reduction in the immunochemical response of β-Lg in fermented milk compared with nonfermented milk. Industrially manufactured sour milk as well as acidophilus milk (pasteurized milk containing *L. acidophilus*) both showed reduced levels of immunoreactivity by ELISA. Fermentation of milk products increases the susceptibility of milk proteins such as β-Lg to peptic digestion and reduces their stability. During fermentation, microbial enzyme proteases cleave milk proteins into peptides, and peptidases further break these down into smaller peptides and amino acids thereby destroying antigenic epitopes that may be present [118]. The action of enzymes produced by microorganisms during fermentation is primarily responsible for the reduction of antigenicity during fermentation of milk [119]. As with enzymatic hydrolysis, fermentation does not completely eliminate allergenicity but only slightly attenuates it [115, 119]. Due to the relative resistance of proteins such as β-Lg to acid hydrolysis and pepsin degradation, portions of these proteins can still remain intact after stomach digestion, making them still potentially highly allergenic [51].

Interestingly, some studies have suggested that consumption of fermented milk may modulate clinical symptoms in atopic infants and children and also reduce the development of atopic allergy in at-risk infants during the first 2 years of life [120]. Prioult et al. [121] reported antiallergy properties of *Bifidobacterium lactis* NCC362. They demonstrated that a cell-free extract of *B. lactis* NCC362 hydrolyzed acidic and basic peptides isolated from a tryptic-chymotryptic hydrolysate of β-Lg. The *B. lactis*-driven hydrolysis released acidic peptides with reduced residual allergenicity. Moreover, these peptides exhibited immunomodulatory properties, such as stimulating interferon (IFN)-γ and IL-10 production and downregulating IL-4 in primed mice suggestive of a pro-Th1/anti-Th2 cell response. An increase in Th1 response and a decrease in Th2 response result in lower IgE antibody production and consequently a reduced allergic response [122, 123]. These studies are interesting, and further investigation should shed more light on the specific mechanisms at play.

8.4.5. Irradiation

Gamma radiation has long been applied to extend the shelf life of food as well as to inactivate microbes and enzymes present in the food. Studies in the late 1950s and early 1960s [124, 125] suggested that radiation of milk protein beyond 28 kGy resulted in extensive molecular alteration of protein structure, which was effective in reducing anaphylactogenic properties. Lee et al. [126] evaluated the effect of gamma radiation on the allergenic and antigenic properties of α-CN and β-Lg at 3, 5, and 10 kGy using IgE antibodies from milk hypersensitive patients and rabbit IgG antibodies in a competitive indirect ELISA. Their results showed that allergenicity and antigenic property of the irradiated milk proteins were modified to different extents depending on the level of irradiation. At higher doses of irradiation, the α-CN and β-Lg were structurally altered and broke down into smaller molecules or coagulated to larger molecules due to interactions between the molecules. As a result of the aggregation, epitopes were masked, which reduced the antigenic response. Similarly, gamma irradiation of BSA was reported to disrupt the ordered structure of the protein and also resulted in degradation, cross-linking, and aggregation of the polypeptide chains [127]. No studies on antigenicity were, however, conducted. The use of food irradiation as a method for food preservation is regulated in many countries. In Canada, the only foods approved for irradiation are potatoes, onions, wheat, flour, whole wheat flour, whole and ground spices, and dehydrated seasoning preparations. Restrictions and bans exist in the European Union (EU) as well as in other countries. Thus, while research on the effect of irradiation on the antigenicity of milk proteins is of academic interest, it may have little industrial relevance.

8.4.6. HP

HP processing is a relatively new technology that is used to improve the nutritional and functional quality of food. The use of HP treatment in foods can have a disruptive effect on protein tertiary and quaternary structure with relatively little influence on secondary structure [128]. Kleber et al. [129] studied the antigenic response of a WPI solution, sweet whey, and skim milk subjected to HP treatment (200–600 MPa) at different temperatures (30–68°C) and times (0–30 minutes) by indirect competitive ELISA. They found that the antigenicity of β-Lg increased with increasing pressure and holding time in all the solutions due to weakening of noncovalent bonds and the unfolding of the protein structure, which resulted in greater accessibility of allergenic epitopes. In combination with low thermal treatment (40 and 50°C), the antigenicity of β-Lg increased, whereas at higher temperatures (60 and 68°C), a marked decrease in antigenicity was observed. Elevated temperature and pressure treatment can result in Maillard reactions, which may lead to destruction of linear epitopes with the consequence of a reduced antigenic response [79]. Izquierdo et al. [130] also showed that HP treatment (350–600 MPa) dramatically increased the susceptibility of β-Lg to hydrolysis by pepsin, trypsin, chymotrypsin, pronase, and thermolysin due to conformational changes in protein structure. Recently, Zeece et al. [131] evaluated the effect of HP treatment on the *in vitro* pepsin digestion of β-Lg. Peptides released were analyzed by mass spectrometry and compared with known sequences of antigenic epitopes. The authors found that increasing the pressure level from 400 to 800 MPa increased the digestibility of β-Lg. The study further showed that most of the peptides in the hydrolysate had low MW and were short in length (e.g., 7–10 residues), which would reduce their ability to cross-link IgE antibodies on sensitized mast cells. Thus, HP treatment may lower β-Lg allergenicity by reducing its structural integrity and promoting unfolding in ways that increase accessibility to digestive enzymes. In contrast to these findings, Chicon et al. [132] reported that treatment of β-Lg and WPI at 200 and 400 MPa increased binding to β-Lg-specific rabbit IgG antibodies and did not affect binding to IgE antibodies from allergic patients. The authors reported that higher digestibility of HP-treated WPI with pepsin did not have an impact on the ability of IgE antibodies to bind the hydrolysates when compared with untreated samples. As with the other processing treatments, further biological and clinical investigation is needed to determine if HP treatment has the ability to produce truly hypoallergenic milk.

8.5. NUTRITIONAL AND FUNCTIONAL ALTERNATIVES TO DAIRY INGREDIENTS

8.5.1. Alternatives to Cow's Milk Protein

The composition and function of the major and minor cow's milk proteins are listed in Table 8.1. Cow's milk contains about 3.3 g of protein/100 mL of milk.

TABLE 8.1 Cow Milk Protein Composition and Functions

Sl. No.	Milk Protein Constituents	Amount in g/100 mL[a]	Molecular Weight (kDa)[b]	*pI*[b]	Function[b]
I	Casein	2.95			
1	α_{s1}-Casein	1.19	23.6	4.9–5.0	Ion carrier for calcium, phosphorus, iron, zinc, and copper, precursors of bioactive peptides
2	α_{s2}-Casein	0.31	25.2	5.2–5.4	
3	β-Casein	0.98	24.0	5.1–5.4	
4	κ-Casein	0.35	19.0	5.4–5.6	
5	γ-Casein	0.12	30	NR	
II	Whey proteins	0.63			
1	β-Lactoglobulin	0.32	18.3	5.3	Retinol carrier, binding fatty acids, possible antioxidant
2	α-Lactalbumin	0.12	14.2	4.8	Lactose synthesis in mammary gland, calcium carrier, immunomodulation, anticarcinogenic
3	Bovine serum albumin (BSA)	0.04	66.4	4.9–5.1	Transport of ligands, antioxidant
III	Immunoglobulins (A, M, and G)	0.08	<150	NR	Immune protection
IV	Lactoferrin	0.10	76.2	8.7	Carrier for iron, antimicrobial
V	Lactoperoxidase	0.003	89	NR	Antimicrobial
VI	Lysozyme	0.00004	18.0	NR	Antimicrobial, synergistic effect with immunoglobulins and lactoferrrin

[a]Reference 133.
[b]References 25, 39, 134, and 135.
pI, isoelectric point; NR, not reported.

In many countries, cow's milk serves as a major source of dietary protein for both children as well as adults because of its high protein quality and physiological properties. The nutritional value of proteins is usually related to their ability to meet nitrogen and amino acid requirements for tissue growth and maintenance [136]. Nitrogen requirements represent the amount of dietary nitrogen necessary to replenish nitrogen and amino acid losses during metabolic activities without affecting protein and energy metabolism [137]. Proteins in milk are of high quality and supply adequate amounts of all nine essential

TABLE 8.2 Protein Quality of Milk and Milk Proteins [138–140]

		Milk and Milk Proteins		
Sl. No.	Quality Attributes	Milk	Casein	Whey Proteins
1	Biological value (BV %)	84.1	79.0	93.4
2	Protein digestibility (PD %)	95.1	97.0	94.8
3	Net protein utilization (NPU %)	78.8	76.9	85.7
4	Protein efficiency ratio (PER)	2.68	2.72	3.02
5	Protein digestibility corrected amino acid score (PDCAAS)	1.21	1.23	1.15

amino acids (lysine, threonine, valine, isoleucine, leucine, methionine, phenylalanine, tryptophan, and histidine) required for growth and development of humans. The quality of cow's milk and some of the different protein fractions is given in Table 8.2.

For milk-allergic patients, however, consumption of cow's milk results in adverse reactions. The only treatment of CMA at the present time is complete avoidance of milk proteins. Alternative foods that provide adequate nutrition and which can serve as nutritional and functional milk protein substitutes are, therefore, needed. Although soy proteins are by themselves allergenic to some consumers, they are the first to be recommended as alternatives because of their hypoallergenicity, low cost, and convenience [141]. The Committee on Nutrition of the American Academy of Pediatrics has recommended formulae based on soy protein isolates for use in infant feeding when their nutritional needs are not being met from breast milk or cow's milk formulae as well as for treatment of IgE-mediated CMA in infants [115, 142]. The protein content of soy (40%) is much higher than that of cereal grains, 3 times richer than eggs, and 11 times richer than milk. Soy protein quality (i.e., protein digestibility corrected amino acid score) has also been shown to be comparable to that of meat and eggs. Zezulka and Calloway [143] investigated the quality of soy protein in adult men and reported that 0.6 g of soy protein per kilogram body weight per day produced a nitrogen balance that was equivalent to 0.4 g of egg white protein per kilogram body weight per day. This indicates that soy protein is capable to meet amino acid requirements of adults and further supplementation with methionine only improves the quality of soy protein [144, 145]. The digestibility and nutritional quality of soy protein is, however, affected by the presence of antinutritional factors such as trypsin inhibitor, phytic acid, and phenolic compounds. Heat processing improves the digestibility through denaturation of protein and inactivation of enzyme inhibitors as well as destruction of antinutritional factors.

Proteins from other legumes and nuts may also serve as good alternatives to milk proteins. Chickpea seeds have been considered a suitable source of dietary protein, due to their good balance of amino acids, their high protein bioavailability, and their relatively low levels of antinutritional factors [146]. The calculated protein efficiency ratio (C-PER) score for isolated chickpea protein was 2.6 and 2.3 for isolates prepared by micellization and isoelectric precipitation, respectively, which is more or less equal to milk CN (C-PER of 2.5) [147]. Proteins from chickpea were reported to contain adequate amounts of most essential amino acids for preschool children and all essential amino acids for adults [147, 148].

Ohlson and Anjou [149] also reported that rapeseed/canola proteins contain high amounts of lysine, methionine, and cysteine and their protein nutritional profile is similar to milk CN and superior to that of other vegetable proteins such as soy, pea, and wheat. Protein efficiency ratio (PER) of rapeseed/canola protein is 2.64 compared with 2.19 for soy proteins. Some other legumes, grains, and nuts of interest are mung bean, pea, cowpea, sesame seed, sunflower seed, and peanut as they are high in protein content and are frequently consumed as alternatives to animal proteins. Further studies are, however, required to determine how they can be used in different food applications and their suitability as a sole source of protein for allergenic children as well as for adults suffering from CMA.

8.5.2. Alternatives to Lipids from Cow's Milk

Cows' milk contains about 4% fat and consists mainly of triglycerides (97%) together with smaller amounts of phospholipids (0.5%) and cholesterol (0.3%). Cow's milk fat typically contains 70% saturated fatty acids, 25% monounsaturated fatty acids, and 5% polyunsaturated fatty acids [150]. Saturated fatty acids in milk are mainly comprised of long-chain fatty acids such as palmitic C16 (25%), stearic C18 (14%), and myristic C14 (3%), and short-chain fatty acids of butyric to capric C4 to C10 (9%) [151]. Milk contains lower amounts of the essential fatty acids (i.e., oleic, linoleic acid, and linolenic acid) [150] needed by humans for various metabolic activities (e.g., synthesis of hormones like prostaglandins). Fortunately, there is an abundant supply of alternative sources of essential and nonessential fatty acids in the diet. Sunflower seed, safflower seed, canola, olive, corn, cottonseed, and soy are good sources of both essential and nonessential fatty acids and can be used as replacements in the diet of CMA patients. Additionally, for those patients who are not allergic to peanuts or tree nuts, these can offer good source of healthy oils in the diet. Since many plant oils are mostly unsaturated, they may also confer additional health benefits [152].

8.5.3. Alternative to Sugars in Cow's Milk

Lactose is a disaccharide composed of glucose and galactose, and milk is the sole source of this naturally available sugar (4.6 g/100 mL of milk). Lactose influences the absorption of calcium and hastens beneficial microbial growth in intestine. In some instances, lactose affects the nutritional value of protein by conjugation via Maillard reaction [153]. There is really no alternative for lactose; however, other sugars can be used to replace milk sugar in food formulation. Nutritionally, the effect of replacement of lactose with partially hydrolyzed rice syrup on the small intestine development in weaned pigs was studied by Kang et al. [154]. They reported that lactose replacement with hydrolyzed rice syrup did not affect alkaline phosphatase activity, mucosal protein/DNA ratio, or intestinal morphology and concluded that partially hydrolyzed rice syrup could be used as a replacement for lactose in manufactured liquid diets for neonatal pigs. Similar observations were made when corn syrup solids were replaced for lactose in the diet of neonatal pigs [155]. Little research has been done on humans and thus the value of such animal studies and their relevance for humans could be questioned. Although lactose plays an important role in improving calcium absorption and promoting bone mineralization in infants, studies indicate that it does not enhance calcium absorption in adults. Other factors such as protein intake, vitamin D, dietary fiber, and genetic disorders also affect calcium bioavailability. For infants, lactose can be supplied by breast milk; hence, the interest in replacing lactose from cow's milk in the diet of humans may be more of a question of taste and physicochemical functionality than a matter of nutritional or physiological benefit. The physicochemical or nutritional role of lactose as a prebiotic could be replaced by oligosaccharides (e.g., stachyose, raffinose, etc.) and fructo-oligosaccharides (e.g., inulin) from Jerusalem artichoke, soybean, rice bran, wheat, barley, and so on. Some of these products themselves may be allergenic; thus, appropriate precautions need to be taken in their use.

A significant number of people also suffer from lactose intolerance, which is distinctly different from milk allergy. Lactose intolerance does not involve the immune system and results from the lack of the enzyme lactase in certain individuals, which limits their ability to effectively digest lactose. Further information on lactose intolerance can be found in the papers written by Shaw and Davies [156] and Vesa et al. [157].

8.5.4. Alternatives to Cow's Milk Vitamins and Minerals

Milk serves as a good source of vitamin A, riboflavin, biotin, and cyanocobalamin (vitamin B_{12}) and contains relatively small amounts of vitamin D, vitamin E, and vitamin K. Although milk has a relatively low concentration of niacin, the value of milk in meeting the dietary needs for niacin is high because of interconversion of tryptophan [153]. Milk contains high amounts of calcium (122 mg/100 mL), phosphorus (92 mg/100 mL), and magnesium; moderate amounts of iodine, iron, and zinc; and traces of potassium, sodium, sulfur, and chloride [153]. The vitamin and mineral composition of milk is presented in Table 8.3. Soy has been suggested as an alternative for milk and it contains high concentrations of potassium, phosphorus, magnesium, sulfur, calcium, chloride, and sodium, which range from 0.2% to 2.1%. Soy also contains minor minerals like selenium, iron, zinc, manganese, copper, and iodine, which range

TABLE 8.3 Vitamins and Minerals Composition of Whole Milk (3.25% fat) per 100 g of Milk [158]

Sl. No.	Nutrients	Amount
I. Vitamins		
1	Thiamine (mg)	0.04
2	Riboflavin (mg)	0.18
3	Niacin (mg)	0.10
4	Vitamin B_6 (mg)	0.03
5	Pantothenic acid (mg)	0.36
6	Folic acid (μg)	0
7	Choline (mg)	14.30
8	Vitamin C (mg)	0
9	Vitamin B_{12} (μg)	0.44
10	Vitamin A (IU)	102
11	Vitamin D (IU)	40
12	Vitamin E (α-tocopherol) (mg)	0.06
13	Vitamin K (μg)	0.20
II. Minerals		
14	Calcium (mg)	113
15	Iron (mg)	0.03
16	Magnesium (mg)	10
17	Phosphorus (mg)	91
18	Potassium (mg)	143
19	Sodium (mg)	40
20	Zinc (mg)	0.40
21	Copper (mg)	0.01
22	Selenium (μg)	3.70

between 0.01 and 140 ppm. Additionally, it is a good source of thiamine, riboflavin, and vitamins A (β-carotene), D, E, and K [152]. Other cereals, legumes, and nuts contain good sources of vitamins and minerals and could be considered as alternative sources for these micronutrients. Phytic acid is one of the major antinutritional factors present in plant foods, which affect the bioavailability of minerals in the body; its concentration in replacement foods should, therefore, be carefully considered. Processing techniques such as heating, enzymatic hydrolysis, and fermentation may be used to reduce the phytate content in many plant foods.

8.5.5. Alternatives to Dairy Products

8.5.5.1. Milk Substitutes. A milk substitute is a nondairy product based mostly on vegetarian milk that could be used to replace cow's milk in the diet. Good milk substitutes must be capable of supplying well-balanced nutrients such as protein, energy, vitamins, and minerals equal to cow's milk; be easy to

Fig. 8.1 Commercially available dairy free milks: (A) rice milk; (B) soymilk; (C) almond milk; (D) vegetable milk.

produce; and be made of ingredients that are easily available and low in cost. The protein advisory group of the Food and Agriculture Organization of the United Nations has recommended that imitation milks should contain 3.5% fat, 3.5% protein, and 5% carbohydrate [159]. Selection of an appropriate milk substitute also depends on whether it is to be used as an ingredient or consumed as a beverage. Three common categories of milk substitutes are grain-based beverages (e.g., rice milk), legume/nut-based beverages (e.g., soymilk, almond milk), and carbohydrate-based beverages (e.g., beverages made from maltodextrin from potatoes) (Fig. 8.1). More recently, alternatives such as hemp milk have also become available.

Among the different plant protein sources, soy is the most commonly used alternative for the production of imitation milk because of its high protein and fat content. Furthermore, soymilk may offer nutritional and health benefits since it contains no cholesterol or lactose and only small quantities of saturated fatty acids [160]. Oral administration of soymilk also provides some beneficial effects in terms of increasing the bifidobacteria population due to the presence of certain types of oligosaccharides, such as raffinose and stachyose, in soy that can be utilized by bifidobacteria and lactobacilli as energy sources [161, 162]. Differences in the nutrient composition of cow's milk and soymilk are presented in Table 8.4.

There have been some concerns about the nutritional quality of soymilk due to the presence of antinutritional factors such as trypsin inhibitors and lectins in soy. Some consumers have also had concerns regarding the taste and flavor of soymilk and soy-derived products. There is very little basis for these concerns today as various studies have clearly demonstrated that processing

TABLE 8.4 Difference in the Nutrient Composition of Cow's Milk and Soymilk per 100 g of Milk [158]

Sl. No.	Nutrients	Cow's Milk	Soymilk[a]
1	Water (g)	88.32	85.61
2	Energy (kcal)	60.00	63.00
3	Carbohydrates (g)	4.52	9.95
4	Protein (g)	3.22	2.26
5	Fat (g)	3.25	1.53
6	SFA (g)	1.86	0.21
7	MUFA (g)	0.81	0.33
8	PUFA (g)	0.19	0.83
9	Cholesterol (mg)	10.00	0.00
10	Fiber (g)	0.00	0.40
11	Ash (g)	0.69	0.65
12	Calcium (mg)	113.00	25.00
13	Iron (mg)	0.03	0.48
14	Magnesium (mg)	10.00	15.00
15	Phosphorus (mg)	91.00	51.00
16	Potassium (mg)	143.00	143.00
17	Sodium (mg)	40.00	53.00
18	Zinc (mg)	0.40	0.34
19	Copper (mg)	0.01	0.20
20	Manganese (mg)	0.003	NR
21	Selenium (μg)	3.70	4.80
22	Thiamine (mg)	0.04	0.02
23	Riboflavin (mg)	0.18	0.26
24	Niacin (mg)	0.10	0.51
25	Vitamin B_6 (mg)	0.03	0.08
26	Panthotenic acid (mg)	0.36	0.09
27	Folic acid (μg)	0.00	0.00
28	Choline (mg)	14.30	23.60
29	Vitamin C (mg)	0.00	0.00
30	Vitamin B_{12} (μg)	0.44	0.70
31	Vitamin A (IU)	102.00	3.00
32	Vitamin D (IU)	40.00	0.00
33	Vitamin E (mg)	0.06	0.01
34	Vitamin K (μg)	0.20	3.00

[a]Soy silk plain.
SFA, saturated fatty acids; MUFA, monounsaturated fatty acids; PUFA, polyunsaturated fatty acids; NR, not reported.

conditions used for making soymilk adequately destroys the antinutritional properties. Furthermore, extensive research has resulted in the production of soymilk products with very little beany taste and which have colors and viscosities comparable to cow's milk. The taste and flavor of soymilk and other milk substitutes can also be improved through flavoring and the addition of sweeteners.

Hackler et al. [163] evaluated the effect of heating at 93 and 121°C for varying periods on the quality of soymilk using weaning rats. The results showed that soymilk cooked at 93°C for 1–6 hours inactivated 90% of trypsin inhibitor without affecting the protein quality and available lysine content. Cooking of soymilk at 121°C for 5–10 minutes gave a product with higher protein quality and available lysine content, however, when compared to products heated for longer periods at 121°C, which caused a decline in protein quality and available lysine content due to Maillard reactions. Extensive thermal treatment may also adversely affect the quality of soymilk, including its flavor, color, and, in some cases, nutritional quality [164].

Caygill et al. [165] studied the quality of imitation milk from mung bean, cowpea, and chickpea after homogenization with soy oil. The imitation milks were nutritionally inferior compared with cow's milk because of lower protein content (1.5–1.9 g/100 g) and lack of essential amino acids (isoleucine and tryptophan). However, potassium and phosphorus levels in imitation milks were comparable to cow's milk, but they contained lower amounts of sodium, calcium, and zinc. Although trypsin inhibitor activity was found to be partially stable to pasteurization, it was lower in all the imitation milks (i.e., milk from mung bean, cowpea, and chickpea) compared with soymilk. To improve the nutritional value of the imitation milks and make them comparable with cow's milk, the authors suggested that the imitation milk prepared from legumes be mixed with cereal proteins to meet the amino acid requirements for humans. Wadud et al. [166] developed soymilk-based baby foods using corn, rice, and wheat and evaluated the protein quality (PER and NPU). The study revealed that PER and NPU scores for soymilk with corn, rice, and wheat were 3.18, 3.12, and 3.00; and 84.30%, 83.80%, and 83.00%, respectively, which was comparable with milk CN (3.25 and 85.00%). The soy-based products were creamish white in color, and had a good taste and fluffy texture and were similar to commercially available baby foods both nutritionally and organoleptically. In other studies, an effort was made to improve the nutritional quality of soymilk by adding SoyTrim® (Archer Daniels Midland, Decatur, IL), a hydrocolloidal mixture of soy flour and oat bran, at the rate of 200 g/kg [167]. The SoyTrim blend incorporated products contained 0.8 g/kg of soluble fiber (β-glucan) and had a reduced fat content (12.2 g/kg) compared with the control soymilk, and no apparent changes in the organoleptic qualities of the SoyTrim-incorporated products were observed [167].

Rice milk is less viscous than soymilk and contains less protein but could be a good alternative to soymilk. Tree nut milk (e.g., almond and cashew) is creamier than rice milk and has a consistency that is similar to soymilk. It has a strong nutty flavor and could be used as an alternative for people who are not allergic to nuts. Hemp milk is a new product that is made from the seeds of hemp. Hemp seeds contain up to 25% protein and hemp meal (defatted hemp seed) contains approximately 31–33% protein. These proteins are of very high quality and contain all 8–10 essential amino acids in the correct proportions needed for human nutrition. Proteins in hemp are also easy to digest. The majority of hemp protein consists of 65% edestin, a globular protein. The high nutritional quality of hemp proteins makes it a promising alternative source of plant protein for use in human food applications [168, 169]. Milk alternatives are also sometimes prepared from oats and barley or from a blend of the different products listed above. Carbohydrate-based alternatives are an interesting addition to the market. They often consist of maltodextrin (from different sources), cereal solids, oil, gums, sweeteners, vitamins, and minerals and are frequently labeled as dairy free and/or gluten free.

There is some controversy regarding use of the word "milk" for nondairy beverages, so the term "drink" or "beverage" is often preferred.

8.5.5.2. Alternatives to Milk Powder. Milk powder is widely used by consumers and the food industry as a substitute for fresh milk and as an ingredient for the formulation of processed food products. Spray drying is the technique commonly used for drying of milks because the very short time of heat contact and the high rate of evaporation gives a relatively high-quality product with relatively low cost. Spray-dried powders have several advantages, which include minimal thermal damage, reduction in volume thus requiring less space for storage, ease of transport, and a longer shelf life [170]. As with fresh milk, spray-dried soy flours, soy protein concentrates (>65% protein content on dry basis), or soy protein isolates (>90% protein content on dry basis) are the most commonly used alternatives to whole milk or whey powder [171]. Dried soymilk powder is often used in confectionaries, meat fillers, and beverages [172]. Rahardjo et al. [173] reported that addition of spray-dried soymilk powder to pork sausage patties improved the textural properties of the patties without causing flavor changes.

The type of processing condition used for making soy protein powders can, however, impact their functionality. Spray-dried cow milk powders, especially those obtained from small-scale spray dryers often have small particle sizes (<50 μm) with poor reconstitution properties [174]. Agglomeration has been suggested as a technique to improve the wettability of spray-dried soy protein powders [175]. Jinapong et al. [176] produced instant soymilk powder using ultrafiltration, spray drying, and fluidized-bed agglomeration. Ultrafiltration increased total solids, protein, and fat contents but decreased carbohydrate and ash contents of soymilk, and the process led to an increase in particle size, wettability, and dispersibility of the resultant spray-dried powders. However, spray-dried soymilk powders without ultrafiltration were very small (~15 μm) and very cohesive leading to poor flowability. Agglomeration of spray-dried powders with maltodextrin as an aqueous binder using a fluidized-bed agglomerator improved the handling and reconstitution properties of the powders. The optimum binder concentration was found to be 10% w/v maltodextrin, which resulted in the largest particle size of the agglomerated powder (260 μm) having a good flowability and low cohesiveness. The wettability of this agglomerated powder was good, but its dispersibility (61%) still required improvement.

Other protein powders, concentrates, and isolates that could be used to replace cow's milk powder are pea protein isolate, canola protein isolate, peanut protein isolate, almond protein isolate, and hemp protein isolate. While commercial sources of these products are available, extensive research is needed in order to determine the appropriate conditions for their use in different food formulations.

8.5.5.3. Milk Hydrolysates. Milk protein hydrolysates are commonly used in infant formulae to combat problems of food allergy. Such hydrolysate formulae must be safe for CMA patients and provide adequate amounts of all nutrients. For infants with food allergies, this is especially important as they need more nutrients than healthy infants because of their impaired ability to utilize nutrients, which may be caused by allergy-induced bowel inflammation [177]. Infant formulae with protein contents of only 1.2–1.3 g/100 mL were shown to ensure normal infant growth [178]. Protein hydrolysates included in hypoallergenic formulae for infants with CMA must have a high content of di- and tripeptides, which are rapidly absorbed by the intestinal epithelium, a low free amino acid content to limit bitterness, a high nutritional value, and a very low antigenicity [179]. This is particularly important as the presence of intact or partially hydrolyzed proteins (containing allergenic epitopes) in the protein formulae can induce immune-mediated allergic reactions. Two main ways to supply tailored amounts of amino acids for children are by using extensively hydrolyzed protein hydrolysates and a mixture of synthetic amino acids [180, 181]. Enzymatic hydrolysis is a preferred method for the preparation of tailor-made peptides because of their large-scale commercial feasibility and their moderate cost [182]. Milk hydrolysates have similar or enhanced nitrogen digestibility and a good nutritional quality compared with native milk proteins. The residual allergenicity of these formulae, however, remains an issue. The use of extensively or partially hydrolyzed milk protein formulae is estimated to have increased 10–50% among formulae-fed infants in countries such as the United States and France [183]. CN and whey protein are considered as the main source to develop hypoallergenic protein hydrolysates. The main drawback of milk hydrolysates is their unpleasant odor and taste as well as inadequate nutrient supply [103], [183] and potential residual allergenicity.

Plant proteins are used in a number of formulated foods as an alternative to proteins from animal sources such as milk. Among plant proteins, soy is the source most widely used to obtain protein hydrolysates, again because of its nutritional quality, lower cost, and higher availability. Moure et al. [184] demonstrated that enzymatic hydrolysis of a soluble protein present in waste liquor from soy processing produced fractions with a similar digestibility as CN. In another study, Moughan et al. [185] reported that it is unlikely that replacement of milk protein with soy-based formulae will cause any major differences in the digestive process. Seppo et al. [186] compared nutrient intake, nutritional status, and growth of cow's milk-allergic infants fed with soy formulae and extensively hydrolyzed whey formulae from cow's milk supplemented with vitamin D and calcium for 2 years. The study revealed no significant difference in the growth (% weight for length) of infants consuming soy formulae and whey-hydrolyzed formulae during the 2 years. Serum transferrin receptor concentrations indicated a greater iron inadequacy in the tissue of infants in the soy formulae group compared with the hydrolyzed whey formulae group. The authors, however, found no significant differences between the groups either in the percentages of mean cell volume, hemoglobin, zinc, and ferritin or in the percentages of high alkaline phosphatase activity. In the study conducted by Lasekan et al. [187], the authors also found no significant differences in anthropometric and blood profile of soy formulae-fed infants and those fed with the cow's milk-based formulae. These studies indicate the comparable safety and effectiveness of the soy formulae as an alternative to the hydrolyzed milk formulae.

Pulse legumes such as chickpea and pea have been successfully used for the production of hydrolysates, but the presence of antinutritional factors and the development of bitterness during processing could lower their utility [147, 188]. Rice has been recognized as the most hypoallergenic cereal and its proteins have been hydrolyzed and used in different food applications. D'Auria et al. [189] reported that rice hydrolysate formulae might be a nutritionally suitable alternative for infants with cows' milk protein allergy. One report indicated that a homemade rice-and-lamb diet reduced the severity of skin lesions and improved weight gain in children with atopic dermatitis and multiple food hypersensitivities [190]. Rice proteins are also allergenic; however, studies suggest that they trigger adverse reactions in <1% of allergic children [191].

8.5.5.4. Alternatives to Fermented Dairy Foods

8.5.5.4.1. Yogurt. Yogurt (or yoghurt) is a semisolid fermented milk product that is produced by heating followed by coagulation of milk using lactic acid bacteria (e.g., *L. delbrueckii* sp. *bulgaricus*, *S. salivarius* sp. *thermophilus*). The coagulation results from an increase in the concentration of lactic acid due to bacterial metabolism of lactose and results in the formation of a three-dimensional gel network, which is capable of retaining the water present in milk. Nutritionally, plain yogurt has a similar composition to that of the milk used and is an excellent source of high-quality protein, riboflavin, niacin, vitamin B_{12}, calcium, phosphorus, magnesium, and zinc [192]. Consumption of fermented dairy products has traditionally been the main source of probiotic and other beneficial bacteria in the diet of humans. Alternatives to dairy-based yogurts should ideally meet this requirement as well as be of good quality and taste. Technically, one of the greatest challenges for the production of non-dairy-based yogurts is the absence of lactose required for fermentation by lactic acid bacteria. Fortunately, several studies have identified bacterial strains that metabolize other oligosaccharides in foods.

Pinthong et al. [193] found that yogurt could be produced from soymilk supplemented with glucose (1% w/v) and yeast extract (0.1% w/v) through fermentation by lactic acid bacteria, including *Lactobacillus bulgaricus* and *Streptococcus thermophilus*. Cheng et al. [194] compared the sensory qualities of soy yogurt and commercial plain cow's milk yogurt using trained human panelists. The soy yogurt (sogurt) was prepared with soymilk, 0.15% calcium acetate, 0.5% gelatin, and lactose (0% or 2%) by fermenting with *L. casei* and *S. thermophilus*. Sogurt with lactose was similar to yogurt in acidity and sourness and the products did not differ in intensity of butter aroma, although the sogurt had a beanier and raisin-like aroma, more bitter and stringent taste, sandier mouthfeel, and was firmer and more yellow in color. The authors suggested that bleaching soy flour and the use of stabilizers could produce a sogurt that was more like yogurt [194].

The influence of initial dry matter content of reconstituted soymilk on the quality of yogurt fermented with two types of starter culture (*L. delbrueckii* spp. *bulgaricus* and *S. thermophilus*) was examined by Denkova and Murgov [195]. They reported that soy yogurt obtained from reconstituted soymilk with 10% dry matter content and 1% starter culture showed good physicochemical and organoleptic qualities. The product had titratable acidity of 65–67°T, 10^{10} cfu/g viable count and could be stored for 3 days at 25–30°C and for more than 2 weeks in refrigeration conditions. Adetunji et al. [196] also studied the effect of preprocessing (soaking in a hot water with and without salts such as sodium hydroxide, sodium carbonate, and sodium hydrogen carbonate) on the nutritional quality of soy yogurt. The soy yogurt produced from sodium carbonate pretreated soybean had the highest mineral content (calcium: 179.7 ppm, potassium: 136.6 ppm, magnesium: 652.2 ppm, and sodium: 495.0 ppm). Yogurt prepared from the sodium hydroxide pretreated soybean scored highest for taste, texture, and overall acceptability. Changes in the nutritional composition of soymilk fermented with *Bifidobacterium longum B6* and *Bifidobacterium infantis CCRC 14633* were investigated by Hou et al. [197]. They observed an increase in crude protein, thiamine, riboflavin, titrable acidity, glucose, fructose, galactose, acidic acid and lactic acid content, and decreased niacin, stachyose, raffinose, and sucrose content. The influence of *Bifidobacterium* sp. on microbial growth of *S. thermophilus* and *L. bulgaricus* and volatile compounds in soymilk and cow's milk was also investigated by Murti et al. [198]. Soy yogurt fermented with mixed cultures (*S. thermophilus*, *L. bulgaricus*, and *Bifidobacterium* sp.) had less acidity and lower *L. bulgaricus* growth and was more viscous with higher amounts of diacetyl and ethanol than that of cow's milk yogurt. Furthermore, yogurt from cow's milk was found to contain higher acetone, which gave a pleasant flavor to the product while *n*-hexanoic acid was found in the soy yogurt, which gave an unpleasant off-flavor to the product and resulted in a lower sensory score for the soy yogurt [198]. Total microbial load was also higher in yogurt produced from whole milk and powdered milk (0.93 and 0.85×10^8 cfu/mL, respectively) than in soy yogurt (0.52×10^8 cfu/mL) [199].

Recently, Farinde et al. [200] found that maize steepwater could be used as an alternative to cow's milk starter culture for the production of yogurt. They observed no significant difference in protein and total viable count of soy yogurt fermented with maize steepwater as compared with commercial culture fermented soy yogurt.

Martensson et al. [201] attempted to formulate an oat-based symbiotic yogurt containing oat milk (16% or 18% dry matter content), xanthan gum (0.03%), and vegetable fat (1.8%). The product was fermented with different mixed yogurt cultures (*L. delbrueckii* sp. *bulgaricus* and *S. salivarius* sp. *thermophilus* at the ratio of 1:1 with and without *Bifidobacterium* sp.). Flavor compounds such as strawberry or mixed berry jam were added after fermentation, and the product was evaluated for quality by subjective analysis. The oat-based yogurt had higher acceptability in terms of acidity, texture, and overall appearance. The products fermented with *Bifidobacterium* had lower pH (3.9) and a more sour taste and fresh aroma typical of yogurt flavor. Addition of xanthan gum (stabilizer) improved the texture of the oat-based yogurt and reduced syneresis. Three of the oat-based yogurt, however, received lower sensory analysis scores for flavor and acceptability compared with the commercial yogurt. β-Glucan content was affected by the type of bacterial strain used [202]. No change in β-glucan content was observed during fermentation with *Lactobacillus reuteri* ATCC 55730 and *L. acidophilus* DSM 20079; however, a decrease in β-glucan content was observed in oat-based yogurt fermented with *Bifidobacterium bifidum* DM 20456 probably due to the β-glucanase activity of the *B. bifidum strain*.

Rao et al. [203] prepared yogurt from cowpea and mung beans fermented with *L. bulgaricus* (ATCC 11842) and *S. thermophilus*. The products were, however, found to be inferior to cow's milk yogurt, although the attributes were scored within the acceptable range. In another study, *L. acidophilus* and *Lactobacillus plantarum* were used as starter organisms for the production of a yogurt-like product from cowpea milk. There was a decrease in the final pH and an increase in the titratable acidity, crude protein, total ash, calcium, potassium, and phosphorus content of the fermented cowpea milk, while a decrease was obtained for lipid, moisture, and crude fiber contents. Consumer acceptability of the cowpea yogurt varied, with preference being given to the banana- and strawberry-flavored samples [204]. In general, the results of various studies suggest that it is possible to produce yogurt-like products from nondairy sources, but the quality and acceptability of the product vary as a function of the ingredients, substrate used for fermentation, type of cultures used, and processing conditions.

8.5.5.4.2. Kefir. Kefir is an acidic and mildly alcoholic fermented dairy product that is produced by milk fermented with kefir grains. It differs from other fermented milk products in that it is not the result of the metabolic activity of a single species of microflora, but is rather the product of fermentation with a mixed group of microflora confined to a matrix of discrete "kefir grains," which are recovered after fermentation [205]. Kefir grains are small, gelatinous, yellowish, and irregularly shaped masses consisting of both lactic acid bacteria and yeasts that are embedded in a slimy polysaccharide matrix named kefiran, which is thought to be produced by the lactobacilli in the grain [206]. Various lactic acid bacteria and yeasts have been identified in kefir grains (e.g., *Lactobacillus breris*, *L. helveticus*, *Lactobacillus kefir*, *L. mesenteroides*, *Kluyveromyces lactis*, *Kluyveromyces marxianus*, and *Pichia fermentans*) [207, 208]. The microorganisms in the kefir grains produce lactic acid, antibiotics, and several kinds of bactericides, which inhibit the proliferation of both degrading and pathogenic microorganisms in kefir milk [207].

Soymilk has been successfully used for the production of kefir. Furthermore, there is evidence that oral administration of soymilk kefir decreased 65% of tumor growth in mice by inhibiting tumor cells and also increased total IgA levels in the small intestine [209]. Soymilk kefir possessed significant antimutagenic and antioxidant activities as evaluated by *in vitro* assays [210]. Antiallergenic properties of soymilk kefir and the effect on gut microflora have also been studied *in vivo* using male BALB/c mice [211]. Oral administration of soymilk kefir for 28 days in egg-allergic mice significantly decreased ($P < 0.05$) serum-specific IgE and IgG1 antibody levels compared with mice fed with soymilk (control) as analyzed by sandwich ELISA. In addition, mice consuming soymilk kefir had higher populations of beneficial microorganisms (*Bifidobacterium* sp. and *Lactobacillus* sp.) and lower populations of pathogenic microorganisms (*Clostridium perfringens*) in the fecal matter than the control mice. Thus, consumption of soymilk kefir enhanced mucosal resistance to gastrointestinal pathogens by increasing growth of beneficial microorganisms and also suppressed IgE and IgG antibody levels in egg-allergic mice. The antiallergenic property of soymilk was not associated with the isoflavones in soymilk [211]. Miyake et al. [212], on the other hand, reported an inverse association between soy isoflavone consumption and the prevalence of rhinitis in pregnant women allergic to Japanese cedar pollinosis. The influence of carbohydrates (glucose, lactose, and sucrose) on microbial growth, ethanol production, and galactosidase activity in soymilk kefir was also investigated by Liu and Lin [213].

8.5.5.4.3. Alternatives to Cheese. Cheese analogs are gaining increasing acceptance with food processors and consumers because of many potential benefits such as cost-effectiveness, simplicity of their manufacture, and the replacement of selected milk ingredients by cheaper vegetable products [214]. Development of cheese analogs involves the use of fat and/or protein sources other than those native to milk, together with a flavor system simulating as closely as possible to that of the natural product [215]. Imitation mozzarella cheese was made using soy isolate and gelatin as the protein instead of caseinate, and its textural and ultrastructural properties were evaluated. Of all the hydrocolloids evaluated, GFS® gum (a mixture of xanthan, locust bean, and guar gum) (Kelco Co., Rahway, NJ) was found to give the best textural properties and melting quality to the analog. Optimum processing conditions were 5 minutes each of dry blending and wet blending of ingredients at 80°C [216]. Furthermore, imitation mozzarella cheese was functionally more stable during refrigerated storage than natural mozzarella cheese. Such stability makes analogs very attractive to the food processing and service industries [217]. Most of the studies on imitation cheese have, however, focused on partial replacement of CN with peanut protein [218]; fermented soymilk powder [219]; and starches from maize, wheat, potato, and rice [220]. Further studies are needed on the use of completely "dairy-free" ingredients. Soy tofu is made from only soy ingredients and salts. It has some semblance to cheese and is becoming more popular as an imitation cheese [221]. The gelation property of soy proteins is the prime reason for their use in imitation cheese [222]. The conventional process of making tofu is to denature the soy protein by heating, and then to coagulate the protein using coagulants and heating [223]. The curd formed consists of aggregated protein particle, which forms a gel capable of retaining water, lipids, sugar, flavors, and other components. Tofu hardness is affected by the ratio of protein to lipid [224]. Recently, many researchers have attempted to produce tofu from soymilk using HP processing technology as an alternative to conventional heat process [225–229]. The texture of these products is different from regular tofu; they have a softer texture with greater elasticity and glossier appearance [230, 231].

8.5.5.5. Other Dairy Products

8.5.5.5.1. Butter. Butter consists of 80% fat and has long been used by consumers in the preparation of bakery products, pastries, confectionery, and convenience foods because of its ability to improve the organoleptic quality of foods. Margarine, which is produced from plant oils, is recognized by many consumers as a good alternative to butter. It is cheap, has high availability, is low in saturated fatty acids (depending on the type of oil used and processing), free of cholesterol, and a suitable carrier for fortification of vitamins A and D [232]. The fatty acid composition of butter and different margarines is provided in Table 8.5. Concerns have been expressed about the use of hydrogenated oils in the production of some types of margarine, and, as a result, food manufacturers are resorting to the use of nonhydrogenated oils in margarine production. Some studies have reported that dietary intake of margarines low in saturated fatty acids decreased serum low-density lipoprotein (LDL) levels [233, 234].

Other alternatives to butter include lard and plant oils depending on the specific application. Spreadability, mouthfeel, and flavor are two important quality attributes of table spreads that need to be taken into consideration in the selection of an appropriate replacement for butter. Spreadability refers to the ease of spreading uniformly over a product surface and depends on the solid fat content and finess of the fat crystal matrix [232]. Butter has better spreadability at 15°C, while hard and soft margarines show better spreadability at 20°C and 0–10°C, respectively [235]. Thus, margarine is spreadable when taken from the refrigerator, while butter cannot be spread easily under these

TABLE 8.5 Fatty Acid Composition of Butter and Margarines (1 tbsp or 15 g of Serving Size) [232]

Product	Energy (kcal)	Cholesterol (mg)	Total Fat (g)	SFA (g)	MUFA (g)	PUFA (g)	TFA (g)
Butter	102	31.10	12	7.20	3.30	0.40	0.30
Hard margarine from corn and soybean oils	101	0	11	2.10	5.20	3.50	1.90
Soft margarine from hydrogenated soybean oil	101	0	11	1.90	5.10	3.80	0.80
Margarine-like spread (40% fat) from hydrogenated soybean oil	50	0	6	0.90	2.40	2.00	0–2.50
Liquid margarine	102	0	11	1.90	4.00	5.10	0

SFA, saturated fatty acids; MUFA, monounsaturated fatty acids; PUFA, polyunsaturated fatty acids; TFA, *trans*-fatty acids.

conditions. Mouthfeel, on the other hand, refers to the rapid melting of the product at mouth temperature (35–37°C). Although there is no difference between the mouthfeel of butter and margarine, the flavor of butter is superior to margarine [232]. Laitinen et al. [236] compared the sensory qualities of sweet rolls, ginger cookies, and cakes prepared with butter and margarine using subjective analysis. They reported that the texture and taste of sweet rolls prepared using margarine was better than sweet rolls made with butter, and no difference in flavor and mouthfeel of sweet rolls made with butter and margarine was observed. Cakes prepared with butter, on the other hand, scored higher for flavor and mouthfeel than cakes made with margarine, and no difference was reported in flavor for ginger cookies prepared with either margarine or butter.

8.5.5.5.2. Cream. Cream is concentrated butter fat (approximately 18–35%) obtained by skimming or mechanical separation of milk. Depending on the fat content and processing, cream can be categorized into clotted cream (heat-treated cream containing 55% fat), double cream (48% fat), whipped cream (35% fat), sterilized cream (23% fat), single cream (18% fat), and half cream (12% fat) [237]. Cream is used as an ingredient in many foods including ice cream, soups, sauces, cakes, custard, and coffee [238]. There is, unfortunately, no adequate substitute for dairy cream, although creams from soymilk and other sources such as coconut are available. Researchers at the company

Danisco have been working on the development of an imitation cream based on tailored blends of emulsifiers and stabilizers [239]. Such imitation creams may offer good stability before and after whipping, along with short whipping time; high overrun; high foam stiffness; smooth surface (lack of visible air cells); good shape retention; good spreadability of foam; and a clean/neutral flavor [239].

8.5.5.5.3. Frozen Desserts. The broad term frozen desserts refers to ice cream and other related products including frozen custard, mellorine (vegetable fat frozen dessert), sherbet, and frozen confections made by mixing of milk components with sweeteners, stabilizers, emulsifiers, and flavorings as well as frozen fruit products [240]. The composition of frozen desserts varies widely depending on the intended market [240]. Fortunately, there are some commercial nondairy frozen desserts that could be suitable for consumption by CMA patients. Most sorbets are made from fruit and should be suitable for consumption by CMA patients if no dairy ingredients are added. Sherbets, on the other hand, may contain dairy ingredients, and the ingredient labeling should be checked carefully. *Parev* ice or Kosher ice cream have been described as being free of dairy fat or any other dairy-derived ingredient [238, 240]. Soy ice cream, hemp ice cream, and hazelnut ice cream are also commercially available but they have not been studied as extensively as dairy ice creams. In the late 1980s, Pitz [241] developed an imitation ice cream product comprising of vegetable fat (coconut or corn oil), hydrogenated vegetable oil (mixture of coconut, palm kernel, and soybean oil), along with water, protein (eggs), sweetener (sucrose), emulsifier (sorbitan monostearate), stabilizer (carrageenan and guar gum), coloring (β-carotene), and flavorings (vanilla). The product was reported to have the taste, appearance, palatability, and aroma of conventional ice cream. It is, however, unclear if this product was ever widely commercialized.

The taste and texture of nondairy frozen desserts may be very different from traditional dairy products. Furthermore, if these products are processed in facilities that also process dairy, they may be contaminated with dairy proteins if appropriate allergen management controls are not in place. Low levels of milk protein contamination have been reported in some commercial soy ice creams [242]. In Canada, dairy protein contamination of some nondairy frozen desserts has resulted in recalls by the Canadian Food Inspection Agency [243].

8.6. CROSS-REACTIVITY

Most allergens that cross-react have similar chemical structure and may or may not be genetically related. Although soy has received a lot of attention as an alternative to cow's milk, some soy proteins are reportedly cross-reactive with milk proteins, and a significant number of patients with milk protein enterocolitis have soy protein enterocolitis [243]. Reports of the prevalence of soy sensitization in CMA infants with proctocolitis and enterocolitis have ranged between 25% and 60% [244–246]. Zeiger et al. [247] reported that 14% of infants with symptoms of CMA reacted adversely to soy (although reports of anaphylaxis to soy were rare). Similar observations were made by Klemola et al. [248] who found 10% of 170 cow's milk-allergic infants reacted to soy and 2% reacted to extensively hydrolyzed formulae in double-blind, placebo-controlled food challenges (DBPCFCs). Ahn et al. [249] found that 18.3% of 244 cow's milk-allergic children (2.5 years) were sensitized to soy proteins and had specific IgE levels of 12.4 ± 20 kU/L (range 0.89–101 kU/L) against soy protein as analyzed by Pharmacia CAP Immunoassay. Rozenfeld et al. [250] identified glycinin A4A5B3 as the first soy protein that shares IgE epitopes with bovine CNs. In addition, the α-subunit of soy β-conglycinin was also recognized by α-CN and β-CN monoclonal antibodies by immunoblot analysis [251].

Additionally, Fiocchi et al. [252] found sensitization to rice in eight children confirmed allergic to cow's milk and soy, but the rice-induced skin response was less intense than the one induced by soy. The identification of high-affinity IgE antibodies and the cross-reactive IgE epitopes in phylogenetically unrelated allergens may constitute the basis to understand multiple food allergies [253]. Further studies to identify cross-reactive epitopes in foods that are unrelated to cow's milk allergens will be useful. It should be mentioned that due to high sequence homology, IgE cross-reactivities exist between milk proteins from cow and those of other species such as ewe and goat [254–256]. These sources of milk, therefore, are not suitable for individuals with CMA. Some studies have suggested that mare and camel milk may be less allergenic [257–259], and further investigation using DBPCFCs will be useful.

8.7. CONCLUSION

Although some food processing techniques may reduce the allergenicity of milk, they are unable to completely eliminate the allergenic potential of milk allergens. For CMA patients, avoidance of cow's milk and dairy products is the only way to prevent allergic reactions. This may be difficult when foods are not properly labeled, or when hidden or undeclared dairy proteins are present in foods. Avoidance of dairy foods and ingredients may also be difficult due to their widespread use. Furthermore, identification of dairy ingredients may sometimes be difficult due to the use of multiple definition for milk ingredients (e.g., butter, CN, caseinates, cheese, delactosed whey, demineralized whey, dried milk, dry milk solids, lactose, lactalbumin, lactate, lactoglobulin, LF, lactoserum, sour cream solids, whey, whey protein concentrate, WPI).

Regulations in many countries such as in Canada and the United States now require that the common name of foods be used on labels in order to make it easier for allergic patients to identify these foods. These regulations should ensure greater security and protection of the allergic consumer. Food manufacturers also have a role to play in providing safe foods to consumers. Foods labeled as "dairy free" or "nondairy" should be processed in a way as to ensure that they do not contain residual milk allergens. In general, manufacturers need to ensure that ingredients being used in processing do not contain dairy proteins. Guidelines for the control of allergens during food processing are discussed in greater detail elsewhere in this book. Adequate training of employees in the food processing and foodservice industries on allergen control will help to ensure that these guidelines are put in place during the production of foods labeled as "dairy free" or "nondairy."

There are no general rules regarding how different allergens respond to physical, chemical, or biochemical processing [260]. Much of the work carried out on dairy allergens has been done on purified proteins rather than whole milk and without much attention to the effect of the food matrix on the antigenicity of the proteins, their detection, and the potential impact on allergenicity. Such studies are needed as well as clinical studies to establish thresholds levels of dairy proteins required for elicitation of allergic reactions. It is anticipated that the market for "allergen-controlled" foods will grow and as such it will also be of value to identify other alternatives to soy in order to expand food choice for dairy-allergic consumers.

Finally, it is important to mention that heavy investments have been made by the food industry and in research to find alternative replacements for several dairy-derived ingredients and foods. These products continue to fill grocery shelves. Adequate caution needs to be exercised when producing these products in lines and/or factories that also process milk-based products in order to avoid cross-contamination in the finished products. As some of these replacement ingredients are themselves potentially allergenic, appropriate vigilance is required as well as ongoing research to identify suitable ingredients that can expand choice for the food-allergic consumer.

REFERENCES

1. Taylor, S. (2006). The basics of food allergy. In: *Detecting Allergens in Food*, Koppelman, S.J., Hefle, S.L. (Eds.). Cambridge, UK: Woodhead Publishing/Boca Raton, FL: CRC Press LLC, pp. 1–20.
2. Sampson, H.A. (2003). Food allergy. *The Journal of Allergy and Clinical Immunology*, *111*, S540–S547.
3. Sicherer, S.H., Sampson, H.A. (2006). Food allergy. *The Journal of Allergy and Clinical Immunology*, *117*, S470–S475.
4. Vierk, K.A., Koehler, K.M., Fein, S.B., Street, D.A. (2007). Prevalence of self-reported food allergy in American adults and use of food labels. *The Journal of Allergy and Clinical Immunology*, *119*, 1504–1510.
5. Branum, A.M., Lukacs, S.L. (2008). Food allergy among U.S. children: trends in prevalence hospitalizations. *NCHS Data Brief*, *10*. Available at www.cdc.gov/nchs/data/databriefs/db10.pdf

6. Singh, A.P., McClements, D.J., Marangoni, A.G. (2004). Solid fat content determination by ultrasonic velocimetry. *Food Research International*, *37*, 545–555.
7. Smith, A.K., Kakuda, Y., Goff, H.D. (2000). Changes in protein fat structure in whipped cream caused by heat treatment and addition of stabilizer to the cream. *Food Research International*, *33*, 697–706.
8. Tamime, A.Y., Robinson, R.K. (1988). Fermented milks and their future trends—part II. Technological aspects. *Journal of Dairy Research*, *55*, 281–307.
9. Lucey, J.A., Tamehana, M., Singh, H., Munro, P.A. (1998). A comparison of the formation, rheological properties and microstructure of acid skim milk gels made with a bacterial culture or glucono-δ-lactone. *Food Research International*, *31*, 147–155.
10. Fox, P.F. (2003). Milk protein: general and historical aspects. In: *Advanced Dairy Chemistry I. Proteins*, 3rd edition, Part A, Fox, P.F., McSweeney, P.L.H. (Eds.). New York: Kluwer Academic/Plenum Publishers, pp. 1–39.
11. Havea, P. (2006). Protein interactions in milk protein concentrate powders. *International Dairy Journal*, *16*, 415–422.
12. Rosenberg, M. (1995). Current and future applications for membrane processes in the dairy industry. *Trends in Food Science and Technology*, *6*, 12–19.
13. Britten, M., Pouliot, J. (1996). Characterization of whey protein isolate obtained from milk microfiltration permeate. *Lait*, *76*, 255–265.
14. Sarkar, B.C.R., Rykala, A.J., Duncan, C.W. (1953). The essential amino acid content of the proteins isolated from milk of the cow, ewe, sow and mare. *Journal of Dairy Science*, *36*, 859–864.
15. Drescher, K., Roos, N., Pfeuffer, M., Seyfert, H.M., Schrezenmeir, J. Hagemeister, H. (1999). Recovery of ^{15}N-lactoferrin is higher than that of ^{15}N-casein in the small intestine of suckline, but not adult miniature pigs. *The Journal of Nutrition*, *129*, 1026–1030.
16. Jones, D.B., Divine, J.P. (1944). The protein nutritional value of soybean, peanut, and cottonseed flours and their value as supplements to wheat flour. *The Journal of Nutrition*, *28*, 41–49.
17. de Oliveira, J.E.D., Scatena, L. (1967). Nutritional value of protein from a soybean milk powder. *Journal of Food Science*, *32*, 592–594.
18. Aguilera, J.M., Stanley, D.W. (1999). *Microstructural Principles of Food Processing Engineering*, 2nd edition. Gaithersburg, MD: Aspen Publishers, p. 299.
19. Wal, J.M. (1998). Cow's milk allergens. *Allergy*, *53*, 1013–1022.
20. Bazinet, L., Ippeersiel, D., Gendron, C., Tetreault, C., Rene-Paradis, J., Beaudry, J., Britten, M., Mahdavi, B., Amiot, J., Lamarche, F. (2001). Comparison between reconstituted and fresh skim milk chemical and electrochemical acidifications. *Journal of the Science of Food and Agriculture*, *82*, 1356–1364.
21. Dalgleish, D.G., Law, A.J.R. (1989). pH-induced dissociation of bovine casein micelles ?? Mineral solubilization and its relation to casein release. *Journal of Dairy Research*, *56*, 727–735.
22. Alvarez, E.M., Risso, P.H., Gatti, C.A., Burgos, M., Sala, V.S. (2006). Calcium induced aggregation of bovine caseins: effect of phosphate and citrate. *Colloid and Polymer Science*, *285*, 507–514.

23. Hunt, J.A., Dalgleish, D.G. (1995). Heat stability of oil-in-water emulsions containing milk proteins: effect of ionic strength and pH. *Journal of Food Science*, *60*, 1120–1131.

24. Erba, D., Ciappellano, S., Testolin, G. (2002). Effect of the ratio of casein phosphopeptides to calcium (w/w) on passive calcium transport in the distal small intestine of rats. *Nutrition*, *18*, 743–746.

25. Shah, N. (2000). Effects of milk-derived bioactives: an overview. *The British Journal of Nutrition*, *84*, S3–S10.

26. Mulvihill, D.M. (1992). Production, functional properties and utilization of milk protein products. In: *Advanced Dairy Chemistry I. Proteins*, Fox, P.F. (Ed.). London: Elsevier Science Publishers, pp. 369–401.

27. Hung, S.C., Zayas, J.F. (1992). Functionality of milk proteins and corn germ protein flour in comminuted meat products. *Journal of Food Quality*, *15*, 139–152.

28. Broccoli, G., Berti, M., Pistolesi, E., Cestaro, B. (2005). Hydrolyzed milk-serum peptides reduce body weight and fat content of dietary obese rats ameliorating their antioxidant status and liver functions. *Panminerva Medica*, *47*, 123–129.

29. Boza, J., Martinez, O., Baro, L., Suarez, M.D., Gil, A. (1995). Influence of casein and casein hydrolysate diets on nutritional recovery of starved rats. *Journal of Parenteral and Enteral Nutrition*, *19*, 216–221.

30. Miclo, L., Perrin, E., Driou, A., Papadopoulos, V., Boujrad, N., Vanderesse, R., Boudier, J.F., Desor, D., Linden, G., Gaillard, J.L. (2001). Characterization of α-casozepine, a tryptic peptide from bovine αs1-casein with benzodiazepine like activity. *The FASEB Journal*, *15*, 1780–1782.

31. Fee, C.J., Chand, A. (2006). Capture of lactoferrin and lactoperoxidase from raw whole milk by cation exchange chromatography. *Separation and Purification Technology*, *48*, 143–149.

32. Zhu, H., Damodaran, S. (1994). Heat-induced conformational changes in whey protein isolate and its relation to foaming properties. *Journal of Agricultural and Food Chemistry*, *42*, 846–855.

33. Boye, J.I., Alli, I., Ismail, A.A., Gibbs, B.F., Konishi, Y. (1995). Factors affecting molecular characteristics of whey protein gelation. *International Dairy Journal*, *5*, 337–353.

34. Walzem, R.L., Dillard, C.J., German, J.B. (2002). Whey components: millennia of evolution create functionalities for mammalian nutrition: what we know and what we may be overlooking. *Critical Reviews in Food Science and Nutrition*, *42*, 353–375.

35. Bounous, G. (2000). Whey protein concentrate (WPC) and glutathione modulation in cancer treatment. *Anticancer Research*, *20*, 4785–4792.

36. Fayed, A., Awainry, Z., Hofi, M.A., Hofi, A.A. (1991). Ultrafiltration whey protein concentrate as a substitute for egg white in pie topping meringue. *Egyptian Journal of Food Science*, *19*, 301–313.

37. Hsu, C.K., Kolbe, E. (1996). The market potential of whey protein concentrate as a functional ingredient in surimi seafood. *Journal of Dairy Science*, *79*, 2146–2151.

38. Anthony, J.C., Yoshizawa, F., Anthony, T.G., Vary, T.C., Jefferson, L.S., Kimball, S.R. (2000). Leucine stimulates translation initiation in skeletal muscle of

postabsorptive rats via a rapamycin sensitive pathway. *The Journal of Nutrition*, *130*, 2413–2419.

39. Monaci, L., Tregoat, V., van Hengel, A.J., Anklam, E. (2006). Milk allergens, their characteristics and their detection in food: a review. *European Journal of Food Research Technology*, *223*, 149–179.
40. Venter, C., Pereira, B., Grundy, J., Clayton, C.B., Roberts, G., Higgins, B., Dean, T. (2006). Incidence of parentally reported and clinically diagnosed food hypersensitivity in the first year of life. *The Journal of Allergy and Clinical Immunology*, *117*, 1118–1124.
41. Eller, E., Kjaer, H.F., Host, A., Andersen, K.E., Bindslev-Jensen, C. (2009). Food allergy and food sensitization in early childhood: results from the DARC cohort. *Allergy*, *64*, 1023–1029.
42. Sakai, K., Yoshino, K., Satter, M.A., Ota, F., Nii, Y., Fukuta, K., Ueda, N., Shimizu, Y., Yamamoto, S. (2000). Effect of pH variation and NaCl on in vitro digestibility of cow's milk proteins in commercially available infant formulas. *Journal of Nutrition Science and Vitaminology*, *46*, 325–328.
43. Venter, C., Pereira, B., Voigt, K., Grundy, J., Clayton, B.C., Roberts, G., Higgins, B., Arshad, S.H., Dean, T. (2008). Prevalence and cumulative incidence of food hypersensitivity in the first 3 years of life. *Allergy*, *63*, 354–359.
44. Skripak, J.M., Matsui, E.C., Mudd, K., Wood, R.A. (2007). The natural history of IgE mediated cow's milk allergy. *The Journal of Allergy and Clinical Immunology*, *120*, 1172–1177.
45. Jost, R. (1988). Physicochemical treatment of food allergens: application to cow's milk proteins. *Nestle Nutrition Workshop Series*, *17*, 187–197.
46. Wal, J.M. (2002). Milk allergy. In: *Food Allergy and Intolerance Current Issues and Concerns*, Emerton, V. (Ed.). Cambridge, UK: Royal Society of Chemistry, p. 31.
47. Wal, J.M. (1998). Strategies for assessment and identification of allergenicity in (novel) foods. *International Dairy Journal*, *8*, 413–423.
48. Chatchatee, P., Jarvinen, K.M., Bardina, L., Beyer, K., Sampson, H.A. (2001). Identification of IgE- and IgG-binding epitopes on αs1-casein: differences in patients with persistent and transient cow's milk allergy. *The Journal of Allergy and Clinical Immunology*, *107*, 379–383.
49. Jarvinen, K.M., Beyer, K., Vila, L., Chatchatee, P., Busse, P.J., Sampson, H.A. (2002). B-cell epitopes as a screening instrument for persistent cow's milk allergy. *The Journal of Allergy and Clinical Immunology*, *110*, 293–297.
50. Vila, L., Beyer, K., Jarvinen, K.M., Chatchatee, P., Bardina, L., Sampson, H.A. (2001). Role of conformational and linear epitopes in the achievement of tolerance in cow's milk allergy. *Clinical and Experimental Allergy*, *31*, 1599–1606.
51. Reddy, I.M., Kella, N.K.D., Kinsella, J.E. (1988). Structural and conformational basis of the resistance of β-lactoglobulin to peptic and chymotryptic digestion. *Journal of Agricultural and Food Chemistry*, *36*, 737–741.
52. Eigel, W.N., Butler, J.E., Ernstrom, C.A., Farrell, H.M., Harwalkar, V.R., Jenness, R., Whitney, R.M. (1984). Nomenclature of proteins of cow's milk: sixth revision. *Journal of Dairy Science*, *67*, 1599–1631.

53. Zimmermann J.G., Braunitzer, G. (1987). Modern aspects of the primary structure and function of β-lactoglobulins. *Milchwissenschaft*, *42*, 294–297.

54. Maynard, F., Weingand, A., Hau, J., Jost, R. (1998). Effect of high pressure treatment on the tryptic hydrolysis of bovine β-lactoglobulin AB. *International Dairy Journal*, *8*, 125–133.

55. Jarvinen, K.M., Chatchatee, P., Bardina, L., Beyer, K., Sampson, H.A. (2001). IgE and IgG binding epitopes on alpha-lactalbumin and beta-lactoglobulin in cow's milk allergy. *International Archives of Allergy and Immunology*, *126*, 111–118.

56. Selo, I., Clement, G., Bernard, H., Chatel, J.M., Cremion, C., Peltre, G., Wal, J.M. (1999). Allergy to bovine β-lactoglobulin: specificity of human IgE to tryptic peptides. *Clinical and Experimental Allergy*, *29*, 1055–1063.

57. Adams, S.L., Barnett, D., Walsh, B.J., Pearce, R.J., Hill, D.J., Howden, M.E. (1991). Human IgE-binding synthetic peptide of bovine β-lactoglobulin and α-lactalbumin: *in vitro* cross-reactivity of the allergens. *Immunology and Cell Biology*, *69*, 191–197.

58. Brew, K., Castellino, F.J., Vanaman, T.C., Hill, R.L. (1970). The complete amino acid sequence of bovine α-lactalbumin. *The Journal of Biological Chemistry*, *245*, 4570–4582.

59. Maynard, F., Jost, R., Wal, J.M. (1997). Human IgE binding capacity of tryptic peptides from bovine α-lactalbumin. *International Archives of Allergy and Immunology*, *113*, 478–488.

60. Carter, D.C., Ho, J.X. (1994). Structure of serum albumin. *Advances in Protein Chemistry*, *45*, 153–203.

61. Restani, P., Ballabio, C., Cattaneo, A., Isoardi, P., Terracciano, L., Fiocchi, A. (2004). Characterization of bovine serum albumin epitopes and their role in allergic reactions. *Allergy*, *59*, 21–24.

62. Beretta, B., Conti, A., Fiocchi, A., Gaiaschi, A., Galli, C.L., Giuffrida, M.G., Ballabio, C., Restani, P. (2001). Antigenic determinants of bovine serum albumin. *International Archives of Allergy and Immunology*, *26*, 188–195.

63. Pierce, A., Colavizza, D., Benaissa, M., Maes, P., Tartar, A., Montreuil, J., Spik, G. (1991). Molecular cloning and sequence analysis of bovine lactotransferrin. *European Journal of Biochemistry/FEBS*, *196*, 177–184.

64. Knight, K.L., Becker, R.S. (1987). Isolation of genes encoding bovine IgM, IgG, IgA and IgE chains. *Veterinary Immunology and Immunopathology*, *17*, 17–24.

65. Heyermann, H., Butler, J.E., Frangione, B. (1992). The heterogeneity of cattle IgG2 V. Differences in the primary structure of cattle IgG2 allotypes. *Molecular Immunology*, *29*, 1147–1152.

66. Natale, M., Bisson, C., Monti, G., Peltran, A., Garoffo, L.P., Valentini, S., Fabris, C., Bertino, E., Coscia, A., Conti, A. (2004). Cow's milk allergens identification by two-dimensional immunoblotting and mass spectrometry. *Molecular Nutrition and Food Research*, *48*, 363–369.

67. Lefranc-Millot, C., Vercaigne-Marko, D., Wal, J.M., Lepretre, A., Peltre, G., Dhulster, P., Guillochon, D. (1996). Comparison of the IgE titers to bovine colostral G immunoglobulins and their F(ab')2 fragments in sera of patients allergic to milk. *International Archives of Allergy and Immunology*, *110*, 156–162.

68. Guo, M.R., Fox, P.F., Flynn, A., Kindstedt, P.S. (1995). Susceptibility of β-lactoglobulin and sodium caseinate to proteolysis by pepsin and trypsin. *Journal of Dairy Science*, *78*, 2336–2344.

69. Takagi, K., Teshima, R., Okunuki, H., Sawada, J. (2003). Comparative study of *in vitro* digestibility of food proteins and effect of preheating on the digestion. *Biological and Pharmaceutical Bulletin*, *26*, 969–973.

70. Peyron, S., Mouecoucou, J., Fremont, S., Sanchez, C., Gontard, N. (2006). Effects of heat treatment and pectin addition on β-lactoglobulin allergenicity. *Journal of Agricultural and Food Chemistry*, *54*, 5643–5650.

71. Wal, J.M. (2003). Thermal processing and allergenicity of foods. *Allergy*, *58*, 727–729.

72. Maleki, S.J., Sathe, S.K. (2006). The effects of processing methods on allergenic properties of food protein. In: *Food Allergy*, Maleki, S.J., Burks, A.W., Helm, R.M. (Eds.). Washington, DC: ASM Press, pp. 309–322.

73. Wroblewska, B., Jedrychowski, L. (1994). Detection and determination of α-lactalbumin in milk using enzyme-linked immunosorbent assay (ELISA). *Polish Journal of Food and Nutrition Science*, *3*, 45–53.

74. Duranti, M., Carpen, A., Iametti, S., Pagani, S. (1991). α-Lactalbumin detection in heat-treated milks by competitive ELISA. *Milchwissenschaft*, *46*, 230–232.

75. Kohno, Y., Honma, K., Saito, K., Shimojo, N., Tsunoo, H., Kamingogawa, S., Niimi, H. (1994). Preferential recognition of primary protein structures of a-casein by IgG and IgE antibodies of patients with milk allergy. *Annals of Allergy*, *73*, 419–422.

76. Takahashi, T., Yamauchi, K., Kaminogawa, S. (1990). Comparison between the antigenicity of native and unfolded β-lactoglobulin. *Agriculture and Biological Chemistry*, *54*, 691–697.

77. Huang, O., Coleman, J.W., Stanworth, D.R. (1985). Investigation of the allergenicity of β-lactoglobulin and its cleavage fragments. *International Archives of Allergy and Immunology*, *78*, 337–344.

78. Ehn, B.M., Ekstrand, B., Bengtsson, U., Ahlstedt, S. (2004). Modification of IgE binding during heat processing of the cow's milk allergen α-lactoglobulin. *Journal of Agricultural and Food Chemistry*, *52*, 1398–1403.

79. Kleber, N., Krause, I., Illgner, S., Hinrichs, J. (2004). The antigenic response of β-lg is modulated by thermally induced aggregation. *European Food Research and Technology*, *219*, 105–110.

80. Davis, P.J., Williams, S.C. (1998). Protein modification by thermal processing. *Allergy*, *53*, 102–105.

81. Rytkönen, J., Karttunen, T.J., Karttunen, R., Valkonen, K.H., Jenmalm, M.C., Alatossava, T., Bjorksten, B., Kokkonen, J. (2002). Effect of heat denaturation on beta-lactoglobulin-induced gastrointestinal sensitization in rats: denatured beta LG induces a more intensive local immunologic response than native beta LG. *Pediatric Allergy and Immunology*, *13*, 269–277.

82. Chen, W.L., Hwang, M.T., Liau, C.Y., Ho, J.C., Hong, K.C., Mao, S.J. (2005). Beta-lactoglobulin is a thermal marker in processed milk as studied by electrophoresis and circular dichroic spectra. *Journal of Dairy Science*, *88*, 1618–1630.

83. Rosendal, A., Barkholt, V. (2000). Detection of potentially allergenic material in 12 hydrolyzed milk formulas. *Journal of Dairy Science*, *83*, 2200–2210.

84. Walter, F.R., Berin, M.C., Arnaboldi, P., Escalante, C.R., Dahan, S., Rauch, J., Jensen-Jarolim, E., Mayer, L. (2008). Pasteurization of milk proteins promotes allergic sensitization by enhancing uptake through Peyer's patches. *Allergy*, *63*, 882–890.

85. Host, A., Samuelsson, E.G. (1988). Allergic reactions to raw, pasteurized, and homogenized/pasteurized cow milk: a comparison—a double-blind placebo-controlled study in milk allergic children. *Allergy*, *43*, 113–118.

86. Norgaard, A., Bernard, H., Wal, J.M., Peltre, G., Skov, P.S., Poulsen, L.K., Bindslev-Jensen, C. (1996). Allergenicity of individual cow milk proteins in DBPCFC-positive milk allergic adults. *The Journal of Allergy and Clinical Immunology*, *97*, 237.

87. Nowak-Wegrzyn, A., Bloom, K.A., Sicherer, S.H., Shreffler, W.G., Noone, S.R.N., Wanich, N., Sampson, H.A. (2008). Tolerance to extensively heated milk in children with cow's milk allergy. *The Journal of Allergy and Clinical Immunology*, *122*, 342–347.

88. Darewicz, M., Dziuba, J. (2001). *The* effect of glycosylation on emulsifying and structural properties of bovine beta-casein. *Nahrung/Food*, *45*, 15–20.

89. Ledl, F., Schleicher, E. (1990). New aspects of the Maillard reaction in foods and in the human body. *Angewandte Chemie International*, *29*, 565–594.

90. Chevalier, F., Chobert, J.M., Dalgalarrondo, M., Choiset, Y., Haertle, T. (2002). Maillard glycation of β-lactoglobulin induces conformation changes. *Narhung/Food*, *46*, 58–63.

91. Sehon, A.H. (1982). Suppression of IgE antibody responses with tolerogenic conjugates of allergens and haptens. *Progress in Allergy*, *32*, 161–202.

92. Kobayashi, K., Hirano, A., Ohta, A., Yoshida, T., Takahashi, K., Hattori, M. (2001). Reduced immunogenicity of beta-lactoglobulin by conjugation with carboxymethyl dextrans differing in molecular weight. *Journal of Agricultural and Food Chemistry*, *49*, 823–831.

93. Hattori, M., Miyakawa, S., Ohama, Y., Kawamura, H., Yoshida, T., To-o, K., Kuriki, T., Takahashi, K. (2004). Reduced immunogenicity of β-lactoglobulin by conjugation with acidic oligosaccharides. *Journal of Agricultural and Food Chemistry*, *52*, 4546–4553.

94. Enomoto, H., Hayashi, Y., Li, C.P., Ohki, S., Ohtomo, H., Shiokawa, M., Aoki, T. (2009). Glycation and phosphorylation of alpha-lactalbumin by dry heating: effect on protein structure and physiological functions. *Journal of Dairy Science*, *92*, 3057–3068.

95. Otte, J., Zakora, M., Qvist, K.B., Olsen, C.E., Barkholt, V. (1997). Hydrolysis of bovine β-lactoglobulin by various proteases and identification of selected peptides. *International Dairy Journal*, *7*, 835–848.

96. Jost, R., Monti, J., Pahud, J. (1987). Whey protein allergenicity and its reduction by technological means. *Food Technology*, *41*, 118–121.

97. Creamer, L.K., Nilsson, H.C., Paulsson, M.A., Coker, C.J., Hill, J.P., Jimenez-Flores, R. (2004). Effect of genetic variation on the tryptic hydrolysis of bovine β-lactoglobulin A, B, and C. *Journal of Dairy Science*, *87*, 4023–4032.

98. Kilara, A., Panyam, D. (2003). Peptides from milk proteins and their properties. *Critical Reviews in Food Science and Nutrition*, *43*, 607–633.
99. Gauthier, S.F., Pouliot, Y. (2003). Functional and biological properties of peptides obtained by enzymatic hydrolysis of whey proteins. *Journal of Dairy Science*, *86*, 78–87.
100. Poms, R.E., Anklam, E. (2004). Effects of chemical, physical and technological processes on the nature of food allergens. *Journal of AOAC International*, *87*, 1466–1474.
101. Asselin, J., Hebert, J., Amiot, J. (1989). Effects of *in vitro* proteolysis on the allergenicity of major whey proteins. *Journal of Food Science*, *54*, 1037–1039.
102. Pahud, J.J., Monti, J.C., Jost, R. (1985). Allergenicity of whey protein: its modification by tryptic *in vitro* hydrolysis of the protein. *Journal of Pediatric Gastroenterology and Nutrition*, *4*, 408–413.
103. Businco, L., Dreborg, S., Einarsson, R., Giampietro, P.G., Host, A., Keller, K.M., Strobel, S., Wahn, U. (1993). Hydrolysed cows milk formulae: allergenicity and use in treatment and prevention. *Pediatric Allergy and Immunology*, *4*, 101–111.
104. van Beresteijin, E.C.H., Peeters, R.A., Kaper, J., Meijer, R.J.G.M., Robben, A.J.P.M., Schmidt, D.G. (1994). Molecular mass distribution, immunological properties and nutritive value of whey protein hydrolysates. *Journal of Food Protection*, *57*, 619–625.
105. Haddad, Z.H., Kalra, V., Verma, S. (1979). IgE antibodies to peptic and peptic-tryptic digests of β-lactoglobulin: significance in food hypersensitivity. *Annals of Allergy*, *42*, 368–371.
106. Görtler, I., Urbanek, R., Forster, J. (1995). Characterization of antigens and allergens in hypo-allergenic infant formulae. *European Journal of Pediatrics*, *154*, 289–294.
107. Schmidt, D.G., Meijer, R.J., Slangen, C.J., van Beresteijn, E.C. (1995). Raising the pH of the pepsin-catalyzed hydrolysis of bovine whey proteins increases the antigenicity of the hydrolysates. *Clinical and Experimental Allergy*, *25*, 1007–1017.
108. Duchateau, J., Michils, A., Lambert, J., Gossart, B., Casimir, G. (1998). Anti-betalactoglobulin IgG antibodies bind to a specific profile of epitopes when patients are allergic to cow's milk proteins. *Clinical and Experimental Allergy*, *28*, 824–833.
109. Ragno, V., Giampietro, P.G., Bruno, G., Businco, L. (1993). Allergenicity of milk protein hydrolysate formulae in children with cow milk allergy. *European Journal of Pediatrics*, *152*, 760–762.
110. Svenning, C., Brynhildsvold, J., Molland, T., Langrud, T., Vegarud, G.E. (2000). Antigenic response of whey proteins and genetic variants of β-lactoglobulin: the effect of proteolysis and processing. *International Dairy Journal*, *10*, 699–711.
111. Bertrand-Harb, C., Baday, A., Dalgalarrondo, M., Chobert, J.M., Haertle, T. (2002). Thermal modifications of structure and denaturation of α-lactalbumin and β-lactoglobulin induce changes of solubility and susceptibility to proteases. *Nahrung/Food*, *46*, 283–289.
112. Bonomi, F., Fiocchi, A., Frokiaer, H., Gaiaschi, A., Iametti, S., Poiesi, C., Rasmussen, P., Restani, P., Rovere, P. (2003). Reduction of immunoreactivity of bovine beta-

lactoglobulin upon combined physical and proteolytic treatment. *Journal of Dairy Research*, *70*, 51–59.

113. Boza, J.J., Jimenez, J., Martinez, O., Suarez, M.D., Gil, A. (1994). Nutritional value and antigenicity of two milk protein hydrolysates in rats and guinea pigs. *The Journal of Nutrition*, *124*, 1978–1986.
114. El-Agamy, E.I. (2007). The challenge of cow milk protein allergy. *Small Ruminant Research*, *68*, 64–72.
115. American Academy of Paediatrics. Committee on Nutrition. (2000). Hypoallergenic infant formulas. *Pediatrics*, *106*, 346–349.
116. Jedrychowski, L., Wroblewska, B. (1999). Reduction of the antigenity of whey proteins by lactic acid fermentation. *Food and Agricultural Immunology*, *11*, 91–99.
117. Maier, I., Okun, M.V., Pittner, F., Lindner, W. (2006). Changes in peptic digestibility of bovine β-lactoglobulin as a result of food processing studied by capillary electrophoresis and immunochemical methods. *Journal of Chromatography B*, *841*, 160–167.
118. Law, J., Haandrikman A. (1997). Proteolytic enzymes of lactic acid bacteria. *International Dairy Journal*, *7*, 1–11.
119. Kleber, N., Weyrich, U., Hinrichs, J. (2006). Screening of lactic acid bacteria with potential to reduce antigenic response of β-lactoglobulin bovine skim milk and sweet whey. *Innovative Food Science and Emerging Technologies*, *7*, 233–238.
120. Cross, M.L., Stevenson, L.M., Gill, H.S. (2001). Anti-allergy properties of fermented foods: an important immunoregulatory mechanism of lactic acid bacteria? *International Immunopharmacology*, *1*, 891–901.
121. Prioult, G., Pechquet, S., Fliss, I. (2005). Allergenicity of acidic peptides from bovine β-lactoglobulin is reduce by hydrolysis with *Bifidobacterium lactis* NCC362 enzymes. *International Dairy Journal*, *15*, 439–448.
122. Coffman, R.L., Carty, J. (1990). A T-cell activity that enhances polyclonal IgE production and its inhibition by interferon-G. *Scandinavian Journal of Immunology*, *20*, 243–251.
123. Paludan, S.R. (1998). Interleukin-4 and interferon-G: the quintessence of a mutual antagonistic relationship. *Scandinavian Journal of Immunology*, *48*, 459–468.
124. Kraybill, H.F., Linder, R.O., Read, M.S., Shaw, T.M., Issac, G.J. (1959). Effect of ionizing radiation on the allergenicity of milk protein. *Journal of Dairy Science*, *42*, 581–588.
125. Grogan, F.T. Jr., Crawford, L.V. (1961). Allergenicity of cow's milk protein III. Effects of irradiation on the antigenicity of cows milk as studied by the agar diffusion technique. *The Journal of Immunology*, *87*, 240–244.
126. Lee, J.W., Kim, J.H., Yook, H.S., Kang, K.O., Lee, S.Y., Hwang, H.J., Byun, M.W. (2001). Effects of gamma radiation on the allergenic and antigenic properties of milk proteins. *Journal of Food Protection*, *64*, 272–276.
127. Gaber, M.H. (2005). Effect of γ-irradiation on the molecular properties of bovine serum albumin. *Journal of Bioscience and Bioengineering*, *100*, 203–206.
128. Puppo, M.C., Chapleau, N., Speroni, F., de Lambellerie-Anton, M., Michel, F., Añón, M.C., Anton, M. (2004). Physicochemical modifications of high-pressure

treated soybean protein isolates. *Journal of Agricultural and Food Chemistry*, *52*, 1564–1571.

129. Kleber, N., Maier, S., Hinrichs, J. (2007). Antigenic response of bovine beta lactoglobulin influenced by ultra high pressure treatment and temperature. *Innovative Food Science and Emerging Technologies*, *8*, 39–45.
130. Izquierdo, F.J., Alli, I., Gomez, R., Ramaswamy, H.S., Yaylayan, V. (2005). Effects of high pressure and microwave on pronase and alpha-chymotrypsin hydrolysis of beta-lactoglobulin. *Food Chemistry*, *92*, 713–719.
131. Zeece, M., Huppertz, T., Kelly, A. (2008). Effect of high pressure treatment on in-vitro digestibility of β-lactoglobulin. *Innovative Food Science and Emerging Technologies*, *9*, 62–69.
132. Chicon, R., Belloque, J., Alonso, E., Lopez-Fandino, R. (2008). Immunoreactivity and digestibility of high pressure treated whey proteins. *International Dairy Journal*, *18*, 367–376.
133. Jensen, G. (1995). *Handbook of Milk Composition*. New York: Academic Press, p. 465.
134. Yoshida, S. (1988). Isolation of lactoperoxidase of 89,000 daltons and a globulin of 81,000 daltons from milk acid whey. *Journal of Dairy Science*, *71*, 2021–2027.
135. Fox, P.F., McSweeney, L.H. (1998). *Dairy Chemistry and Biochemistry*. New York: Kluwer Academic/Plenum Publisher, pp. 317–346.
136. Millward, D.J., Pacy, P.J. (1995). Postprandial protein utilisation and protein quality assessment in man. *Clinical Science*, *88*, 597–606.
137. Manatt M.W., Garcia P.A. (1992). Nitrogen balance: concepts and techniques. In: *Modern Methods in Protein Nutrition and Metabolism*, Nissen, S. (Ed.). San Diego, CA: Academic Press, pp. 9–66.
138. Tomarelli, R.M., Bernhart, F.W. (1962). Biological assay of milk and whey protein compositions for infant feeding. *The Journal of Nutrition*, *78*, 44–50.
139. Van Dael, P., Kastenmayer, P., Clough, J., Jarret, A.R., Barclay, D.V., Maire, J.C. (2005). Substitution of casein by β-casein or of whey protein isolate by α-lactalbumin does not affect mineral balance in growing rats. *The Journal of Nutrition*, *135*, 1438–1443.
140. European Dairy Association. (1997). *Nutritional Quality of Proteins*. Brussels, Belgium: European Dairy Association.
141. Brady, M.S., Rickard, K.A., Fitzgerald, J.F., Lemons, J.A. (1986). Specialized formulas and feedings for infants with malabsorption or formula intolerance. *Journal of the American Dietetic Association*, *86*, 191–200.
142. American Academy of Pediatrics. Committee on Nutrition. (1998). Soy protein based formulas: recommendations for use in infant feeding. *Pediatrics*, *101*, 148–153.
143. Zezulka, A.Y., Calloway, D.H. (1976). Nitrogen retention in men fed varying levels of amino acids from soy protein with or without added L-methionine. *The Journal of Nutrition*, *106*, 212–221.
144. Young, V.R. (1991). Soy protein in relation to human protein and amino acid nutrition. *Journal of the American Dietetic Association*, *91*, 828–835.
145. Friedman, M., Brandon, D.L. (2001). Nutritional and health benefits of soy proteins. *Journal of Agricultural and Food Chemistry*, *49*, 1069–1086.

146. Friedman, M. (1996). Nutritional value of proteins from different food sources: a review. *Journal of Agricultural and Food Chemistry*, *44*, 6–29.

147. Paredes-Lopez, O., Ordorica-Falomir, C., Olivares-Vazquez, M.R. (1991). Chickpea protein isolates: physicochemical, functional and nutritional characterization. *Journal of Food Science*, *56*, 726–729.

148. Clemente, A., Vioque, J., Sanchez-Vioque, R., Pedroche, J., Bautista, J., Millan, F. (1999). Protein quality of chickpea (*Cicer arietinum* L.) protein hydrolysates. *Food Chemistry*, *67*, 269–274.

149. Ohlson, R., Anjou, K. (1979). Rapeseed protein products. *Journal of the American Oil Chemists' Society*, *56*, 431–437.

150. Moore, J.H. (1978). Cows milk fat and human nutrition. *Proceedings of the Nutrition Society*, *37*, 231–240.

151. Jensen, R.O., Ferris, A.M., Lammi-Keefe, C.J. (1991). The composition of milk fat. *Journal of Dairy Science*, *74*, 3228–3243.

152. Liu, K. (1997). *Soybeans: Chemistry, Technology and Utilization*. New York: Chapman and Hall, pp. 28–35.

153. Walsh, M.K., Dunccan, S.E., McMahon, D.J. (2000). Milk. In: *Food Chemistry: Principles and Applications*, Christen, G.L., Smith, J.S. (Eds.). West Sacramento, CA: Science Technology System, pp. 291–310.

154. Kang, P., Yin, Y.L., Ruan, Z., Pan, J., Hu, Q., Deng, Z.Y., Xiong, H., Xie, M.Y. (2008). Effect of replacement of lactose with partially hydrolysed rice syrup on small intestine development in weaned pigs from 7 to 21 days. *Journal of the Science of Food and Agriculture*, *88*, 1932–1938.

155. Oliver, W.T., Mathews, S.A., Phillips, O., Jones, E.E., Odle, J., Harrell, R.F. (2002). Efficacy of partially hydrolyzed corn syrup solids as replacement for lactose in manufactured liquid diets for neonatal pigs. *Journal of Animal Science*, *80*, 143–153.

156. Shaw, A.D., Davies, G.J. (1999). Lactose intolerance: problems in diagnosis and treatment. *Journal of Clinical Gastroenterology*, *28*, 208–216.

157. Vesa, T.H., Marteau, P., Korpela, R. (2000). Lactose intolerance. *Journal of the American College of Nutrition*, *19*, 165S–175S.

158. USDA. (2008). United States Department of Agriculture Nutrient Database for Standard Reference, Release 21. Available at http://www.nal.usda.gov/fnic/foodcomp/search/

159. Protein Advisory Group. (1972). Guideline for the preparation of milk substitutes of vegetable protein. *PAG Guideline No. 13*. Rome: Food and Agriculture Organisation of the United Nations.

160. Liu, J.R., Lin, C.W. (2000). Production of kefir from soymilk with or without added glucose, lactose or sucrose. *Journal of Food Science*, *65*, 716–719.

161. Cheng, I.C., Shang, H.F., Lin, T.F., Wang, T.H., Lin, H.S., Lin, S.H. (2005). Effect of fermented soy milk on the intestinal bacterial ecosystem. *World Journal of Gastroenterology*, *11*, 1225–1227.

162. Scalabrini, P., Rossi, M., Spettoli, P., Matteuzzi, D. (1998). Characterization of *Bifidobacteria* strains for use in soymilk fermentation. *International Journal of Food Microbiology*, *39*, 213–219.

163. Hackler, L.R., van Buren, J.P., Streinkraus, K.H., El-Rawi, I., Hand, D.B. (1965). Effect of heat treatment on nutritive value of soymilk protein fed to weanling rats. *Journal of Food Science*, *30*, 723–728.

164. Kwok, K.C., Niranjan, K. (1995). Review: effect of thermal processing on soymilk. *International Journal of Food Science and Technology*, *30*, 263–295.

165. Caygill, J.C., Jones, J.A., Ferber, C.E.M. (1981). Imitation milks from *Cicer arietinum* (L.), *Vigna unguiculata* (L.) Walpers and *Vigna radiata* (L.) Wilczek and other legumes. *Journal of the Science of Food and Agriculture*, *32*, 601–607.

166. Wadud, S., Abid, H., Ara, H., Kosar, S., Shah, W.H. (2004). Production, quality evaluation and storage stability of vegetable protein-based baby foods. *Food Chemistry*, *85*, 175–179.

167. Inglett, G.E., Carriere, C.J., Maneepun, S., Boonpunt, T. (2002). Nutritional value and functional properties of a hydrocolloidal soybean and oat blend for use in Asian foods. *Journal of the Science of Food and Agriculture*, *83*, 86–92.

168. Silversides, F.G., Lefrancois, M.R. (2005). The effect of feeding hemp seed meal to laying hens. *British Poultry Science*, *46*, 231–235.

169. Gibb, D.J., Shah, M.A., Mir, P.S., McAllister, T.A. (2005). Effect of full-fat hemp seed on performance and tissue fatty acids of feedlot cattle. *Canadian Journal of Animal Science*, *85*, 223–230.

170. Hall, C.W., Hedrick, T.I. (1972). *Drying of Milk and Milk Products*, 2nd edition. Westport, CT: AVI Publishing Company, pp. 59–93, 232–255.

171. Snyder, H.E., Kwon, T.W. (1987). *Soybean Utilization*. New York: The AVI Publishing Company, pp. 163–289, 309–327.

172. Wilson, L. (1995). Soy foods. In: *Practical Handbook of Soybean Processing and Utilization*, Erickson, D.R. (Ed.). Champaign, IL: AOCS Press, pp. 428–459.

173. Rahardjo, R., Wilson, L.A., Sebranek, J.G. (1994). Spray dried soymilk used in reduced fat pork sausage patties. *Journal of Food Science*, *59*, 1286–1290.

174. Masters, K. (1991). *Spray Drying Handbook*, 5th edition. Essex: Longman Scientific and Technical, p. 30.

175. Wijeratne, D.B.T. (1993). Spray drying of soymilk. PhD Thesis, University of Illinois, Urbana-Champaign.

176. Jinapong, N., Suphantharika, M., Jamnong, P. (2008). Production of instant soymilk powders by ultrafiltration, spray drying and fluidized bed agglomeration. *Journal of Food Engineering*, *84*, 194–205.

177. Majamaa, H., Miettinen, A., Laine, S., Isolauri, E. (1996). Intestinal inflammation in children with atopic eczema: faecal eosinophil cationic protein and tumor necrosis factor-α as noninvasive indicators of food allergy. *Clinical Experimental and Allergy*, *26*, 181–187.

178. Raiha, N., Minoli, I., Moro, G. (1989). Milk protein intake in the term infant I. Metabolic responses and effects on growth. *Acta Paediatrica*, *75*, 881–886.

179. Siemensma, A.D., Weijer, W.J., Bak, H.J. (1993). The importance of peptide lengths in hypoallergenic infant formulas. *Trends in Food Science and Technology*, *4*, 16–21.

180. Adler Nissen, J. (1986). *Enzymic Hydrolysis of Food Proteins*. New York: Elsevier, pp. 9–24.

181. Herrmann, G., Schwarz, A., Wandrey, C., Kula, M.R., Kanup, G., Drauz, K.H., Berndt, H. (1991). Scale-up of enzymatic peptide synthesis in an enzyme membrane reactor. *Biotechnology and Applied Biochemistry*, *13*, 346–353.

182. Clemente, A. (2000). Enzymatic protein hydrolysates in human nutrition. *Trends in Food Science and Technology*, *11*, 102–109.

183. Hernell, A., Lönnerdal, B. (2003). Nutritional evaluation of protein hydrolysate formulas in healthy term infants: plasma amino acids, haematology, and trace elements. *The American Journal of Clinical Nutrition*, *78*, 296–301.

184. Moure, A., Dominguez, D., Parajo, J.C. (2005). Fractionation and enzymatic hydrolysis of soluble protein present in waste liquors from soy processing. *Journal of Agricultural and Food Chemistry*, *53*, 7600–7608.

185. Moughan, P.J., Pedraza, M., Smith, W.C., William, M., Wilson, M.N. (1990). An evaluation with piglets of bovine milk, hydrolyzed bovine milk, and isolated soybean proteins included in infant milk formulas I. Effect on organ development, digestive enzyme activities, and amino acid digestibility. *Journal of Pediatric Gastroenterology and Nutrition*, *10*, 385–394.

186. Seppo, L., Korpela, R., Lonnerdal, B., Metsaniitty, L., Juntunen-Backman, K., Klemola, T., Paganus, A., Vanto, T. (2005). A follow up study of nutrient intake, nutritional status and growth of infants with cow milk allergy fed either a soy formula or an extensively hydrolyzed whey formula. *The American Journal of Clinical Nutrition*, *82*, 140–145.

187. Lasekan, J.B., Ostrom, K.M., Blatter, M.M., Ndife, L.I., Gooch, W.M., Cho, S. (1999). Growth of newborn, term infants fed soy formulas for 1 year. *Clinical Pediatrics*, *38*, 563–571.

188. Humiski, L.M., Aluko, R.E. (2007). Physicochemical and bitterness properties of enzymatic pea protein hydrolysates. *Journal of Food Science*, *72*, S605–S611.

189. D'Auria, E., Sala, M., Lodi, F., Radaelli, G., Riva, E., Giovannini, M. (2003). Nutritional value of rice hydrolysate formula in infants with cow's milk protein allergy: a randomized pilot study. *The Journal of International Medical Research*, *31*, 215–222.

190. Martino, F., Bruno, G., Aprigliano, D., Agolini, D., Guido, F., Giardini, O., Businco, L. (1998). Effectiveness of a homemade meat-based formula (the Rezza-Cardi diet) as a diagnostic tool in children with food-induced atopic dermatitis. *Pediatric Allergy and Immunology*, *9*, 192–196.

191. Helm, R.M., Burks, A.W. (1996). Hypoallergenicity of rice protein. *Cereal Food World*, *41*, 839–843.

192. Buttriss, J. (1997). Nutritional properties of fermented milk products. *International Journal of Dairy Technology*, *50*, 21–27.

193. Pinthong, R., Macrae, R., Rothwell, J. (1980). The development of a soy based yoghurt I. Acid production by lactic acid bacteria. *International Journal of Food Science and Technology*, *15*, 647–652.

194. Cheng, Y.J., Thompson, L.D., Brittin, H.C. (1990). Sogurt, a yogurt-liked soybean product: development and properties. *Journal of Food Science*, *55*, 1178–1179.

195. Denkova, Z.R., Murgov, I.D. (2005). Soy milk yoghurt. *Biotechnology and Biotechnology Equipment*, *19*, 193–195.

196. Adetunji, O.A., Betikum, E., Ojo, A., Solomon, B.O. (2006). Effect of some selected processing routes on the nutritional value of soy yoghurt. *Journal of Applied Sciences*, *6*, 527–530.

197. Hou, J.W., Yu, R.C., Chou, C.C. (2000). Changes in some components of soymilk during fermentation with bifidobacteria. *Food Research International*, *33*, 393–397.

198. Murti, T.W., Bouillanne, C., Landon, M., Desmazeaud, M.J. (1992). Bacterial growth and volatile compounds in yoghurt type products from soymilk containing *Bifidobacterium* spp. *Journal of Food Science*, *58*, 153–157.

199. Muhammad, B.F., Abubakar, M.M., Adegbola, T.A., Oyawoye, E.O. (2005). Effects of culture concentration and inoculation temperature on physicochemical, microbial and organoleptic properties of yoghurt. *Nigerian Food Journal*, *23*, 156–165.

200. Farinde, E.O., Obatolu, V.A., Fasoyiro, S.B., Adeniran, A.H., Agboola, E.R. (2008). Use of alternative raw materials for yoghurt production. *African Journal of Biotechnology*, *7*, 3339–3345.

201. Martensson, O., Andresson, C., Andresson, K., Oste, R., Holst, O. (2001). Formulation of an oat based fermented product and its comparison with yoghurt. *Journal of the Science of Food and Agriculture*, *81*, 1314–1321.

202. Martensson, O., Oste, R., Host, O. (2002). The effect of yoghurt culture on the survival of probiotic bacteria in oat based non dairy products. *Food Research International*, *35*, 775–784.

203. Rao, D.R., Pulusani, S.R., Chawan, C.B. (1988). Preparation of a yoghurt like product from cowpeas and mung beans. *International Journal of Food Science and Technology*, *23*, 195–198.

204. Sanni, A.I., Onilude, A.A., Adeleke, E.O. (1999). Preparation and characteristics of lactic acid fermented cow pea milk. *Zeitschrift-für-Lebensmitteluntersuchung-und-Forschung A*, *208*, 225–229.

205. Marshall, V.M., Cole, W.M. (1985). Methods for making kefir and fermented milks based on kefir. *Journal of Dairy Research*, *52*, 451–452.

206. La Riviere, J.W.M., Kooiman, P., Schmidt, K. (1967). Kefiran, a noval polysaccharide produced in the kefir grain by *Lactobacillus brevis*. *Archives of Microbiology*, *59*, 269–278.

207. Angulo, L., Lopez, E., Lema, C. (1993). Microflora present in kefir grains of the Galician region (north-west of Spain). *Journal of Dairy Research*, *60*, 263–267.

208. Lin, C.W., Chen, C.L., Liu, J.R. (1999). Identification and characterisation of lactic acid bacteria and yeasts isolated from kefir grains in Taiwan. *Australian Journal of Dairy Technology*, *54*, 14–18.

209. Liu, J.R., Wang, S.Y., Lin, Y.Y., Lin, C.W. (2002). Antitumor activity of milk-kefir and soymilk-kefir in tumor-bearing mice. *Nutrition and Cancer*, *44*, 182–187.

210. Liu, J.R., Chen, M.J., Lin C.W. (2005). Antimutagenic and antioxidant properties of milk-kefir and soymilk-kefir. *Journal of Agricultural and Food Chemistry*, *53*, 2467–2474.

211. Liu, J.R., Wang, S.Y., Chen, M.J., Yueh, P.Y., Lin, C.W. (2006). The anti-allergenic properties of milk kefir and soymilk kefir and their beneficial effects on the intestinal microflora. *Journal of the Science of Food and Agriculture*, *86*, 2527–2533.

212. Miyake, Y., Sasaki, S., Ohya, Y., Miyamoto, S., Matsunaga, I., Yoshida, T., Hirota, Y., Oda, H. (2005). Soy, isoflavones, and prevalence of allergic rhinitis in Japanese women: the Osaka maternal and child health study. *The Journal of Allergy and Clinical Immunology*, *115*, 1176–1183.

213. Liu, J.R., Lin, C.W. (2000). Production of kefir from soymilk with or without added glucose, lactose or sucrose. *Journal of Food Science*, *65*, 716–719.

214. Eymery, O., Pangborn, R.M. (1988). Influence of fat, citric acid and sodium chloride on texture and taste of a cheese analog. *Science des Aliments*, *8*, 15–32.

215. Shaw, M. (1984). Cheese substitutes: threat or opportunity? *International Journal of Dairy Technology*, *37*, 27–31.

216. Yang, C.S.T., Taranto, M.V., Cheryan, M. (1983). Optimization of textural and morphological properties of a soy gelatine mozzarella cheese analog. *Journal of Food Processing and Preservation*, *7*, 41–65.

217. Kiely, L.J., McConnell, S.L., Kindstedt, P.S. (1991). Observations on the melting behavior of imitation Mozzarella cheese. *Journal of Dairy Science*, *74*, 3568–3572.

218. Santos, B.L., Resurreccion, A.V.A., Garcia, V.V. (1989). Quality characteristics and consumer acceptance of a peanut-based imitation cheese spread. *Journal of Food Science*, *54*, 468–471, 494.

219. Hofmann, C.J., Marshall, W.E. (1985). Lactic fermentation of ground soybean for use in imitation cream cheese products. *Journal of Food Science*, *50*, 325–329.

220. Mounsey, J.S., O'Riordan, E.D. (2001). Characteristics of imitation cheese containing native starches. *Journal of Food Science*, *66*, 586–591.

221. Anderson, J.J.B., Anthony, M., Messina, M., Garner, S.C. (1999). Effects of phyto-oestrogens on tissues. *Nutrition Research Reviews*, *12*, 75–116.

222. Garcia, M.C., Torre, M., Marina, M.L., Laborda, F. (1997). Composition and characterization of soybean and related products. *Critical Reviews in Food Science and Nutrition*, *37*, 361–391.

223. Yamauchi, F., Yamagishi, T., Iwabuchi, S. (1991). Molecular understanding of heat-induced phenomena of soybean protein. *Food Reviews International*, *7*, 283–322.

224. Taira, H. (1990). Quality of soybeans for processed foods in Japan. *Japan Agricultural Research Quarterly*, *24*, 224–230.

225. Kajiyama, N., Isobe, S., Uemura, K., Noguchi, A. (1995). Changes of soy protein under ultra-high hydraulic pressure. *International Journal of Food Science and Technology*, *30*, 147–158.

226. Molina, E., Defaye, A.B., Ledward, D.A. (2002). Soy protein pressure induced gels. *Food Hydrocolloids*, *16*, 625–632.

227. Apichartsrangkoon, A. (2003). Effects of high pressure on rheological properties of soy protein gels. *Food Chemistry*, *80*, 55–60.

228. Préstamo, G., Lesmes, M., Otero, L., Arroyo, G. (2000). Soybean vegetable protein (tofu) preserved with high pressure. *Journal of Agricultural and Food Chemistry*, *48*, 2943–2947.

229. Prestamo, G., Fontecha, J. (2007). High pressure treatment on the tofu fatty acids and acylglycerols content. *Innovative Food Science and Emerging Technologies*, *8*, 188–191.

230. Saowapark, S., Apichartsrangkoon, A., Bell, A.E. (2008). Viscoelastic properties of high pressure and heat induced tofu gels. *Food Chemistry*, *107*, 984–989.

231. Farr, D. (1990). High pressure technology in the food industry. *Trends in Food Science and Technology*, *1*, 4–16.

232. Morris, D.H., Vaisey-Genser, M. (2003). Margarine dietary importance. *Encyclopaedia of Food Science and Nutrition*, 3719–3725. Available at doi:10.1016/B0-12-227055-X/00741-0

233. Denke, M.A., Adams-Huet, B., Nguyen, A.T. (2000). Individual cholesterol variation in response to a margarine or butter based diet. *The Journal of the American Medical Association*, *284*, 2740–2747.

234. Zock, P.L., Katan, M.B. (1997). Butter, margarine and serum lipoproteins. *Atherosclerosis*, *131*, 7–16.

235. Board, P.W., Aicken, K., Kuskis, A. (1980). Measurement of the spreadability of margarine and butter using a single pin maturometer. *International Journal of Food Science and Technology*, *14*, 277–283.

236. Laitinen, J., Hyvonen, T., Maliranta, H. (1993). Sensory evaluation of sweet rolls, ginger cookies and cakes made with butter, oil or margarine. *Food Quality and Preference*, *4*, 94.

237. The Cheese and Cream Regulations. (1995). UK Statutory Instruments and Explanatory Memorandum, Statutory Instrument No. 3240. Available at http://www.opsi.gov.uk/si/si1995/Uksi_19953240_en_1.htm

238. Varnam, A.H., Sutherland, J.P. (2001). *Milk and Milk Products Chemistry, Technology and Microbiology*, Vol. *1*. New York: Springer LLC, pp. 183–223.

239. Danisco Australia Private Limited. (2007). Imitation cream. Available at http://www.foodprocessing.com.au/products/6314-Imitation-cream

240. Marshall, R.T., Arbuckle, W.S. (1996). *Ice Cream*, 5th edition. New York: Chapman and Hall, 1–9.

241. Pitz, D.E. (1987). Lactose free synthetic ice creams. United States Patent 4643906. Available at http://www.freepatentsonline.com/4643906.html

242. Raja, V., Rampilli, M., Cattaneo, T.P.M., Toppino, P.M. (1999). Identification of cow's milk components in soy ice cream. *Rivista-di-Scienza-dell'Alimentazione*, *28*, 25–30.

243. Canadian Food Inspection Agency. (2009). Allergy alert—undeclared milk in so good brand creamy vanilla non-dairy frozen dessert. Available at http://www.inspection.gc.ca/english/corpaffr/recarapp/2009/20090408be.shtml

244. Burks, A.W., Casteel, H.B., Fiedorek, S.C., Williams, L.W., Pumphrey, C.L. (1994). Prospective oral food challenge study of two soybean protein isolates in patients with possible milk or soy protein enterocolitis. *Pediatric Allergy and Immunology*, *5*, 40–45.

245. Whitington, P.F., Gibson, R. (1977). Soy protein intolerance: four patients with concomitant cow's milk intolerance. *Pediatrics*, *59*, 730–732.

246. Powell, G.K. (1986). Food protein-induced enterocolitis of infancy: differential diagnosis and management. *Comprehensive Therapy*, *12*, 28–37.

247. Zeiger, R.S., Sampson, H.A., Bock, S.A., Burks, A.W. Jr., Harden, K., Noone, S., Martin, D., Leung, S., Wilson, G. (1999). Soy allergy in infants and children with IgE-associated cow's milk allergy. *The Journal of Pediatrics*, *134*, 614–622.

248. Klemola, T., Vanto, T., Juntunen-Backman, K., Kalimo, K., Korpela, R., Varjonen, E. (2002). Allergy to soy formula and to extensively hydrolysed whey formula in infants with cow's milk allergy: a prospective, randomized study with a follow-up to the age of 2 years. *The Journal of Pediatrics*, *140*, 219–224.
249. Ahn, K.M., Han, Y.S., Nam, S.Y., Park, H.Y., Shin, M.Y., Lee, S.I. (2003). Prevalence of soy protein hypersensitivity in cow's milk protein-sensitive children in Korea. *Journal of Korean Medical Science*, *18*, 473–477.
250. Rozenfeld, P., Docena, G.H., Anon, M.C., Fossati, C.A. (2002). Detection and identification of a soy protein component that cross-reacts with caseins from cow's milk. *Clinical and Experimental Allergy*, *130*, 49–58.
251. Curciarello, R., Lareu, J.F., Fossati, C.A., Docena, G.H., Petruccelli, S. (2008). Immunochemical characterization of *Glycine max* L. Merr. var Raiden as a possible hypoallergenic substitute for cows milk allergic patients. *Clinical and Experimental Allergy*, *38*, 1559–1565.
252. Fiocchi, A., Travaini, M., D'Auria, E., Banderali, G., Bernardo, L., Riva, E. (2003). Tolerance to a rice hydrolsate formula in children allergic to cow's milk and soy. *Clinical and Experimental Allergy*, *33*, 1576–1580.
253. Elsayed, S., Hill, D.J., Do, T.V. (2004). Evaluation of the allergenicity and antigenicity of bovine-milk alpha-s1-casein using extensively purified synthetic peptides. *Scandinavian Journal of Immunology*, *60*, 486–493.
254. Spuergin, P., Walter, M., Chiltz, E., Deichmann, K., Forster, J., Mueller, H. (1997). Allergenicity of alpha-caseins from cow, sheep and goat. *Allergy*, *52*, 293–298.
255. Restani, P., Gaiaschi, A., Plebani, A., Beretta, B., Cavagni, G., Fiocchi, A., Poiesi, C., Velona, T., Ugazio, A.G., Galli, C.L. (1999). Cross reactivity between milk proteins from different animal species. *Clinical and Experimental Allergy*, *29*, 997–1004.
256. Pessler, F., Morris, N. (2004). Anaphylactic reaction to goat's milk in a cow's milk allergic infant. *Pediatric Allergy and Immunology*, *15*, 183–185.
257. Businco, L., Giampietro, P.G., Lucenti, P., Lucaroni, F., Pini, C., Di Felice, G., Iacovacci, P., Curadi, C., Orlandi, M. (2000). Allergenicity of mare's milk in children with cow's milk allergy. *The Journal of Allergy and Clinical Immunology*, *105*, 1031–1034.
258. Shabo, Y., Barzel, R., Margoulis, M., Yagil, R. (2005). Camel milk for food allergies in children. *The Israel Medication Association Journal*, *7*, 796–798.
259. El-Agamy, E.I., Nawai, M., Shamsia, S.M., Awad, S., Haenlein, G.F.W. (2009). Are camel milk proteins convenient to the nutrition of cow milk allergic children? *Small Ruminant Research*, *82*, 1–6.
260. Thomas, K., Herouet-Guicheney, C., Ladics, G., Bannon, G., Cockburn, A., Crevel, R., Fitzpatrick, J., Mills, C., Privalle, L., Vieths, S. (2007). Evluating the effect of food processing on the potential human allergenicity of novel proteins: international workshop report. *Food and Chemical Toxicology*, *45*, 1116–1122.

9

PROCESSING OF EGG-FREE FOODS

Valéry Dumont and Philippe Delahaut

9.1. INTRODUCTION

Thanks to a perfect balance of its components, egg is recognized universally as a basic commodity. In Europe, the average annual consumption of eggs is 210 per person. Production and consumption worldwide have tripled since the 1960s and continue to increase [1].

Egg is an important ingredient of many food products because of its functional assets. This makes it difficult for allergic patients to avoid it. Egg allergy is generally accepted as one of the most common food allergies [2–4], with an estimated prevalence varying from 0.2% to 7% [5]. After a presentation of egg and egg allergy, this chapter addresses the task of managing the allergen risk to produce egg-free foods.

9.2. GENERALITIES ON EGG AND EGG ALLERGY

9.2.1. Clinical Aspect of Egg Allergy

Egg allergy is one of the most common food allergies, affecting 1–2% of young children. In a study by Rancé et al. [6], of 544 children with a positive prick test and/or specific immunoglobulin E (IgE) and a positive food challenge, 36% displayed egg allergy (Fig. 9.1). The condition usually develops within the first 2 years of life [4]. According to some studies, 50–58% of allergic children outgrow their allergy after the age of 5 years [1, 7], but data from Savage

Allergen Management in the Food Industry, Edited by Joyce I. Boye and Samuel Benrejeb Godefroy

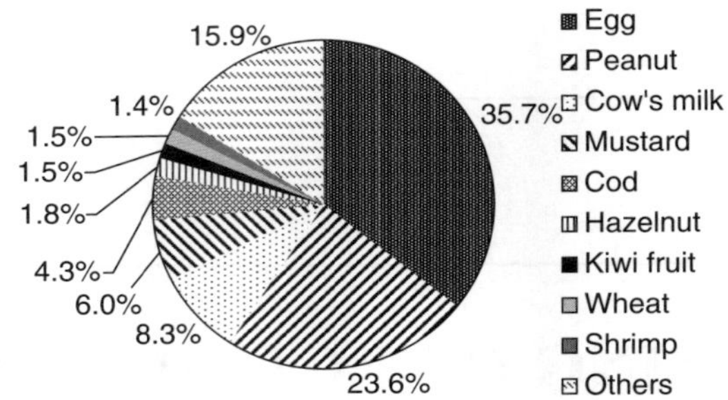

Fig. 9.1 Frequency of food allergy to various foods in 544 children with positive prick tests and/or specific IgE and positive food challenge [6].

et al. [8] suggest that this does not happen so early. The authors found that 4% developed tolerance by age 4. By age 6, 12% of egg-allergic children have outgrown their allergy. This increases to 37% by age 10, and 68% by age 16. Hen's egg allergy is more significant than allergy to eggs from other birds like goose and duck.

9.2.2. Egg Composition

Eggs are composed of 56–61% egg white and 27–32% egg yolk [9–11]. The relative proportion of each component varies according to the age of the hen and the environmental conditions. Egg white represents approximately two-thirds of the total weight of the egg without its shell, with water accounting for 87–89% of that weight. The remaining weight of the egg white comes from protein (9–11%), carbohydrates (0.8%), trace minerals (0.5%), lipids (0.1%), and vitamins. Egg yolk contains nearly 50% water, 32–35% lipid, and 16% protein. Minerals and carbohydrates are present in the egg yolk at 1.6% and 0.5%, respectively. Except for riboflavin and niacin, most vitamins are present in the egg yolk.

9.2.3. Egg Proteins Responsible for Allergy

Although the major egg allergens are mainly contained in egg white, clinically relevant allergenic egg proteins have been identified in both egg white and yolk [12, 13].

Table 9.1 shows the molecular and biological properties of identified egg allergens. Although ovalbumin seems to be one of the most studied antigens, it does not appear to be the most allergenic molecule [13]. Aabin et al. [15] reported the frequencies of reactivity of 53% for ovotransferrin, 38% for ovomucoid, 32% for ovalbumin, and 15% for lysozyme. In other studies, ovomucoid has been shown to be the dominant allergen in egg white [16].

TABLE 9.1 Molecular and Biological Properties of Identified Egg Allergens [14]

Protein Name	MW (kDa)	Rel % (w/w)	Biological Function
Egg white proteins			
Ovomucoid (Gal d1)	28	11	Trypsin inhibitor activity
Ovalbumin (Gal d2)	44	54	Not really known
Ovotransferrin (Gal d3)	78	12	Antimicrobial activity Activation of immune system Antioxidant properties
Lysozyme (Gal d4)	14	3.4	Bacteriolytic activity
Ovomucin	165	3.5	Antiviral activity
Egg yolk proteins			
Phosvitin	35	13.4	Cation-binding property Antibacterial activity Antioxidant activity
α-Livetin (Gal d5)	69	9.3	Albumin
Apovitellenins	9.5–170	37	Lipid-binding activity

MW, molecular weight; Rel % (w/w), relative proportion in weight compared with egg white

This discrepancy may be partly due to the use of impure proteins in some studies. Moreover, according to Walsh et al. [17], the existence of different groups of patients reacting in a similar way to four discrete sets of purified proteins may explain why various researchers have reported different allergens to be important in egg hypersensitivity. The four sets are as follows: set 1: the egg white proteins lysozyme and ovalbumin; set 2: the egg white protein ovomucoid; set 3: the egg white protein ovomucin; and set 4: the egg white protein ovotransferrin and the egg yolk proteins apovitellenin I, apovitellenin VI, and phosvitin.

9.2.3.1. Ovomucoid. Ovomucoid or Gal d1 forms about 11% of the egg white. It is a highly glycosylated protein, with 25% of its mass being attributed to carbohydrates. The protein consists of three structurally independent homologous domains [18]. An epitope mapping study has shown that specific immunoglobulin G (IgG) and IgE from patient sera bind to the whole ovomucoid molecule comprising all three domains [19]. It has been shown that the antigenicity and allergenicity of ovomucoid remain after fragmentation by pepsin digestion [20, 21]. This protein is also thermolabile, and this has an impact on allergenicity: patients may be tolerant to cooked eggs while having severe allergic reactions to raw eggs [22].

9.2.3.2. Ovalbumin. The most abundant and most studied protein in egg white is ovalbumin [13, 23]. This protein, also called Gal d2, is a phosphoglycoprotein. Its biological function is not really known. It has homology with the serine protease inhibitor family, although it lacks any protease inhibitor activity [24]. Pellegrini et al. [25] showed that proteolytic peptide fragments of ovalbumin display antimicrobial activity, whereas the whole protein itself is not bactericidal.

9.2.3.3. Ovotransferrin. Also called Gal d3 or conalbumin, ovotransferrin belongs to the homologous group of transferrins, which are single-chain bilobe proteins possessing an Fe^{+}-binding site in each lobe [26, 27]. Its antimicrobial activity due to its iron-binding properties has been well investigated [28]. Other biological functions have been described, including activation of the immune system [29] and antioxidant properties [30].

9.2.3.4. Lysozyme. Lysozyme or Gal d4 is a single polypeptide chain of 129 amino acids cross-linked by four disulfide bridges [31, 32]. It can catalyze the hydrolysis of specific kinds of polysaccharides comprising the cell walls of bacteria [33–36]. It is thus widely used as a preservative in several foods and drugs.

9.2.3.5. Ovomucin. Ovomucin is a heavily glycosylated protein consisting of two subunits [37]. It maintains the structure and viscosity of the egg white [38]. It also displays properties such as antiviral activity [39] and immunoactivator activity [40]. Walsh et al. published the first report on the *in vitro* allergenicity of this protein in 1988 [41]. This was also the first time egg yolk proteins were described as allergens.

9.2.3.6. Egg Yolk Proteins. Although egg yolk proteins were previously believed to have no allergenicity potential, some studies have highlighted minor allergens among them [12, 17, 41]. Phosvitin is a highly phosphorylated molecule with a great cation-chelating capacity [42]. More than 90% of the iron in eggs is bound to phosvitin. Because of this property, phosvitin shows antibacterial [43] and antioxidant activities [44]. The protein α-livetin or Gal d5 is an egg yolk allergen also implicated in the bird-egg syndrome [45]. The third set of egg yolk allergens consists of apovitellinin-containing lipoproteins. Apovitellenins I and VI are reported to show IgE-binding activity [17].

9.3. EGG: A MULTIFUNCTIONAL INGREDIENT IN THE FOOD INDUSTRY

Egg and egg components are widely used in many different food products, such as bakery products, sauces, ice cream, processed meat products, and pasta. They are also found in cosmetics, shampoos, and pharmaceuticals. Egg is a multifunctional ingredient: binder, coagulant, emulsifier, preservative, and so on. Egg components may be used individually for specific applications in food preparation. For example, lysozyme is used as a preservative, lecithin as an emulsifier, and provitamin A as a coloring. Yet, these egg-derived products are sources of allergens for egg-sensitive individuals.

Egg is available in different forms: whole egg powder, pouches of liquid eggs, only egg yolk powder or egg white powder, egg-derived products (lecithin, lysozyme), egg proteins, and so on.

This can make it fairly complex to manage eggs or egg-derived products in an allergen control program.

9.4. MANAGING THE ALLERGEN RISK TO PRODUCE EGG-FREE FOODS

In order to make egg-free foodstuffs, food producers have to know how eggs end up in the final product. Eggs may be incorporated into a product voluntarily or involuntarily. In the first case, it is necessary to know what function egg fulfills in the product, in order to find substitutes. For instance, half a small, ripe, mashed banana could bring moisture equivalent to that afforded by an egg, and cooked oatmeal or potato starch could serve as a binder in the same manner as egg.

Another way to avoid a severe allergic reaction to egg-containing products is to render egg products hypoallergenic through food processing. Methods such as heat treatment, enzymatic fragmentation, irradiation, or a combination of these have been shown to reduce allergenicity [14, 45–47]. The main issue here is the risk of altering the functional properties of egg proteins. Moreover, this kind of treatment merely reduces allergenicity of egg-containing products but does not eliminate it. The only truly effective approach to eliminating egg allergy is total avoidance of the offending compound. This means developing food products that do not contain any trace of egg.

In dealing with food allergens like egg, the main task is to assess where the risk arises. Involuntary introduction of egg into a product can come from different sources: human error, the raw material supplier, cross-contamination on the product lines, and so on. Contamination can be of two kinds: primary and secondary. Primary contamination occurs when a component used voluntarily in a recipe is introduced into a product that should not contain it as, for example, when a processed meat product is prepared in a mixer after another product containing egg, without any cleaning step between the two products. The second product is thus contaminated with egg, and this is a primary contamination. Secondary contamination occurs when an allergen is involuntarily present in a component that is itself introduced involuntarily into a product. On a product line, for instance, a chocolate biscuit might be produced directly after an almond biscuit, and egg powder might have contaminated the almond paste used for the almond biscuit. In this example, the presence of egg in the almond biscuit and the presence of almond in the chocolate biscuit are primary contaminations, but the fortuitous presence of egg in the chocolate biscuit via the almond paste is a secondary contamination.

Contamination can also be heterogeneous or homogeneous. Heterogeneous contamination is the most difficult contamination to assess and to manage. For instance, egg powder transported on employee clothing and falling into a non-egg-containing product is a heterogeneous contamination. Only a small amount of products is contaminated by batch. It could also be, for instance, only one cream injector for cakes that has not been well cleaned after egg-containing cream cakes have been produced. On a product line with eight injectors, the cakes from the seven well-cleaned injectors will not have been contaminated. The first cakes from the contaminated injector will contain egg. After a while, the injector will be "cleaned" by dilution with the non-egg-containing cream and the last cakes could be without egg. The batch of those cakes presents heterogeneity of the contamination. Less than 12.5% of the production is contaminated by egg. Homogeneous contamination such as egg contaminating dough in a blender could be evaluated by analysis more easily than heterogeneous contamination. Indeed, each product of the batch contains the same amount of egg. A second example is an ingredient contaminated with egg used for the preparation of products. All products of the batch are contaminated with egg.

Human error and employee ignorance about allergens are often the main causes of the fortuitous presence of allergens in food. The only solution is training and making each employee aware of what an allergen is. Employees have to know that allergens can be present in various forms: powder, liquid, small pieces. For some consumers, allergens are delicious foods, but for others they are life threatening. Each employee should know how to avoid accidental introduction of allergens into products upon leaving a cafeteria, for instance, or when shifting between two different production areas.

The second most important step in food allergen management is the handling of raw materials. Knowledge is of the essence. It is important to know what is in the finished product and what enters the factory. This means first obtaining a food allergen declaration from each supplier of raw materials. This declaration could be a questionnaire transmitted with each raw material. It has to be checked via an audit and analyses. With complete information, raw materials could be segregated per allergen in dedicated and clearly labeled bags or containers. Tools used to handle allergenic products should also be identified by color coding and/or label. One possibility is to dedicate a color to each allergen. For instance, red could be used for each utensil, bag, and tool used to handle egg or an egg-containing product, blue for milk, yellow for peanut, and so on.

The research and development department is another important place to assess and manage the risk of egg contamination. Each recipe should be evaluated in order to know the function of each component and the advantages of using it. It should be advantageous to avoid the use of allergenic food ingredients if other ingredients work just as well. As already mentioned, some substitutes for eggs exist, like banana, tofu, or cooked oatmeal. One should avoid using allergenic ingredients in such slight amount that they have no functional effect on the finished product. This happens sometimes, because some recipes have been used for many years without ever really being assessed. Eggs are perhaps added to the product to color it a little, and another food coloring could replace them easily. The same approach should be used when developing a new recipe. In conclusion, reformulation of old recipes and allergen-conscious development of new ones could even improve the final product.

Engineering of the fittings and system design should also be considered in a control program. Whenever possible, one should favor accessible, "visible" fittings. If egg is a binder for the components of a product, it can also bind onto equipment walls, which can be adequately cleaned only if they are accessible. Flows of raw materials, semifinished product, rework, and staff must be drawn. This is the best way to see if there are intersections between product lines, these being potential sources of contamination. The logistics should be precise in order to avoid wrong labeling and packaging—staff might transport an allergen from one production line to another; containers that are supposed to carry one specific product might be reused for another product; the same line might be used to package different products. The system design must take into account the allergen problem. The best way to avoid cross-contamination is to make production as rectilinear as possible, without any intersections between production lines, from reception of raw materials to packaging of the final product.

For consumers, reading the label is the only way to know what is in the product that they would like to eat. Food control agencies and public health commissions lay down labeling regulations such as the European Directive 2003/89 and its subsequent amendments, in order to inform allergic consumers of the presence of allergens in prepackaged foodstuffs. Yet for these consumers, the problem arises from cross-contamination, for which there is no legislation. The majority of food producers prefer using precautionary labeling such as "may contain" instead of devising a real allergen control program; but autocontrol system and quality standards require a more documented risk assessment and management program for avoiding cross-contamination by food allergens.

Control of packaging and labeling is very important. Any change in the recipe should be mentioned clearly on the label, especially if a new allergen has been introduced. Discussion with the research and development department may be necessary to delay production of the new product in order to sell out all the old batches.

The name used for the component should clearly refer to the allergen. For instance, ovalbumin or lysozyme should be followed by a reference to egg. Finally, the right product has to be in the right package with the right label. At the end of production, for instance, a clear separation should be established between egg-containing products and others. If only one packaging line is available, products without eggs should be packaged first. Before packaging of the other products begins, label and package changes should be checked.

Production scheduling may be the best way to avoid cross-contamination. Most producers do not have one production line for each product they make, so there are intersections between flows. There are several easy ways to decrease the risk of contamination. One is to minimize changeovers. Long runs of allergenic products should be favored. This decreases the number of cleaning steps and increases profitability. Another easy measure is to group together products containing the same source of allergens. For instance, all egg-containing products could be made, then all milk-containing products, and so on, if possible on different days. Production can be planned so as to start production of allergen-free products right after the cleaning step, these being followed by products with few allergens and finally by products containing the most allergens. Finally, to minimize and even eliminate the risk of contaminating allergen-free products, it is best, if possible, to dedicate processing, personnel, containers, tools, and so on, to products containing specific allergens.

Recycling of semifinished products like dough or cutting wastes is common in the food industry, as avoiding it is too expensive. This rework should be viewed as raw material and processed as such. This applies to labeling, color coding, and storage. In this way, it should be easy to reintroduce the rework into production in a "like into like" manner, or better, in an "exact into exact" manner.

The last step in a food allergen management program is cleanup. Even if good planning makes it possible to minimize this step, no allergen control program can be set up without cleaning. Compared with the work and equipment needed, time for cleaning is the limiting factor. Cleaning is time–consuming, and time is money. The cleaning step should be determined according to the process used and products made. It is important to make the right choice of a cleanup system: wet or dry, in-place or out-of-place, scraping or cleaning. Push-through to "rinse" the production line could be a good solution to close the cleaning step, but it has to be tested to evaluate how much is necessary.

9.5. CONCLUSIONS

Allergen-free and particularly egg-free food processing is a complex issue. The omnipresence of egg and egg-derived components in cooked or manufactured food products makes it hard to set up a food allergen control program. Yet the only efficient way for allergic consumers to avoid an allergic reaction is complete elimination of the allergen in their diet. The food industry has a key role in developing foodstuffs safe for human consumption without neglecting the nutritional contribution and the diversity of diet. Overuse of precautionary labeling could have a negative impact on the behavior of allergic consumers: they might marginalize the labeling and decide despite everything to eat the products. To protect consumers, it is necessary to devise assessment and management schemes to avoid the risk of cross-contamination by food allergens.

REFERENCES

1. Nys, Y., Sauveur, B. (2004). Valeur nutritionnelle des oeufs. *INRA Productions and Animales*, *17*, 385–393.
2. Sampson, H.A. (2004). Update on food allergy. *The Journal of Allergy and Clinical Immunology*, *113*, 805–819.
3. Eggesbo, M., Botten, G., Halvorsen, R., Magnus, P. (2001). The prevalence of allergy to egg: a population-based study in young children. *Allergy*, *56*, 403–411.
4. Heine, R.G., Laske, N., Hill, D.J. (2006). The diagnosis and management of egg allergy. *Current Allergy and Asthma Reports*, *6*, 145–152.
5. Rona, R.J., Keil, T., Summers, C., Gislason, D., Zuidmeer, L., Sodergren, E., Sigurdardottir, S.T., Lindner, T., Goldhahn, K., Dahlstrom, J., McBride, D., Madsen, C. (2007). The prevalence of food allergy: a meta-analysis. *The Journal of Allergy and Clinical Immunology*, *120*, 638–646.
6. Rancé, F., Kanny, G., Dutau, G., Moneret-Vautrin, D.A. (1999). Food hypersensitivity in children: clinical aspects and distribution of allergens. *Pediatric Allergy and Immunology*, *10*, 33–38.
7. Boyano-Martinez, T., Garcia-Ara, C., Diaz-Pena, J.M., Martin-Esteban, M. (2002). Prediction of tolerance on the basis of quantification of egg white-specific IgE antibodies in children with egg allergy. *The Journal of Allergy and Clinical Immunology*, *110*, 304–309.
8. Savage, J.H., Matsui, E.C., Skripak, J.M., Wood, R.A. (2007). The natural history of egg allergy. *The Journal of Allergy and Clinical Immunology*, *120*, 1413–1417.
9. Mine, Y., Rupa, P. (2004). Immunological and biochemical properties of egg allergens. *World's Poultry Science Journal*, *60*, 321–330.
10. Blum, J.C., Sauveur, B. (1996). Caractéristiques et qualité de l'oeuf de poule. *Cahiers de Nutrition et de Diététique*, *31*, 369–378.
11. Kovacs-Nolan, J., Phillips, M., Mine, Y. (2005). Advances in the value of eggs and egg components for human health. *Journal of Agricultural and Food Chemistry*, *53*, 8421–8431.
12. Anet, J., Back, J.F., Baker, R.S., Barnett, D., Burley, R.W., Howden, M.E. (1985). Allergens in the white and yolk of hen's egg—a study of IgE binding by egg proteins. *International Archives of Allergy and Applied Immunology*, *77*, 364–371.
13. Poulsen, L.K., Hansen, T.K., Norgaard, A., Vestergaard, H., Stahl Skov, P., Bindslev-Jensen, C. (2001). Allergens from fish and egg. *Allergy*, *56*, 39–42.
14. Mine, Y., Yang, M. (2008). Recent advances in the understanding of egg allergens: basic, industrial, and clinical perspectives. *Journal of Agricultural and Food Chemistry*, *56*, 4874–4900.
15. Aabin, B., Poulsen, L.K., Ebbehoj, K., Norgaard, A., Frokiaer, H., Bindslev-Jensen, C., Barkholt, V. (1996). Identification of IgE-binding egg white proteins: comparison of results obtained by different methods. *International Archives of Allergy and Immunology*, *109*, 50–57.
16. Bernhisel-Broadbent, J., Dintzis, H.M., Dintzis, R.Z., Sampson, H.A. (1994). Allergenicity and antigenicity of chicken egg ovomucoid (Gal d III) compared with ovalbumin (Gal d I) in children with egg allergy and in mice. *The Journal of Allergy and Clinical Immunology*, *93*, 1047–1059.

17. Walsh, B.J., Hill, D.J., Macoun, P., Cairns, D., Howden, M.E. (2005). Detection of four distinct groups of hen egg allergens binding IgE in the sera of children with egg allergy. *Allergologia et Immunopathologia*, *33*, 183–191.

18. Kato, I., Schrode, J., Kohr, W.J., Laskowski, M. Jr. (1987). Chicken ovomucoid: determination of its amino acid sequence, determination of the trypsin reactive site, and preparation of all three of its domains. *Biochemistry*, *26*, 193–201.

19. Mine, Y., Wei Zhang, J. (2002). Identification and fine mapping of IgG and IgE epitopes in ovomucoid. *Biochemical and Biophysical Research Communications*, *292*, 1070–1074.

20. Matsuda, T., Gu, J., Tsuruta, K., Nakamura, R. (1985). Immunoreactive glycopeptides separated from peptic hydrolysate of chicken egg white ovomucoid. *Journal of Food Science*, *50*, 592–594.

21. Kovacs-Nolan, J., Zhang, J.W., Hayakawa, S., Mine, Y. (2000). Immunochemical and structural analysis of pepsin-digested egg white ovomucoid. *Journal of Agricultural and Food Chemistry*, *48*, 6261–6266.

22. Eigenmann, P.A. (2000). Anaphylactic reactions to raw eggs after negative challenges with cooked eggs. *The Journal of Allergy and Clinical Immunology*, *105*, 587–588.

23. Raikos, V., Hansen, R., Campbell, L., Euston, S.R. (2006). Separation and identification of hen egg protein isoforms using SDS–PAGE and 2D gel electrophoresis with MALDI-TOF mass spectrometry. *Food Chemistry*, *99*, 702–710.

24. Huntington, J.A., Stein, P.E. (2001). Structure and properties of ovalbumin. *Journal of Chromatography. B, Biomedical Sciences and Applications*, *756*, 189–198.

25. Pellegrini, A., Hulsmeier, A.J., Hunziker, P., Thomas, U. (2004). Proteolytic fragments of ovalbumin display antimicrobial activity. *Biochimica et Biophysica Acta*, *1672*, 76–85.

26. Mizutani, K., Mikami, B., Hirose, M. (2001). Domain closure mechanism in transferrins: new viewpoints about the hinge structure and motion as deduced from high resolution crystal structures of ovotransferrin N-lobe. *Journal of Molecular Biology*, *309*, 937–947.

27. Keung, W.M., Azari, P., Phillips, J.L. (1982). Structure and function of ovotransferrin I. Production of iron-binding fragments from iron-ovotransferrin by the action of immobilized subtilisin—purification and characterization of the fragments. *The Journal of Biological Chemistry*, *257*, 1177–1183.

28. Giansanti, F., Rossi, P., Massucci, M.T., Botti, D., Antonini, G., Valenti, P., Seganti, L. (2002). Antiviral activity of ovotransferrin discloses an evolutionary strategy for the defensive activities of lactoferrin. *Biochemistry and Cell Biology*, *80*, 125–130.

29. Xie, H., Huff, G.R., Huff, W.E., Balog, J.M., Rath, N.C. (2002). Effects of ovotransferrin on chicken macrophages and heterophil-granulocytes. *Developmental and Comparative Immunology*, *26*, 805–815.

30. Ibrahim, H.R., Hoq, M.I., Aoki, T. (2007). Ovotransferrin possesses SOD-like superoxide anion scavenging activity that is promoted by copper and manganese binding. *International Journal of Biological Macromolecules*, *41*, 631–640.

31. Canfield, R.E. (1963). The amino acid sequence of egg white lysozyme. *The Journal of Biological Chemistry*, *238*, 2698–2707.

32. Canfield, R.E., Liu, A.K. (1965). The disulfide bonds of egg white lysozyme (muramidase). *The Journal of Biological Chemistry*, *240*, 1997–2002.
33. Hughey, V.L., Johnson, E.A. (1987). Antimicrobial activity of lysozyme against bacteria involved in food spoilage and food-borne disease. *Applied Environmental Microbiology*, *53*, 2165–2170.
34. Hughey, V.L., Wilger, P.A., Johnson, E.A. (1989). Antibacterial activity of hen egg white lysozyme against *Listeria monocytogenes* Scott A in foods. *Applied Environmental Microbiology*, *55*, 631–638.
35. Proctor, V.A., Cunningham, F.E. (1988). The chemistry of lysozyme and its use as a food preservative and a pharmaceutical. *Critical Reviews in Food Science and Nutrition*, *26*, 359–395.
36. Salton, M.R. (1957). The properties of lysozyme and its action on microorganisms. *Bacteriological Reviews*, *21*, 82–100.
37. Itoh, T., Miyazaki, J., Sugawara, H., Adachi, S. (1987). Studies on the characterization of ovomucin and chalaza of the hen's egg. *Journal of Food Science*, *52*, 1518–1521.
38. Mine, Y., Kovacs-Nolan, J. (2004). Biologically active hen egg components in human health and disease. *The Journal of Poultry Science*, *41*, 1–29.
39. Tsuge, Y., Shimoyamada, M., Watanabe, K. (1997). Bindings of ovomucin to Newcastle disease virus and anti-ovomucin antibodies and its heat stability based on binding abilities. *Journal of Agricultural and Food Chemistry*, *45*, 4629–4634.
40. Tanizaki, H., Tanaka, H., Iwata, H., Kato, A. (1997). Activation of macrophages by sulfated glycopeptides in ovomucin, yolk membrane, and chalazae in chicken eggs. *Bioscience, Biotechnology, and Biochemistry*, *61*, 1883–1889.
41. Walsh, B.J., Barnett, D., Burley, R.W., Elliott, C., Hill, D.J., Howden, M.E. (1988). New allergens from hen's egg white and egg yolk: *in vitro* study of ovomucin, apovitellenin I and VI, and phosvitin. *International Archives of Allergy and Applied Immunology*, *87*, 81–86.
42. Greengard, O., Sentenac, A., Mendelsohn, N. (1964). Phosvitin, the iron carrier of egg yolk. *Biochimica et Biophysica Acta*, *90*, 406–407.
43. Choi, I., Jung, C., Seog, H., Choi, H. (2004). Purification of phosvitin from egg yolk and determination of its physicochemical properties. *Food Science and Biotechnology*, *13*, 434–437.
44. Lu, C.L., Baker, R.C. (1986). Characteristics of egg yolk phosvitin as an antioxidant for inhibiting metal-catalyzed phospholipid oxidations. *Poultry Science*, *65*, 2065–2070.
45. Quirce, S., Maranon, F., Umpierrez, A., de las Heras, M., Fernandez-Caldas, E., Sastre, J. (2001). Chicken serum albumin (Gal d 5) is a partially heat-labile inhalant and food allergen implicated in the bird-egg syndrome. *Allergy*, *56*, 754–762.
46. Lee, J.W., Lee, K.Y., Yook, H.S., Lee, S.Y., Kim, H.Y., Jo, C., Byun, M.W. (2002). Allergenicity of hen's egg ovomucoid gamma irradiated and heated under different pH conditions. *Journal of Food Protection*, *65*, 1196–1199.
47. Hildebrandt, S., Kratzin, H.D., Schaller, R., Fritsche, R., Steinhart, H., Paschke, A. (2008). *In vitro* determination of the allergenic potential of technologically altered hen's egg. *Journal of Agricultural and Food Chemistry*, *56*, 1727–1733.

10

FISH AND SHELLFISH ALLERGENS

ANGELINA O. DANQUAH, JOYCE I. BOYE, AND
BENJAMIN K. SIMPSON

10.1. INTRODUCTION

Food allergy is an abnormal reaction of the body's immune system to certain foods. Substances in foods causing such reactions are referred to as allergens/antigens and are naturally occurring proteins [1–5]. Other substances such as food additives, which may not be protein in nature such as dust, mold, or pollen, may also provoke allergic reactions in some individuals [6]. Food allergy is the most frequent cause of anaphylaxis treated in emergency rooms [7, 8]. Adverse reactions to foods occur at all ages. Epidemiological data suggest that food allergies affect about 5–8% of young children and 3–4% of adults in developed countries [2, 9–12].

There are only very few foods responsible for the majority of food allergies [11]. Foods and ingredients known to cause hypersensitivity include fish (i.e., both saltwater and freshwater finfish) and fish products, crustacean (e.g., shrimp, prawns, crab, lobster, and crayfish) and their products, mollusks (e.g., snails, oysters, clams, squid, octopus, and cuttlefish), eggs and egg products, milk and milk products, cereals containing gluten (e.g., wheat, rye, barley, and their hybridized strains and products), peanuts, soybeans and their products, tree nuts (e.g., almonds, walnuts, pecans, cashews, Brazil nuts, hazelnuts, pistachios, pine nuts, macadamia nuts, chestnuts, and hickory nuts) and their products, and seeds (e.g., sesame seeds, poppy seeds, sunflower seeds, cottonseeds, peas, and lentils) [1]. Any one or a few of the many proteins in each of these foods

Allergen Management in the Food Industry, Edited by Joyce I. Boye and
Samuel Benrejeb Godefroy

is capable of acting as an allergen. In all, over 170 foods are known to provoke allergic reactions in humans [1].

Food allergens from animal (including fish and shellfish) and plant origins have several biochemical and physicochemical properties in common [12]. These include high thermostability conferred by intramolecular disulfide bonds, posttranslational N-glycosylation, resistance to proteolysis, and an enhanced capacity to bind ligands like cations, lipids, or steroids. Well-known food allergens include the major seafood allergens parvalbumin and tropomyosin, and cow's milk allergens casein, β-lactoglobulin, and α-lactalbumin. Other allergens include Sin a 1, a storage 2S albumin from mustard; 11S and 7S storage globulins of soy [12]; Mal d 1, the 18-kD protein linked with pollen allergies in apple [13, 14]; Pru av 1, the major allergen from cherry; Api g 1 from celery [15–17]; Pyr c 1 from pear [18]; Cor a 1 from hazelnut [19]; Dau c 1 from carrot [20]; and Ara h 1 and Ara h 2 from peanut [21]. The subject of food allergy is, thus, quite vast. This chapter will specifically focus on allergies elicited by seafood consumption and seafood processing and will highlight the allergenic proteins involved as well as the major foods and ingredients of most concern to seafood-allergic individuals.

10.2. FISH AND SHELLFISH ALLERGY

Throughout recorded history, the aquatic environment (especially the marine environment) has served as a major source of food (and nutrients) for the human population and other animals [22]. Examples of economically important and most frequently consumed fish and shellfish species in North America include tuna, trout, salmon, cod, halibut, Alaska pollock, catfish, shrimp, lobster, the snow crab, oysters, and clams (Tables 10.1–10.3). In addition, a variety of imitation surimi-type products, such as imitation crab meat, shrimp, sausages, sauces, and flavors are produced from different species of fish and shellfish. Consumption of seafood has increased in recent years due to its high nutritive value and perceived superior health benefits.

It is now common knowledge that omega-3 fatty acids, mostly found in fish, have beneficial health effects. These fatty acids have been implicated in preventing age-related macular degeneration [23]. High levels of docosahexaenoic acid (DHA) are found in the retina, specifically in the disk membrane of the outer segments of the receptor cells, suggesting that DHA plays an important role in the retina function [24]. In animals, dietary deprivation of DHA results in abnormal electroretinograms and visual impairment accompanied by lower retinal DHA levels, further suggesting that DHA may be beneficial in reducing the risk of eye diseases that are associated with vascular changes of the retina or choroids [23]. Several beneficial effects have been attributed to long-chain polyunsaturated fatty acid (LC-PUFA) consumption in general. In particular, LC-PUFAs from fish, shellfish, and their supplements have been found to reduce mortality and cardiovascular disease risk [25]. LC-PUFAs have also

TABLE 10.1 Some Economically Important Fish Species of North America

Marine Fishes/Market Name	Common Name	Scientific Name
Alaska pollock	Walleye pollock	*Theragra chalcogramma*
Anchovy (and sardines)	Northern anchovy	*Engraulis mordax*
Bluefish	Bluefish	*Pomatomus saltatrix*
Bonitos	Pacific bonito	*Sarda chiliensis*
Butterfishes	Butterfish	*Peprilus triacanthus*
Capelin	Capelin	*Mallotus villosus*
Cods	Atlantic cod	*Gadus morhua*
Dogfish (and sharks)	Spiny dogfish	*Squalus acanthias*
Flounders	Blackback/winter	*Pseudopleuronectes americanus*
Grenadiers	Rock grenadier	*Coryphaenoides rupestris*
Groupers	Red grouper	*Epinephelus morio*
Haddock	Haddock	*Melanogrammus aeglefinus*
Hakes	White hake	*Urophycis tenuis*
Halibuts	Pacific halibut	*Hippoglossus stenolepis*
Herrings	Pacific herring	*Clupea pallasii*
Jack mackerel	Jack mackerel	*Trachurus symmetricus*
Mackerels	Atlantic mackerel	*Scomber scombrus*
Menhadens	Atlantic menhaden	*Brevoortia tyrannus*
Monkfishes	Goosefish	*Lophius americanus*
Mullets	Striped mullet	*Mugil cephalus*
Ocean catfishes	Atlantic wolffish	*Anarhichas lupus*
Ocean perch	Deepwater redfish	*Sebastes mentalla*
Plaice	American plaice	*Hippoglossoides platessoides*
Pollock	Pollock	*Pollachius virens*
Porgies/scup	Scup	*Stenotomus chrysops*
Rockfishes	Yelloweye rockfish	*Sebastes ruberrimus*
Sablefish	Sablefish	*Anoplopoma fimbria*
Salmon	Atlantic salmon	*Salmo salar*
Salmon	Chinook salmon	*Oncorhynchus tshawytscha*
Salmon	Chum salmon	*Oncorhynchus keta*
Salmon	Coho salmon	*Oncorhynchus kisutch*
Salmon	Pink salmon	*Oncorhynchus gorbuscha*
Salmon	Sockeye salmon	*Oncorhynchus nerka*
Sea basses (and basses)	Black sea bass	*Centropristis striata*
Sea trouts	Gray sea trout	*Cynoscion regalis*
Skates	Thorny skate	*Amblyraja radiata*
Snappers	Red snapper	*Lutjanus campechanus*
Soles	Rock sole	*Lepidopsetta bilineata*
Swordfish	Swordfish	*Xiphias gladius*
Tilefishes	Golden tilefish	*Lopholatilus chamaeleonticeps*
Tunas	Bluefin tuna	*Thunnus thynnus*
Turbot	Greenland turbot	*Reinhardtius hippoglossoides*
Whitings	Pacific whiting	*Merluccius productus*

TABLE 10.2 Some Economically Important Freshwater Fish Species of North America

Freshwater Fishes/Market Name	Common Name	Scientific Name
Basses	Bigmouth bass	*Micropterus salmoides*
Catfishes	Channel catfish	*Ictalurus punctatus*
Char	Arctic char	*Salvelinus alpinus*
Eels	American eel	*Anguilla rostrata*
Trout (inconnu)	Rainbow trout	*Stenodus leucichthys*
Pike	Northern pike	*Esox lucius*
Smelts	Rainbow smelt	*Osmerus mordax*
Sturgeons	Lake sturgeon	*Acipenser fulvescens*
Trouts	Lake trout	*Salvelinus namaycush*
Walleye	Walleye pike	*Sander vitreus*
Whitefishes	Lake whitefish	*Coregonus clupeaformis*
Yellow perch	Yellow perch	*Perca flavescens*

TABLE 10.3 Some Economically Important Shellfish Species of North America

Shellfishes/Market Name	Common Name	Scientific Name
Abalones	Red abalone	*Haliotis rufescens*
Clam	Soft-shell clam	*Mya arenaria*
Clams	Northern quahog/hard-shell	*Mercenaria mercenaria*
Clams (Atlantic)	Atlantic surf clam	*Spisula solidissima*
Clams (Pacific)	Geoduck clam	*Panopea abrupta*
Crabs	Blue crab	*Callinectes sapidus*
Crabs	Snow crab	*Chionoecetes opilio*
Crabs	Dungeness crab	*Cancer magister*
Crayfishes	Red swamp crayfish	*Procambarus clarkii*
Lobsters	American lobster	*Homarus americanus*
Mussels	Blue mussel	*Mytilus edulis*
Oysters	Eastern oyster	*Crassostrea virginica*
Scallops	Sea scallop	*Placopecten magellanicus*
Shrimps (pandalid)	Northern (pink) shrimp	*Pandalus borealis*
Shrimps (penaeid)	Brown shrimp	*Farfantpenaeus aztecus*
Squids	Northern shortfin squid	*Illex illecebrosus*

been suggested to alleviate attention deficit/hyperactivity disorder (AD/HD), autism, dyspraxia, dyslexia, aggression [26], and inflammatory diseases [27].

Increasing awareness of the health benefits of omega-3 fatty acids has resulted in more food manufacturers and consumers looking for ways to increase fish in their formulations and diets. Furthermore, public health concerns with certain muscle foods (e.g., beef and its higher cholesterol content

TABLE 10.4 Fish and Shellfish Species Known to Cause Allergic Reactions in Humans

Phylum	Class	Species
Mollusks	Gastropods	Abalone, alikreukel, snails
	Bivalves	Clams, cockles, mussels, oysters, scallops
	Cephalopods	Cuttlefish, octopus, squid
Arthropods	Crustacea	Crabs, lobsters, prawns, shrimps, crayfish (freshwater)
Chordates	Osteichthyes	Bonito, cod, flounder, grouper, haddock, halibut, hake, herring, mackerel, pike, plaice, salmon, sole, snapper, sprat, trout, tuna
	Condrichthyes	Rays/skates, sharks

as well as the mad cow disease or bovine spongiform encephalopathy scare, pork and related swine flu, poultry and avian flu) have augured well for the increased consumption of fish, shellfish, and their products. Apart from commercial fishing, the aquaculture industry also supplies an additional amount of fish and shellfish to boost the natural harvest from the wild.

Consumption, handling, or exposure of certain individuals to fish and shellfish, however, induces allergic-type reactions in these individuals. In fact, seafood (encompassing fish, shellfish, and edible seaweeds) is one of the most common food sources that may trigger allergic reactions in consumers [28]. Studies suggest that seafood can cause severe adverse reactions in sensitized individuals after direct exposure or otherwise to minute amounts [29–31]. Examples of fish and shellfish known to produce allergic-type responses are listed in Table 10.4.

Fish and shellfish handling/processing include practices involving crustacean species (crabs, lobsters, prawns, and shrimp) such as "tailing" (lobsters), "cracking," butchering and degilling (crabs), semi-mechanized peeling, heading and deveining and "blowing" pressurized water or compressed air (shrimps), or cooking, boiling, steaming, cutting, mincing, scrubbing or washing of these products; shucking (oysters), depuration, chopping, dicing and slicing in the case of mollusks; and heading, gutting or evisceration, skinning, mincing, trimming, filleting, cooking, salting, milling and bagging with regular finfish and cartilaginous fish [32]. These handling and processing operations invariably create an environment that bring people in the workplace in direct or indirect contact with fish components that could produce allergic reactions. In a study by Caiaffa et al. [33], it was reported that a severe IgE-mediated fish allergy could be triggered by various kinds of indirect contact such as inhalation of airborne cooking fish particles and kissing (both passionate and platonic). The study further indicated that ingestion of chicken by a person not allergic to chicken could elicit anaphylaxis shock if the chicken had been fed with fish-based fodder.

Lehrer et al. [22] evaluated seafood-induced allergies in 30 shrimp-sensitive and 37 fish-allergic individuals and observed that generalized itching, urticaria, and swelling of the lips and tongue were among the symptoms experienced. In addition, others have reported pulmonary, gastrointestinal symptoms, and anaphylactic shock [34–37]. Thus, food-induced allergic reactions elicited by consumption of seafood products are similar to those experienced as a result of the consumption of a variety of other food allergy-causing products [22, 38].

The health benefits associated with the consumption of fish and shellfish have resulted in a steady increase in per capita consumption of these products. This increase is expected to continue in view of growing consumer concerns about beef, pork, and sometimes poultry. Consequently, new food products that incorporate fish and shellfish are being developed [39], which raises concerns about fish- and shellfish-induced allergic reactions in sensitized individuals.

10.2.1. Seafood Allergens

Researchers have identified a few proteins in seafoods that are considered as major allergens. A major allergen denotes a substance originating from an appropriate source material that will induce an IgE response in more than 50% of a sample of patients exposed to the allergen source material and who have clinical symptoms on contact with that source material [22, 40]. Parvalbumin, Gad c 1, a 12-kDa molecular mass protein, is reported as a major allergen in codfish (*Gadus callarias*) [41]. The muscle protein, tropomyosin, with a molecular mass of ~36 kDa [35–37], is reported as the major allergen in shrimp [42, 43]. Tropomyosin is also present in lobster [44, 45], crab [46], and mollusks such as squid, oyster, snail, mussels, clam, and scallops [47–49].

Both parvalbumin in fish and tropomyosin in shellfish are mostly found in muscle cells. Lehrer et al. [22] reported that tropomyosin is a major muscle protein that is present in all living animals. Structurally, the molecules appear to be highly conserved based on the substantial amino acid sequence identity of tropomyosin from unrelated species. The portion of a food protein that may cause allergic reaction may be a simple chain of a few amino acids on the primary structure (linear epitope) or a unique three-dimensional conformation of the protein structure (conformational epitope) [50]. About 60% of the amino acid sequence of shrimp tropomyosin is similar to those of vertebrates. Tropomyosin is composed of two polypeptide chains, each in α-helix formation coiled around one another in the coiled-coil formation [51]. Although the structure is well known, there is a paucity of information about the IgE-binding epitopes and no information about any T-cell epitopes.

Reese et al. [52] screened a unidirectional expression cDNA library from the tail muscle of the shrimp *Penaeus aztecus*, with a Pen a 1-specific (i.e., the shrimp muscle protein tropomyosin) antibody and identified four possible clones. All four recombinant proteins were recognized in immunoblot analysis by serum IgE from shrimp-allergic patients. They purified one of the clones and used it to construct a peptide library, which was then screened with individual serum from shrimp-allergic subjects to identify four IgE-reactive peptides that were between 13 and 21 amino acids long. All the peptides were found in the second half of the molecule including the carboxyl terminus. Similarly, areas of the tropomyosin molecule that contain important IgE-binding regions were also identified by Ayuso et al. [53, 54] using synthetic overlapping peptides. In all, five major IgE-binding regions representing eight epitopes have been identified in shrimp tropomyosin [22]. Studies by Leung et al. [47] indicated that the allergic epitopes on tropomyosin are conserved among invertebrates including insects, suggesting that persons sensitive to shrimp could be sensitized to or have cross-reactivities to some insects.

10.2.2. Occupational Seafood Allergy

The increase of fish and crustacean allergy is mainly attributed to the increased consumption of seafood. However, workers in the seafood processing industry are likely to develop an occupational seafood allergy due to direct skin contact during handling seafood or inhalation of seafood aerosols. This may occur during cooking or when cleaning storage tanks with pressured water [32, 55]. Seitz et al. [55] reported of a truck driver who acquired fish and crustacean allergy by direct skin and mucosal contact due to unprotected handling of fresh seafood. The clinical symptoms of his allergy started as contact urticaria and progressed to generalized urticaria and later anaphylaxis and occupational asthma.

Occupational reactions have been reported among a variety of seafood workers. These include fishermen, fish and prawn workers, seafood processing workers, canners, restaurant cooks, and other workers in the seafood industry [22, 56–60]. The snow crab industry is one of the best investigated seafood industries for occupational allergic reactions. Cartier and coworkers [56] demonstrated that workers in the snow crab industry were exposed to occupational allergens through direct contact with seafood products as well as inhalation of bits of seafood or water droplets generated during processing. Out of the 303 crab workers investigated, 18% reported rhinitis or conjunctivitis, about 24% reported some sort of skin rash, and over a third reported asthma. In Germany, about 9500 cases of occupational skin diseases were among a total of 25,000 occupational diseases reported in 2005 [55]. The most important heat- and ingestion-resistant fish allergens are parvalbumins, for example, the 12-kDa muscle protein Gad c 1 [61]. About 70% of all patients with fish allergy develop symptoms to several fish species, while the remaining 30% react to only one fish species. Thus, even in jobs with limited direct exposure, workers not using appropriate skin protection are at risk for developing occupational seafood allergy, which may require termination of employment [55].

10.3. FOOD PRODUCTS CONTAINING FISH AND FISH INGREDIENTS

Apart from allergic reactions elicited by direct ingestion of fish/shellfish as food and occupational/workplace hazards, food products fabricated with fish ingredients could also be important sources of fish allergens. Food products of concern in this regard include stocks and soups, and dishes such as paella (a Spanish dish made from rice that may contain lobster and/or shrimp), bouillabaisse (a traditional Provençal stew containing vegetables and different kinds of cooked fish and shellfish), fritto misto (mixed fried fish and shellfish dish popular in Italy and coastal Mediterranean countries), Asian foods (may contain fish sauce and/or chopped pieces of different kinds of fish/shellfish), gumbo (a traditional stew usually served with rice and vegetables along the Gulf coast of the United States), surimi-type products (made from white fish and/or shellfish extracts), Caesar salad dressing and Worcestershire sauce (both of which may contain anchovies), kedgeree (a dish prepared from rice and fish), fish sauce and fish paste (produced by fermentation, common examples are nuoc mam and nam pla), and patum peprium (a dish prepared with anchovies).

Products derived from fish and shellfish, such as glucosamine, chitinous polymers, some calcium supplements, fish gelatin, and fish oils, may also produce allergic reactions in some individuals depending on the concentration of residual fish proteins. Glucosamine is produced from the exoskeletons of shellfish and is used for the treatment of arthritis. If not of high purity, the glucosamine may have proteins coextracted with it, which could pose allergic hazards to some individuals. A similar situation could occur for chitinous polymers (chitin and chitosan) used in commercial fat absorbers and in skin moisturizers. Calcium supplements derived from shellfish may also contain residual shells with proteins, which may have the potential to induce allergic reactions in certain individuals.

An area of significant concern today is the potential allergenic risk of fish oils and fish oil capsules. Fish oil has well-documented health benefits by virtue of its high omega-3 PUFA content as previously indicated [62]. Interest in its incorporation into foods has, therefore, grown tremendously. Consumers have expressed concerns about the potential threat this could pose to individuals who are allergic to fish/shellfish products. As with most oils, adverse reactions can occur if residual proteins are present. Highly sensitized patients can react to these minute amounts of allergenic proteins. Refined oils generally have little or no residual proteins and are considered to be of less risk. However, cold-pressed oils, which are considered to be of higher quality, may contain much higher levels of protein and, therefore, pose greater risks. Fish oil is obtained by cooking body tissues of fatty fish and extracting the oil. The residual protein content will depend on the specific unit operations used to refine the oil, and thus they may, or may not, contain relevant concentrations of allergenic proteins.

Interestingly, several research studies have suggested that consumption of fish oils could reduce allergic sensitization. Olsen et al. [62] compared the effects of marine fish oil with olive oil intake by women in late pregnancy on the risk of asthma in their offspring, and found that increasing intake of omega-3 PUFAs had prophylactic effects in relation to the occurrence of asthma in offspring. Similar observations were made by Salam et al. [63] and Romieu et al. [64] in their studies on the effects of fish consumption by mothers during pregnancy and early childhood asthma. Sausenthaler et al. [65] and Calvani et al. [66] reported reduced allergic sensitization in offsprings of pregnant mothers fed with fish diets high in omega-3 PUFAs versus their counterparts who received fats/oils from other sources (e.g., butter or margarine). Marine omega-3 PUFAs with their high content of eicosapentaenoic acid (EPA) and DHA may impact fetal immune function via various mechanisms, that is, via incorporation of EPA and DHA in cell membranes at the expense of arachidonic acid; arachidonic acid formation is further reduced via competition for cyclooxygenase and lipoxygenase enzymes by the EPA and DHA, both processes effectively diminishing the synthesis of the proinflammatory eicosanoids derived from arachidonic acid [62]. Marine omega-3 PUFAs may also form resolvins and neuroprotectins that have anti-inflammatory and neuroprotective effects [67, 68]. Omega-3 PUFAs are also known to lower production of prostaglandin E2 (PGE2), which can alter the balance of Th1 and Th2 cytokines and the production of immunoglobulin E, which is needed for binding with allergens to elicit allergic reactions.

Fish gelatin is another product that has spurred much interest. Fish gelatin is produced from fish collagen and is used in a variety of products in the food and pharmaceutical industries. Some uses include as a gelling agent in cooking, for making gelatin desserts (trifles, aspic, marshmallows) and confectioneries (gummy bears, jelly beans), as stabilizers or texturizers (in jams, yogurt, and cream cheese), in fat-reduced foods to impart "mouthfeel," as a bulking agent to create volume without adding calories, as coatings or shells of pharmaceutical capsules, as a carrier agent for imparting color to beverages, as a hydrating agent in cosmetics, and as a biological substrate to culture adherent cells. Traditionally, bovine and porcine gelatins have been mostly used. The bovine spongiform encephalopathy (BSE) scare and religious beliefs have made the use of gelatin from these sources not acceptable to some consumers, thus, the interest in the use of fish as source of gelatin. Unlike gelatin from bovine and porcine sources, fish gelatin is also considered suitable for kosher and halal diets.

Fish gelatin is made from the skins of edible fish, mostly codfish. The successive steps for making fish gelatin of high purity involve rinsing/washing of the skins with water, draining of the water, and acidification to a final pH of 4.5–5.0 with acetic acid, followed by heating and agitation at 190–200°C for 3–6 hours. This is followed by filtration to clarify the extract and separation of the hot extract by ion exchange chromatography (IEX) to remove residual salts and heavy metals. The purified liquor is concentrated by evaporation followed by drying and grinding into a fine powder. Fish gelatin produced this way is typically devoid of residual fish flesh that may have been associated with the raw material. Thus, the process should remove the major fish allergen parvalbumin because of its high water solubility. Although there were no reports of clinical reactions to fish gelatin in processed foods, it was useful to investigate the possible allergenicity of this product, since it was derived from the skins of fish species (like cod) known to produce allergic responses in sensitized individuals. Hansen et al. [69] conducted a study to verify the allergenicity of codfish gelatin, and observed no adverse allergic response in a group of 30 patients receiving a cumulative dose of 3.61 g of fish gelation. However, when the gelatin level was increased to a cumulative dose of 7.61 g of fish gelatin, one patient out of the 30 showed a mild allergic response. Based on these findings, the authors concluded that highly purified fish gelatin does not appear to pose health risks to individuals with fish allergy. These observations are consistent with previous findings [70] in which oral dosage of tuna fish gelatin at a level of 5 g did not produce adverse effects in patients who had shown IgE binding to tuna skin gelatin.

Isinglass is a related substance to collagen (gelatin). It is produced from the swim bladders of fish species like sturgeon and cod and is used for the clarification of alcoholic beverages. Although no direct investigation of the possible allergenicity of isinglass was found in the literature, Hansen et al. [69] suggested that the findings from their study with codfish skin gelatin was likely relevant to all forms of fish gelatins and to isinglass, since collagen from various fish species display a high degree of structural homology.

10.4. EFFECTS OF PROCESSING ON ALLERGENICITY

Proteins are known to undergo modifications when subjected to various physical and chemical treatments. Some of the changes that occur include conformational changes in the higher orders of protein structure, unfolding and aggregation, as well as chemical modifications. Such changes in the protein structure could potentially affect existing epitopes on protein molecules or generate new ones [50]. Thus, processing operations including thermal processing, irradiation, as well as high-pressure and chemical treatments that cause conformational changes in food proteins, may influence the allergenicity of foods. In general, the antigenicity/immunogenicity of heat-labile proteins is reduced by heat treatment, while that of thermostable allergenic proteins is affected less or not at all by thermal processing [71]. An example of this is the codfish allergen, Gad c 1, which is resistant to heat, and its allergenicity is not alleviated by heat treatment. The situation appears to be different for tuna and salmon, which have both been shown to be more allergic in the raw form compared with the heat processed (canned) forms [72].

Enzymatic hydrolysis also appears to have mixed effects on food allergens. For example, the codfish allergen, Gad c 1, is quite resistant to proteolysis, and a combination of four proteases was necessary to achieve destruction of IgE binding [73]. However, allergenicity induced by rice was reduced by hydrolysis with actinase [74], bromelain treatment reduced allergenicity in wheat [75], while combination treatment with trypsin, elastase, and pepsin reduced allergenicity of hazelnut [76]. Physical treatments like ultracentrifugation was shown to reduce allergenicity of cow's milk proteins [77]; however, similar studies have not been reported for fish/shellfish allergens. As indicated above, highly refined oils produced by solvent extraction are believed to pose no threat to persons who suffer allergic reactions, since the proteins in the raw material are removed [78]. However, cold-pressed oils and other minimally processed oils may pose health risks as they could contain some residual protein [79].

Food processing may additionally be associated with other chemical changes that can influence allergenicity. For example, nonenzymatic glycation reactions can take place between carbonyl groups of sugars and the primary amino groups of proteins, peptides, or amino acids to give rise to a plethora of compounds collectively known as Amadori products. The cross-linkages that occur during these reactions can affect allergenic epitopes modifying their immunogenic properties. Studies on the IgE cross-reactivity of bread revealed enhanced IgE reactivity by the Amadori products [80]. Similar studies are, however, lacking for seafood allergens.

10.5. CONCLUDING REMARKS

Food allergy is an important public health problem, which is particularly bothersome because of its prevalence, potential severity, and lack of curative therapies. Of the priority allergens, fish and shellfish are of particular concern because these allergies persist in the adult population and can potentially induce life-threatening responses. Unfortunately, there is relatively less information on seafood allergy compared with the other priority allergens. While some therapeutic approaches have been described to manage food allergy in general, they are not without their limitations. The current standard of practice is to educate patients to avoid ingesting foods that trigger allergic reactions and initiate therapy quickly if ingestion should occur. This prescription is reactive rather than proactive. The growing use of seafood and seafood ingredients in food production increases concerns about the management of seafood allergy risks in the general population. Questions persist about the safety of ingredients, processing aids, functional foods, and nutraceutical products such as gelatin, isinglass, and omega-3 capsules. While some answers may be gleaned from the existing literature, they do not provide adequate information to address questions raised by consumers and the food industry. Thus, sustained research is needed to generate relevant data that will allow the food industry and regulatory agencies to effectively address some of the gaps identified.

REFERENCES

1. Taylor, S.L. (2000). Emerging problems with food allergens. In: *Food, Nutrition and Agriculture–26*. Rome: FAO Food and Nutrition Division, pp. 14–29. Available at http://www.fao.org/docrep/003/X7133M/x7133m03.htm#P0_0.
2. Kanny, G., Moneret-Vautrin, D.A., Flabbee, J., Beaudouin, E., Morisset, M., Thevenin, F. (2001). Population study of food allergy in France. *The Journal of Allergy and Clinical Immunology*, *108*, 133–140.
3. Sampson, H. (1992). Food hypersensitivity: manifestation, diagnosis, and natural history. *Food Technology*, *46*, 141–144.
4. Melcalfe, D. (1989). Diseases of food hypersensitivity. *The New England Journal of Medicine*, *321*, 255–257.
5. Bock, S. (1982). The natural history of food hypersensitivity. *The Journal of Allergy and Clinical Immunology*, *69*, 173–177.
6. Breneman, J.C. (1987). *Handbook of Food Allergies*. New York: Marcel Dekker.
7. Lehrer, S.B., Laurianne, G.W., Bost, K.L., Sorensen, R.U. (1999). Immunotherapy for food hypersensitivity. *Immunology and Allergy Clinics of North America*, *19*, 563–580.
8. Yocum, M.W., Khan, D.A. (1994). Assessment of patients who have experienced anaphylaxis: a three year survey. *Mayo Clinic Proceedings*, *69*, 16–23.
9. Sicherer, S.H., Munoz-Furlong, A., Sampson, H.A. (2004). Prevalence of seafood allergy in the United States determined by a random telephone survey. *The Journal of Allergy and Clinical Immunology*, *114*, 159–165.
10. Munoz-Furlong, A., Sampson, H.A., Sicherer, S.H. (2004). Prevalence of self-reported seafood allergy in the U.S. *The Journal of Allergy and Clinical Immunology*, *113*, S100.
11. Sampson, H.A. (2004). Update on food allergy. *The Journal of Allergy and Clinical Immunology*, *113*, 805–819.
12. Breiteneder, H., Mills, E.N.C. (2005). Molecular properties of food allergens. *The Journal of Allergy and Clinical Immunology*, *115*, 14–23.
13. Vanek-Krebitz, M., Hoffmann-Sommergruber, K., Laimer da Camara, M., Susani, M., Ebner, C., Kraft, D., Scheiner, O., Breiteneder, H. (1995). Cloning and sequencing of Mal d 1, the major allergen from apple (*Malus domestica*), and its immunological relationship to Bet v 1, the major birch pollen allergen. *Biochemical and Biophysical Research Communications*, *214*, 538–551.
14. Son, D.Y., Scheurer, S., Haustein, D., Vieths, S. (1999). Pollen-related food allergy: cloning and immunological analysis of isoforms and mutants of Mal d 1, the major apple allergen, and Bet v 1, the major birch pollen allergen. *European Journal of Nutrition*, *38*, 201–215.
15. Scheurer, S., Metzner, K., Haustein, D., Vieths, S. (1997). Molecular cloning, expression and characterization of Pru av 1, the major cherry allergen. *Molecular Immunology*, *34*, 619–629.
16. Scheurer, S., Son, D.Y., Bohm, M., Karamloo, F., Franke, S., Hoffmann, A., Haustein, D., Vieths, S. (1999). Cross-reactivity and epitope analysis of Pru a 1, the major cherry allergen. *Molecular Immunology*, *36*, 155–167.

17. Breiteneder, H., Hoffmann-Sommergruber, K., O'Riordain, G., Susani, M., Ahorn, H., Ebner, C., Kraft, D., Scheiner, O. (1995). Molecular characterization of Api g 1, the major allergen of celery (*Apium graveolens*), and its immunological and structural relationships to a group of 17 kDa tree pollen allergens. *European Journal of Biochemistry/FEBS*, *233*, 484–489.
18. Karamloo, F., Scheurer, S., Wangorsch, A., May, S., Haustein, D., Vieths, S. (2001). Pyr c 1, the major allergen from pear, is a new member of the Bet v 1 allergen family. *Journal of Chromatography*, *756*, 281–293.
19. Luttkopf, D., Muller, U., Skov, P.S., Ballmer-Weber, B.K., Wuthrich, B., Skamstrup Hansen, K., Poulsen, L.K., Kastner, M., Haustein, D., Vieths, S. (2002). Comparison of four variants of a major allergen in hazelnut (*Corylus avellana*) Cor a 1.04 with the major hazel pollen allergen Cor a 1.01. *Molecular Immunology*, *38*, 515–525.
20. Hoffmann-Sommergruber, K. (2000). Plant allergens and pathogenesis-related proteins. What do they have in common? *International Archives of Allergy and Immunology*, *122*, 155–166.
21. Gruber, P., Vieths, S., Wangorsch, A., Nerkamp, J., Hofmann, T. (2004). Maillard reaction and enzymatic browning affect the allergenicity of Pru av 1, the major allergen from cherry (*Prunus avium*). *Journal of Agricultural and Food Chemistry*, *52*, 4002–4007.
22. Lehrer, S.B., Ayuso, R., Reese, G. (2003). Seafood allergy and allergen: a review. *Marine Biotechnology*, *5*, 339–348.
23. Hodge, W.G., Schachter, H.M., Barnes, D., Pan, Y., Lowcock, E.C., Zhang, L., Sampson, M., Morrison, A., Tran, K., Miguelez, M., Lewin, G. (2005). Efficacy of ω-3 fatty acids in preventing age-related macular degeneration. *Ophthalmology*, *113*, 1165–1173.
24. Rodriguez de Turco, E.B., Gordon, W.C., Bazan, N.G. (1991). Rapid and selective uptake, metabolism, and cellular distribution of docosahexaenoic acid among rod and cone photoreceptor cells in the frog retina. *The Journal of Neuroscience*, *11*, 3667–3678.
25. Griffin, B.A. (2008). How relevant is the ratio of dietary n-6 to n-3 polyunsaturated fatty acids to cardiovascular disease risk? Evidence from the OPTILIP study. *Current Opinion in Lipidology*, *19*, 57–62.
26. Kidd, P.M. (2007). Omega-3 DHA and EPA for cognition, behaviour, and mood: clinical findings and structural-functional synergies with cell membrane phospholipids. *Alternative Medicine Review*, *12*, 207–227.
27. Galatowicz, G., Ajayi, Y., Stern, M.E., Calder, V.L. (2007). Ocular anti-allergic compounds selectively inhibit human mast cell cytokines *in vitro* and conjunctival cell infiltration *in vivo*. *Clinical and Experimental Allergy*, *37*, 1648–1656.
28. Taylor, S.L., Bush, R.K. (1988). Allergy by ingestion of seafoods. In: *Marine Toxins and Venoms*, Vol. 3, Tu, A.T. (Ed.). New York: Marcel Dekker, pp. 149–183.
29. Bahna, S.L. (2004). You can have fish allergy and eat it too! *The Journal of Allergy and Clinical Immunology*, *114*, 125–126.
30. Bahna, S.L. (1994). Exquisite food allergy without eating. *Allergy*, *49*, 129–130.

31. Crespo, J.F., Pascual, C., Dominguez, C., Ojeda, I., Munoz F.M., Martin-Esteban, M. (1995). Allergic reactions associated with airborne fish particles in IgE-mediated fish hypersensitive patients. *Allergy*, *50*, 257–261.

32. Jeebhay, M.F., Robins, T.G., Lehrer, S.B., Lopata, A.L. (2001). Occupational seafood allergy: a review. *Occupational and Environmental Medicine*, *58*, 553–562.

33. Caiaffa, M.F., De Serio, A., Lotti, A., Kourtis, G., Macchia, L. (2008). Indirect food contact in severe fish allergy. *Revue Française d'Allergologie et d'Immunologie Clinique*, *48*, 111–112.

34. O'Neil, C., Helbling, A.A., Lehrer, S.B. (1993). Allergic reactions to fish. *Clinical Reviews in Allergy*, *11*, 183–200.

35. Daul, C.B., Morgan, J.E., Lehrer, S.B. (1993). Hypersensitivity reactions to crustacean and molluscs. *Clinical Reviews in Allergy*, *11*, 201–222.

36. Daul, C.B., Slattery, M., Morgan, J.E., Lehrer, S.B. (1993). Common crustacean allergens: identification with shrimp-specific allergens. In: *Molecular Biology and Immunology of Allergens*, Kraft, D., Sehon, A. (Eds.). Baton Rouge, LA: CRC Press, pp. 291–293.

37. Daul, C.B., Slattery, M., Reese, G., Lehrer, S.B. (1994). Identification of the major brown shrimp (*Penaeus aztecus*) as the muscle protein tropomyosin. *International Archives of Allergy and Clinical Immunology*, *105*, 49–55.

38. Lehrer, S.B., Reese, G. (1998). Food allergens: implications for biotechnology. In: *Biotechnology and Safety Assessment*, 2nd edition, Thomas, J.A. (Ed.). New York: Taylor and Francis, pp. 127–150.

39. Lehrer, S.B., Helbling, A., Daul, C.B. (1992). Seafood allergy: prevalence and treatment. *Journal of Food Safety*, *13*, 61–76.

40. Lofkvist, T., Agrell, B., Dreborg, S., Svesson, G. (1994). Effects of immunotherapy with a purified standardized allergen preparation of *Dermatophagoides farinae* in adults with perennial allergic rhinoconjunctivitis. *Allergy*, *49*, 100–107.

41. Elsayed, S., Bennich, H. (1975). The primary structure of allergen M from cod. *Scandinavian Journal of Immunology*, *4*, 205–208.

42. Shanti, K.N., Martin, B.M., Nagpal, S., Metcalfe, D.D., Sabba-Rao, P.V. (1993). Identification of tropomyosin as the major shrimp allergen and characterization of its IgE binding epitopes. *Journal of Immunology*, *151*, 5354–5363.

43. Leung, P.S., Chu, K.H., Chow, W.K., Aftab, A., Bandea, C.I., Kwan, H.S., Nagy, S.M., Gershwin, M.E. (1994). Cloning, expression, and primary structure of *Metapenaeus ensis* tropomyosin, the major heat-stable shrimp allergen. *The Journal of Allergy and Clinical Immunology*, *92*, 837–845.

44. Leung, P.S., Chen, Y.C., Mykles, D.L., Chow, W.K., Li, C.P., Chu, K.H. (1998). Molecular identification of the lobster muscle protein tropomyosin as a seafood allergen. *Molecular Marine Biology and Biotechnology*, *7*, 12–20.

45. Mykles, D.L., Cotton, J.L., Taniguchi, H., Sano, K., Maeda, Y. (1998). Cloning of tropomyosins from lobster (*Homarus americanus*) striated muscles, fast and slow isoforms may be generated from the same transcript. *Journal of Muscle Research and Cell Motility*, *19*, 105–115.

46. Leung, P.S., Chen, Y.C., Gershwin, M.R., Wong, S.H., Kwan, H.S., Chu, K.H. (1998). Identification and molecular characterization of *Charybdis feriatus* tropomyosin,

the major crab allergen. *The Journal of Allergy and Clinical Immunology*, *102*, 847–852.

47. Leung, P.S., Chow, W.K., Duffey, S., Kwan, H.S., Gershwin, M.E., Chu, K.H. (1996). IgE reactivity against a cross-reactive allergen in crustacean and mollusca: evidence for tropomyosin as the common allergen. *The Journal of Allergy and Clinical Immunology*, *98*, 954–961.
48. Ishikawa, M., Ishida, M., Shimakura, K., Nagashima, Y., Shiomi, K. (1998). Purification and IgE-binding epitopes of a major allergen in the gastropod *Turbo cornutus*. *Bioscience, Biotechnology, and Biochemistry*, *62*, 1337–1343.
49. Chu, K.H., Wong, S.H., Leung, P.S. (2000). Tropomyosin is the major mollusc allergen: reverse transcriptase polymerase chain reaction, expression and IgE reactivity. *Marine Biotechnology*, *2*, 499–509.
50. Sathe, S.K., Teuber, S.S., Roux, K.H. (2005). Effects of food processing on the stability of food allergens. *Biotechnology Advances*, *23*, 423–429.
51. Smillie, L.B. (1979). Structure and functions of tropomyosins from muscle and non-muscle sources. *Trends in Biochemical Sciences*, *4*, 151–155.
52. Reese, G., Jeoung, B.J., Daul, C.B., Lehrer, S.B. (1997). Characterization of recombinant shrimp allergen Pen a 1 (tropomyosin). *International Archives of Allergy and Immunology*, *113*, 240–242.
53. Ayuso, R., Reese, G., Lehrer, S.B. (1999). IgE-antibody recognition of Pen a 1 (shrimp tropomyosin): localization of continuous, antigenic sites by synthetic overlapping peptides. *The Journal of Allergy and Clinical Immunology*, *103*, S866.
54. Ayuso, R., Lehrer, S.B., Reese, G. (2002). Identification of continuous, allergenic regions of the major shrimp allergen Pen a 1(tropomyosin). *International Archives of Allergy and Immunology*, *127*, 27–37.
55. Seitz, C.S., Brocker, E.B., Trautmann, A. (2008). Occupational allergy due to seafood delivery: case report. *Journal of Occupational Medicine and Toxicology*, *3*, 11–13.
56. Cartier, A., Malo, J.L., Forest, F., Lafrance, M., Pineau, L., St-Aubin, J.J., Dubois, J.Y. (1984). Occupational asthma in snow crab-processing workers. *The Journal of Allergy and Clinical Immunology*, *74*, 261–269.
57. Cartier, A., Malo, J.L., Ghezzo, H., McCants, M., Lehrer, S.B. (1986). IgE sensitization in snow crab-processing workers. *The Journal of Allergy and Clinical Immunology*, *78*, 344–348.
58. Desjardins, A., Malo, J.L., L'Archeveque, J., Cartier, A., McCants, M., Lehrer, S.B. (1995). Occupational IgE-mediated sensitization and asthma caused by clam and shrimp. *The Journal of Allergy and Clinical Immunology*, *96*, 608–617.
59. Lemiere, C., Desjardins, A., Lehrer, S., Malo, J.L. (1996). Occupational asthma to lobster and shrimp. *Allergy*, *51*, 272–273.
60. O'Neil, C., Lehrer, S.B. (1996). Occupational reactions to food allergens. In: *Food Allergy: Adverse Reactions to Foods and Food Additives*, Metcalfe, D.D., Sampson, H.A., Simon, R.A. (Eds.). Oxford: Blackwell Publishing, pp. 311–335.
61. Van Do, T., Elsayed, S., Florvaag, E., Hordvik, I., Endresen, C. (2005). Allergy to fish parvalbumins: studies on the cross-reactivity of allergens from 9 commonly consumed fish. *The Journal of Allergy and Clinical Immunology*, *116*, 1314–1320.
62. Olsen, S.F., Østerdal, M.L., Salvig, J.D., Mortensen, L.M., Rytter, D., Secher, N.J., Henriksen, T.B. (2008). Fish oil intake compared with olive oil intake in late

pregnancy and asthma in the offspring: 16 y of registry-based follow-up from a randomized controlled trial. *The American Journal of Clinical Nutrition*, *88*, 167–175.

63. Salam, M.T., Li, Y.F., Langholz, B., Gilliland, F.D. (2005). Maternal fish consumption during pregnancy and risk of early childhood asthma. *The Journal of Asthma*, *42*, 513–518.
64. Romieu, I., Torrent, M., Garcia-Esteban, R., Ferrer, C., Ribas-Fitó, N., Antó, J.M., Sunyer, J. (2007). Maternal fish intake during pregnancy and atopy and asthma in infancy. *Clinical and Experimental Allergy*, *37*, 518–525.
65. Sausenthaler, S., Koletzko, S., Schaaf, B., Lehmann, I., Borte, M., Herbarth, O., Von Berg, A., Wichmann, H.E., Heinrich, J. (2007). Maternal diet during pregnancy in relation to eczema and allergic sensitization in the offspring at 2 y of age. *The American Journal of Clinical Nutrition*, *85*, 530–537.
66. Calvani, M., Alessandri, C., Sopo, S.M., Panetta, V., Pingitore, G., Tripodi, S., Zappalà, D., Zicari, A.M. (2006). Consumption of fish, butter and margarine during pregnancy and development of allergic sensitizations in the offspring: role of maternal atopy. *Pediatric Allergy and Immunology*, *17*, 94–102.
67. Serhan, C.N., Gotlinger, K., Hong, S., Arita, M. (2004). Resolvins, docosatrienes, and neuroprotectins, novel omega-3 derived mediators, and their aspirin-triggered endogenous epimers: an overview of their protective roles in catabasis. *Prostaglandins and Other Lipid Mediators*, *73*, 155–172.
68. Fritsche, K. (2006). Fatty acids as modulators of the immune response. *Annual Review of Nutrition*, *26*, 45–73.
69. Hansen, T.K., Poulsen, L.K., Skov, P.S., Hefle, S.L., Hlywka, J.J., Taylor, S.L., Bindslev-Jensen, U., Bindslev-Jensen, C. (2004). A randomized, double-blinded, placebo-controlled oral challenge study to evaluate the allergenicity of commercial, food-grade fish gelatin. *Food and Chemical Toxicology*, *42*, 2037–2044.
70. André, F., Cavagna, S., André, C. (2003). Gelatin prepared from tuna skin: a risk factor for fish allergy or sensitization? *International Archives of Allergy and Immunology*, *130*, 17–24.
71. Clare Mills, E.N., Sancho, A.I., Rigby, N.M., Jenkins, J.A., Mackie, A.R. (2009). Impact of food processing on the structural and allergenic properties of food allergens. *Molecular Nutrition and Food Research*, *53*, 963–969.
72. Bernhisel-Broadbent, J., Scanlon, S.M., Sampson, H.A. (1992). Fish hypersensitivity I. *In vitro* and oral challenge results in fish-allergic patients. *The Journal of Allergy and Clinical Immunology*, *89*, 730–737.
73. Aas, K., Elsayed, S.M. (1969). Characterization of a major allergen (cod): effect of enzymic hydrolysis on the allergenic activity. *The Journal of Allergy*, *44*, 333–343.
74. Watanabe, M., Miyakawa, J., Ikezawa, Z., Suzuki, Y., Hirao, T., Yoshizawa, T., Arai, S.I. (2006). Production of hypoallergenic rice by enzymatic decomposition of constituent proteins. *Journal of Food Science*, *55*, 781–783.
75. Tanabe, S., Arai, S., Yanagihara, Y., Mita, H., Takahashi, K., Watanabe, M. (1996). A major wheat allergen has a Gln-Gln-Gln-Pro-Pro motif identified as an IgE-binding epitope. *Biochemical and Biophysical Research Communications*, *219*, 290–293.

76. Wigotzki, M., Steinhart, H., Paschke, A. (2000). Influence of varieties, storage and heat treatment on IgE-binding proteins in hazelnuts (*Corylus avellana*). *Food and Agricultural Immunology*, *12*, 217–229.

77. Halken, S., Host, A., Hansen, L.G., Osterballe, O. (1993). Safety of a new, ultrafiltrated whey hydrolysate formula in children with cow milk allergy: a clinical investigation. *Pediatric Allergy and Immunology*, *4*, 53–59.

78. Bush, R.K., Taylor, S.L., Nordlee, J.A., Busse, W.W. (1985). Soybean oil is not allergenic to soybean-sensitive individuals. *The Journal of Allergy and Clinical Immunology*, *76*, 242–245.

79. Hoffman, D.R., Collins-Williams, C. (1994). Cold-pressed peanut oils may contain peanut allergen. *The Journal of Allergy and Clinical Immunology*, *93*, 801–802.

80. Nakamura, A., Watanabe, K., Ojima, T., Ahn, D.H., Saeki, H. (2005). Effect of Maillard reaction on allergenicity of scallop tropomyosin. *Journal of Agricultural and Food Chemistry*, *53*, 7559–7564.

11

PROCESSING FOODS WITHOUT PEANUTS AND TREE NUTS

SAHUL H. RAJAMOHAMED AND JOYCE I. BOYE

11.1. INTRODUCTION

Peanuts and tree nuts have been part of the human diet dating as far back as 7000 BC [1]. In many ancient societies, peanuts and tree nuts were used as ingredients in the preparation of stews, soups, and various sauces. Today, peanuts and tree nuts constitute one of the major components in many ethnic cuisines both as a delicacy and for core nutrition. Although peanuts and tree nuts are from different families, they are collectively termed as edible nuts (Fig. 11.1), and this chapter will attempt to cover both. The peanut (*Arachis hypogea*) is an annual herbaceous plant belonging to the Leguminosae family, which includes green peas, lentils, soybeans, and kidney beans. In contrast to tree nuts, peanut pods containing the peanut seeds grow under the ground. They are also sometimes called earthnut, groundnut, or monkey nut. Tree nuts, on the other hand, are edible seeds that grow on trees. They include almonds (*Prunus dulcis*), cashews (*Anacardium occidentale*), Brazil nuts (*Bertholetia excelssa*), hazelnuts (*Corylus avellana*), macadamia nuts (*Macadamia integrifolia*), pecans (*Carya illinoinensis*), pine nuts (*Pinus pinea*), pistachios (*Pistachia vera*), and walnuts (*Juglans regia*).

Peanuts and tree nuts are widely consumed by people in many different forms. A survey conducted by the United States Department of Agriculture (USDA) [3] in 1994–1996 on the consumption pattern of peanut and peanut

Allergen Management in the Food Industry, Edited by Joyce I. Boye and Samuel Benrejeb Godefroy

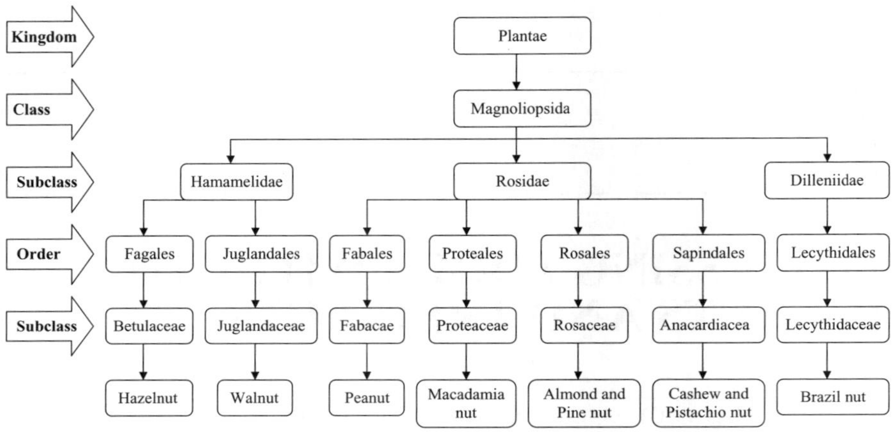

Fig. 11.1 Taxonomic classification of peanuts and tree nuts based on information (*Source*: USDA-NRCS) [2].

products with 14,262 free living individuals (including men, women, and children) revealed that 24% of respondents included peanuts or peanut products in their diet. Higher consumption rates were reported in a later study conducted in 2001–2004 with 17,306 individuals (>2 years of age) in which 34% of respondents indicating that they consumed peanuts and tree nuts either as a snack (6%), peanut butter (8%), or as an ingredient in food recipes (21%) [4]. Of the foods that contributed peanuts and tree nuts as ingredients in recipes or foods, candy was the top contributor at 46%, followed by baked items/desserts (24%), cookies (17%), ready-to-eat cereals (9%), and entrees (4%). In another study conducted in 10 western European countries [5], it was reported that 4.4% and 2.3% of the population (36,994 individuals), respectively, consumed whole tree nuts and peanut, and among tree nuts, walnuts, almonds, and hazelnuts were the most commonly consumed.

Although peanuts and tree nuts are very nutritious and provide a lot of culinary pleasure, their consumption by some people unfortunately results in allergic reactions, some of which can be fatal [6]. Allergic reactions to peanuts and tree nuts represent some of the most severe allergies known. As there are currently no effective treatments for peanut and tree nut allergy, the avoidance of these foods is the only safe treatment for sensitized individuals. This chapter will summarize the major allergenic proteins present in peanuts and tree nuts and will provide a list of some alternative foods and ingredients that could be used to replace peanuts and tree nuts in the formulation of foods targeted for peanut- and tree nut-allergic individuals.

11.2. PEANUT AND TREE NUT ALLERGY

11.2.1. Prevalence of Peanut and Tree Nut Allergies

Approximately 1.5% and 0.8% of children (3–4 years) in the United Kingdom and United States, respectively, have peanut allergy [7, 8], and the prevalence rate especially among children appears to be increasing [7–9]. Family history, prenatal maternal consumption of peanuts, and early exposure to peanut allergen either by ingestion or skin application of peanut oil have been suggested as possible etiological factors for peanut sensitization among children [10–12].

The prevalence of tree nut allergy in the general population, on the other hand, is unclear. A survey in the U.S. population (13,493 respondents) reported rates of 0.6% [8]. Various studies have indicated high coprevalence of tree nut allergy in children and adults with peanut allergy. In a survey of 5149 voluntary registrants, Sicherer et al. [13] reported that 68%, 9%, and 23% of the registrants, primarily children younger than 18 years, were allergic to peanut, tree nut, and both, respectively. An earlier study by the same authors found that 34% of 102 peanut-allergic patients had allergic reactions to at least one tree nut and among these walnut, almond, pecan, and cashew were the most common tree nuts responsible for the allergic reactions [6].

Prevalence rates for peanut and tree nut allergy appear to be location specific with rates being lower in countries outside of North America and Europe. The prevalence of peanut and tree nut allergies in Israel, as an example, is much lower (0.04% and 0.02%, respectively) than in Europe and the United States probably due to lower consumption of peanut products (1.4 kg/person/year) as well as differences in cooking and preparation methods [14–17]. In China, the prevalence rate of peanut allergy is also comparatively lower (~0.3% in young children) than in the United States even though the consumption level of peanut is roughly the same (2.3–2.7 kg/person in China and >2.7 kg/person in the United States). This further supports the notion that the differences in prevalence rates may be due to different cooking and preparation methods [16] or to some other physiological or biosocial factors.

11.2.2. Clinical Features of Peanut and Tree Nut Allergies

Allergic symptoms following the ingestion of peanuts and tree nuts can occur from within minutes to a few hours after consumption and may be characterized by a variety of symptoms including oral pruritus, nausea, vomiting, urticaria, and angioedema to bronchospasm [18]. In severe cases, anaphylaxis with angioedema, bronchitis, and hypotension are manifested, and symptoms may sometimes recur 1–8 hours after the initial symptoms have resolved [19].

A voluntary peanut and tree nut registry [6] indicated that 72% of respondents experienced allergic reaction to peanuts at the median age of 24 months (range 6–108 months), and the median time for onset of symptoms was 3 minutes (range 1–45 minutes). Fifty percent of the peanut-allergic patients showed isolated skin symptoms, 17% noticed skin and respiratory symptoms, and 21% observed skin, respiratory, and gastrointestinal (GI) symptoms, while only 2% experienced respiratory and GI symptoms alone. Similar reactions were reported by tree nut-allergic patients with a median age of first allergic reactions at 62 months and reaction occurring after a 2-minute exposure (range 1–30 minutes).

The minimum amount of protein required to trigger an allergic reaction in peanut- and tree nut-sensitized individuals is highly variable. Very small amounts, less than one kernel, of peanuts and tree nuts can induce allergic reactions in peanut and tree nut-allergic patients [20]. Flinterman et al. [21] examined the threshold dose required to provoke allergic symptoms in 27 peanut-sensitive children (median age of 7.5 years) using a double-blind, placebo-controlled food challenge (DBPCFC). One milligram of peanut flour corresponding to 2 mg of whole peanut was the no observed adverse effect level (NOAEL). Eliciting doses for subjective (oral allergic syndrome [OAS], nausea, and/or abdominal pain) and objective symptoms (urticaria, rhinoconjunctivitis, vomiting, diarrhea, hoarseness, stridor, bronchoconstriction, and/or tachycardia) were 10 mg to 3 g and 100 mg to 3 g, respectively. As little as 100 μg of peanut protein has been reported to provoke symptoms in some subjects with peanut allergy [22]. The Threshold Working Group of the Department of Health and Human Services of the Food and Drug Administration (FDA) reported threshold levels of 0.25–10 mg and 0.02–7.5 mg of peanut and tree nut proteins, respectively [20]. Other workers have reported lowest provoking doses for peanut-allergic patients ranging between 100 μg and 125 mg of peanut protein [23, 24].

Variability in the reported data could be due to differences in the degree of sensitization as well as matrix effects. Grimshaw et al. [25] found that matrix effects markedly influenced threshold levels. When patients were fed peanut in a high-fat recipe (31.5%), 12–31 times higher doses of peanut were required to elicit allergic reactions compared with a low-fat recipe (22.9%). This may be due to concealing of allergenic epitopes by the fat, which masked allergen detection in the oral mucosa and/or slower release of allergen during digestion in the stomach and small intestine due to the high fat content.

11.2.3. Peanut and Tree Nut Allergens

11.2.3.1. Peanut Allergens. Eight peanut allergens have so far been identified (Table 11.1). Of these, Ara h 1, Ara h 2, and Ara h 3 are recognized by over 90% of peanut-allergic patients and are classified as major allergens [26–28]. Ara h 4, Ara h 5, Ara h 6, Ara h 7, and Ara h 8 are classified as minor allergens due to their lower sensitizing rate among peanut-allergic patients [29, 30]. There is some disagreement surrounding this classification, however. In a DBPCFC study on immunoglobulin E (IgE) reactivity to peanut allergens

TABLE 11.1 Major and Minor Peanut Allergens

Peanut Allergens	MW (kDa)	Protein Family	Identified Epitopes	Allergenicity* (%)	References
Ara h 1	63.5	Vicilin	23	100	26, 33
Ara h 2	17.0	Conglutin	10	100	27, 31
Ara h 3	57.0	Glycinin	4	44–77	39
Ara h 4	35.9	Glycinin	NR	53	44
Ara h 5	14.0	Profilin	NR	13	44
Ara h 6	14.5	Conglutin	NR	38	44
Ara h 7	15.8	Conglutin	NR	43	44
Ara h 8	16.9	PR-10 protein	NR	NR	48

*Percentage of human sera that bound to this allergen.
MW, molecular weight; NR, not reported; PR, pathogenesis-related protein family.

with 20 peanut-allergic children (3–15 years) [31], it was reported that a larger number of subjects (16 children) more frequently recognized Ara h 2 and Ara h 6 than Ara h 1 and Ara h 3 (recognized by only 10 children) at the first time point (before DBPCFC). Twenty months after the food challenge (second time point), Ara h 2 was still recognized by all the children (100%), followed by Ara h 6 (85%), Ara h 3 (55%), and Ara h 1 (45%). The authors for this study, therefore, concluded that Ara h 2 and Ara h 6 were more potent peanut allergens for peanut-sensitive children. Further research on the identification and classification of the major allergens in peanuts will, therefore, be useful. Details on the molecular properties of some of the allergenic proteins in peanut are provided below.

11.2.3.1.1. Ara h 1. Ara h 1 is a glycoprotein with a molecular weight (MW) of 63.5 kDa. It contains 521–523 amino acids and accounts for 12–16% of the total storage protein in peanut. Ara h 1 has sequence homology with the vicilin seed storage proteins belonging to the cupin superfamily. It consists of 23 linear IgE-binding epitopes, and among these sequences, amino acid residues 25–34, 65–74, 89–98, and 498–507 are considered to be immunodominant [32]. Single amino acid changes particularly in the center of each epitope results in loss of IgE reactivity [32]. The IgE-binding epitopes of Ara h 1 are clustered into two main regions despite their more even distribution in the primary sequence of the protein. As a result, they are able to bind more efficiently with antibodies, which results in more efficient release of chemical mediators from mast cells and, as a consequence, often leads to the provocation of more severe clinical symptoms in peanut-allergic patients [33]. During heating, Ara h 1 is capable of forming a stable homotrimer and larger complexes that are resistant to proteolytic digestion with pepsin, trypsin, and chymotrypsin [34, 35].

11.2.3.1.2. Ara h 2. Ara h 2 is also a glycoprotein with a MW of 17.5 kDa and constitutes about 5.9% of the total protein in peanut. The three-dimensional structure of Ara h 2 is stabilized by four disulfide bonds, and it has a structural homology with the conglutin (2S) family of seed storage proteins. Ten IgE-binding epitopes have been identified in the primary sequence of Ara h 2, and among them, three are immunodominant (i.e., 27–36, 57–66, and 65–74 amino acid residues). The IgE-binding epitopes are evenly distributed along the linear sequence of the molecule, and mutational alteration of epitopes with a single amino acid particularly in the center results in loss of IgE reactivity [27]. Ara h 2 is poorly digested by proteases (i.e., pepsin, trypsin, and chymotrypsin) in the GI tract; on digestion, it produces a 10-kDa protein fragment, which is resistant to further digestion [36]. The 10-kDa fragment begins at amino acid 23 at the N-terminal and contains approximately 90 amino acids, 2–7 IgE-binding epitopes, 6 of 8 cysteine residues, and 11 trypsin digestion sites [36]. Ara h 2 is more susceptible to trypsin digestion if it is reduced or denatured, which indicates that disulfide bonds play a major role in its digestibility [36, 37]. Ara h 2 has structural homology to bifunctional trypsin/α-amylase inhibitors (RBI) from ragi seed, and the inhibitor activity increases during roasting [37].

11.2.3.1.3. Ara h 3. Ara h 3 has a high structural homology with the legumin (glycinin) seed storage proteins, which belong to the cupin superfamily of proteins [38, 39]. Ara h 3 was identified as a 14-kDa protein by peanut-allergic patients using immunoblot, but cDNA clone encoding of the protein revealed its MW to be ~57 kDa. Glycinin is a major seed storage protein (11S) consisting of both acidic (~40 kDa) and basic (~20 kDa) subunits linked together by disulfide bonds [40]. Each pair of acidic and basic polypeptides is encoded by a single gene and cleaved posttranslationally [41]. The 14-kDa protein corresponds to the N-terminal portion of a peanut glycinin subunit located at the 5′-end of the cDNA clone proteins, which is posttranslationally cleaved by endopeptidases [28, 39, 42]. Four IgE-binding epitopes have been detected (i.e., amino acid residues 33–47, 240–254, 274–293, and 303–317), and alteration of the peptide sequence with single amino acids reduced the IgE-binding activity significantly [39]. Ara h 3 is pepsin labile; heating slightly decreases the pepsin digestibility; however, it does not appear to affect IgE-binding activity [43].

11.2.3.1.4. Other Peanut Allergens. Ara h 4 is a 35.9-kDa protein that shows over 90% structural homology with Ara h 3 by cDNA clone encoding (which has resulted in some researchers suggesting that Ara h 3 and Ara h 4 are the same allergen) [42, 44, 45]. The International Union of Immunological Societies (IUIS), however, lists these two proteins as different with a biochemical name listing of cupin (legumin-type, 11S globulin, glycinin) for Ara h 3 (MW 60 kDa) and cupin (legumin-type, 11S, glycinin) for Ara h 4 with a MW of 37 kDa (http://www.allergen.org/Allergen.aspx). Ara h 5 is a 14- to 15-kDa protein containing 131 amino acids and belongs to the plant profilin family. The frequency of sensitization to Ara h 5 by peanut-allergic individuals is reportedly low (13%); however, it is a prominent cross-reactive allergen with birch pollen, grass, and wheat [44, 46]. Ara h 5 shows 76% amino acid identity to the grass (*Phleum pratense*) profilins and 83% identity to the soybean profilins [47]. Ara h 6 and Ara h 7 have MWs of 14.5 and 15.8 kDa, respectively, and are both classified as belonging to the conglutin family of seed storage proteins. Ara h 6 and Ara h 7 exhibit 59% and 35% amino acid sequence identity with Ara h 2, respectively, whereas these two allergens reveal only 35% amino acid identity with each other. Reported frequency of sensitization of Ara h 6 and Ara h 7 by peanut-allergic individuals was 38% and 43%, respectively [44]. Ara h 8 is a 16.9-kDa protein containing 157 amino acids and belongs to the pathogenesis-related protein family. The Ara h 8-specific IgE-binding epitopes are cross-reactive with the major birch pollen antigen (*Betula verrucosa*—Bet v 1), which shares 45.9% identity in their amino acid sequence and shows secondary structure similarities by far ultraviolet circular dichroism (UVCD) spectroscopy [48].

11.2.3.2. Tree Nut Allergens. A voluntary survey report on tree nut allergy estimated that 54% of tree nut-allergic individuals (464) had allergic reactions to single tree nuts, which included walnut (34%), cashew (20%), almond (15%), pecan (9%), pistachio (7%), hazelnut, Brazil nut, macadamia nut, pine nut, and hickory (less than 5% each), and 46% to multiple tree nuts [13]. A limited number of allergens from tree nuts have been identified. The majority of tree nut allergens belong to seed storage proteins such as vicilins (7S), legumins (11S), and profilins (2S) [49]. The major allergens identified in tree nuts and their biochemical characteristics are summarized in Table 11.2. Details on the molecular properties of these allergens are provided below.

11.2.3.2.1. Cashew. Ana o 1 (50 kDa), Ana o 2 (33 and 53 kDa), and Ana o 3 (12 kDa) are major allergens in cashew, which are recognized by more than 50% of cashew-allergic patients [50–52]. Ana o 1 has a structural homology (52–62% amino acid sequence similarity) with some other vicilins belonging to the cupin superfamily and contains 11 IgE-binding epitopes among which amino acid residues 1–15, 57–71, and 521–535 are immunodominant [51]. Ana o 2 is a member of the legumin protein family, which also belongs to the cupin superfamily; probing of overlapping synthetic peptides with pooled sera from cashew-allergic patient identified 22 IgE-reactive amino acid residues [53]. Ana o 3 belongs to the 2S albumin family and contains 16 IgE-binding epitopes with the amino acid residues 33–44, 57–68, 72–83, and 102–113 being immunodominant [52]. Ana o 1, Ana o 2, and Ana o 3 are stable to heat, and there is apparently no change in their IgE-binding activity after heating when measured by enzyme-linked immunosorbent assay (ELISA) [54].

11.2.3.2.2. Walnuts. The first allergen identified in walnut was a 14-kDa protein (Jug r 1) belonging to the 2S albumin family. Jug r 1 is considered a major

TABLE 11.2 Major Tree Nut Allergens and Their Characteristics

Tree Nuts	Allergen	MW (kDa)	Protein Family	Allergen Type	Identified Epitopes	Allergen Stability	References
Cashew	Ana o 1	50	Vicilin (7S)	Major	11	Thermostable	51, 54
	Ana o 2	33 and 53	Legumin (11S)	Major	22	Thermostable	53, 54
	Ana o 3	12	Albumin (2S)	Major	16	Thermostable	52, 54
Walnut	Jug r 1	14	Albumin (2S)	Major	3	NR	55, 56
	Jug r 2	44–47	Vicilin (7S)	Major	NR	NR	57
	Jug r 3	9	LTP	NR	NR	NR	58
	Jug r 4	NR	Legumin (11S)	NR	NR	NR	58, 76
Hazelnut	Cor a 1	18	PR-10	Major	NR	Thermolabile	59, 61
	Cor a 2	14	Profilin	Major	NR	NR	59
	Cor a 8	9	LTP	Major	NR	NR	63
	Cor a 9	35–40	Legumin (11S)	Major	NR	NR	63
	Cor a 11	47	Vicilin (7S)	NR	NR	NR	63
Brazil nut	Ber e 1	9	Albumin (2S)	Major	NR	Thermostable and resistant to proteolysis	64, 65
	Ber e 2	22 and 35	Legumin (11S)	Minor	NR	NR	67
Almond	NR	45	Vicilin (7S)	Major	NR	NR	70
	NR	20–22 and 38–42	Legumin (11S)	Major	NR	NR	68, 69
	NR	12	Albumin (2S)	Major	NR	NR	70

MW, molecular weight; NR, not reported; PR, pathogenesis-related protein family; LTP, lipid transfer protein.

walnut allergen and is recognized by 75% of walnut-allergic patients [55]. Three IgE-binding epitopes were identified in Jug r 1 by overlapping synthetic peptide analysis and mutations of critical core amino acid positions located at 36–39 and 42 resulted in loss of IgE-binding activity [56]. A vicilin-like protein (44–47 kDa), identified as a second major allergen in walnut (Jug r 2), was recognized by 9 of 15 allergic patient sera by immunoblot analysis. Although Jug 2 shows 70% amino acid sequence identity with Ara h 1 by deduced sequence analysis, it does not cross-react with Ara h 1 [57]. Two other allergens, Jug r 3 (9 kDa) and Jug r 4, belonging to the lipid transfer protein and legumin protein families, respectively, have also been identified in walnut [49, 58], but they have been less studied.

11.2.3.2.3. Hazelnut. An 18-kDa protein (Cor a 1) present in both hazelnut kernel and pollen is considered as a major hazelnut allergen; four isoforms have been identified, which include Cor a 1.01, 1.02, and 1.03 in hazel pollen, and 1.04 in hazelnut [59]. This probably explains why some hazelnut-sensitive individuals were reported to react to both pollen- and non-pollen-related allergens depending on the route of sensitization [59]. Lüttkopf et al. [60] studied the molecular characteristics of a major hazelnut allergen (Cor a 1.04) and its immunoreactivity using sera of hazelnut-allergic patients. Cor a 1.04 expressed at least four sub-isoforms (Cor a 1.0401, Cor a 1.0402, Cor a 1.0403, and Cor a 1.0404) by cDNA encoding and showed 97–99% identity among each other by deduced amino acid sequence analysis; however, only 63% identity was found with the hazel pollen Cor a 1. Enzyme allergosorbent tests (EASTs) with sera of hazelnut-allergic patients (43) showed 95%, 93%, 91%, and 74% IgE-binding activity to Cor a 1.0401, Cor a 1.0402, Cor a 1.0403, and Cor a 1.0404, respectively. Cor a 1.04 is labile to heat treatment and is digested by proteases in the GI tract [61, 62]. Cor a 2 is a 14-kDa protein that belongs to the profilin protein family. Forty-one percent of hazelnut-allergic patients recognized Cor a 2 by immunoblot, and it is considered a major allergen in hazelnut-allergic patients [61]. Cor a 1 and Cor a 2 have homology with major allergens from birch (Bev v 1 and Bev v 2, respectively) [59 ,63]. Immunoblotting of raw hazelnut extract with the sera of 65 hazelnut-allergic patients also resulted in the identification of 47-, 35-, and 9-kDa proteins, which belong to the vicilin, legumin, and lipid transfer protein families, respectively, and were labeled as major hazelnut allergens [63].

11.2.3.2.4. Other Tree Nut Allergens. The 2S albumin (9–12 kDa) in Brazil nut is used to improve the nutritional quality of methionine-deficient crops by genetic engineering because of its higher methionine content (18%) [64]. The protein has been reported as a major allergen (Ber e 1) and was recognized by the sera of eight out of nine Brazil nut-allergic patients [64]. Researchers [64] also evaluated the immunoreactivity of 2S albumin from transgenic soybean carrying Brazil nut 2S gene and nontransgenic soybean by EAST. Seven out of nine Brazil nut-allergic patients' sera reacted to transgenic soybean; however, none of the sera reacted to nontransgenic soybean by EAST. The study clearly showed how food allergens could be transferred to other nonrelated foods by genetic engineering. Ber e 1 is thermostable and resistant to hydrolysis by pepsin, trypsin, and chymotrypsin due to its high content of sulfur-rich amino acids (8% cysteine, 18% methionine) [65, 66]. Immunoblotting of raw Brazil nut extract with the sera of 11 Brazil nut-allergic patients further identified allergenic proteins with varying MWs as follows: less than 5 kDa (18.2%), 14.4 kDa (36.6%), 22 kDa (36.6%), 35 kDa (36.6%), and 48 and 55 kDa (27.3%). The 35- and 22-kDa proteins (Ber e 2) are α- and β-subunits of legumin (12S or 11S) seed storage protein [67]; much less is known about the other proteins.

Information about almond allergen is scanty. Studies indicate that the 45-, 20–22, 38–48, and 12-kDa proteins corresponding to vicilins (7S), legumins (11S), and albumins (2S), respectively, from almond extract reacted with IgE from sera of almond-allergic patients by immunoblot analysis [68–70]. The IUIS lists Pru du 4 as a 14-kDa protein with two isoforms (Pru du 4.0101 and Pru du 4.0102) and Pru du 5 with a MW of 10 kDa and having one isoform (Pru du 5.0101) as almond allergens (http://www.allergen.org/Allergen.aspx). Five allergens are also reported for pistachio, which are identified as Pis v 1, Pis v 2, Pis v3, Pis v 4, and Pis v 5 with MWs of 7, 32, 55, 25.7, and 36 kDa, respectively.

11.2.4. Cross-Reactivity of Peanut and Tree Nut Allergens

Clinical studies indicate that 35–40% of peanut-allergic patients show multiple sensitivities to tree nuts [6, 71]. This may be due to the presence of cross-reactive epitopes in both peanut and tree nut allergens [72]. Cross-reactivity between allergens occurs when IgE antibodies originally raised against one allergen recognizes and binds to a structurally similar protein from a different source upon exposure to the cross-reacting agent [73].

Cor a 9, a hazelnut allergen, has 67% amino acid sequence homology of an IgE-binding epitope with Ara h 3 [74], and cross-reactivities between the two have been reported. Eigenmann et al. [38] also reported *in vitro* IgE cross-reactivity between peanut allergen Ara h 3 and soy glycinin, which resulted from homology of three IgE-binding epitopes (suggesting that allergens from the legumin protein family [glycinin 11S] may be cross-reactive) [75, 76]. The major peanut allergen, Ara h 1, is a member of the vicilin seed storage protein family, and studies indicate that it shares structurally related IgE-binding epitopes with walnut (Jug r 2), hazelnut (Cor a 11), and cashew nut (Ana o 1) [77]. Wang et al. [51], however, did not find significant sequence homology between epitopes of Ara h 1 and Ana o 1 allergens. This latter finding is supported by Rance et al. [78] who also reported no cross-reactivity between the Ana o 1 cashew nut allergen and the peanut allergen Ara h 1. Strong *in vitro* cross-reactivity between cashew nut and pistachio nut was, however, observed, probably because they both belong to the same botanical family (Anacardiaceae family). Two out of five (40%) cashew-allergic patients reportedly exhibited

IgE-binding activity to pistachio even though they had never eaten pistachio [79]; this is likely due to the high degree of structural homology of the IgE-binding epitopes (80% amino acid identity and 90% sequence similarity) of Pis v 3 (pistachio) and Ana o 1 (cashew).

Other studies have reported multiple tree nut sensitivities in tree nut-allergic patients to walnut (Jug r 4), cashew (Ana o 2), and hazelnut (Cor a 9) [53, 58, 76]. Although these studies suggest the presence of multiple cross-reactive IgE antibodies to peanuts and tree nuts, the clinical relevance of cross-reactivity is not fully known or understood. de Leon et al. [80] demonstrated biological cross-reactivity of peanut-specific IgE antibodies with tree nuts using *in vitro* basophil activation assay. They noticed effector cell activation due to cross-reactivity between peanut-specific IgE antibodies with tree nut allergens (almond, Brazil nut, and hazelnut), which suggests that IgE cross-reactivity might contribute to the manifestation of tree nut allergy in peanut-allergic individuals upon exposure.

11.2.5. Effect of Processing on Allergenicity of Peanuts and Tree Nuts

Food products are subjected to a variety of processing treatments from the time they are harvested until they reach the consumer. Factors such as food composition, time, temperature, and conditions of harvest, cooking, and storage can alter the properties of proteins in foods and render them more or less allergenic [81]. Important characteristics of food allergens include their stability to processing and their resistance to digestion by proteolytic enzymes. Reduction of the allergenic properties of foods by processing could be beneficial to peanut- and tree nut-allergic individuals. Allergenicity of foods can potentially be reduced either by removal, destruction, proteolytic modificatin, or masking of IgE-binding sites to make the allergenic proteins unavailable [82].

Thermal processing is commonly applied to improve the quality of food in terms of its nutritional value (e.g., by enhancing digestibility) and sensory quality (e.g., through improvements of appearance, texture, flavor, and taste), as well as safety and storage stability (e.g., by inactivation of microbes and toxins). Thermal processing induces various physical, chemical, and biochemical changes in foods and is dependent on the intrinsic characteristics of the food, such as pH, temperature, and duration of processing [83]. For food allergens, changes to the allergenic properties during thermal processing can be influenced by the method employed to process the food (e.g., dry or wet). Since the majority of studies conducted on the effect of processing have been done on peanuts, greater emphasis will be paced on peanut in this section.

The influence of thermal processing such as frying, roasting, and boiling on the allergenicity of peanut proteins from two peanut varieties (Florunner and Valencia) was studied by Beyer et al. [16]. Sodium dodecyl sulfate–polyacrylamide gel electrophoresis (SDS-PAGE) electrophoretic pattern revealed alterations in the molecular structure of the protein fractions from both varieties on processing. The changes in structure were, however, comparable after frying in vegetable oil for 5–10 minutes and boiling at 100°C for 20 minutes but were different after roasting at 170°C for 20 minutes. Furthermore, immunoblot analysis with peanut-allergic patients' sera showed lower IgE-binding intensities of Ara h 1, Ara h 2, and Ara h 3 levels in the fried and boiled peanuts compared with the roasted peanuts. The authors suggested that occurrence of irreversible alterations in protein structure at higher temperature (e.g., Maillard reaction) may have caused chemical modifications, which increased the allergenicity of peanuts during roasting. Maleki et al. [35] also observed a 90-fold increase in allergenic properties of roasted peanut compared with raw peanut, which was mainly attributed to chemical modifications resulting from Maillard reactions. Chung and Champagne [84] investigated the effect of Maillard reaction products (MRPs) on allergenicity of peanut protein. A solution of lectin (32 kDa) (5 mg/mL) was reacted with sugar (150 mM glucose or fructose) and incubated at 50°C at pH 8 for 5 weeks. SDS-PAGE analysis showed three to four additional adducts (between 48 and 77 kDa, 32–36 kDa, and 26 kDa) when compared with lectin without sugar, which strongly recognized IgE from the sera of peanut-allergic patients by competitive immunoblotting. Furthermore, roasted mature peanuts exhibited higher IgE inhibitory effect than roasted immature peanuts. The results, therefore, suggests that MRPs are potentially allergenic, and products formed during roasting increases allergenicity of the peanuts. Various MRPs such as advanced glycation end products (AGEs), N-carboxymethyl lysine (CML), malondialdehyde (MDA), and 4-hydroxynonenal (HNE) are formed during roasting of peanuts. AGE and MDA products were found to be present at higher levels in roasted peanuts, and AGE exhibited higher level of IgE binding by competitive ELISA; higher levels of AGE adducts in roasted peanuts may, therefore, be associated with an increase in allergenicity of peanuts [85].

Mondoulet et al. [86] examined the effect of thermal processing on the IgE-binding capacity of whole peanut protein extract and purified peanut allergens (Ara h 1 and Ara h 2) using EAST. The IgE-binding capacity of peanut extracts prepared from boiled peanuts (30 minutes) was twofold lower than in raw and roasted peanut extracts; however, no significant difference between IgE-binding capacity of extracts prepared from raw and roasted peanuts was observed. Immunoreactivity of Ara h 1 and Ara h 2 was found to be higher in roasted peanuts than raw and boiled peanuts by competitive EAST. The decrease in allergenicity of boiled peanuts could potentially be due to the transfer of low-MW allergen (Ara h 2) into the water during cooking.

Ara h 2 from roasted peanuts reportedly had threefold higher trypsin inhibitory activity (TIA) compared with the native Ara h 2 due to the formation of intramolecular cross-links (disulfide bonds), which contributed to higher structural stability of the Ara h 2 against denaturation [37]. The structural integrity of Ara h 2 masked trypsin digestion sites resulting in increased TIA on roasting [36]. Moreover, SDS-PAGE analysis on digestibility of Ara h 1 in the presence of Ara h 2 by trypsin revealed that Ara h 2 from roasted peanuts affects the digestibility of native Ara h 1 (raw peanut); both bands were detectable even after overnight incubation with trypsin. Ara h 1 was easily digested within 30–60 minutes of incubation with trypsin in the absence of Ara h 2 or when Ara h 2 was denatured. Apparently, Ara h 2 provides additional protection to Ara h 1 against trypsin digestion in roasted peanuts [37]. The results suggest that structural integrity and stability of Ara h 2 during roasting reduces its digestibility by proteolytic enzymes in the GI tract and its presence further protects Ara h 1 from digestion, leading to larger amounts of intact Ara h 1 and Ara h 2 being released in the GI tract, which are easily recognized by peanut-sensitive patients. This may explain why the allergenicity of peanut increases upon roasting compared with other thermal process like boiling and frying.

There are some contrasting studies; however, Koppelman et al. [34] found that the allergenicity of Ara h 1 was not affected by heating, although the native Ara h 1 underwent heat-induced denaturation at the molecular level. In another study, dry roasting of hazelnut at 140°C for 40 minutes reportedly reduced the allergenicity of hazelnuts significantly compared with raw hazelnut [61]. Approximately 100-fold higher concentration of roasted hazelnut extract was required for 50% inhibition of IgE-binding activity compared with raw hazelnut extract. Five out of 17 hazelnut-allergic individuals experienced OAS after ingestion of the roasted hazelnut at a median dose of 7 g by DBPCFC, whereas all subjects reacted to raw hazelnut at a median dose of 2.0 g. The response to roasted hazelnut might be due to sensitization by the thermostable low-MW allergens other than Cor a 8 and residual activity of Cor a 1 or Cor a 2 in the roasted hazelnuts [61].

Masking of allergenic IgE epitopes by cross-linking may be a way to reduce allergenicity of peanuts and tree nuts. Mouecoucou et al. [87] evaluated the effect of polysaccharide addition on *in vitro* allergenicity of peanut. Peanut protein isolate (PPI) (15 mg/mL) treated with 1.7% w/v polysaccharides (gum arabic, xylan, and low-methylated pectin) at 1:1 ratio by volume and left to incubate overnight at 4°C modified the digestibility of the PPI, and the hydrolyzed products were found to react less with rabbit anti-immunoglobulin G (IgG) and IgE antibodies from peanut allergic sera. The decrease in immunoreactivity was attributed to potential nonspecific interactions between polysaccharides and peptides from PPI. Among different polysaccharides, xylan showed higher reduction in *in vitro* allergenicity of PPI compared with the other polysaccharides. Similarly, peanut extract (5 mg/mL) treated with peroxidase (10 mg/mL) resulted in lower IgE binding of Ara h 1 and Ara h 2 in roasted peanuts compared with raw peanuts. Cross-linking of proteins by peroxidase is dependent on the presence of tyrosine residues. At higher temperatures, peanut proteins get denatured and expose the buried tyrosine residues that were before inaccessible to peroxidase cross-linking [88].

An eightfold decrease in the allergenic property of peanut butter extract (IC_{50} = 4 μg/mL) treated with phytic acid (2 mM w/v) compared with the untreated peanut extract (IC_{50} = 0.5 μg/mL) was observed by competitive

ELISA [89]. Phytic acid reacted predominantly with the major peanut allergens (Ara h 1 and Ara h 2) and formed insoluble complexes with the raw peanut extract but not with the roasted peanut extract due to the lack of free amino groups (lysine). At pH 8.5, a significantly higher amount of peanut allergens was detected by SDS-PAGE compared with pH 3 and 7, which indicates that pH also plays a significant role in the phytic acid–protein complex formation with peanut extract. Recently, Chung et al. [90] demonstrated that the use of nonthermal pulsed UV light could modify the immunogenic properties of peanut proteins. A six- to sevenfold lower IgE-binding activity (IC_{50} = 4 μg/mL) was observed in peanut butter treated by pulsed UV light (220.8 J/cm^2 energy level, 3 pulses/second, 360 μs pulse width) compared with the untreated peanut butter (IC_{50} = 0.6 μg/mL).

11.2.6. Effect of Refining on Allergenicity of Peanut and Tree Nut Oils

Refining is a physical and chemical process applied to remove certain compounds present in oils and involves acid and alkali treatment, bleaching, and deodorization. Depending on the degree of refining, the oil may or may not contain allergenic proteins [91]. Hourihane et al. [92] found that refined peanut oil did not pose a threat to peanut-sensitive subjects, although 10% reacted to crude peanut oil in a crossover DBPCFC study (i.e., 6/60 individuals; 4/6 had subjective symptom and 2/6 had objective symptoms). Thus, refined peanut oil may be safe for peanut-sensitive individuals, but "gourmet oils" (i.e., refined oil blended with crude peanut oil) may pose a risk. Similar results were reported by Taylor et al. [93] who observed no adverse reactions in peanut-allergic patients challenged with refined peanut oil. Hoffman and Collins-Williams [94] observed specific IgE reactivities with cold-pressed crude peanut oil but not with refined oil. The presence of 1.32 mg/kg of peanut proteins in cold-pressed oil was reported by Martin-Hernandez et al. [95]; however, no proteins were detected in refined oil. Major proteins identified from the cold-pressed unrefined peanut oil by SDS-PAGE analysis had MWs of 65, 40, 31, 23, and 18 kDa. Interestingly, Olszewski et al. [96] recorded the presence of protein (0.1–0.2 μg of protein/g of refined oil) in commercially available refined peanut oils, which reacted with IgE antibodies in the sera of peanut-allergic patients. An 18-kDa protein present in refined peanut oil, which had a *pI* close to the *pI* of Ara h 2 (*pI* 4.5), was found to be responsible for the allergic reaction in peanut-sensitive individuals. Pons et al. [97] also reported that a purified oleosin protein extracted from crude peanut oil with a MW of 18 kDa reacted with the sera of 3 out of 14 peanut-allergic patients. Very few studies have been conducted on tree nut oils.

Differences in unit operations and temperatures employed to process refined oil play a significant role in determining the allergenicity of peanut and tree nut oils. Teuber et al. [98] evaluated the allergenicity of gourmet nut oils processed by different methods. Their results showed the oils that underwent less processing at lower temperature had higher residual protein and greater

IgE binding (Table 11.3). As questions still linger about the allergenicity of refined and cold-pressed oils, further clinical investigation will be useful to confirm some of the findings reported in order to ensure the safety of peanut and tree nut oils for consumption by peanut- and tree nut-allergic patients.

TABLE 11.3 Immunoreactivity of Commercially Available Peanut and Tree Nut Oil Processed at Different Temperatures

Oil Type[a]	Protein Concentration (μg/mL)	Maximum Processing Temperature (°C)	Immunoassay Relative Reactivity[b]
Peanut			
Flora	10.5	54	+ + +
Hain	3.0	230–260	+/–
Spectrum	10.7	65–93	+ +
Hollywood	5.7	230–260	–
Nut extract	9001	4	+ + + +
Almond			
Flora	62.2	57	+/–
Hain	2.2	230–260	–/+
Spectrum	16.7	230–260	–/+
Gourmet Int.	12.7	90; 190	+ +
Nut extract	5300	4	+ + +
Walnut			
Flora	16.5	71	+
Hain	9.2	230–260	–
Spectrum	7.0	230–260	+/–
Gourmet Int.	20.4	90; 190	+ +
Diamond	7.0	230–260	+
Nut extract	4671	4	+
Hazelnut			
Gourmet Int.	62.0	90; 190	+ +
Nut extract	3900	4	+ + +
Macadamia			
Oil of Aloha	21.3	65–93	–/+
Loriva	81.1	65–93	+
Nut extract	3000	4	+ +
Pistachio			
Gourmet Int.	36.0	90; 190	+
Nut extract	5000	4	+ + +

Source: Reference 98—Reprinted from Teuber, S.S., Brown, R.L., Happanen, L.A.D. (1997). Allergenicity of gourmet nut oils processed by different methods. *Journal of Allergy and Clinical Immunology*, *99*, 502–507, with permission from Elsevier.

[a]Flora and spectrum oils were unrefined oil; spectrum almond and walnut oils were refined; Hain and Hollywood oils were refined oil; Gourmet International (Int.) was a blend of refined and unrefined oils.

[b]Relative reactivity was based on a subjective visual scale with "+ + + +" means most reactive protein extract, "–/+" means barely visible band, and "+/–" means very light band.

11.3. NUTRITIONAL AND FUNCTIONAL PROPERTIES OF PEANUTS AND TREE NUTS

11.3.1. Nutritional Properties of Peanuts and Tree Nuts

11.3.1.1. Nutrient Composition. The nutrient composition of peanuts and tree nuts is summarized in Table 11.4. Peanuts and tree nuts provide useful nutrients in the diet, and, in order to find adequate alternatives for nut-allergic patients, the role these foods play in the diet needs to be carefully understood. Peanuts and tree nuts are good sources of protein and mono- (MUFAs) and polyunsaturated fatty acids (PUFAs), and they also contain significant amounts of vitamins and minerals. Peanuts generally contain higher amounts of protein (25.80 g/100 g) than tree nuts (7.91–21.22 g/100 g), and among tree nuts, almond and pistachio contain higher amounts of protein (~21 g/100 g), followed by pecans (~9 g/100 g) and macadamia nuts (~8 g/100 g) [99]. The quality of protein in peanuts and tree nuts is somewhat low due to the lack of some essential amino acids (EAAs) such as methionine, lysine, and tryptophan. Some nuts, such as Brazil nut, contain higher methionine content (~9 g/100 g of total protein) than other tree nuts and peanut, and may have higher nutritional value. As presented in Table 11.5, however, in general, there is not much difference in the EAA composition of peanuts and tree nuts, although their nutritional quality may differ. Protein digestibility corrected amino acid score (PDCAAS) for peanut is 0.52, which is comparatively superior to tree nuts, which have lower PDCAAS (0.40–0.43) [100, 101].

The fat content of peanuts and tree nuts is high (44–76 g/100 g). Tree nuts, especially macadamia nuts, pecans, pine nuts, and Brazil nuts, are particularly high in fat (76, 72, 68, and 66 g/100 g, respectively) compared with peanut (49 g/100 g). Walnuts and pine nuts contain higher amounts of the PUFAs (47 and 34 g/100 g of total fat, respectively), while macadamia nut, hazelnut, pecan, and almond are high in MUFAs (59, 46, 41, and 31 g/100 g of total fat, respectively) [99]. Additionally, peanuts and tree nuts contain moderate amounts of carbohydrate (16 and 18 g/100 g, respectively) and fiber (9 and 8 g/100 g, respectively). They also contain non-nutrient components like phytates and tannins. Tree nuts have higher phytate content (0.23 g/100 g) and lower tannins (0.20 g/100 g) than peanuts (0.17 and 0.29 g/100 g, respectively). These compounds are considered as antinutrients, but recent research studies suggest that phytates and tannins may have bioactive properties that may be of nutritional interest [103, 104]. Fortunately, no detectable amounts of TIA and hemagglutinating activity have been reported in tree nuts or peanuts [102].

11.3.1.2. Nutritional Significance and Health Benefits of Peanut and Tree Nut Consumption. There is increasing evidence that peanut and tree nut consumption may be associated with nutritional and health benefits. Consumption of 42 g (1.5 oz) of peanut or tree nuts per day provides greater

TABLE 11.4 Nutrient Content of Peanuts and Tree Nuts (100 g of Edible Portion) [99]

Nutrient	Peanut	Almond	Brazil Nut	Cashew	Hazelnut	Macadamia Nut	Pecan	Pine Nut	Pistachio	Walnut
Energy (kcal)	567	575	656	553	628	718	691	673	557	654
Protein (g)	25.80	21.22	14.32	18.22	15.0	7.91	9.17	20.6	20.61	15.23
Total fat (g)	49.24	49.42	66.43	43.85	60.7	75.77	71.97	68.3	44.44	65.21
Saturated fat (g)	6.83	3.73	15.10	7.78	4.46	12.06	6.18	4.89	5.44	6.12
MUFA (g)	24.42	30.89	24.55	23.79	45.65	58.87	40.80	18.76	23.32	8.93
PUFA (g)	15.56	12.05	20.57	7.84	7.92	1.50	21.61	34.07	13.45	47.17
ω-6 Fatty acids (g)	15.56	12.05	20.54	7.78	7.83	1.30	20.62	33.15	13.20	38.00
ω-3 Fatty acids (g)	—	—	0.03	0.06	0.09	0.20	0.99	0.16	0.25	9.08
Carbohydrate (g)	16.13	21.67	12.30	30.20	16.70	13.82	13.86	13.08	27.97	13.71
Dietary fiber (g)	8.50	12.20	7.50	3.30	9.70	8.60	9.60	3.70	10.00	6.70
Vitamin B_1 (mg)	0.64	0.21	1.00	0.20	0.64	1.20	0.66	0.36	0.87	0.34
Vitamin B_2 (mg)	0.14	1.01	0.12	0.20	0.11	0.16	0.13	0.22	0.16	0.15
Niacin (mg)	12.06	3.39	1.62	1.40	1.80	2.47	1.17	4.38	1.30	1.12
Vitamin B_6 (mg)	0.35	0.14	0.25	0.26	0.56	0.28	0.21	0.09	1.70	0.53
Pantothenic acid (mg)	1.77	0.47	0.23	1.21	0.92	0.76	0.86	0.33	0.52	0.57
Folate (μg)	240.00	50.00	22.00	69.00	113.00	11.00	22.00	34.00	51.00	98.00
Vitamin A (IU)	—	1.00	—	—	20.00	—	56.00	29.00	553.00	20.00
Vitamin E (mg ATE)	8.33	26.18	5.73	0.57	15.0	0.54	4.05	9.33	2.30	0.70
Calcium (mg)	92.00	264.00	160.00	37.00	114.00	85.00	70.00	16.00	107.00	98.00
Iron (mg)	4.58	3.72	2.43	6.70	4.70	4.00	2.53	5.53	4.15	3.00
Magnesium (mg)	168.00	268.00	376.00	292.00	163.00	130.00	121.00	251.00	121.00	158.00
Phosphorus (mg)	376.00	484.00	725.00	593.00	290.00	188.00	277.00	575.00	490.00	346.00
Potassium (mg)	706.00	705.00	659.00	660.00	680.00	368.00	410.00	597.00	1025.00	441.00
Sodium (mg)	18.00	1.00	3.00	12.00	—	5.00	—	2.00	1.00	2.00
Zinc (mg)	3.27	3.08	4.06	5.78	2.45	1.30	4.53	6.45	2.20	3.09
Copper (mg)	1.14	1.00	1.74	2.20	1.73	0.76	1.20	1.32	1.30	1.59
Manganese (mg)	1.93	2.28	1.22	1.66	6.18	4.13	4.50	8.80	1.20	3.41
Selenium (μg)	7.20	2.50	1917.00	19.70	2.40	3.60	3.80	0.70	7.00	4.90
Phytosterols (mg)	220.00	172.00	NR	NR	96.00	116.00	102.00	141.00	214.00	72.00

MUFA, monounsaturated fatty acid; PUFA, polyunsaturated fatty acid; IU, international unit; ATE, α-tocopherol equivalent; NR, not reported.

TABLE 11.5 Essential Amino Acid (EAA) Content of Peanuts and Tree Nuts in g/100 g of Protein

	Peanuts and Tree Nuts									
EAA	Peanut	Almond	Brazil Nut	Cashew Nut	Hazelnut	Macadamia Nut	Pecan	Pine Nut	Pistachio	Walnut
Lysine (5.8/1.6)	3.88	3.06	2.95	4.59	2.93	4.10	3.17	3.54	4.64	2.71
Leucine (6.6/1.9)	7.03	7.19	7.89	8.00	7.40	6.55	7.51	7.30	7.56	7.76
Valine (3.5/1.3)	3.95	4.41	4.71	5.65	4.66	4.31	4.72	4.52	5.69	4.61
Isoleucine (2.8/1.3)	3.45	3.79	3.21	4.15	3.69	3.26	4.08	3.65	4.10	4.00
Threonine (3.4/0.9)	2.21	2.60	2.27	3.22	2.95	2.81	2.90	2.43	2.97	3.00
Phenylalanine (6.3/1.9)	5.38	5.46	4.06	4.83	4.54	3.34	5.09	3.59	4.92	4.63
Methionine (2.5/1.7)	1.31	0.81	8.98	2.27	1.90	2.15	2.52	2.93	1.88	2.14
Tryptophan (1.1/0.5)	0.73	0.70	0.71	1.31	0.98	0.59	0.47	0.84	0.78	0.55
Histidine (1.9/1.6)	2.54	2.97	2.92	2.68	2.65	2.45	2.80	2.23	2.38	2.43

Source: Reference 102—Reprinted with permission from Venkatachalam, M., Sathe, S.K. (2006). Chemical composition of selected nut seeds. *Journal of Agricultural and Food Chemistry*, *54* (13), 4705–4714, American Chemical Society.

Values in parenthesis represents Food and Agriculture Organization of the United Nations (FAO)/World Health Organization (WHO) recommended EAA requirements for preschool children (2–5 years) and adults, respectively.

than 10% of the recommended dietary allowances (RDAs) for protein, fat, thiamine, niacin, pyridoxine, pantothenic acid, vitamin E, vitamin K, iron, magnesium, phosphorus, zinc, copper, selenium, and folate for an adult male (31–50 years) [4]. A study on the diet quality of free living men (4751), women (4572), and children (4939 boys and girls) showed that respondents who consumed peanuts as part of their meal (24%) had higher intake of nutrients especially vitamin A, vitamin E, calcium, magnesium, and zinc than nonconsumers [3]. Although peanut consumers showed higher energy intake (2569, 1727, and 2052 kcal/day for men, women, and children, respectively) compared with nonpeanut consumers (2214, 1546, and 1952 kcal/day for men, women, and children, respectively), the mean body mass index (BMI) for peanut users was lower in all age groups (25.7 kg/m^2) than nonpeanut users (26.2 kg/m^2).

Another study reported that obese subjects following a moderate-fat diet (35% energy from fat) containing several nuts, peanut butter, and olive oil for 18 months had greater and more sustained weight loss than obese subjects prescribed a low-fat diet (20% total energy from fat), which did not contain peanuts and tree nuts [105]. Furthermore, Johnston and Buller [106] demonstrated that the addition of peanut products to a high glycemic load meal significantly reduced postprandial glycemia by rapid stimulation of insulin release and glucose uptake. Hu et al. [107] also observed a negative association between nut consumption and BMI among 86,000 females (34–59 years). The substitution of 1 oz of nuts for the equivalent energy from carbohydrate in an average diet was similarly reported to reduce coronary heart disease (CHD) risk by 30%, and the substitution of nut fat for saturated fat reduced CHD risk by 45% [108]. These studies were corroborated by Fraser and Shavlik [109] who found that the consumption of nuts (peanuts and tree nuts) five times per week reduced the risk of death from CHD by 31%.

Phytosterols present in peanuts and tree nuts lower serum low-density lipoprotein (LDL) cholesterol by inhibiting dietary cholesterol absorption [110–113]. The phytosterol content in tree nuts ranges from 95 mg/100 g in Brazil nuts to 280 mg/100 g in pistachios and is comparable to that found in peanuts (90–300 mg/100 g) [114]. The rich sources of several essential vitamins and minerals, MUFA and PUFA, fiber, polyphenols, and phytochemicals in peanuts and tree nuts could promote health benefits and reduce the risk of chronic diseases [115]. Many different phytochemicals are present in peanuts and tree nuts, which have been associated with an array of bioactivities, including antioxidant, antiviral, antiproliferative, and anti-inflammatory actions potentially capable of affecting the initiation and progression of several pathogenic process [116]. In 2003, FDA approved a qualified health claim for nuts (peanuts and tree nuts), which associates the consumption (1.5 oz or 42 g) of nuts with reduced risk of cardiovascular diseases [117]. Thus, the avoidance of nuts in the diet may have health implications, and it is of critical importance that alternatives identified to replace nuts in the diet provide equal or better nutritional and health benefits.

11.3.2. Functional Properties of Peanuts and Tree Nuts

Apart from their nutritional importance, proteins from peanuts and tree nuts may also play a vital role in conferring specific functional properties to foods. Functional properties such as solubility, emulsion, gelation, and water and oil absorption capacities are essential in many food product formulations. A brief summary of some of these functional properties is provided below.

11.3.2.1. Solubility. Among the functional properties solubility is a prime requirement for other functional properties like foaming, gelation, emulsion, and whipping [118]. Eighty percent of the proteins from defatted peanut flour are soluble in the neutral pH range (6.5–7.0) [119]. The highest solubility occurs in the pH range of 1–2 and 8–10, and the lowest solubility is observed at pH 3.5–5.0, which is the isoelectric point (*pI*) of peanut protein [119, 120]. Solubility profiles of most tree nut proteins are similar to peanut. Cashew nut protein isolate (CPI) prepared by isoelectric precipitation was reported to have higher solubility at pH values above and below the isoelectric point (*pI* 5) [121]. Similar results have been reported for almond, walnut, and Brazil nut protein isolates, which all had higher solubilities at pH values below or above the *pI* (pH 5.0) [122–124].

11.3.2.2. Water and Oil Absorption Capacities. Interactions of water and oil with proteins are very important in food systems because of their effects on flavor and food texture. Ihekoronye [125] observed higher water absorption capacity (WAC) for defatted peanut flour (409.30 g/100 g) compared with ovalbumin (399.75 g/100 g) and peanut protein concentrate (PPC) (231.16 g/100 g). The authors attributed this to the higher carbohydrate content in defatted peanut flour, which was probably able to bind more water. PPC, on the other hand, had a higher fat absorption capacity (FAC) (254.60 g/100 g) than ovalbumin (216.69 g/100 g) and defatted peanut flour (186.40 g/100 g). CPI was found to have lower WAC (1.45 mL/g of protein) compared with casein (2.81 mL/g of protein), but its FAC was superior (1.40 mL/g of protein) to ovalbumin (0.84 mL/g of protein) [121]. Brazil nut and macadamia nut protein isolates were found to have higher WAC (2.0 and 1.62 mL/g protein, respectively) and FAC (1.40 and 1.22 mL/g, respectively) than CPI [124, 126].

11.3.2.3. Foaming Property. Foaming properties are essential in the formulation and production of foods such as cakes, ice creams, meringues, whipped cream, marshmallows, and some beverages. Ihekoronye [125] observed a 90% increase in foam volume of PPC (3% w/v solution) compared with a 58% increase for defatted peanut flour at the same concentration. For both ingredients, the foams were reported to remain stable over a 2-hour period. Foams made from PPC were found to be soft flowing and cream colored with medium-sized air bubbles compared with ovalbumin, which formed white foam with small-sized air bubbles. Similarly, CPI was reported to have higher foaming capacity (39.8 mL of foam/5 mg of protein) than ovalbumin (18.5 mL of foam/5 mg of protein) at pH 7.0; however, foam stability was lower in CPI (25.8 seconds) compared with ovalbumin (147.1 seconds) [121]. In other studies, the foaming capacity of almond protein isolate (API 1% w/v, pH 6.46) was found to be comparable to soy protein isolate (SPI) (58 and 64 mL of foam, respectively), and it had better foam stability at acidic pH (≤5.0) compared with the SPI [127].

11.3.2.4. Gelation. A gel is a three-dimensional ordered matrix that is capable of holding significant amounts of fluid. This property is useful in the development of products such as jellies, puddings, comminuted meat and fish products, doughs, and other bakery and confectionery applications. Defatted peanut flour and PPC were reported to form very weak gels at 5% (w/v) concentration and stronger gels at 9% (w/v) concentration [125]. Peanut flour from heat-pressed peanut cake, however, formed firm gels at a lower concentration (7% w/v) compared with flour from cold-pressed peanut cake, which formed weaker gels at the same concentration [119]. Least gelation capacity (LGC), which is the minimum concentration required to form a gel, was higher for CPI (13.5% w/v) than cashew nut protein concentrate (CPC) (10.0%) and defatted cashew nut powder (6.5%) [128]. API solutions (4% w/v, pH 8.2) reportedly formed firm gels after heating at above 90°C for 5 minutes [129].

11.4. NUTRITIONAL AND FUNCTIONAL ALTERNATIVES TO PEANUTS AND TREE NUTS

There are currently no cures for food allergy; thus, the only way to avoid allergic reactions is for food-allergic patients to avoid consumption of foods to which they are sensitized. Food manufacturers interested in developing specific "allergen-free" foods must, therefore, identify appropriate alternatives that could be used to replace these ingredients in food formulations. Alternative ingredients should ideally confer similar nutritional and functional properties in order not to compromise the health status of allergic individuals or taste.

Ingredients from peanuts and tree nuts are used extensively in foods such as chocolate, cereals, soups, and seasonings. Production of "free-from" foods is still in its infancy, and as such, there is limited research on appropriate alternatives that could be used for different food applications. Some suggestions on alternatives that could be considered to replace peanut and tree nut products in "nut-free" foods are presented in Table 11.6 and summarized below.

11.4.1. Alternatives to Roasted Peanuts and Tree Nuts

Roasted sesame and sunflower seeds have nutrient profiles that are comparable to roasted peanuts and tree nuts and could be used as alternatives. Roasted soybeans (soy nuts) are also available commercially and could also serve as a replacement. Soy nuts contain higher protein (35–40%) compared

TABLE 11.6 Nutrient Compositions of Alternatives to Peanuts and Tree Nuts and Its Products for 100 g of Edible Portion [99]

	Alternatives to Raw Seeds			Alternatives to Dry Roasted without Salt				
Nutrients	SF	SE	SO	PA	CA	SF	SE	SO
Energy (kcal)	584	573	446	585	574	582	565	471
Protein (g)	20.78	17.73	36.49	23.68	15.31	19.33	16.96	35.22
Total fat (g)	51.46	49.67	19.94	49.66	46.35	49.80	48.0	25.40
SFA (g)	4.45	6.95	2.88	6.89	9.15	5.21	6.00	3.67
MUFA (g)	18.52	18.75	4.40	24.64	27.31	9.50	18.12	5.61
PUFA (g)	23.13	21.77	11.25	15.69	7.83	32.88	21.03	14.33
Ash (g)	3.02	4.45	4.87	3.60	3.95	5.60	4.00	3.88
Carbohydrate (g)	20.00	23.45	30.16	21.51	32.69	24.07	25.74	33.55
Fiber (g)	8.60	11.80	9.30	8.00	3.00	11.10	14.00	17.70
Thiamine (mg)	1.48	0.79	0.87	0.43	0.20	0.10	1.20	0.10
Riboflavin (mg)	0.35	0.24	0.87	0.09	0.20	0.24	0.46	0.14
Niacin (mg)	8.33	4.51	1.62	13.52	1.40	7.04	5.43	1.41
Vitamin B_6 (mg)	1.34	0.79	0.37	0.25	0.25	0.80	0.14	0.20
Pantothenic acid (mg)	1.13	0.05	0.79	1.39	1.21	7.04	0.68	0.45
Folate (μg)	227	97	375	145	69	237	96	211
Choline (mg)	55.10	25.60	115.90	55.30	61.00	55.10	25.60	NR
Vitamin E (mg)	33.23	0.25	0.85	6.93	0.92	26.10	0.25	0.91
Calcium (mg)	78	975	277	54	45	70	989	138
Iron (mg)	5.25	14.55	15.70	2.26	6.00	3.80	14.76	3.90
Magnesium (mg)	325	351	280	176	260	129	356	145
Phosphorus (mg)	660	629	704	358	490	1155	638	363
Potassium (mg)	645	468	1797	658	565	850	475	1470
Sodium (mg)	9.00	11.00	2.00	6.00	16.00	3.00	11.00	4.00
Zinc (mg)	5.00	7.75	4.89	3.31	5.60	5.29	7.16	3.14
Copper (mg)	1.80	4.08	1.65	0.67	2.22	1.83	2.47	0.82
Manganese (mg)	1.95	2.46	2.51	2.08	0.82	2.11	2.49	2.15
Selenium (μg)	53.00	5.70	17.00	7.50	11.70	79.30	5.80	19.10

[a]Partially defatted.
[b]Nutrient composition of soy butter is not available in the USDA database.
SF, sunflower seeds; SE, sesame seeds; SO, soybeans; PA, peanuts; CA, cashew nuts; SFA, saturated fatty acid; MUFA, monounsaturated fatty acid; PUFA, polyunsaturated fatty acid; NR, not reported.

with peanuts and tree nuts, and they are also rich in carbohydrate, fiber, and calcium but have much lower fat content (17–23%). Similarly, roasted or extruded pulses, such as peas and chickpeas, have been consumed for centuries in places like Turkey and in other parts of the world and could be used as replacement snack or to replace roasted peanuts and tree nuts as ingredients in foods. These alternatives also contain allergenic proteins, and since some peanut- and tree nut-allergic patients may also be allergic to sesame, sunflower,

Alternatives to Defatted Flours				Alternatives to Oil				Alternatives to Butter without Salt[b]			
PA	SF[a]	SE[a]	SO	PA	SF	SE	SO	PA	CA	SF	SE
327	326	382	330	884	884	884	884	588	587	579	595
52.20	48.06	40.32	47.01	—	—	—	—	21.93	17.56	19.66	17.00
0.55	1.61	11.89	1.22	100	100	100	100	49.54	49.41	47.73	53.76
0.06	0.13	1.63	0.13	16.90	9.00	14.20	15.65	9.51	9.76	5.00	7.52
0.22	0.25	4.40	0.20	46.20	57.35	39.70	22.78	20.72	29.12	9.11	20.30
0.14	0.87	5.11	0.53	32.00	28.96	41.70	57.74	11.33	8.35	31.52	23.56
4.75	7.04	6.05	6.15	—	—	—	—	3.00	2.50	3.95	5.00
34.70	35.83	35.14	38.37	—	—	—	—	23.98	27.57	27.42	21.19
15.80	5.20	NR	17.5	—	—	—	—	5.70	2.00	NR	9.30
0.70	3.18	2.53	0.69	—	—	—	—	0.10	0.31	0.32	1.22
0.48	0.26	0.27	0.25	—	—	—	—	0.10	0.18	0.28	0.47
27.0	7.31	12.60	2.61	—	—	—	—	13.16	1.59	5.32	5.45
0.50	0.75	0.15	0.57	—	—	—	—	0.55	0.25	0.80	0.14
2.74	6.59	2.76	1.99	—	—	—	—	1.04	1.20	7.03	0.69
248	222	29	305	—	—	—	—	35	68	237	98
108.70	NR	NR	11.30	0.10	0.20	0.20	0.20	65.70	NR	NR	25.80
0.05	NR	NR	0.12	15.69	41.08	1.40	—	5.97	NR	NR	0.25
140	114	150	241	—	—	—	—	54	43	122	426
2.10	6.62	14.30	9.24	0.03	0.03	-	0.05	2.16	5.03	4.75	8.95
370	346	362	290	—	—	—	—	179	258	369	95
760	689	810	674	—	—	—	—	335	457	736	732
1290	67	425	2384	—	—	—	—	592	546	72	414
180.00	3.00	41.00	20.00	—	—	—	—	476	15.00	3.00	115.00
5.10	4.95	10.70	2.46	0.01	—	—	0.01	2.67	5.16	5.29	4.62
1.80	1.71	1.43	4.06	—	—	—	—	0.57	2.19	1.83	1.61
4.90	1.97	1.40	3.01	—	—	—	—	1.36	0.81	2.11	1.45
7.10	58.2	NR	1.7	—	—	—	—	40.4	11.50	NR	1.70

soy, or pulses [71, 130], appropriate precautions and adequate labeling is required in order to facilitate decision making.

11.4.2. Alternatives to Peanut and Tree Nut Flours

Defatted peanut and tree nut flour is a by-product from the oil industry. The flours generally contain approximately 50% protein and are often used in bakery products and confectioneries due to their high nutritional value and good functionality. Sesame, sunflower, and soy flours contain similar protein contents (48.06, 40.32, and 47.01 g/100 g, respectively) and could be considered as replacements. An added advantage is that the protein quality (biological value) of these ingredients is higher than for peanuts and tree nuts (except cashew). Sen and Bhattacharyya [131, 132] evaluated the nutritional quality of sesame and sunflower seed protein extract using animal experiment. Their results indicated that rats fed with decorticated sesame and sunflower protein extract for 28 days showed similar gain in body weight and food protein efficiency ratio compared with soybean and casein. Furthermore, rats fed with soybean protein, sesame protein, and sunflower protein had lower plasma lipid profiles than rats fed with casein.

Functional property studies of sesame and soybean protein isolates [133] revealed no significant differences in properties such as solubility, WAC and FAC, emulsifying capacity, foaming capacity, and whipping property. Although the study did not compare the properties of the sesame protein isolate and SPI with those of peanut and tree nut proteins, the results, nevertheless, provide some useful information on how these alternative ingredients are likely to perform when used in foods.

Other novel ingredients that could be considered are high-protein flours or concentrates from canola and pulses such as peas, chickpeas, and lentils. Paredes-Lopez et al. [134] found that the *in vitro* digestibility and calculated protein efficiency ratio of chickpea protein isolate prepared by a micellization process was higher (94.1% and 2.6, respectively) than casein (90.0% and 2.5, respectively) and showed good functional properties. Proteins from pea and canola have also been shown to have good nutritional and functional properties [135, 136]. As a nutritional and functional alternative, high-protein dairy flours, concentrates, or isolates could also be considered. As these proteins are equally allergenic, appropriate precautions need to be taken in their labeling.

11.4.3. Alternatives to Peanut and Tree Nut Oils

Peanut and tree nut seeds contain 40–50% oil depending on variety and growing conditions. Peanut and tree nut oil extracted from shelled seeds are used in culinary preparations for their pleasant flavor and also for therapeutic purposes. Due to their high PUFA and vitamin E content, peanut and tree nut oils play a significant role in a healthy diet. Vegetable oils such as soybean, canola, sunflower, palm, coconut, sesame, safflower, olive, and corn oils can be used to replace peanut and tree nut oils in cooking and in food processing. The nutritional benefits of oils such as olive oil and canola oil are commonly known. Sesame and sunflower seeds contain similar quantities of oil as in peanut and tree nut oils. Sesame oil, however, contains higher PUFA levels (41.7%) and lower MUFA levels (39.7%) than peanut oil (32% and 46%, respectively), while sunflower oil is rich in MUFA and has lower PUFA content (57.3% and 28.9%, respectively). Although soybean seeds contain lower oil content (~20%), soy oil is higher in PUFA (57.74%) and lower in MUFA

(22.78%) than peanut oil. Peanut oil is widely used for frying of foods due to its high smoke point (244°C). The smoke point of soybean, canola, sunflower, palm, coconut, sesame, safflower, olive, and corn oils range between 220 and 242°C [137]; depending on the food in question, these oils could be used as effective replacements for peanut and tree nut oils in the manufacture of "nut-free" foods. For use as a salad oil, pumpkin seed, flaxseed, grape seed, avocado, and hemp seed oils are potentially good alternatives.

11.4.4. Alternatives to Peanut and Tree Nut Butters

Peanut and tree nut butters are pastes made from roasted peanut or tree nuts with or without addition of salt, sweetener, and oil. They are often consumed in a manner similar to butter from cow's milk (i.e., as a spread) and are also commonly used in snacks and chocolate preparations because of their nutritional quality and functionality. Suggested alternatives to peanut and tree nut butters include sesame, soy, pea, hemp, pumpkin seed, and sunflower butters (Fig. 11.2). Sesame, sunflower, soy, and hemp butters contain approximately

Fig. 11.2 Commercially available alternatives to peanut and tree nut butter. (A) Soybean seed butter; (B) sunflower seed butter; (C) pumpkin seed butter; (D) sesame seed butter; (E) pea seed butter.

17–19% protein and 48–54% fat, which is comparable to peanut and tree nut butters (approximately 15–22% protein and 50–60% fat). Pea and pumpkin seed butters contain less protein (12–14%) and fat (30–35% fat). Sesame, sunflower, hemp, pea, and soy butters are commercially available; however, utilization of these butters in confectionery and bakery applications as an alternative to peanut and tree nut butters has not been extensively studied.

11.5. HIDDEN AND UNINTENTIONAL SOURCES OF PEANUT AND TREE NUT ALLERGENS

Complete avoidance of peanuts and tree nuts is often difficult due to hidden and undeclared sources of peanut and tree nut allergens in foods (Table 11.7). Numerous reports have indicated that convenience food products such as snacks, breakfast cereals, cookies, bakery items, and ice creams are often contaminated with peanuts and tree nuts due to improper handling, use of contaminated raw materials, cross-contamination in the production line, and so on [138, 139]. This poses a potential risk for peanut- and tree nut-allergic individuals.

TABLE 11.7 Hidden and Unintentional Sources of Peanuts and Tree Nuts in Foods

Hidden sources
Baked foods—biscuits, cakes, cookies, egg rolls
Cereal-based products
Sauces
Ice cream
Milk formulas
Soups
Candy
Vegetable meatballs
Pasta products
Chili
Vegetable oil
Margarine
Ointment containing peanut and tree nut oil
Drugs containing peanut and tree nut oil
Unintentional sources
Using contaminated raw materials
Sharing of utensils
Common packaging materials
Lack of equipment cleanliness
Improper handling of food

Sicherer et al. [6] reported that 55% of peanut-allergic patients experienced accidental ingestion of peanut with an average of two accidents per patients over a period of 5.4 years, which most commonly happened in schools (29%), home (22%), and restaurants (8%), especially in Asian restaurants. The presence of hidden peanut ingredients in processed foods, cross-contamination, sharing food with friends, and skin contact with peanut butter in school classroom projects were suggested as possible modes for accidental ingestion of peanut allergens.

Flinterman et al. [21] reported the case of four peanut-sensitive children (mean age of 7.5 years) who developed allergic symptoms after ingestion of food products that contained hidden sources of peanut. The culprit foods were vegetarian meatball, chocolate ice cream, chocolate candy bar, and chocolate lollipop. Some reports have also suggested accidental ingestion of peanuts and tree nuts due to the use of containers that were previously used for carrying foods containing peanuts. Furlong et al. [140] reported higher chances of accidental exposure to peanuts and tree nuts (67% and 24%, respectively) in food establishments, and desserts were the most common sources of exposure to peanuts and tree nuts (43%) followed by entrées (35%), appetizers (13%), and others (9%). The inadvertent ingestion of peanuts and tree nuts was mainly from sauces, dressings, or egg rolls, as well as cross-contact due to sharing of equipment and utensils and improper handling. Candy, chocolate products, cookies, chili, dry soup mixtures, pasta, sandwiches, and Vietnamese dishes have also been frequently reported to be contaminated with peanuts and tree nuts [141, 142].

Moneret-Vautrin et al. [12] detected the presence of peanut oil in 11 milk formulas, which resulted in the development of allergic reactions in peanut-sensitive infants (4–13 months). Undeclared peanut (50 ppm) was also found in 19 out of 49 industrially produced snack products including cereals, cereal bars, cookies, and various types of snacks [138]. The Western Australian Food Monitoring Programme (WAFMP) reported the presence of undeclared almond and hazelnut in different food products including biscuits, cakes, chocolates, and ice creams [139]. Fifty-five percent of the food products were contaminated with almond and hazelnut, and among these, 14% contained greater than 1 ppm of almond and hazelnut allergens [139].

Food manufacturers and providers need to exercise greater caution in order to prevent the inadvertent presence of peanuts and tree nuts in foods labeled as being "free from" specific nuts. In general, the presence of undeclared peanuts and nuts in various types of foods can be attributed to a wide variety of factors including the use of contaminated raw materials; common transport containers for peanuts, tree nuts, and other foods; lack of separate production lines and equipment for processing "nut-free" foods; processing of "nut-free" products immediately after processing nut-containing products; and insufficient cleaning steps as well as unsafe rework managements [138, 139]. Potential sources of contamination and cross-contact, therefore, need to be carefully identified and properly controlled as discussed elsewhere in this book.

It is important to mention here that peanut oil is sometimes used in the development of drugs, ointments, and massage creams, and exposure to these products could provoke allergic reactions in peanut-sensitive individuals [143, 144]. Transgenic plants could also result in the introduction of allergens in the food chain as was demonstrated with the case of three Brazil nut-allergic patients who had positive skin reactions to transgenic soybean carrying the Brazil nut 2S gene [64]. Transgenic foods, therefore, need to be carefully screened and appropriately labeled if potential allergenic proteins are present.

For the allergic consumer, an important step in successful allergen avoidance is verification of food labels for the presence of peanuts and tree nuts when purchasing prepacked food items. In response to strict food labeling regulations in many countries, manufacturers are increasingly using precautionary labels such as "may contain traces of peanuts and tree nuts" or "made in a factory that processes peanuts and tree nuts" to help allergic patients. These kinds of labels unintentionally create greater confusion rather than provide guidance about the safety of foods for the allergic consumer [138]. Flinterman et al. [21] examined the dietary management of peanut-allergic children (median age 7.5 years) during a 12-month follow-up period. Three peanut-sensitive children out of 11 developed mild to moderate allergic symptoms after the intake of foods with a "may contain" peanut label. The study indicated that "may contain" labeling can affect food choices of allergic individuals and may strongly influence compliance in a negative way. Furthermore, the amount of allergens present in the peanut- and tree nut-contaminated food items is often unknown.

11.6. ANALYTICAL TOOLS FOR DETECTION OF PEANUT AND TREE NUT ALLERGENS

Reliable analytical tools are necessary to detect the presence of hidden peanut and tree nut allergens in foods in order to ensure compliance with food labeling regulations and to improve consumer protection. Various immunoassay and DNA-based methods are available for the detection of hidden peanuts and tree nuts in foods. Immunoassays such as radioallergosorbent test (RAST), EAST, ELISA, and immunoblotting techniques target total protein or specific peanut and tree nut allergens. The DNA-based assays are based on the amplification of a specific DNA segment of the allergen by the polymerase chain reaction (PCR) [145]. Among protein-based assays, ELISA is widely used to detect food allergens because of its high precision, sensitivity, speed, and ease of handling. Even though DNA marker-based assays may, in some instances, be more sensitive and reproducible than ELISA techniques, due to their higher cost, complicated procedures, and longer duration, their applicability in allergen detection has been limited [146].

ELISA tests are based on interactions between antibodies and antigens and consist mainly of three processes, namely sample preparation, extraction, and determination. Extraction of proteins from allergenic foods is a prerequisite for allergen detection by ELISA test. Currently available ELISA kits detect soluble or buffer extractable peanut and tree nut proteins in foods. An inefficient extraction process can, therefore, affect the detectability of allergens and give false-negative results. Nogueira et al. [147] reported that the yield of proteins extracted from raw peanuts was higher than from roasted peanuts using the Bradford and bicinchoninic acid (BCA) protein assays. The concentration of purified Ara h 1 and Ara h 2 also ranged from 2.55 to 0.30 μg/μL, depending on the degree of roasting [147]. Poms et al. [148] reported that the extraction efficiency of soluble protein from oil and dry roasted peanuts (140–190°C) reduced by 50% and 75–80%, respectively, compared with raw peanuts using Tris-buffered saline (TBS). Addition of 6 M urea (denaturing agent) to the TBS extraction buffer increased the protein yield for the dry and oil roasted peanuts. Cong et al. [149] observed no difference in ELISA response of fried, boiled, and roasted peanuts extracted with a urea buffer compared with raw peanuts. There was, however, a remarkable difference (decrease) in Ara h 1 content for the fried and boiled peanuts when only phosphate buffer saline (PBS) at pH 7.6 was used as the extraction buffer. Reduced extractability in the absence of a denaturing agent may be due to heat-induced conformational changes in the peanut protein structure (hydrophobic interactions), which make the proteins less soluble in aqueous extraction buffers. A standardized buffer system for extraction of protein from processed peanuts and tree nuts will, therefore, be very useful.

The detection of peanut and tree nut allergens in chocolate by ELISA is particularly challenging due to the inherent matrix effect. Grimshaw et al. [25] observed difficulty in extracting peanut proteins from high-fat recipes such as chocolate. Moreover, chocolate contains high amounts of tannins and other phenolic compounds, which effectively bind with proteins and may also interfere with the ELISA test by binding with antibodies [150]. Addition of 5% nonfat dry milk (w/v) for 2.5 hours at room temperature or 15 minutes at 60°C increased the extraction of peanut allergen (Ara h 1) from 0.16% to 0.33% in peanut chocolate [151]. Trucksess et al. [152] studied the variation of analytical results for peanut contamination in energy bars and milk chocolate using commercially available ELISA kits. The sampling and subsampling steps were the major source of variability contributing 60–90% of total variance compared with the analytical variability associated with testing the energy bars and milk chocolate for peanut contamination. Increasing sample size from 1 to 4 bars and using 20 g of subsample instead of 5 g and duplicate analyses of each extract reduced total variability for detecting peanuts in energy bars or similar snack foods (candy bars/breakfast bars). For powdered milk chocolate, 20 g of subsample and duplicate analyses of each extract were recommended.

The type of antibody used in an ELISA is also important for accurate detection of peanut and tree nut allergens. Most commercial ELISA kits use either polyclonal antibodies (IgG) from rabbit, sheep, or goat, or monoclonal antibodies from mice. Antibodies raised against raw or minimally processed peanuts

TABLE 11.8 Commercially Available Test Kits for Detection of Peanuts and Tree Nuts Allergen

Name of the Nut	Detection Method	Method	Type	LOD (ppm)	Name of Kit	Manufacturer	References
Peanut							
1	Protein	ELISA	Quantitative	0.1	BIOKITS Peanut Assay	Tepnel Life Sciences	www.tepnel.com
2	Protein	ELISA	Quantitative	1.0	Veratox® for Peanut Allergen	Neogen	www.neogen.com
3	Protein	ELISA	Quantitative	1.5	RIDASCREEN® FAST Peanut	R-Biopharm	www.r-biopharm.com
4	Protein	ELISA	Quantitative	<2.5	RIDASCREEN® Peanut	R-Biopharm	www.r-biopharm.com
5	Protein	ELISA	Quantitative	0.5–1.0	Peanut Protein Residue, Peanut Residue Mass	ELISA Systems	www.elisasystems.net
6	DNA	PCR	Quantitative	<10	SureFood® Peanut PCR-ELISA	R-Biopharm	www.r-biopharm.com
7	Protein	LFD	Qualitative	<0.1	BIOKITS Rapid Peanut Test	Tepnel Life Sciences	www.tepnel.com
8	Protein	LFD	Qualitative	0.5	TECRA PEANUT VIA™	Biotrace International	http://www.tecra.net; www.biotrace.com
9	Protein	LFD	Qualitative	5.0	Reveal® for Peanut Allergen	Neogen	www.neogen.com
10	Protein	ELISA	Qualitative	5.0	Alert® for Peanut Allergen	Neogen	www.neogen.com
11	Protein	LFD	Qualitative	5.0	BIOKITS RAPID 3-D™ Peanut Test	Tepnel Life Sciences	www.tepnel.com
12	DNA	PCR	Qualitative	10–50	SureFood® Allergen Peanut Real-Time PCR	R-Biopharm	www.r-biopharm.com
13	DNA	PCR	Qualitative	<10	BIOKITS Peanut PCR Mastermix Pod	Tepnel Life Sciences	www.tepnel.com
Almond							
1	Protein	ELISA	Qualitative	5.0	Alert® for Almond Allergen	Neogen	www.neogen.com
2	Protein	ELISA	Quantitative	1.7	RIDDASCREEN® FAST Mandel/Almond	R-Biopharm	www.r-biopharm.com
3	Protein	LFD	Qualitative	1–50	BIOKITS RAPID 3-D™ Almond Test	Tepnel Life Sciences	www.tepnel.com
4	DNA	PCR	Qualitative	<10	SureFood® Allergen Almond Real-Time PCR	R-Biopharm	www.r-biopharm.com
Hazelnut							
1	Protein	ELISA	Quantitative	1.5	RIDASCREEN® FAST Hazelnut	R-Biopharm	www.r-biopharm.com
2	Protein	ELISA	Qualitative	5.0	RIDA® QUICK Hazelnut	R-Biopharm	www.r-biopharm.com
Walnut							
1	Protein	ELISA	Quantitative	0.25	BIOKITS Walnut Assay	Tepnel Life Sciences	www.tepnel.com
2	DNA based	PCR	Qualitative	<10	SureFood® Allergen Walnut Real-Time PCR	R-Biopharm	www.r-biopharm.com

ELISA, enzyme-linked immunosorbent assay; PCR, polymerase chain reaction; LFD, lateral flow device; LOD, limit of detection.

and tree nuts may not recognize highly processed peanuts and tree nuts. Hence, the selection of appropriate sample size, suitable buffer, and right antibodies is an important factor that needs to be considered as this may influence accurate detection of peanut and tree nut allergens in foods. Three commercial ELISA kits (Biokits, Tepnel Biosystems, Gainesville, FL; Veratox, Neogen Corp., Lansing, MI; and RIDASCREEN®, FAST, R-Biopharm, Darmstadt, Germany) have been validated by the Association of Official Analytical Chemists (AOAC) for the detection of peanuts at levels of 5 μg/g in processed foods including breakfast cereals, cookies, ice cream, and milk chocolate [153]. Several other test kits are commercially available for the detection of peanut and tree nut allergens in food products and are listed in Table 11.8.

11.7. GUIDELINES FOR PROCESSING FOODS "FREE FROM" PEANUTS AND TREE NUTS

To ensure the safety of peanut- and tree nut-allergic consumers, food manufacturers must carefully assess the potential risks of peanut and tree nut contamination during processing. This includes evaluation of all ingredients used in processing as well as possible contamination from rework, processing, and cleaning aids. Cleaning of equipment is an essential step before processing foods "free from" peanuts and tree nuts. Inadequate cleaning may lead to traces of peanuts and tree nuts in the equipment, which increases the risk of cross-contact. Stephan et al. [154] found that cleaning water from peanut mush equipment contained 98.5–1379 μg/mL of peanut protein when analyzed by sandwich ELISA and Bradford assays. However, when the equipment was washed with wash water using acid (P3-horolith CIP, conductance 5.5 mS/cm) and alkali solutions (P3 mip, conductance 22 mS/cm), results for the presence of peanut protein were negative. Evaluation of cleaning efficiency by measuring total protein content in washing water and residual proteins on surfaces could lower the risks of cross-contact.

Additionally, multiple production lines operating in a single facility may present potential pitfalls for peanut and tree nut contamination [139]. Proper scheduling of production lines can help to prevent cross-contact. As an example, in shared facilities, food products containing peanuts and tree nuts can be produced at the end of the week or at the end of the production shift. The equipment can then be dismantled for cleaning and completely washed to remove hidden peanuts and tree nuts. Reviewing and verifying equipment cleanliness before processing food products that are "free from" peanuts and tree nuts will reduce the chances of contamination. Creating awareness among employees about hidden and unintentional sources of peanuts and tree nuts in food products coupled with adequate training can help to prevent the risk of cross-contact and the inadvertent presence of allergens. The use of certified raw materials and thoroughly cleaned individual containers for transport will also help to improve the safety of food products labeled as "free from" peanuts and tree nuts. As the risk of cross-contact may be high in a shared facility, dedicated facilities may be the most suitable option for the production of foods labeled as "free from" peanuts and tree nuts.

11.8. CONCLUSION

Due to increasing prevalence of peanut and tree nut allergy in the general population, there is a growing demand for food products labeled as "free from" peanuts and tree nuts. Food manufacturers need guidance on how to process such foods. They also need to understand how cross-contact occurs, the sources of contamination, and the effect of processing treatments on allergen detection and allergenicity. Research is needed to improve our understanding of the effects of different processing treatments on the allergenicity of peanut and tree nut proteins and to identify potential techniques that can be used to reduce the allergenicity of these proteins.

Cross-contamination of peanuts and tree nuts in food products pose allergic risks to peanut- and tree nut-sensitive individuals. The use of good agricultural practices (GAPs) and good manufacturing practices (GMPs) during production and processing may help to reduce the risks of cross-contact in peanut- and tree nut-free foods. Furthermore, appropriate GAP and GMP may help to reduce the use of broad labeling such as "may contain" or "made in a factory that processes peanuts and tree nuts" and will enhance the capacity of the food industry to more accurately indicate the level of contamination likely to be present in their food products as well as increase their capacity to produce food products that are truly "free from" peanuts and tree nuts. Additionally, new research studies on threshold dose determination for peanuts and tree nuts such as that conducted by Taylor et al. [155] will be useful to establish guidelines for the industry.

REFERENCES

1. Rainey, C., Nyquist, L. (1997). Nuts: nutrition and health benefits of daily use. *Nutrition Today*, *32*, 157–163.
2. United States Department of Agriculture Natural Resources Conservation Services (USDA-NRCS). (2009). Plants classification report. Available at http://plants.usda.gov/classification.html
3. Griel, A.E., Eissenstat, B., Juturu, V., Hsieh, G., Kris-Etherton, P.M. (2004). Improved diet quality with peanut consumption. *Journal of the American College of Nutrition*, *23*, 660–668.
4. King, J.C., Blumberg, J., Ingwersen, L., Jenab, M., Tucker, K.L. (2008). Tree nuts and peanuts as components of a healthy diet. *The Journal of Nutrition*, *138*, 1736S–1740S.

5. Jenab, M., Sabate, J., Slimani, N., Ferrari, P., Mazuir, M., Casagrande, C., Deharveng, G., Tjonneland, A., Olsen, A., Overvad, K., Boutron-Ruault, M.C., Clavel-Chapelon, F., Boeing, H., Weikert, C., Linseisen, J., Rohrmann, S., Trichopoulou, A., Naska, A., Palli, D., Sacerdote, C., Tumino, R., Mattiello, A., Pala, V., Bueno-de-Mesquita, H.B., Ocké, M.C., Peeters, P.H., Engeset, D., Skeie, G., Jakszyn, P., Ardanaz, E., Quirós, J.R., Chirlaque, M.D., Martinez, C., Amiano, P., Berglund, G., Palmqvist, R., van Guelpen, B., Bingham, S., Key, T., Riboli, E. (2006). Consumption and portion sizes of tree nuts, peanuts and seeds in the European Prospective Investigation into Cancer and Nutrition (EPIC) cohorts from 10 European countries. *The British Journal of Nutrition*, *96*, S12–S23.
6. Sicherer, S.H., Burks, A.W., Sampson, H.A. (1998). Clinical features of acute allergic reactions to peanut and tree nuts in children. *Pediatrics*, *102*, 1–6.
7. Grundy, J.R., Matthews, S., Bateman, B., Dean, T., Arshad, S.H. (2002). Rising prevalence of allergy to peanut in children: data from 2 sequential cohorts. *The Journal of Allergy and Clinical Immunology*, *110*, 784–778.
8. Sicherer, S.H., Munoz-Furlong, A., Sampson, H.A. (2003). Prevalence of peanut and tree nut allergy in the United States determined by means of a random digit dial telephone survey: a 5-year follow-up study. *The Journal of Allergy and Clinical Immunology*, *112*, 1203–1207.
9. Sampson, H.A. (2004). Update on food allergy. *The Journal of Allergy and Clinical Immunology*, *113*, 805–819.
10. Hourihane, J.O., Dean, T.P., Warner, J.O. (1996). Peanut allergy in relation to hereditary, maternal diet, and other atopic diseases: results of a questionnaire survey, skin prick testing, and food challenges. *British Medical Journal*, *313*, 518–521.
11. Zimmerman, B., Forsyth, S., Gold, M. (1989). Highly atopic children: formation of IgE antibody to food protein, especially peanut. *The Journal of Allergy and Clinical Immunology*, *83*, 764–770.
12. Moneret-Vautrin, D.A., Hatahet, R., Kanny, G. (1994). Risks of milk formulas containing peanut oil contaminated with peanut allergens in infants with atopic dermatitis. *Pediatric Allergy and Immunology*, *5*, 184–188.
13. Sicherer, S.H., Furlong, T.J., Munoz-Furlong, A., Burks, A.W, Sampson H.A. (2001). A voluntary registry for peanut and tree nut allergy: characteristics of the first 5149 registrants. *The Journal of Allergy and Clinical Immunology*, *108*, 128–132.
14. Dalal, I., Binson, I., Reifen, R., Amitai, Z., Shohat, T., Rahmanmi, S., Levine, A., Ballin A., Somekh, E. (2002). Food allergy is a matter of geography after all: sesame as a major cause of severe IgE mediated food allergic reactions among infants and young children in Israel. *Allergy*, *57*, 362–365.
15. Sheskin, A., Regew, A. (2001). *Israel Agriculture–Facts and Figures*, 2nd edition. Beit Dagan, Israel: Publication of State of Israel/Ministry of Agriculture and Rural Development, Foreign Trade Centre and Foreign Relationship Department.
16. Beyer, K., Morrow, E., Li, X.M., Bardina, L., Bannon, G.A., Burks, A.W., Sampson, H.A. (2001). Effects of cooking methods on peanut allergenicity. *The Journal of Allergy and Clinical Immunology*, *107*, 1077–1081.
17. Levy, Y., Broides, A., Segal, N., Danon, Y.L. (2003). Peanut and tree nut allergy in children: role of peanut snacks in Israel? *Allergy*, *58*, 1206–1207.

18. Sampson, H.A. (2002). Clinical practice—peanut allergy. *The New England Journal of Medicine*, *346*, 1294–1299.
19. Stark, B.J., Sullivan T.J. (1986). Biphasic and protracted anaphylaxis. *The Journal of Allergy and Clinical Immunology*, *78*, 76–83.
20. United States Food and Drug Administration (FDA), Department of Health and Human Services, Threshold Working Group. (2006). Approaches to establish thresholds for major food allergens and for gluten in food. Available at http://www.fda.gov/Food/LabelingNutrition/FoodAllergensLabeling/GuidanceComplianceRegulatoryInformation/ucm106108.htm
21. Flinterman, A.E., Pasmans, S.G., Hoekstra, M.O., Meijer, Y., Van Hoffen, E., Knol, E.F., Hefle, S.L., Bruijnzeel-Koomen, C.A., Knulst, A.C. (2006). Determination of no-observed adverse effect levels and eliciting doses in a representative group of peanut sensitized children. *The Journal of Allergy and Clinical Immunology*, *117*, 448–454.
22. Hourihane, J.O., Kilburn, S.A., Nordlee, J.A., Hefle, S.L., Taylor, S.L., Warner, J.O. (1997). An evaluation of the sensitivity of subjects with peanut allergy to very low doses of peanut protein: a randomized, double-blind, placebo-controlled food challenge study. *The Journal of Allergy and Clinical Immunology*, *100*, 596–600.
23. Taylor, S.L., Hefle, S.L., Bindslev-Jensen, C., Bock, S.A., Burks, A.W. Jr., Christie, L., Hill, D.J., Host, A., Hourihane, J.O., Lack, G., Metcalfe, D.D., Moneret-Vautrin, D.A., Vadas, P.A., Rance, F., Skrypec, D.J., Trautman, T.A., Yman, I.M, Zeiger, R.S. (2002). Factors affecting the determination of threshold doses for allergenic foods: how much is too much? *The Journal of Allergy and Clinical Immunology*, *109*, 24–30.
24. Wensing, M., Penninks, A.H., Hefle, S.L., Koppelman, S.J., Bruijnzeel-Koomen, C.A., Knulst, A.C. (2002). The distribution of individual threshold doses eliciting allergic reactions in a population with peanut allergy. *The Journal of Allergy and Clinical Immunology*, *110*, 915–920.
25. Grimshaw, K.E.C., King, R.M., Nordlee, J.A., Hefle, S.L., Warner, J.O., Hourihane, J.O. (2003). Presentation of allergen in different food preparations affects the nature of the allergic reaction—a case series. *Clinical and Experimental Allergy*, *33*, 1581–1585.
26. Burks, A.W., Williams, L.W., Helm, R.M., Connaughton, C., Cockrell, G., O'Brien, T. (1991). Identification of a major peanut allergen, Ara h I, in patients with atopic dermatitis and positive peanut challenge. *The Journal of Allergy and Clinical Immunology*, *88*, 172–179.
27. Stanley, J.S., King, N., Burks, A.W., Huang, S.K., Sampson, H., Cockrell, G., Helm, R.M., West, C.M, Bannon, G.A. (1997). Identification and mutational analysis of the immunodominant IgE binding epitopes of the major peanut allergen Ara h 2. *Archives of Biochemistry and Biophysics*, *342*, 244–253.
28. Burks, W., Sampson, H.A., Bannon G.A. (1998). Peanut allergens. *Allergy*, *53*, 725–730.
29. Scurlock, A.M., Burks, A.W. (2004). Peanut allergenicity. *Annals of Allergy, Asthma & Immunology*, *93*, S12–S18.
30. Wen, H.W., Borejsza-Wysocki, W., Decory, R.T., Baeumner A., Durst, R.A. (2007). Peanut allergy, peanut allergens, and methods for the detection of peanut con-

tamination in food products. *Comprehensive Reviews in Food Science and Food Safety*, *6*, 47–58.

31. Flinterman, A.E., Van Hoffen, E., Jager, C.F.D.H., Koppelman, S., Pasmans, S.G., Hoekstra, M.O., Bruijnzeel-Koomen, C.A., Knulst, A.C., Knol, E.F. (2007). Children with peanut allergy recognize predominantly Ara h 2 and Ara h 6 which remains stable over time. *Clinical and Experimental Allergy*, *37*, 1221–1228.
32. Burks, A.W., Shin, D., Cockrell, G., Stanley, J.S., Helm, R.M., Bannon, G.A. (1997). Mapping and mutational analysis of the IgE-binding epitopes on Ara h 1, a legume vicilin protein and a major allergen in peanut hypersensitivity. *European Journal of Biochemistry/FEBS*, *245*, 334–339.
33. Shin, D.S., Compadre, C.M., Maleki, S.J., Kopper, R.A., Sampson, H., Huang, S.K., Burks, A.W, Bannon, G.A. (1998). Biochemical and structural analysis of the IgE binding sites on Ara h 1, an abundant and highly allergenic peanut protein. *The Journal of Biological Chemistry*, *273*, 13753–13759.
34. Koppelman, S.J., Bruijnzeel-Koomen, C.A., Hessing, M., De Jongh, H.H. (1999). Heat induced conformational changes of Ara h 1, a major peanut allergen, do not affect its allergenic properties. *The Journal of Biological Chemistry*, *274*, 4770–4777.
35. Maleki, S.J., Chung, S.Y., Champagne, E.T., Raufman, J.P. (2000). The effects of roasting on the allergenic properties of peanut proteins. *The Journal of Allergy and Clinical Immunology*, *106*, 763–768.
36. Sen, M., Kopper, R., Pons, L., Abraham, E.C., Burks, A.W., Bannon, G.A. (2002). Protein structure plays a critical role in peanut allergen stability and may determine immunodominant IgE-binding epitopes. *The Journal of Immunology*, *169*, 882–887.
37. Maleki, S.J., Viquez, O., Jacks, T., Dodo, H., Champagne, E.T., Chung, S.Y., Landry, S.J. (2003). The major peanut allergen, Ara h 2, functions as a trypsin inhibitor, and roasting enhances this function. *The Journal of Allergy and Clinical Immunology*, *112*, 190–195.
38. Eigenmann, P.A., Burks, A.W., Bannon, G.A., Sampson, H.A. (1996). Identification of unique peanut and soy allergens in sera adsorbed with cross reacting antibodies. *The Journal of Allergy and Clinical Immunology*, *98*, 969–978.
39. Rabjohn, P., Helm, E.M., Stanley, J.S., West, C.M., Sampson, H.A., Burks, A.W., Bannon, G.A. (1999). Molecular cloning and epitope analysis of the peanut allergen Ara h 3. *The Journal of Clinical Investigation*, *103*, 535–542.
40. Staswick, P.E., Hermodson, M.A., Nielsen, N.C. (1984). Identification of the cystines which link the acidic and basic components of the glycinin subunits. *The Journal of Biological Chemistry*, *259*, 13431–13435.
41. Tumer, N.E., Thanh, V.H., Nielsen, N.C. (1981). Purification and characterization of mRNA from soybean seeds. *The Journal of Biological Chemistry*, *256*, 8756–8760.
42. Koppelman, S.J., Knol, E.F., Vlooswijk, R.A., Wensing, M., Knulst, A.C., Hefle, S.L., Gruppen, H., Piersma, S. (2003). Peanut allergen Ara h 3: isolation from peanuts and biochemical characterization. *Allergy*, *58*, 1144–1151.
43. van Boxtel, E.L., van den Broek, L.A.M., Koppelman, S.J, Gruppen, H. (2008). Legumin allergens from peanuts and soybeans: effects of denaturation and aggregation on allergenicity. *Molecular Nutrition and Food Research*, *52*, 674–682.

44. Kleber-Janke, T., Crameri, R., Appenzeller, U., Schlaak, M., Becker, W.M. (1999). Selective cloning of peanut allergens, including profilin and 2S albumins, by phage display technology. *International Archives of Allergy and Immunology*, *119*, 265–274.
45. Boldt, A., Fortunato, D., Conti, A., Petersen, A., Ballmer-Weber, B., Lepp, U., Reese, G., Becker, W.M. (2005). Analysis of the composition of an immunoglobulin E-reactive high-molecular-weight protein complex of peanut extract containing Ara h 1 and Ara h 3/4. *Proteomics*, *5*, 675–686.
46. Valenta, R., Duchene, M., Ebner, C., Valent, P., Sillaber, C., Deviller, P., Ferreira, F., Tejkl, M., Edelmann, H., Kraft, D. (1992). Profilins constitute a novel family of functional plant pan-allergens. *Journal of Experimental Medicine*, *175*, 377–385.
47. Pastorello, E.A., Incorvaia, C., Pravettoni, V., Ortolani, C. (1997). Cross reactions in food allergy. *Clinical Reviews in Allergy and Immunology*, *15*, 415–427.
48. Mittag, D., Akkerdaas, J., Ballmer-Weber, B.K., Vogel, L., Wensing, M., Becker, W.M., Koppelman, S.J., Knulst, A.C., Helbling, A., Hefle, S.L., Van Ree, R., Vieths, S. (2004). Ara h 8, a Bet v 1-homologous allergen from peanut, is a major allergen in patients with combined birch pollen and peanut allergy. *The Journal of Allergy and Clinical Immunology*, *114*, 1410–1417.
49. Crespo, J.F., James, J.M., Fernandez-Rodriguez, C., Rodriguez, J. (2006). Food allergy: nuts and tree nuts. *The British Journal of Nutrition*, *96*, S95–S102.
50. Teuber, S.S., Sathe, S.K., Peterson, W.R., Roux, K.H. (2002). Characterization of the soluble allergenic proteins of cashew nut (*Anacardium occidentale* L.). *Journal of Agricultural and Food Chemistry*, *50*, 6543–6549.
51. Wang, F., Robotham, J.M., Teuber, S.S., Tawde, P., Sathe, S.K., Roux, K.H. (2002). Ana o 1, a cashew (*Anacardium occidental*) allergen of the vicilin seed storage protein family. *The Journal of Allergy and Clinical Immunology*, *110*, 160–166.
52. Robotham, J.M., Wang, F., Vest, V.M., Teuber, S.S., Sathe, S.K., Beyer, K., Roux, K.H. (2005). Ana o 3 an important cashew nut allergen of the 2S albumin family. *The Journal of Allergy Clinical Immunology*, *115*, 1284–1290.
53. Wang, F., Robotham, J.M., Teuber, S.S., Sathe, S.K., Roux, K.H. (2003). Ana o 2, a major cashew (*Anacardium occidentale* L.) nut allergen of the legumin family. *International Archives of Allergy and Immunology*, *132*, 27–39.
54. Venkatachalam, M., Monaghan, E.K., Kshirsagar, H.H., Robotham, J.M., O'Donnell, S.E., Gerber, M.S., Roux, K.H., Sathe, S.K. (2008). Effects of processing on immunoreactivity cashew nut (*Anacardium occidentale* L.) seed proteins. *Journal of Agricultural and Food Chemistry*, *56*, 8998–9005.
55. Teuber, S.S., Dandekar, A.M., Peterson, W.R., Sellers, C.L. (1998). Cloning and sequencing of a gene encoding a 2S albumin seed storage protein precursor from English walnut (*Juglans regia*), a major food allergen. *The Journal of Allergy and Clinical Immunology*, *101*, 807–814.
56. Robotham, J.M., Teuber, S.S., Sathe, S.K., Roux, K.H. (2002). Linear IgE epitope mapping of the English walnut (*Juglans regia*) major food allergen, Jug r 1. *The Journal of Allergy and Clinical Immunology*, *109*, 143–149.
57. Teuber, S.S., Jarvis, K.C., Dandekar, A.M., Peterson, W.R, Ansari A. (1999). Identification and cloning of a complimentary DNA encoding a vicilin-like proprotein, Jug r 2, from English walnut kernel (*Juglans regia* L.) a major food allergen. *The Journal of Allergy and Clinical Immunology*, *104*, 1311–1320.

58. Asero, R., Mistrello, G., Roncarolo, D., Amato, S. (2004). Walnut induced anaphylaxis with cross reactivity to hazelnut and Brazil nut. *The Journal of Allergy and Clinical Immunology*, *113*, 358–360.
59. Hirschwehr, R., Valenta, R., Ebner, C., Ferreira, F., Sperr, W.R., Valent, P., Rohac, M., Rumpold, H., Scheiner, O., Kraft, D. (1992). Identification of common allergenic structures in hazel pollen and hazelnuts: a possible explanation for sensitivity to hazelnuts in patients allergic to tree pollen. *The Journal of Allergy and Clinical Immunology*, *90*, 927–936.
60. Lüttkopf, D., Muller, U., Skov, P.S., Ballmer-Weber, B.K., Wuthrich, B., Skamstrup Hansen, K., Poulsen, L.K., Kastner, M., Haustein, D., Vieths, S. (2002). Comparison of four variants of a major allergen in hazelnut (*Corylus avellana*) Cor a 1.04 with the major hazel pollen allergen Cor a 1.01. *Molecular Immunology*, *38*, 515–525.
61. Hansen, K.S., Ballmer-Weber, B.K., Luttkopf, D., Skov, P.S., Wuthrich, B., Bindslev-Jensen, C., Vieths, S., Poulsen, L.K. (2003). Roasted hazelnuts—allergenic activity evaluated by double-blind, placebo-controlled food challenge. *Allergy*, *58*, 132–138.
62. Vieths, S., Reindl, J., Muller, U., Hoffmann, A., Haustein, D. (1999). Digestibility of peanut and hazelnut allergens investigated by a simple *in vitro* procedure. *European Food Research and Technology*, *209*, 379–388.
63. Pastorello, E.A., Vieths, S., Pravettoni, V., Farioli, L., Trambaioli, C., Fortunato, D., Luttkopf, D., Calamari, M., Ansaloni, R., Scibilia, J., Ballmer-Weber, B.K., Poulsen, L.K., Wutrich, B., Hansen, K.S., Robino, A.M., Ortolani, C., Conti, A. (2002). Identification of hazelnut major allergens in sensitive patients with positive double-blind, placebo-controlled food challenge results. *The Journal of Allergy and Clinical Immunology*, *109*, 563–570.
64. Nordlee, J.A., Taylor, S.L., Townsend, J.A., Thomas, L.A., Bush, R.K. (1996). Identification of a Brazil-nut allergen in transgenic soybeans. *The New England Journal of Medicine*, *334*, 688–692.
65. Alcocer, M.J., Murtagh, G.J., Wilson, P.B., Progias, P., Lin, J., Archer, D.B. (2004). The major human structural IgE epitope of the Brazil nut allergen Ber e1: a chimaeric and protein microarray approach. *Journal of Molecular Biology*, *343*, 759–776.
66. Moreno, F.J., Mellon, F.A., Wickham, M.S., Bottrill, A.R., Mills, E.N.C. (2005). Stability of the major allergen Brazil nut 2S albumin (Ber e 1) to physiologically relevant *in vitro* gastrointestinal digestion. *The FEBS Journal*, *272*, 341–352.
67. Pastorello, E.A., Farioli, L., Pravettoni, V., Ispano, M., Conti, A., Ansaloni, R., Rotondo, F., Incorvaia, C., Bengtsson, A., Rivolta, F., Trambaioli, C., Previdi, M., Ortolani, C. (1998). Sensitization to the major allergen of Brazil nut is correlated with the clinical expression of allergy. *The Journal of Allergy and Clinical Immunology*, *102*, 1021–1027.
68. Sathe, S.K., Teuber, S.S., Gradziel, T.M., Roux, K.H. (2001). Electrophoretic and immunological analyses of almond (*Prunus dulcis* L.) genotypes and hybrids. *Journal of Agricultural and Food Chemistry*, *49*, 2043–2052.
69. Roux, K.H., Teuber, S.S., Robotham, J.M., Sathe, S.K. (2001). Detection of the major almond allergen in foods. *Journal of Agricultural and Food Chemistry*, *49*, 2131–2136.

70. Poltronieri, P., Cappello, M.S., Dohmae, N., Conti, A., Fortunato, D., Pastorello, E.A., Ortolani, C., Zacheo, G. (2002). Identification and characterisation of the IgE-binding proteins 2S albumin and conglutin gamma in almond (*Prunus dulcis*) seeds. *International Archives of Allergy and Immunology*, *128*, 97–104.

71. Ewan, P.W. (1996). Clinical study of peanut and nut allergy in 62 consecutive patients: new features and associations. *British Medical Journal*, *312*, 1074–1078.

72. de Leon, M.P., Glaspole, I.N., Drew, A.C., Rolland, J.M., O'Hehir, R.E., Suphioglu, C. (2003). Immunological analysis of allergenic cross reactivity between peanut and tree nuts. *Clinical and Experimental Allergy*, *33*, 1273–1280.

73. Aalberse, R.C. (2007). Assessment of allergen cross reactivity. *Clinical and Molecular Allergy*, *5*, 1–6.

74. Beyer, K., Grishina, G., Bardina, L., Grishin, A., Sampson, H.A. (2002). Identification of an 11S globulin as a major hazelnut food allergen in hazelnut-induced systemic reactions. *The Journal of Allergy and Clinical Immunology*, *110*, 517–523.

75. Barre, A., Jacquet, G., Sordet, C., Culerrier, R., Rouge, P. (2007). Homology modelling and conformational analysis of IgE binding epitopes of Ara h 3 and other legumin allergens with cupin fold from tree nuts. *Molecular Immunology*, *44*, 3243–3255.

76. Wallowitz, M.L., Teuber, S., Beyer, K., Sampson, H.A., Roux, K.H., Sathe, S.K., Wang, F., Robotham, J. (2004). Cross reactivity of walnut, cashew and hazelnut legumin proteins in tree nut allergenic patients. *The Journal of Allergy and Clinical Immunology*, *113*, S156.

77. Barre, A., Sordet, C., Culerrier, R., Rance, F., Didier, A., Rouge, P. (2008). Vicilin allergens of peanut and tree nuts (walnut, hazelnut and cashew nut) share structurally related IgE binding epitopes. *Molecular Immunology*, *45*, 1231–1240.

78. Rance, F., Bidat, E., Bourrier, T., Sabouraud, D. (2003). Cashew allergy: observation of 42 children without associated peanut allergy. *Allergy*, *58*, 1311–1314.

79. Willison, L.N., Tawde, P., Robotham, J.M., Penney, R.M., Teuber, S.S., Sathe, S.K., Roux, K.H. (2008). Pistachio vicilin, Pis v 3, is immunoglobulin E reactive and cross reacts with the homologous cashew allergen, Ana o 1. *Clinical and Experimental Allergy*, *38*, 1229–1238.

80. de Leon, M.P., Drew, A.C., Glaspole, I.N., Suphioglu, C., Rolland, J.M., O'Hehir, R.E. (2005). Functional analysis of cross reactive immunoglobulin E antibodies: peanut specific immunoglobulin E sensitizes basophils to tree nut allergens. *Clinical and Experimental Allergy*, *35*, 1056–1064.

81. Maleki, S.J, Hurlburt, B.K. (2004). Structural and functional alterations in major peanut allergens caused by thermal processing. *Journal of AOAC International*, *87*, 1475–1479.

82. Soler-Rivas, C., Wichers, H.J. (2001). Impact of biochemical and physical procedures on food allergen stability. *Allergy*, *56*, 52–55.

83. Wal, J.M. (2003). Thermal processing and allergenicity of foods. *Allergy*, *58*, 727–729.

84. Chung, S.Y., Champagne, E.T. (1999). Allergenicity of Maillard reaction products from peanut proteins. *Journal of Agricultural Food Chemistry*, *47*, 5227–5231.

85. Chung, S.Y., Champagne, E.T. (2001). Association of end-product adducts with increased IgE binding of roasted peanuts. *Journal of Agricultural Food Chemistry*, *49*, 3911–3916.
86. Mondoulet, L., Paty, E., Drumare, M.F., Ah-Leung, S., Scheinmann, P., Willemot, R.M., Wal, J.M., Bernard, H. (2005). Influence of thermal processing on the allergenicity of peanut proteins. *Journal of Agricultural and Food Chemistry*, *53*, 4547–4553.
87. Mouecoucou, J., Fremont, S., Sanchez, C., Villaumes, C, Mejean, L. (2004). *In vitro* allergenicity of peanut after hydrolysis in the presence of polysaccharides. *Clinical and Experimental Allergy*, *34*, 1427–1437.
88. Chung, S.Y., Maleki, S.J., Champagne, E.T. (2004). Allergenic properties of roasted peanut allergens may be reduced by peroxidase. *Journal of Agricultural Food Chemistry*, *52*, 4541–4545.
89. Chung, S.Y., Champagne, E.T. (2007). Effects of phytic acid on peanut allergens and allergenic properties of extracts. *Journal of Agricultural Food Chemistry*, *55*, 9054–9058.
90. Chung, S.Y., Yang, W., Krishnamurthy, K. (2008). Effects of pulsed UV light on peanut allergens in extracts and liquid peanut butter. *Journal of Food Science*, *73*, C400–C405.
91. Poms, R.E., Anklam, E. (2004). Effects of chemical, physical and technological processes on the nature of food allergens. *Journal of AOAC International*, *87*, 1466–1474.
92. Hourihane, J.O., Bedwani, S.J., Dean, T.P., Warner, J.O. (1997). Randomised, double blind, crossover challenge study of allergenicity of peanut oils in subjects allergic to peanuts. *British Medical Journal*, *314*, 1084–1088.
93. Taylor, S.L., Busse, W.W., Sachs, M.I., Parker, J.L., Yunginger, J.W. (1981). Peanut oil is not allergenic to peanut-sensitive individuals. *The Journal of Allergy and Clinical Immunology*, *68*, 372–375.
94. Hoffman, D.R., Collins-Williams, C. (1994). Cold-pressed peanut oils may contain peanut allergen. *The Journal of Allergy and Clinical Immunology*, *93*, 801–802.
95. Martin-Hernandez, C., Benet, S., Obert, L. (2008). Determination of proteins in refined and nonrefined oils. *Journal of Agricultural and Food Chemistry*, *56*, 4348–4351.
96. Olszewski, A., Pons, L., Moutete, F., Aimone-Gastin, I., Kanny, G., Moneret-Vautrin, D.A., Gueant, J.L. (1998). Isolation and characterization of proteic allergens in refined peanut oil. *Clinical and Experimental Allergy*, *28*, 850–859.
97. Pons, L., Chery, C., Romano, A., Namour, F., Artesani, M.C., Gueant, J.L. (2002). The 18 kDa peanut oleosin is a candidate allergen for IgE-mediated reactions to peanuts. *Allergy*, *57*, 88–93.
98. Teuber, S.S., Brown, R.L, Happanen, L.A.D. (1997). Allergenicity of gourmet nut oils processed by different methods. *The Journal of Allergy and Clinical Immunology*, *99*, 502–507.
99. United States Department of Agriculture (USDA). (2008). Agricultural Research Service Nutrient Database for Standard Reference, Release 21. Available at http://www.nal.usda.gov/fnic/foodcomp/search/

100. FAO. (1991). Protein quality evaluation. Report of the Joint FAO/WHO Expert Consultation. *FAO Food and Nutrition Paper No. 51*. Rome: Food and Agricultural Organization of the United Nations.
101. Ahrens, S., Venkatachalam, M., Mistry, A., Lapsley, K., Sathe, S.K. (2005). Almond (*Prunus dulcis* L.) protein quality. *Plant Foods for Human Nutrition*, *60*, 123–128.
102. Venkatachalam, M., Sathe, S.K. (2006). Chemical composition of selected nut seeds. *Journal of Agricultural and Food Chemistry*, *54*, 4705–4714.
103. van Dokkum, W., Frølich, W., Saltmarsh, M., Gee, J. (2008). The health effects of bioactive plant components in food: results and opinions of the EU COST 926 action. *Nutrition Bulletin*, *33*, 133–139.
104. Dilis, V., Trichopoulou, A. (2009). Nutritional and health properties of pulses. *Mediterranean Journal of Nutrition and Metabolism*, *1*, 149–157.
105. McManus, K., Antinoro, L., Sacks, F. (2001). A randomized controlled trial of a moderate-fat, low-energy diet compared with a low fat, low-energy diet for weight loss in overweight adults. *International Journal of Obesity*, *25*, 1503–1511.
106. Johnston, C.S., Buller, A.J. (2005). Vinegar and peanut products as complementary foods to reduce postprandial glycemia. *Journal of Dietetic Association*, *105*, 1934–1942.
107. Hu, F.B., Stampfer, M.J., Manson, J.E., Rimm, E.B., Colditz, G.A., Rosner, B.A., Speizer, F.E., Hennekens, C.H., Willett, W.C. (1998). Frequent nut consumption and risk of coronary heart disease in women: prospective cohort study. *British Medical Journal*, *317*, 1341–1345.
108. Hu, F.B., Stampfer, M.J. (1999). Nut consumption and risk of coronary heat disease: a review of epidemiologic evidence. *Current Atherosclerosis Reports*, *1*, 204–209.
109. Fraser, G.E., Shavlik, D.J. (1997). Risk factors for all-cause coronary heart disease mortality in the oldest-old—the Adventist Health Study. *Archives of International Medicine*, *157*, 2249–2258.
110. Bruce, B., Spiller, G.A., Farquhar, J.W. (1997). Effects of a plant-based diet rich in whole grains, sun-dried raisins and nuts on serum lipoproteins. *Vegetarian Nutrition: An International Journal*, *1*, 58–63.
111. Spiller, G.A., Jenkins, D.A.J., Bosello, O., Gates, J.E., Cragen, L.N., Bruce, B. (1998). Nuts and plasma lipids: an almond-based diet lowers LDL-C while preserving HDL-C. *Journal of the American College of Nutrition*, *17*, 285–290.
112. Kris-Etherton, P.M., Yu-Poth, S., Sabate, J., Ratcliffe, H.E., Zhao, G., Etherton, T.D. (1999). Nuts and their bioactive constituents: effects on serum lipids and other factors that affect disease risk. *The American Journal of Clinical Nutrition*, *70*, 504S–511S.
113. Bruce, B., Spiller, G.A., Klevay, L.M., Gallagher, S.K. (2000). A diet high in whole and unrefined foods favorably alters lipids, antioxidant defenses, and colon function. *Journal of the American College of Nutrition*, *19*, 61–67.
114. Phillips, K.M., Ruggio, D.M., Ashraf-Khorassani, M. (2005). Phytosterol composition of nuts and seeds commonly consumed in the United States. *Journal of Agricultural and Food Chemistry*, *53*, 9436–9445.
115. Dreher, M.L., Maher, C.V., Kearney, P. (1996). The traditional and emerging role of nuts in healthful diets. *Nutrition Reviews*, *54*, 241–245.

116. Jiang, R., Jacobs, D.R. Jr., Mayer-Davis, E., Szklo, M., Herrington, D., Jenny, N.S., Kronmal, R., Barr, R.G. (2006). Nut and seed consumption and inflammatory markers in the multi-ethnic study of atherosclerosis. *American Journal of Epidemiology*, *163*, 222–231.
117. United States Food and Drug Administration (FDA), Office of Nutritional Products, Labelling and Dietary Supplements. (2003). Qualified health claims: letter of enforcement discretion—nuts and coronary heart disease. Docket No. 02P-0505, July 14.
118. Halling, P.J. (1981). Protein stabilized foams and emulsions. *Critical Reviews in Food Science and Nutrition*, *15*, 155–203.
119. Juliana, K.R., Zhengxing, C. (2008). Effects of processing methods on the physico-functional properties of peanut flour (*Arachis hypogaea* L.). *Biotechnology*, *7*, 168–174.
120. Yu, J., Myrick, P., Ahmedna, M., Rehrah, D., Goktepe, I. (2005). Functional properties of peanut protein concentrates from defatted peanut flour. IFT Annual Meeting, New Orleans, LA, July 15–20, 71B-18.
121. Neto, V.Q., Narain, N., Silva, J.B., Bora, P.S. (2001). Functional properties of raw and heat processed cashew nut (*Anacardium occidentale* L.) kernel protein isolates. *Nahrung/Food*, *45*, 258–262.
122. Sathe, S.K. (1993). Solubilization, electrophoretic characterization, and *in vitro* digestibility of almond (*Prunus amygdalus*) proteins. *Journal of Food Biochemistry*, *16*, 249–264.
123. Sze-Tao, K.W.C., Sathe, S.K. (2000). Walnuts (*Juglans regia* L): proximate composition, protein solubility, protein amino acid composition and protein *in vitro* digestibility. *Journal of the Science of Food and Agriculture*, *80*, 1393–1401.
124. Ramos, C.M.P., Bora, P.S. (2003). Extraction and functional characteristics of Brazil nut (*Bertholletia excelsa* HBI) globulin. *Food Science and Technology International*, *9*, 265–269.
125. Ihekoronye, A.I. (1986). Functional properties of meal products of the Nigerian "Red Skin" groundnut (*Arachis hypogaea* L.). *Journal of the Science of Food and Agriculture*, *37*, 1035–1041.
126. Bora, P.S., Ribeiro, D. (2004). Influence of pH on the extraction yield and functional properties of macadamia (*Macadamia integrofolia*). *Food Science and Technology International*, *10*, 263–267.
127. Sze-Tao, K.W.C., Sathe, S.K. (2000). Functional properties and *in vitro* digestibility of almond (*Prunus dulcis* L.) protein isolate. *Food Chemistry*, *69*, 153–160.
128. Ogunwolu, S.O., Henshaw, F.O., Mock, H.P., Santros, A., Awonorin, S. (2009). Functional properties of protein concentrates and isolates produced from cashew (*Anacardium occidentale* L.) nut. *Food Chemistry*, *115*, 8522–8528.
129. Sathe, S.K., Sze, K.W.C. (1997). Thermal aggregation of almond protein isolate. *Food Chemistry*, *59*, 95–99.
130. Kewalramani, A., Maleki, S.J., Cheng, H., Teuber, S. (2006). Cross reactivity among almond, peanut and other tree nuts in almond allergic patients. *The Journal of Allergy and Clinical Immunology*, *117*, S32.
131. Sen, M., Bhattacharyya, D.K. (2000). Nutritional quality of sunflower seed protein fraction extracted with isopropanol. *Plant Foods for Human Nutrition*, *55*, 265–278.

132. Sen, M., Bhattacharyya, D.K. (2001). Nutritional quality of sesame seed protein fraction extracted with isopropanol. *Journal of Agricultural and Food Chemistry*, *49*, 2641–2646.

133. Kanu, P.J., Kerui, Z., Ming, Z.H., Haifeng, Q., Kanu, J.B., Kexue, Z. (2007). Sesame protein 11: functional properties of sesame (*Sesamum indicum* L.) protein isolate as influenced by pH, temperature, time and ratio of flour to water during its production. *Asian Journal of Biochemistry*, *2*, 289–301.

134. Paredes-Lopez, O., Ordorica-Falomir, C., Olivares-Vazquez, M.R. (1991). Chickpea protein isolates: physicochemical, functional and nutritional characterization. *Journal of Food Science*, *56*, 726–729.

135. Periago, M.J., Vidal, M.L., Ros, G., Rincón, F., Martínez, C., López, G., Rodrigo, J., Martínez, I. (1998). Influence of enzymatic treatment on the nutritional and functional properties of pea flour. *Food Chemistry*, *63*, 71–78.

136. Ghodsvali, A., Khodaparast, M.H.H., Vosoughi, M., Diosady, L.L. (2005). Preparation of canola protein materials using membrane technology and evaluation of meals functional properties. *Food Research International*, *38*, 223–231.

137. Canola Council. (2009). Food oil smoking point. Available at http://www.canolacouncil.org/food_oil_smoke.aspx

138. Schappi, G.F., Konrad, V., Imhof, D., Etter, R., Wuthrich, B. (2001). Hidden peanut allergens detected in various foods: findings and legal measures. *Allergy*, *56*, 1216–1220.

139. Western Australian Food Monitoring Programme (WAFMP). (2004–2005). Report on undeclared tree nut allergens in food. Government of Western Australia. Available at www.public.health.wa.gov.au/cproot/1544/2/Undeclared_Tree_Nut_Allergens_in_Foods.pdf

140. Furlong, T.J., DeSimone, J., Sicherer, S.H. (2001). Peanut and tree nut allergic reactions in restaurants and other food establishments. *The Journal of Allergy and Clinical Immunology*, *108*, 867–870.

141. Hefle, S.L., Bush, R.K., Yunginger, J.W., Chu, F.S. (1994). A sandwich enzyme-linked immunosorbent assay (ELISA) for the quantification of selected peanut proteins in foods. *Journal of Food Protection*, *57*, 419–423.

142. McKenna, C., Klontz, K.C. (1997). Systemic allergic reaction following ingestion of undeclared peanut flour in a peanut-sensitive woman. *Annals of Allergy, Asthma & Immunology*, *79*, 234–236.

143. de Montis, G., Gendrel, D., Chemillier-Truong, M., Dupont, C. (1993). Sensitization to peanut and vitamin D oily preparations (letter). *The Lancet*, *341*, 1411.

144. Weeks R. (1996). Peanut oil in medications. *The Lancet*, *348*, 759–760.

145. Koppelman, S.J., Bleeker-Marcelis, H., Duijn, G., Hessing, M. (1996). Detecting peanut allergens: the development of an immunochemical assay for peanut proteins. *World of Ingredients*, *12*, 35–38.

146. Poms, R.E., Anklam, E., Kuhn, M. (2004). Polymerase chain reaction techniques for food allergen detection. *Journal of AOAC International*, *87*, 1391–1397.

147. Nogueira, M.C.L., McDonald, R., Westphal, C. (2004). Can commercial peanut assay kits detect peanut allergens? *Journal of AOAC International*, *87*, 1480–1484.

148. Poms, R.E., Capelletti, C., Anklam, E. (2004). Effect of roasting history and buffer composition on peanut protein extraction efficiency. *Molecular Nutrition and Food Research*, *48*, 459–464.

149. Cong, Y.J., Lou, F., Xue, W.T., Li, L.F., Wang, J., Zhang, H. (2007). The effect of cooking methods on the allergenicity of peanut. *Food and Agricultural Immunology*, *18*, 53–65.

150. Keck-Gassenmeier, B., Benet, S., Rosa, C., Hischenhuber, C. (1999). Determination of peanut traces in food by a commercially available ELISA test. *Food and Agricultural Immunology*, *11*, 243–250.

151. Pomes, A., Vinton, R., Chapman, M.D. (2004). Peanut allergen (Ara h 1) detection in foods containing chocolate. *Journal of Food Protection*, *67*, 793–798.

152. Trucksess, M.W., Whitaker, T.B., Slate, A.B., Williams, K.M., Brewer, V.A., Whittaker, P., Heeres, J.T. (2004). Variation of analytical results for peanuts in energy bars and milk chocolate. *Journal of AOAC International*, *87*, 943–949.

153. Park, D.L., Coates, S., Brewer, V.A., Garber, E.A., Abouzied, M., Johnson, K., Ritter, B., McKenzie, D. (2005). Performance tested method multiple laboratory validation study of ELISA-based assays for the detection of peanuts in food. *Journal of AOAC International*, *88*, 156–160.

154. Stephan, O., Weisz, N., Vieths, S., Weiser, T., Rabe, B., Vatterott, W. (2004). Protein quantification, sandwich ELISA and real-time PCR used to monitor industrial cleaning procedures for contamination with peanut and celery allergens. *Journal of AOAC International*, *87*, 1448–1457.

155. Taylor, S.L., Crevel, R.W.R., Sheffield, D., Kabourek, J., Baumert, J. (2009). Threshold dose for peanut: risk characterization based upon published results from challenges of peanut-allergic individuals. *Food and Chemical Toxicology*, *47*, 1198–1204.

12

PROCESSING GLUTEN-FREE FOODS

ELKE K. ARENDT AND MARIA HELENA B. NUNES

12.1. INTRODUCTION

Over the last few decades, celiac disease (CD) has become one of the most common food intolerances worldwide. CD is also known as celiac sprue or gluten-sensitive enteropathy and is defined as a chronic inflammatory disease of the small intestine, which results from the interaction between gluten proteins and immune, genetic, and environmental factors [1].

CD is triggered by certain components present in gluten proteins, such as gliadin and glutenin of wheat, secalins of rye, and hordeins of barley. These proteins are resistant to enzymatic digestion in the intestinal tract of celiac patients [2]. When these components are ingested, an immune response occurs, promoting an inflammatory reaction in the mucosa of the small intestine. The disease is associated to the genotypes HLA-DQ2 or DQ8 and is only generated when a person has alleles that encode these types of HLA genes [1, 3]. The presence of DQ2 and DQ8 is typically found in the genome of celiac patients and has also been associated to other autoimmune diseases such as type 1 diabetes, thyroid disease, and lupus [4]. Untreated patients with CD are at increased risk of anemia, edema, osteoporosis, infertility, T-cell lymphoma, and other malignancies [2]. A gluten-free diet not only alleviates the symptoms and enhances nutrition and bone density, but it can also reduce the risk of developing other autoimmune conditions. Thus, the diet leads to a complete healing of the intestinal mucosa, although the histological recovery may never be complete in some patients [5].

Allergen Management in the Food Industry, Edited by Joyce I. Boye and Samuel Benrejeb Godefroy

CD is not only recognized as the most common food disease throughout Europe, but also in the Middle East, Asia, Australia, America, and North Africa [1]. CD occurs in adults and in children with rates approaching 1% of the population [5, 6].

12.2. OVERVIEW OF AVAILABLE GLUTEN INGREDIENTS AND TERMINOLOGIES

The cereals known to cause reactions in celiac patients are wheat, rye, and barley. According to various plant classifications (taxonomy) [7], these cereals belong to the grass family (Poaceae) and are all included in the same tribe, namely Triticeae. Cereals classified in this tribe are considered highly probable to be toxic for celiac patients [7]. The storage proteins in Triticeae (gluten) are composed of alcohol-soluble prolamins, which consist of gliadin fractions and alcohol-insoluble glutelins, which consist of glutenin fractions, portions that trigger the disease [2]. The prolamin fractions are the most toxic for celiac patients and are known as gliadin in wheat, secalin in rye, and hordein in barley. Another cereal, closely related to the ones mentioned above, is oat (Aveneae), which contains avenin. The gluten-free status of oats is currently under discussion. Other cereals such as corn, sorghum, millet, and rice also contain prolamins; however, these are considered safe for consumption by celiac patients. The so-called pseudocereals such as buckwheat, amaranth, and quinoa are not part of the grass family but are considered safe for celiac patients and are commonly used for the production of gluten-free foods.

The toxicity of gluten proteins was extensively evaluated by many authors [2, 8, 9], who concluded that both prolamins and glutelins of wheat, rye, barley, and, possibly, oats appear to be involved in activating celiac disease. Furthermore, for wheat gluten, the most toxic sequences occur in the repetitive N-terminal domain of α/β-gliadins. For rye and barley, no tests were yet performed, but these cereals present homologous structures to the ones of wheat [2].

Wheat is the most important cereal in the world used for human consumption; however, it is the cereal with the highest content of storage proteins, contributing 80–85% of the total wheat protein. Due to this considerable high gluten content, wheat is the ideal candidate for a wide range of cereal products.

The gluten proteins in wheat have unique properties, such as good water absorption capacity, cohesiveness, viscosity, and elastic properties. In a dough system, gliadin contributes to the viscous properties, while glutenin (glutelin fraction) contributes to the elastic properties. A proper mixture of both fractions is essential to impart the viscoelastic properties so well known in wheat-based dough [10]. The adequate mixture of these fractions is only found in wheat, making this cereal the most valuable of all the food grains, being an important tool to provide texture or ingredient binding in many food applications.

12.2.1. Definition of Gluten-Free

The only therapy to treat celiac disease is a strict lifelong adherence to a gluten-free diet. The term "gluten-free" does not refer to the total absence of gluten. In the definition of gluten-free, some residual amount of gluten is allowed; however, this amount is strictly regulated.

Worldwide, foods produced for international trade should follow the standard levels regulated by the Codex Alimentarius. The current Codex Alimentarius Standard for gluten-free foods was established in 1976 and was amended in 1983. The definition of gluten-free is, however, under revision since 1992 by the Codex Committee on Nutrition and Foods for Special Dietary Uses. The last draft of the Codex Commission is from July 2008 and defines gluten-free foods as

1. consisting of or made only from one or more ingredients, which do not contain wheat (i.e., all *Triticum* species, such as durum wheat, spelt, and kamut), rye, barley, oats (determined at a national level), or their crossbred varieties, and the gluten level does not exceed 20 mg/kg in total, based on the food as sold or distributed to the consumer (e.g., bread made from gluten-free cereals);
2. consisting of one or more ingredients from wheat (i.e., all *Triticum* species, such as durum wheat, spelt, and kamut), rye, barley, oats (determined at a national level), or their crossbred varieties, which have been specially processed to remove gluten, and the gluten level does not exceed 20 mg/kg in total, based on the food as sold or distributed to the consumer (e.g., distilled products such as vinegar or liquors, wheat starch) [11].

Decisions on the marketing of products described in this section may be determined at the national level.

The establishment of a standard limit for the gluten content has caused disagreements within the international food authorities. Disagreements regarding the decision on the inclusion of oats and on the exact limit of gluten level in gluten-free foods became remarkable.

In the United States, gluten-free foods are regulated by the U.S. Food and Drug Administration (FDA), who defines the term "gluten-free" as a food that does not contain any ingredient from any species of the grains wheat, rye, barley, or a crossbred hybrid of these grains. Oats are not included in the "prohibited grains." In the case of trace amounts of gluten, these should not exceed the 20 mg/kg. In Canada, a "gluten-free" food must not contain wheat, including spelt and kamut, or oats, barley, rye, triticale, or any part thereof [12]. Recently, the Canadian Celiac Association has revised the position statement on oats, establishing specifications on growth, harvesting, and processing of pure and uncontaminated oats. These are presently considered safe within a daily threshold of 50–70 g for adults and 20–25 g for children [13, 14]. In

Europe, only Finland and the United Kingdom considered oat products as safe in a gluten-free diet, although some precautions and recommendations were made, for example, the celiac patient should consult a doctor before starting a diet containing oats [15].

The safe threshold of gluten in a food product is another controversial issue. Over the last few years, the level of gluten traces has been continuously reduced by the Codex Committee and yet studies have proven that 100 mg/kg gluten is not a [13] suitable limit, since 50 mg/kg of gluten intake demonstrated to be harmful for certain individuals [16]. These authors affirmed that the establishment of a gluten threshold depends not only on the minimum toxic dose but also on the amount of gluten-free products consumed by celiac patients. Due to fact that the clinical research did not reach a consensus of the exact threshold, disagreements are also found regarding the labeling of gluten-free foods. The acceptance of a standard method by the Codex Commission, and results of ongoing research on tolerance levels will allow the commission to move toward adopting a new revised definition of "gluten-free."

12.3. IDENTIFYING NUTRITIONAL AND FUNCTIONAL ALTERNATIVES TO WHEAT, BARLEY, AND RYE

A gluten-free diet affects inevitably the quality of life of individuals undergoing this type of treatment. Lee and Newman [17] reported that the restrictive nature of the diet can be a limiting factor affecting decisions on social activities, contributing negatively to their ability of dining out and traveling and, at a less extent, to their family life or work ability.

Cereals and cereal products are one of the major sources for human nutrition worldwide. Therefore, the avoidance of wheat in particular becomes a problem when the patient has no control over the food provided to them. Moreover, the purchase of products labeled as gluten-free is a concern when prices and variety available is comparatively small to the commercialized conventional products. This problem might be even more evident in countries where a gluten-free diet is still not so common.

In order to create new and healthier options leading to a reduction of price of the gluten-free products, many studies have been performed to investigate the potential of grains other than wheat, barley, and rye for the production of food.

12.3.1. Functional Properties

For baking applications, consideration must be given to certain properties, particularly related to starch, such as the amylose–amylopectin ratio or pasting behavior. Starch is the main component in all the cereal grains, and it may determine the quality and functionality of the cereals when these are submitted to thermal processing. Proteins are the second highest component in cereals, and depending on the type of protein fractions present in the grain, it may also determine their functional and nutritional value.

Rice is considered one of the most suitable grains for celiac patients due to attributes such as low protein and sodium content and the presence of approximately 80% of easily digestible carbohydrates. Rice of good eating quality shows low amylose (19.5%) and low protein content (7.3%). Rice starch gelatinizes at around 65°C and presents a high peak viscosity due to its small size granules [18]. The rice proteins are mainly composed of glutelins, but these have relatively poor functional properties with regard to food processing, due to their hydrophobic nature, which makes them extremely insoluble. Due to these characteristic, rice flour on its own is not suitable for bread production, although it can be used in many other applications in bakery products, such as baby foods or extruded cooked products [19].

Corn contains between 75% and 87% of starch and 6–8% of protein [20]. The storage proteins of corn are prolamins called zeins; they represent 60% of the protein content [21] and are characterized by their insolubility. Cornstarch has a gelatinization temperature ranging from 52.9 to 66.5°C [22]. Corn is widely used in products such as tortillas and extruded snacks. Moreover, the isolation of starch from corn is itself a major food ingredient.

In *sorghum*, starch constitutes an average of 70.8% of the grain [23] and has similar properties to cornstarch; however, the peak viscosity and the water-binding capacity of sorghum have been reported to be lower than that of cornstarch [24]. Starch from waxy sorghum (<1% amylose) is known for its rapid cooking ability, paste clarity, and resistance to gel formation and retrogradation [23]. Sorghum proteins constitute an average of about 11% and are mainly composed of the prolamin fractions, called kafirins. The sorghum proteins are characterized by a high resistance to enzymatic digestion and to disruption during processes such as extrusion, and by a decrease in digestibility upon cooking [25]. This may be associated with the formation of more protein cross-links during cooking [26]. Sorghum is traditionally used to produce porridges, steam-cooked products, fermented or unfermented breads, alcoholic or nonalcoholic beverages, and snacks [27].

The more important *millets* include pearl, foxtail, proso, and finger millets. There are other millets of local significance, including teff, kodo, barnyard, fonio, and little millet. Among millets, pearl millet is the most widely grown, and it is known to have the higher protein content and the more balanced amino acid composition. The protein content of pearl millet ranges from 8% to 19% and its major constituents are prolamins and glutelins. Starch accounts for 56–65% of the kernel, from which 20–22% is amylose [28]. The water-holding capacity of pearl millet starch is higher than that of sorghum and lower than that of maize. Pearl millet starch appears to have a higher swelling power and solubility than other starches [24]. Millets as well as sorghum are mainly used for the production of porridges and fermented beverages [27].

Pseudocereals (buckwheat, amaranth, and quinoa) are dicotyledonous plants, while cereals are monocotyledons and are classified as legume-like [7].

In comparison with cereal grains, pseudocereals have a higher nutritional value, which is mainly connected to their proteins [29]. Pseudocereals have their reserve compounds located at different places in the kernels, but their physicochemical characteristics are comparable to cereal grains.

Buckwheat has attracted increasing attention from food scientists for its positive effects on a number of chronic diseases [30]. Buckwheat contains mostly carbohydrates (63%), where starch represents 55.8% of the grain. The proteins contribute to 11–15% to the weight of the grain. However, in cereals, usually 10–20% of the protein is associated with the embryo, while 80–90% is found in the endosperm. In buckwheat, 55% of the protein is located in the embryo, 35% in the endosperm, while the remainder is found in the hull [31]. Buckwheat is rich in micronutrients, and the amino acid structure and composition are distinctly different from other cereals, as its limiting amino acids are Thr and Met. A variety of buckwheat-based foods have been reported in the literature [32], for example, wine, sauces, cakes, and noodles. Furthermore, its use in the production of gluten-free malt beers [33–37] and bread [38] is also reported.

Amaranth is another pseudocereal that has shown great economic potential and also has a high nutritional value. The storage proteins of the *Amaranthus* species are located in the embryo and endosperm cells and range from approximately 13–18%, where albumins and globulins are the major seed proteins [39, 40]. Starch is stored in the perisperm and is the most abundant carbohydrate component in the amaranth grain. The extremely small starch granules of amaranth provide unique functional properties in food products, acting, for example, as thickener or as fat replacement [41]. Due to the low amylose levels typically found in amaranth starch, this pseudocereal performs poorly in bread and cake formulations [42]. However, species of the amaranth family with higher amylose content may improve its performance as an ingredient in bread.

Quinoa (*Chenopodium quinoa*) is nutritionally interesting because it contains high levels of lysine-rich proteins, polyunsaturated fatty acids, micronutrients, and vitamins C and E [43, 44]. Quinoa contains a slightly higher protein content (13–14%) and a much greater fat content than (5–9.7%) most cereal grains [45, 46]. The starch content ranges from 52% to 69%, which is lower than what has been found in other cereals but comparable to the remaining pseudocereals [44, 46]. Quinoa starch is rich in amylopectin, which leads to low gelatinization temperatures comparable to wheat and barley [47]. Quinoa starch exhibits much higher pasting viscosity than wheat or amaranth starch [48] but has excellent freeze–thaw stability [49], which is due to the high content of amylopectin and low content of amylose. Quinoa has been mainly used in the production of extruded products and as a nutrient complement in wheat breads.

12.3.2. Nutritional Value

It has been shown in a number of studies that the adherence to a gluten-free diet can lead to deficiency in certain vitamins and minerals as well as in fiber.

Thompson [50], for example, reported that gluten-free cereal products generally provide lower amounts of folate and iron and more dietary fiber than their enriched/fortified, gluten-containing counterparts. Also, a malabsorption of vitamin D and calcium is a common manifestation of untreated celiac disease that can lead to a number of degenerative diseases. Patients should be encouraged to consume whole grain and enriched/fortified, gluten-free flours, breads, pastas, and cereals. Considering the deficiencies of a gluten-free diet, some studies have investigated the incorporation of dairy ingredients [51] or dietary fibers in bakery products. Patients should also be encouraged to increase their consumption of noncereal sources of folate, iron, and fiber, which are found in vegetables, fruits, and animal sources [50, 52]. From a nutritional point view, the pseudocereals as well as oats are good sources of fiber and iron and should therefore be incorporated where possible. The protein from pseudocereals can also be used as a supplement when other crops are used in gluten-free food products. Shewry and Halford [53] affirmed that the nutritional deficiencies in the essential amino acids might be overcome with the combination of cereals and legume seeds (pseudocereals). These two types of seeds are essentially complementary in their compositions of essential amino acids: cereals tend to be rich in sulfur-containing amino acids and low in lysine, whereas legume seeds are rich in lysine and low in sulfur-containing amino acids.

12.3.3. Oats

Oat is a very versatile cereal with a mild nutty flavor, and it has been promoted as a cereal with high nutritional value. In a recent review [54], it has been suggested that the daily intake of adequate amount of oats dietary fiber (β-glucan) reduces the serum cholesterol levels as well as the postprandial glycemic and insulinemic response.

Oats have been widely used in products such as snacks, breads, and porridges, as well as additional ingredient to certain products like ice cream or yogurt [55]. Compared with other cereals, oats are characterized by lower carbohydrate content, with higher protein and fat content [56].

In the past, the oat consumption was believed to be harmful for celiac patients. This was mainly due to the fact that there is a high risk of contamination with gluten-containing cereals during the processing of oats and also due to the little information available on the toxicity of pure oats. Recently, a significant amount of studies have been performed to evaluate the inclusion of oats in the gluten-free diet [57–60].

The toxicity of oats became increasingly controversial not only due to medical studies but also to the improved knowledge acquired on the oat proteins. In oats, the prolamin fraction of the storage proteins is avenin, and it belongs to the tribe Aveneae, whose origin differs from the Triticeae cereals (i.e., wheat). Moreover, the major storage protein in oats is globulin, varying from 40–50% to 70–80%, while prolamin fraction accounts for only about 15% of the total protein. Regarding the small amount of prolamins in oats, it has been suggested that a far greater quantity of oats would have to be consumed to cause the same adverse effects as wheat, rye, or barley [61]. This characteristic is also related to the oats' inability to form a viscoelastic dough comparable to wheat [62].

The amino acid composition of oats also reflects its intermediate position in cereal taxonomy, presenting characteristics similar to Triticeae and to other tribes considered safe for celiac disease diet. In oats, the high degree of amidation of glutamic and aspartamic acids correspond to that of wheat, rye, and barley. However, a higher content of lysine and relatively low proline residues present in the avenin peptides contribute to relevant differences between oats and Triticeae cereals. This lack of proline residues creates susceptibility of the oat proteins to be degraded by proteases in the gastrointestinal tract [9].

In patients with celiac disease, T-cell reactivity has been found against a large panel of epitopes derived from α/β-gliadin, γ-gliadin, and glutelin proteins [10]. The most toxic amino acid sequence known is found in prolamins of wheat, rye, and barley, but not in oats, where the amount of N-terminal sequences is much smaller. Together with the fact that oats present no antigenic similarity to wheat and related cereals, this could explain the possibility of oats being suitable for consumption by celiac patients [63, 64]. Previous review studies [15, 60] on oat toxicity summarized several medical interventions carried out from 1995 to 2007. Thompson [60] reported that without exception, the studies [65–70] concluded that there were no adverse effects associated with the regular consumption of moderate amounts of oats (≤62.5 g/day [mean]). In the most recent review carried out by Salovaara et al. [15], almost 300 celiac patients following a gluten-free diet for at least the previous 3 months consumed up to 100 g/day of oats. From these studies, only four patients consuming ≤50 g/day of pure oats were found to develop mucosal changes and dermatitis, this being directly related to the oats ingestion [71]. In other cases reported by Arentz-Hansen et al. [72], T cells were found to be specifically reactive to avenin. Nevertheless, it is proven that most celiac patients can tolerate the moderate consumption of oats (50–70 g/day) over a period of 6–12 months [73]. In 2002, one of the largest intervention studies was performed [68], where 92 patients were initially incorporated. The results of this study revealed that the consumption of pure oats (free from cross-contaminations) can be tolerated by celiac disease patients and no complication in the intestinal mucosa could be detected up to a period of 5 years.

Although there were only very few studies suggesting that a small number of adult patients with celiac disease are intolerant to oats, these have not as such been declared gluten-free. For oat-tolerant celiac patients who included oats in their diet, it was also found that these had more abdominal complaints than those not eating oats [59]. In these cases, commercial oat products were consumed, where cross-contamination may not be excluded. Oats are still not recommended worldwide as a grain to be consumed by celiac disease patients. This is mainly due to the fact that most of the commercial oat products available may contain some cross-contamination with gluten from other grains. Production of oats is predominantly performed in the proximity of wheat, leading to increase in the probability of contamination during harvesting, transporting, and milling processes.

Nevertheless, oats are considered a cereal that may improve the nutritional content of the diet and overall quality of life [59]. Oats can be tolerated by most, but not all, celiac disease patients. Therefore, the present Codex Alimentarius claims that the allowance of oats, not contaminated with wheat, rye, or barley in gluten-free foods, may be determined at a national level [11].

12.4. HIDDEN AND UNINTENTIONAL SOURCES OF GLUTEN

One of the main concerns regarding the use of oats in the production of gluten-free foods is the doubt of oats and oat products being heavily "contaminated" with barley or wheat [74], since these grains are grown in the same regions. Gélinas et al. [75] concluded that the safest foods for celiac patients were based on cereals like rice, corn, or quinoa. On the other hand, foods made with oats or buckwheat encountered contaminations from wheat and barley gluten. However, Dahinden et al. [76] reported no contaminations in all the tested products, which were manufactured in bakeries specialized in gluten-free foods.

Other sources of gluten may also be found in foods derived especially from wheat and barley. These products are obtained by processes such as fermentation (e.g., vinegar, soy sauce, enzymes), hydrolyzation (e.g., wheat starch and protein), and distillation (e.g., ethanol from fermented wheat), or by sprouting (e.g., malt and malt extracts from barley). However, some of these represent a threat to celiac patients.

Dietetic gluten-free food produced for CD patients underlie the regulation of the Codex Alimentarius Standard for Gluten-Free Foods, which recently proposed a maximum level of 20 mg of gluten/kg for naturally gluten-free food [11]. To ensure that the gluten-free products on the market are totally safe for consumption by the celiac patients, numerous analytical methods have been developed in order to detect gluten toxic fractions in cereals and its derivates. The ideal method should detect with equal sensitivity and specificity all toxic fractions in raw ingredients as well as in foods that have been submitted to technological processes, which may lead to the loss of the original epitopes. However, the development and validation of a standard detection method revealed several difficulties, such as the protein profile dependency on the variety and environment conditions; cross-reactions with nontoxic cereals [77]; and the reliability of the measurement of gluten extracted from foodstuff [78] or to accurately achieve the minimum detection limit, which should be of 10 mg gluten/kg as established by the Codex Alimentarius in the "Methods of analysis and sampling" section [11].

The methods currently used for the detection of gluten are mostly based on immunochemical assays such as various enzyme-linked immunosorbent assay (ELISA) tests [79], R5 ELISA [80], or competitive ELISA [81]. Other methods have been tested, such as mass spectroscopy [82] or polymerase chain reaction (PCR) [76, 83] and R5 ELISA in combination with real-time PCR [84]. In 2006, the Codex Committee on Methods of analysis and sampling endorsed the (ELISA) R5 Mendez method as a type I method.

12.5. GUIDELINES FOR GLUTEN-FREE PROCESSING

Gluten is often termed the "structural" protein for bread making. The properties of gluten become evident when flour is hydrated, and an extensible dough with good gas-holding capacity and a bread with good crumb structure are obtained. The major problem associated with gluten-free cereals is the fact that these grains lack proteins with the ability to form networks upon hydration. The absence of gluten often results in a liquid batter rather than a dough [51].

The processing of wheat bread differs in many stages to that of gluten-free (Fig. 12.1). The methods used over the last few years for the production of gluten-free products have been modified and optimized in order to establish the adequate ingredient combination, water levels and mixing, and proofing and baking times.

To simulate the typical network occurring in a wheat dough, many studies [85–92] have investigated not only the possibility of using more than one gluten-free flour in the formulation, but also to use this flour in combination with additional chemical or physical methods and treatments. Studies [88–100] reported the effect of adding different starches, hydrocolloids, sourdough, and enzymes to the formulation in order to create gluten-free breads with structure and good appearance.

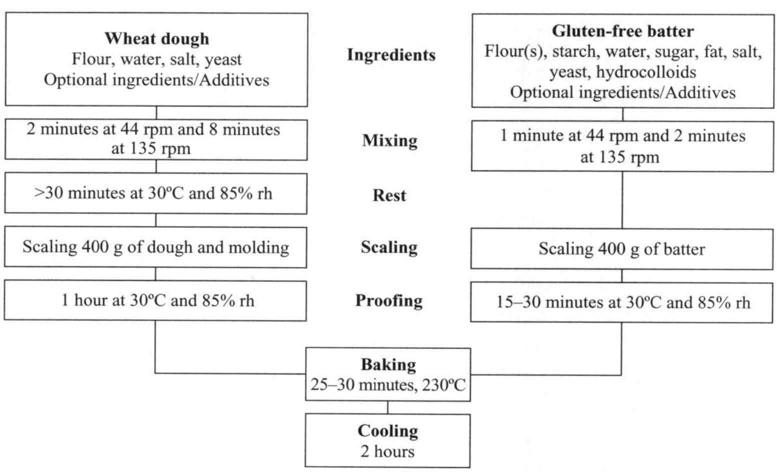

Fig. 12.1 Wheat- and gluten-free bread-making processes.

12.5.1. Starch

Starch has unique properties that determine its functionality in many food applications, particulary bakery products where it contributes to texture, appearance, and overall acceptability of cereal-based foods [101]. When a starch suspension is heated in water above a specific temperature, it undergoes disruption of molecular order, loss of crystallinity, and irreversible granule swelling. The network structure of the gelatinized starch molecules plays a vital role in the formation of the bread crumb [102]. However, the incorporation of starch in bread may also become a problem. The retrogradation of starch granules during the storage period leads to an increase in the staling rate. This is an issue, in particular, while producing gluten-free breads, where certain amounts of pure starches are often combined with gluten-free cereals as a replacement for wheat. Gluten-free breads typically exhibit a crumbling texture, short shelf life, and poor sensorial characteristics. Therefore, other ingredients/additives have been introduced as important tools for the production of gluten-free breads.

12.5.2. Hydrocolloids

Hydrocolloids or gums are used in foods to improve texture, to reduce the starch retrogradation, to increase moisture retention, and to extend the overall quality of the product over time [103]. The incorporation of hydrocolloids in gluten-free breads have been earlier reported and extensively investigated. Many types of hydrocolloids, such as hydroxypropyl methylcellulose (HPMC), locust bean gum, guar gum, carageenan, xanthan gum, pectin, carboxymethyl cellulose (CMC), konjac gum gelatin, agarose, and β-glucan, were evaluated, mainly in rice-based formulations. These studies revealed that all hydrocolloids interact with water, reducing its diffusion and stabilizing its presence [98, 104]. Interactions between hydrocolloids and starch granules may also occur, leading to changes in the starch mobility and enhancing the structure [105] and the viscoelasticity of gluten-free batters and breads [96, 99, 104]. The studies mentioned above came to the conclusion that among all the hydrocolloids reported, HPMC and xantham gum appear to be the best options. Further studies demonstrated that an adequate combination of different hydrocolloids (e.g., HPMC and CMC) act more efficiently as gluten replacement [87, 100].

12.5.3. Dairy Ingredients

The addition of dairy ingredients to gluten-free foods may not only improve the nutritional value, but also the flavor and structure. However, the dairy-based ingredients used in gluten-free products should be low in, or free from, lactose, since a significant amount of celiac patients are intolerant, particularly during the early stages after being diagnosed [106].

Several studies evaluating the quality, rheological properties, and shelf life of gluten-free products have incorporated dairy protein in the formulations. Sodium caseinate [96, 107], skim milk powder [88, 89, 98, 107], sweet and demineralized whey, fresh milk solids, and protein isolate [107] have been evaluated in gluten-free bread formulations. Gallagher et al. [107] studied the effect of a range of dairy powders, which improved the volume, color, crumb and crust appearance, and sensory properties of breads. These authors concluded that, in general, powders with high protein/low lactose content gave better breads, as well as doubled the protein content.

12.5.4. Sourdough

Sourdough is a mixture of flour (e.g., wheat, rye), water, and other ingredients (e.g., NaCl) that is fermented by naturally occurring lactic acid bacteria (LAB) and yeasts. The use of sourdough in bakery products has long been recognized as a tool to improve the texture and palatability of whole grain and fiber-rich products, and it may stabilize or increase the levels of bioactive compounds [108]. Very little literature can be found regarding the incorporation of sourdough in gluten-free breads. Moore et al. [91] reported that the growth of a selected LAB in gluten-free batters was found to be similar to that reported previously for wheat sourdoughs. Sourdough fermentation caused an increase in the dough elasticity as well as a decrease in staling. Another study [90] concluded that the addition of sourdough in gluten-free batters could effectively retard the mold growth, improving significantly the shelf life of the gluten-free breads. Schober et al. [109] demonstrated that the problems associated with sorghum bread production (i.e., flat top and tendency toward a hole in the crumb) could be overcome by the sourdough fermentation of the total sorghum flour.

Nevertheless, research has been carried out to access the possibility of producing sourdough bread that is tolerated by celiac patients. The results of these studies revealed that the use of selected lactobacilli, nontoxic flours, and a long fermentation time is a novel tool for decreasing the level of gluten intolerance in humans [110]. Furthermore, in a study carried out by De Angelis et al. [111], it was also demonstrated that a pool of selected sourdough LAB showed the capacity to degrade prolamins contained in rye flour, and therefore, this methodology could be used as a tool to decrease the risk of rye contamination of gluten-free products for celiac patients [111].

12.5.5. Enzymes

Enzymes used in bakery products have demonstrated the potential to modify protein, starch, and fiber fractions and therefore improve product quality. Several studies have investigated the possibility to use enzymes as a way to improve the quality of gluten-free breads. The main enzymes studied were transglutaminases (TGases) and oxidases. The protein source introduced as substrate is a key element determining the impact of the enzyme on the final product quality. The addition of different protein sources to microbial TGase has been investigated. Marco and Rosell [97] reported that the use of pea, soybean, and whey proteins treated by TGase significantly decreased the peak and final viscosity of the dough. The TGase cross-linked rice proteins resulted in a dough with improved elastic and viscous behavior, as well as an increased emulsifying activity. The resultant breads presented higher volume and crumb strength. Moore et al. [89] evaluated the impact of TGase in gluten-free bread in conjunction with different protein sources such as soya, skim milk, or egg powder. The most pronounced effect reported was the reduction of bread volume due to network formation and the improvement of bread structure. Bread containing skim milk powder and 10 units of TGase per gram of protein showed the most compact structure, and the authors concluded that network formation in gluten-free bread is dependent on the TGase level and type of protein used.

Renzetti et al. 2008 [92] investigated the impact of TGase on a range of gluten-free cereals (brown rice, corn, oats, sorghum, teff, and buckwheat). The optimal substrates for TGase application were buckwheat and brown rice. A significant increase in the pseudoplastic behavior of buckwheat and brown rice batters was observed when 10 units of TGase per gram of protein were used. Results that reflected significant improvements in the textural and structural characteristics of the breads were obtained (Fig. 12.2).

Three-dimensional confocal laser scanning microscopy (CLSM) image elaborations confirmed the formation of protein complexes by TGase action. Results obtained with corn flour showed a negative impact of TGase on the

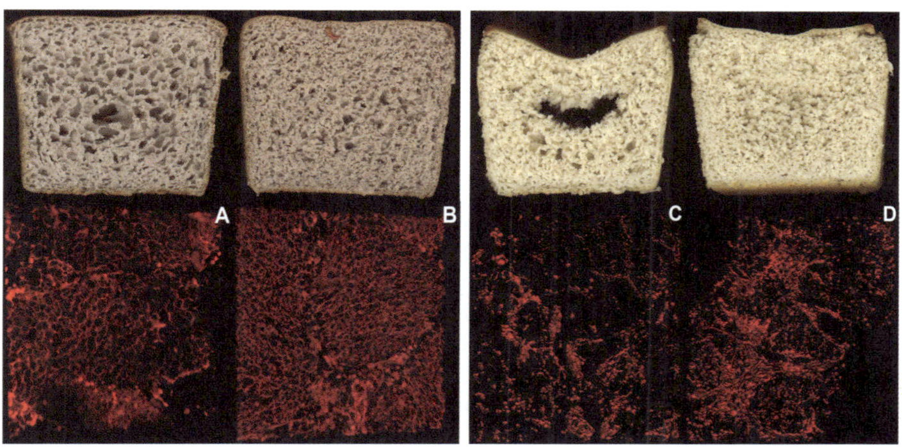

Fig. 12.2 Bread slices and 3D elaboration of CLSM images of buckwheat (BW) and brown rice (BR) bread crumbs (40× magnification): (A) BW control bread (0 U of enzyme); (B) BW 10 U of TGase bread; (C) BR control bread (0 U of enzyme); (D) BR 10 U of TGase bread.

pseudoelastic behavior of the batters, but the resulting breads were significantly improved in terms of increased specific volume and decreased crumb hardness and chewiness. Regarding the oats, sorghum, and teff flours, it was only observed a slight effect by the addition of the enzyme.

Overall, it was concluded that TGase may be successfully applied to gluten-free flours to improve their bread-making potential by promoting network formation; nevertheless, the protein source is a key element determining the impact of the enzyme.

Recently, the same authors investigated the activity and specificity of TGase on the protein fractions of buckwheat flour [112]. This study showed the presence of high-molecular-weight (HMW) protein polymers in the TGase-treated albumin and globulin fraction of buckwheat, indicating that these were cross-linked after TGase treatment. The increase in the average molecular weight confirmed the network formation previously observed, as well as the improved functionality of buckwheat flour in terms of bread-making potential.

Furthermore, several studies have reported that the use of enzymes in combination with hydrocolloids can have a synergistic effect. Gujral and Rosell [94, 95] demonstrated that the use of glucose oxidase and TGase modified the proteins of rice flour. The addition of these enzymes to optimal levels of HPMC promoted an increase in the elastic and viscous modulus enhancing the dough strength and subsequent bread volume. Other studies reported the use of cyclodextrin glycoxyltransferase (CGTase) [113, 114] and α-amylase [115] in combination with different hydrocolloids. These studies revealed that the enzymes decreased crumb firmness and delayed the staling of rice bread.

12.6. CONCLUSION

Celiac disease is a food intolerance characterized by the inflammation of the small intestinal mucosa. It is triggered by specific components present in gluten proteins and thus the avoidance of these types of proteins is the only therapy known. The establishment of gluten consumption thresholds and methods for its detection are essential regarding the safety of celiac disease patients. Gluten has unique properties in the production of high-quality cereal products. Despite the wide range of gluten-free grains, the replacement of wheat by gluten-free cereals proves to be a major challenge for food scientists. The production of high-quality gluten-free bread has been under more rigorous studies than any other product regarding celiac disease. Over the last years, research has focused on the combination of gluten-free flours with other ingredients that may improve the structure as well as the sensory characteristics and the nutritional value of the gluten-free products. The use of several ingredients such as enzymes, sourdough, low-lactose dairy ingredients, and hydrocolloids have been investigated, and their applications have shown they can play an important role in the development of good quality gluten-free products and particularly in gluten-free bread.

REFERENCES

1. Green, P.H.R., Cellier, C. (2007). Celiac disease. *The New England Journal of Medicine*, *357*, 1731–1743.
2. Weiser, H., Koehler, P. (2008). The biochemical basis of celiac disease. *Cereal Chemistry*, *85*, 1–13.
3. Lundin, K.E. (1993). Gliadin-specific, HLA-DQ restricted T cells isolated from the small intestine mucosa of CD patients. *The Journal of Experimental Medicine*, *178*, 187–196.
4. Ventura, A., Neri, E., Ughi, C., Leopaldi, A., Citta, A., Not, T. (2000). Gluten-dependent diabetes-related and thyroid-related autoantibodies in patients with coeliac disease. *The Journal of Pediatrics*, *137*, 253–255.
5. Mulder, C.J.J., Cellier, C. (2005). Coeliac disease: changing views. *Best Practice and Research Clinical Gastroenterology*, *19*, 313–321.
6. Malekzadeh, R., Sachdev, A., Ali, F. (2005). Coeliac disease in developing countries: Middle East, India and North Africa. *Best Practice and Research Clinical Gastroenterology*, *19*, 351–368.
7. Kasarda, D.D. (2001). Grains in relation to celiac disease. *Cereal Foods World*, *46*, 209–210.
8. Shewry, P.R., Tatham, A.S. (1990). The prolamin storage of cereal seeds: structure and evolution. *The Biochemical Journal*, *297*, 1–12.
9. Vader, L.W., Stepniak, D.T., Bunnik, E.M., Kooy, Y.M., de Haan, W., Drijfhout, J.W., Van Veelen, P.A., Koning, F. (2003). Characterization of cereal toxicity for coeliac disease patients based on protein homology in grains. *Gastroenterology*, *125*, 1105–1113.
10. Wieser, H. (2006). Chemistry of gluten proteins. *Food Microbiology*, *24*, 115–119.
11. Codex Alimentarius Commission. (2007). Draft revised standard for gluten-free foods at September 7, ALINORM 08/31/26. In: *Report of the 29th Session of the Codex Committee on Nutrition and Foods for Special Dietary Uses*, Joint FAO/WHO Food Standards Programme, November 12–16, Bad Neuenahr-Ahrweiler, Germany, pp. 42–64.
12. Canadian Celiac Association. (2008). Defining gluten-free in Canada, B.24.018 CsFaDR. Available at http://www.calgaryceliac.com/awareness/Definition%20of%20Gluten-Free.pdf
13. Health Canada. (2007). Celiac disease and the safety of oats: Health Canada's position on the introduction of oats to the diet of individuals diagnosed with celiac disease (CD). Available at http://www.hc-sc.gc.ca/fn-an/securit/allerg/cel-coe/oats_cd-avoine_e.html
14. Canada Celiac Association. (2008). Position statement on oats. Available at http://www.celiac.ca/Articles/PABoats.html
15. Salovaara, H., Kanerva, P., Kaukinen, K., Sontag-Strohm, T. (2009). Oats—an overview from a coeliac disease point of view. In: *The Science of Gluten-Free Foods and Beverages*, Arendt E.F. (Ed.), Proceedings of the First International Conference on Gluten-Free Cereal Products and Beverages. St. Paul, MN: American Association of Cereal Chemists International, Inc. pp. 69–87.

16. Catassi, C., Fabiani, E., Iacono, G., D'Agate, C., Francavilla, R., Biagi, F., Volta, U., Accomando, S., Picarelli, A., De Vitis, I., Pianelli, G., Gesuita, R., Carle, F., Mandolesi, A., Bearzi, I., Fasano, A. (2007). A prospective, double-blind, placebo-controlled trial to establish a safe gluten threshold for patients with coeliac disease. *The American Journal of Clinical Nutrition*, *85*, 160–166.

17. Lee, A.M., Newman, J. (2003). Coeliac disease: its impact on quality of life. *Journal of the American Dietetic Association*, *103*, 1533–1535.

18. Zhou, Z., Robards, K., Helliwell, S., Blanchard, C. (2002). Composition and functional properties of rice. *International Journal of Food Science and Technology*, *37*, 849–868.

19. Rosell, C., Gómez, M. (2006). Rice. In: *Bakery Products: Science and Technology*, Hui, Y.H., Corke, H., De Leyn, I., Nip, W.-K., Cross, N. (Eds.). Oxford: Blackwell Publishing, pp. 123–133.

20. Shukla, R., Cheryan, M. (2001). Zein: the industrial protein from corn. *Industrial and Crops Products*, *13*, 171–192.

21. Lending, C.R., Larkins, B.A. (1989). Changes in the zein composition of protein bodies during maize endosperm development. *The Plant Cell*, *1*, 1011–1023.

22. Seetharaman, K., Tziotis, A., Borras, F., White, P.J., Ferrer, M., Robutti, J. (2001). Thermal and functional characterization of starch from Argentinean corn. *Cereal Chemistry*, *78*, 379–386.

23. Serna-Saldivar, S., Rooney, L.W. (1995). Structure and chemistry of sorghum and millets. In: *Sorghum and Millets—Chemistry and Technology*, American Association of Cereal Chemists Monograph Series, Dendy, D.A.V. (Ed.). St. Paul, MN: American Association of Cereal Chemists, pp. 69–124.

24. Abd Allah, M.A., Mahmoud, R.M., El-Kalyoubi, M.H., Abou Arab, A.A. (1987). *Physical* properties of starches isolated from pearl millet, yellow corn, sorghum, sordan and pearl millet. *Starch/Stärke*, *39*, 9–12.

25. Hamaker, B.R., Kirleis, A.W., Mertz, E.T., Axtell, J.D. (1986). Effect of cooking on the protein profiles and in vitro digestibility of sorghum and maize. *Journal of Agricultural and Food Chemistry*, *34*, 647–649.

26. Duodu, K.G., Taylor, J.R.N., Belton, P.S., Hamaker, B.R. (2003). Factors affecting sorghum protein digestibility. *Journal of Cereal Science*, *38*, 117–131.

27. Murty, D.S., Kumar, K.A. (1995). Traditional uses of sorghum and millets. In: *Sorghum and Millets: Chemistry and Technology*, American Association of Cereal Chemists Monograph Series, Dendy, D.A.V. (Ed.). St. Paul, MN: American Association of Cereal Chemists, pp. 185–221.

28. Sullivan, T.W., Douglass, J.H., Andrews, D.J., Bond, P.L., Hemcock, J.D., Bramel-Cox, P.J., Stegmeir, W.D., Brethour, J.R. (1990). Nutritional value of pearl millet for food and feed. In: *Proceedings of an International Conference on Sorghum Nutritional Quality*, Ejeta, G., Mertz, E.T., Rooney, L., Schaffert, R., Yphe, J. (Eds.), February 26–March 1, Purdue University, West Lafayette, IN, pp. 83–94.

29. Shoenlechner, R., Siebenhandl, S., Berghofer, E. (2008). Pseudocereals. In: *Gluten-Free Cereal Products and Beverages*, Arendt, E.K., Dal Bello, F. (Eds.). Amsterdam: Academic, pp. 149–190.

30. Li, S.Q., Zhang, Q.H. (2001). Advances in the development of functional foods from buckwheat. *Critical Reviews in Food Science and Nutrition*, *41*, 451–464.

31. Aufhammer, W. (2000). *Pseudogetreidearten: Buchweizen, Reismelde und Amarant/ Herkunft, Nutzung und Anbau*. Stuttgart: Verlag Eugen Ulmer GmbH und Co. (in German).
32. Kreft, S., Kreft, M. (2000). Localization and morphology of the buckwheat embryo. *Fagopyrum*, *17*, 15–19.
33. Nic Phiarais, B.P., Schehl, B.D., Oliveira, J.C., Arendt, E.K. (2006). Use of response surface methodology to investigate the effectiveness of commercial enzymes on buckwheat malt for brewing purposes. *Journal of the Institute of Brewing*, *112*, 324–332.
34. Nic Phiarais, B.P., Wijngaard, H.H., Arendt, E.K. (2005). The impact of kilning on enzymatic activity of buckwheat malt. *Journal of the Institute of Brewing*, *111*, 290–298.
35. Wijngaard, H.H., Arendt, E.K. (2006). Optimisation of a mashing program for 100% malted buckwheat. *Journal of the Institute of Brewing*, *112*, 57–65.
36. Wijngaard, H.H., Ulmer, H.M., Arendt, E.K. (2005). The effect of germination temperature on malt quality of buckwheat. *Journal of the American Society of Brewing Chemists*, *63*, 31–36.
37. Wijngaard, H.H., Ulmer, H.M., Neumann, M., Arendt, E.K. (2005). The effect of steeping time on the final malt quality of buckwheat. *Journal of the Institute of Brewing*, *111*, 275–281.
38. Rakkar, P.S. (2007). Development of a gluten-free commercial bread. M.S. Thesis, Auckland University of Technology, Auckland.
39. Coimbra, S., Salema, R. (1994). *Amaranthus hypochondriacus*: seed structure and localization of seed reserves. *Annals of Botany*, *74*, 373–379.
40. Singhal, R.S., Kulkarni, P.R. (1988). *Amaranthus*—an underutilized resource. *International Journal of Food Science and Technology*, *23*, 125–140.
41. Breene, W.M. (1991). Food uses for grain amaranth. *Cereal Foods World*, *36*, 426–430.
42. Stone, L.A., Lorenz, K. (1984). *The* starch of *Amaranthus*—physicochemical properties of functional characteristics. *Starch/Stärke*, *36*, 232–237.
43. Chauhan, G.S., Eskin, N.A.M., Tkachuk, R. (1992). Nutrients and antinutrients in quinoa seed. *Cereal Chemistry*, *69*, 85–88.
44. Guzmán-Maldonado, S.H., Paredes-López, O. (1998). Functional products of plants indigenous to Latin America: amaranth, quinoa, common beans, and botanicals. In: *Functional Foods: Biochemical and Processing Aspects*, Mazza, G. (Ed.). Lancaster, PA: Technomic Publishing Company, pp. 293–328.
45. Risi, J.C., Galwey, N.W. (1989). Chenopodium grains of the Andes: a crop for temperate latitudes. In: *New Crops for Food and Industry*, Wickens, G.E., Haq, N., Day, P. (Eds.). London: Chapman and Hall, pp. 222–234.
46. Ruales, J. (1992). Development of an infant food from quinoa (*Chenopodium quinoa*, Willd): technological aspects and nutritional consequences. PhD Thesis, University of Lund, Lund, Sweden.
47. Hoseney, R.C. (1994). *Principles of Cereal Science and Technology*, 2nd edition. St. Paul, MN: American Association of Cereal Chemists, pp. 40, 147–148, 194–195, 197, 342–343.

48. Qian, J.Y., Kuhn, M. (1999). Characterization of *Amaranthus cruentus* and *Chenopodium quinoa* starch. *Starch/Stärke*, *51*, 116–120.
49. Ahamed, N.T., Singhal, R.S., Kulkarni, P.R., Pal, M. (1996). Physicochemical and functional properties of *Chenopodium quinoa* starch. *Carbohydrate Polymers*, *31*, 99–103.
50. Thompson, T. (2000). Folate, iron, and dietary fibre contents of the gluten-free diet. *Journal of the American Dietetic Association*, *100*, 1389–1396.
51. Gallagher, E., Kunkel, A., Gormley, T.R., Arendt, E.K. (2003). The effect of dairy and rice powder addition on loaf and crumb characteristics, and on shelf life (intermediate and long term) of gluten-free breads stored in a modified atmosphere. *European Journal of Food Research*, *218*, 44–48.
52. Pagano, A.E. (2007). The gluten-free vegetarian. *Practical Gastroenterology*, *9*, 96–106.
53. Shewry, P.R., Halford, N.G. (2002). Cereal seed storage proteins: structures, properties and role in grain utilization. *Journal of Experimental Botany*, *53*, 947–958.
54. Wood, P.J. (2007). Cereal β-glucans in diet and health. *Journal of Cereal Science*, *46*, 230–238.
55. Sontag-Strohm, T., Lehtinen, P., Kaukovirta-Norja, A. (2008). Oat products and their current status in coeliac diet. In: *Gluten-Free Cereal Products and Beverages*, Arendt, E.K., Dal Bello, F. (Eds.). Amsterdam: Academic, pp. 191–202.
56. Zhou, M., Robards, K., Glennie-Holmes, M., Helliwel, S. (1998). Structure and pasting properties of oat starch. *Cereal Chemistry*, *75*, 273–281.
57. Garsed, K., Scott, B.B. (2007). Can oats be taken in a gluten-free diet? A systematic review. *Scandinavian Journal of Gastroenterology*, *42*, 171–178.
58. Hollén, E., Peterson, K.H., Sundqvist, T., Grodzinsky, E., Högberg, L., Laurin, P., Stenhammar, L., Fälth-Magnusson, K., Magnusson, K.E. (2006). Coeliac children on a gluten-free diet with or without oats display equal anti-avenin antibody titres. *Scandinavian Journal of Gastroenterology*, *41*, 42–47.
59. Peraaho, M.K.K., Mustalahti, K., Mustalahti, K., Vuolteenaho, N., Mäki, M., Laippala, P., Collin, P. (2004). Effect of an oats-containing gluten-free diet on symptoms and quality of life in coeliac disease: a randomized study. *Scandinavian Journal of Gastroenterology*, *39*, 27–31.
60. Thompson, T. (2003). Oats and the gluten-free diet. *Journal of the American Dietetic Association*, *103*, 376–379.
61. Kumar, P.J., Farthing, M.G.J. (1995). Oats and celiac disease. *The New England Journal of Medicine*, *333*, 1075–1076.
62. Peterson, D.M., Brinegar, A.C. (1986). Oat storage proteins. In: *Oats: Chemistry and Technology*, Webster, F.H. (Ed.). St. Paul, MN: American Association of Cereal Chemists, pp. 153–197.
63. Janatuinen, E.K., Pikkarainen, P.H., Kemppainen, T.A., Kosma, V.M., Jarvinen, R.M.K., Uusitupa, M.I.J., Julkunen, R.J.K. (1995). A comparison of diets with and without oats in adults with coeliac disease. *The New England Journal of Medicine*, *333*, 1033–1037.
64. Kasarda, D.D. (1994). Toxic cereal grains in coeliac disease. In: *Gastrointestinal Immunology and Gluten-Sensitive Disease, Proceedings of 6th the International*

Symposium on Coeliac Disease, Feighery, C., O'Farrelly, F. (Eds.). Dublin: Trinity College, Oak Tree Press, pp. 203–220.

65. Hardman, C.M., Garioch, J.J., Leonard, J.N., Thomas, H.J.W., Walker, M.M., Lortan, J.E., Lister, A., Fry, L. (1997). Absence of toxicity of oats in patients with dermatitis herpetiformis. *The New England Journal of Medicine*, *337*, 1884–1887.
66. Hoffenberg, E.J., Haas, J., Drescher, A., Barnhurst, R., Osberg, I., Bao, F., Eisenbarth, G. (2000). A trial of oats in children with newly diagnosed celiac disease. *The Journal of Pediatrics*, *137*, 361–366.
67. Janatuinen, E.K., Kemppainen, T.A., Pikkarainen, P.H., Holm, K.H., Kosma, V.M., Uusitupa, M.I., Mäki, M., Julkunen, R.J. (2000). Lack of cellular and humoral immunological responses to oats in adults with coeliac disease. *Gut*, *46*, 327–331.
68. Janatuinen, E.K., Kemppainen, T.A., Julkunen, R.J., Kosma, V.M., Mäki, M., Heikkinen, M., Uusitupa, M.I. (2002). No harm from five year ingestion of oats in coeliac disease. *Gut*, *50*, 332–335.
69. Reunala, T., Collin, P., Holm, K., Pikkarainen, P., Miettinen, A., Vuolteenaho, N., Mäki, M. (1998). Tolerance to oats in dermatitis herpetiformis. *Gut*, *43*, 490–493.
70. Srinivasan, U., Leonard, N., Jones, E., Kasarda, D.D., Weir, D.G., O'Farrelly, C., Feighery, C. (1996). Absence of oats toxicity in adult coeliac disease. *British Medical Journal*, *313*, 1300–1301.
71. Lundin, K.E., Nilsen, E.M., Scott, H.G., Løberg, E.M., Gjøen, A., Bratlie, J., Skar, V., Mendez, E., Løvik, A., Kett, K. (2003). Oats induced villous atrophy in coeliac disease. *Gut*, *52*, 1649–1652.
72. Arentz-Hansen, H., Fleckenstein, B., Molberg, Ø., Scott, H., Koning, F., Jung, G., Roepstorff, P., Lundin, K.E., Sollid, L.M. (2004). The molecular basis for oat intolerance in patients with celiac disease. *PLoS Medicine*, *1*, 084–092.
73. Farrell, R.J., Kelly, P.C. (2002). Celiac sprue. *The England Journal of Medicine*, *346*, 180–188.
74. Thompson, T., Dennis, M., Higgins, L., Lee, A., Sharrett, M.K. (2005). Gluten-free diet survey: are Americans with coeliac disease consuming recommended amounts of fibre, iron, calcium and grain foods? *Journal of Human Nutrition and Dietetics*, *18*, 163–169.
75. Gélinas, P., McKinnon, C.M., Mena, M.C., Méndez, E. (2008). Gluten contamination in cereal foods in Canada. *International Journal of Food Science and Technology*, *43*, 1245–1252.
76. Dahinden, I., von Buren, M., Luthy, J. (2001). A quantitative competitive PCR system to detect contamination of wheat, barley or rye in gluten-free food for coeliac patients. *European Food Research and Technology*, *212*, 228–233.
77. Denery-Papini, S., Nicolas, Y., Popineau, Y. (1999). Efficiency and limitations of immunochemical assays for the testing of gluten-free foods. *Journal of Cereal Science*, *30*, 121–131.
78. Hischenhunber, C., Crevel, R., Jarry, B., Mäki, M., Moneret-Vautrin, D.A., Romano, A., Troncone, R., Ward, R. (2006). Review article: safe amounts of gluten for patients with wheat allergy or coeliac disease. *Alimentary Pharmacology and Therapeutics*, *23*, 559–575.

79. Skerritt, J.H., Hill, A.S. (1992). How free is gluten-free? Relationship between kjeldahl nitrogen values and gluten protein content for wheat starches. *Cereal Chemistry*, *69*, 110–112.

80. Valdes, I., Garcia, E., Llorente, M., Mendez, E. (2003). Innovative approach to low-level gluten determination in foods using a novel sandwich enzyme-linked immunosorbent assay protocol. *European Journal of Gastroenterology and Hepatology*, *15*, 465–474.

81. Redondo, M.C.B., Griffin, P.B., Ransanz, M.G., Ellis, H.J., Ciclitira, P.J., O'Sullivan, C.K. (2005). Monoclonal antibody-based competitive assay for the sensitive detection of coeliac disease toxic prolamins. *Analytica Chimica Acta*, *551*, 105–114.

82. Camafeita, E., Solís, J., Alfonso, P., López, J.A., Sorell, L., Méndez, E. (1998). Selective identification by matrix-assisted laser desorption/ionization time-of-flight mass spectrometry of different types of gluten in foods made with cereal mixtures. *Journal of Chromatography A*, *823*, 299–306.

83. Henterich, N., Osman, A.A., Méndez, E., Mothes, T. (2003). Assay of gliadin by real-time immunopolymerase chain reaction. *Nahrung/Food*, *47*, 345–348.

84. Olexová, L., Dovicovicová, L., Svec, M., Siekel, P., Kuchta, T. (2004). Detection of gluten-containing cereals in flours and gluten-free bakery products by polymerase chain reaction. *Food Control*, *17*, 234–237.

85. Ribotta, P.D., Ausar, S.F., Morcillo, M.H., Perez, G.T., Beltramo, D.M., Leon, A.E. (2004). Production of gluten-free bread using soybean flour. *Journal of the Science of Food and Agriculture*, *84*, 1969–1974.

86. Schober, T.J., O'Brien, C.M., McCarthy, D., Darnedde, A., Arendt, E.K. (2003). Influence of gluten-free flour mixes and fat powders on the quality of gluten-free biscuits. *European Food Research and Technology*, *216*, 369–376.

87. Haque, A., Morris, E.R. (1994). Combined use of ispaghula and HPMC to replace or augment gluten in breadmaking. *Food Research International*, *27*, 379–393.

88. Moore, M.M., Schober, T.J., Dockery, P., Arendt, E.K. (2004). Textural comparison of gluten-free and wheat based doughs, batters and breads. *Cereal Chemistry*, *81*, 567–575.

89. Moore, M.M., Heinbockel, M., Dockery, P., Ulmer, H.M., Arendt, E.K. (2006). Network formation in gluten-free bread with the application of transglutaminase. *Cereal Chemistry*, *83*, 28–36.

90. Moore, M.M., Dal Bello, F., Arendt, E.K. (2008). Sourdough fermented by *Lactobacillus plantarum* FST 1.7 improves the quality and shelf life of gluten-free bread. *European Food Research and Technology*, *226*, 1309–1316.

91. Moore, M.M., Juga, B., Schober, T.J., Arendt, E.K. (2007). Effect of lactic acid bacteria on the properties of gluten-free sourdoughs, batters and the quality and ultrastructure of gluten-free bread. *Cereal Chemistry*, *84*, 357–364.

92. Renzetti, S.D., Dal Bello, F., Arendt, E.K. (2008). Microstructure, fundamental rheology and baking characteristics of batters and breads from different gluten-free flours treated with a microbial transglutaminase. *Journal of Cereal Science*, *48*, 33–45.

93. Nunes, M.H.B., Moore, M.M., Ryan, L.A.M., Arendt, E.K. (2009). Impact of emulsifiers on the quality and rheological properties of gluten-free breads and batters. *European Food Research and Technology*, *228*, 633–642.

94. Gujral, H.S., Rosell, C.M. (2004). Functionality of rice flour modified with a microbial transglutaminase. *Journal of Cereal Science*, *39*, 225–230.
95. Gujral, H.S., Rosell, C.M. (2004). Improvement of the breadmaking quality of rice flour by glucose oxidase. *Food Research International*, *37*, 75–81.
96. Lazaridou, A., Duta, D., Papageorgiou, M., Belc, N., Biliaderis, C.G. (2007). Effects of hydrocolloids on dough rheology and bread quality parameters in gluten-free formulations. *Journal of Food Engineering*, *79*, 1033–1047.
97. Marco, C., Rosell, C.M. (2007). Effect of different protein isolates and transglutaminase on rice flour properties. *Journal of Food Engineering*, *84*, 132–139.
98. McCarthy, D.F., Gallagher, E., Gormley, T.R., Schober, T.J., Arendt, E.K. (2005). Application of response surface methodology in the development of gluten-free bread. *Cereal Chemistry*, *82*, 609–615.
99. Sivaramakrishnan, H.P., Senge, B., Chattopadhyay, P.K. (2004). Rheological properties of rice dough for making rice bread. *Journal of Food Engineering*, *62*, 37–45.
100. Cato, L., Gan, J.J., Rafael, L.G.B., Small, D.M. (2004). Gluten-free breads using rice flour and hydrocolloid gums. *Food Australia*, *56*, 75–78.
101. Ward, F.M., Andon, S.A. (2002). Hydrocolloids as film formers, adhesives and gelling agents for bakery and cereal products. *Cereal Food World*, *47*, 52–55.
102. Sluimer, P. (2005). *Principles of Breadmaking—Functionality of Raw Materials and Process Steps*. St. Paul, MN: American Association of Cereal Chemists, pp. 207.
103. Rojas, J.A., Rosell, C.M., de Barber, C.B. (1999). Pasting properties of different wheat flour-hydrocolloid systems. *Food Hydrocolloids*, *13*, 27–33.
104. Anton, A., Artfield, S. (2008). Hydrocolloids in gluten-free breads: a review. *International Journal of Food Science and Nutrition*, *59*, 11–23.
105. Ahlborn, G.J., Pike, O.A., Hendrix, S.B., Hess, W.M., Huber, C.S. (2005). Sensory, mechanical, and microscopic evaluation of staling in low-protein and gluten-free breads. *Cereal Chemistry*, *82*, 328–335.
106. Ortolani, C., Pastorello, E.A. (1997). Symptoms of food allergy and food intolerance. In: *Study of Nutritional Factors in Food Allergies and Food Intolerance*. Luxembourg: CEC, pp. 26–45.
107. Gallagher, E., Gormley, T.R., Arendt, E.K. (2003). Crust and crumb characteristics of gluten-free breads. *Journal of Food Engineering*, *56*, 153–161.
108. Katina, K., Arendt, E., Liukkonen, K.H., Autio, K., Flander, L., Poutanen, K. (2005). Potential of sourdough for cereal products. *Trends in Food Science and Technology*, *16*, 104–112.
109. Schober, T.J., Bean, S.R., Boyle, D.L. (2007). Gluten-free sorghum bread improved by sourdough fermentation: biochemical, rheological, and microstructural background. *Journal of Agricultural and Food Chemistry*, *55*, 5137–5146.
110. Di Cagno, R., De Angelis, M., Auricchio, S., Greco, L., Clarke, C., De Vincenzi, M., Giovannini, C., D'Archivio, M., Landolfo, F., Parrilli, G., Minervini, F., Arendt, E., Gobbetti, M. (2004). Sourdough bread made from wheat and nontoxic flours and started with selected lactobacilli is tolerated in celiac sprue patients. *Applied and Environmental Microbiology*, *70*, 1088–1096.
111. De Angelis, M., Coda, R., Silano, M., Minervini, F., Rizzello, C.G., Di Cagno, R., Vicentini, O., De Vincenzi, M., Gobbetti, M. (2006). Fermentation by selected

sourdough lactic acid bacteria to decrease coeliac intolerance to rye flour. *Journal of Cereal Science*, *43*, 301–314.

112. Renzetti, S., Behr, J., Vogel, R.F., Arendt, E.K. (2008). Transglutaminase polymerization of buckwheat (*Fagopyrum esculentum* Moench) proteins. *Journal of Cereal Science*, *48*, 747–754.
113. Gujral, H.S., Guardiola, I., Carbonell, J.V., Rosell, C.M. (2003). Effect of cyclodextrin glycoxyl transferase on dough rheology and bread quality from rice flour. *Journal of Agricultural and Food Chemistry*, *51*, 3814–3818.
114. Gujral, H.S., Haros, M., Rosell, C.M. (2003). Starch hydrolyzing enzymes for retarding the staling of rice bread. *Cereal Chemistry*, *80*, 750–754.
115. Gujral, H.S., Haros, M., Rosell, C. (2004). Improving the texture and delaying staling in rice flour chapati with hydrocolloids and alpha-amylase. *Journal of Food Engineering*, *65*, 89–94.

13

PROCESSING FOODS WITHOUT SOYBEAN INGREDIENTS

JOYCE I. BOYE, LAMIA L'HOCINE, AND SAHUL H. RAJAMOHAMED

13.1. INTRODUCTION

The quality of soybean as food for human consumption has been shown to be equal to that of many animal-based products. Today, soybean is one of the world's most economical and valuable agricultural commodities. A distinct advantage of soybean in the agricultural and industrial sectors is that it can be processed into oil, protein products, nutraceuticals, chemicals, and high nutritional foods [1]. In the food market, soybean is processed into beverages, meat and simulated meat products, confectionery, and bakery and cereal products. Furthermore, soy protein and soy protein products have been approved by the Food and Drug Administration (FDA) of the United States and Japan (Foods for Specified Health Use [FOSHU]) as a "functional food," making it one of the most valuable vegetable proteins in the world today.

Despite its high nutritional quality, soybean is listed by the Food and Agriculture Organization of the United Nations (FAO) as one of the eight priority allergens [2]. Prevalence rate for soybean allergy ranges between 0.3% and 0.4% [3, 4]. Clinical reactivity levels, also referred to as threshold dose or eliciting dose, vary significantly and ranges from as low as 0.0013 to 500 mg of soy protein [5, 6]. Identification and avoidance of soy ingredients is, therefore, important especially for highly sensitive soy-allergic patients. This chapter will provide an overview of the composition and molecular properties of soybean including the known allergenic proteins, a summary of the different types of

Allergen Management in the Food Industry, Edited by Joyce I. Boye and Samuel Benrejeb Godefroy

soy foods and soy ingredients currently available on the market, potential sources of hidden soy allergens in foods, and a list of alternate foods and ingredients that could be used as nutritional or functional replacements in the food processing and foodservice industries.

13.2. COMPOSITION AND MOLECULAR PROPERTIES OF SOYBEAN AND SOYBEAN COMPONENTS

Table 13.1 provides the nutrient composition of soybean. Proximate chemical composition of soybean varies depending on the varietal type and growing conditions; reasonable average figures are 35–40% protein, 17–23% lipid, 31% carbohydrate, and 4–5% ash on a dry weight (d.w.) basis with the water content of stored mature soybean ranging between 12% and 13% [8]. Molecular properties of these components are important in conferring specific nutritional and functional properties to soybean and soy products, and some of which are discussed in detail below.

13.2.1. Major Components of Soybean

13.2.1.1. Proteins. Soy proteins are of high nutritional quality when compared with other plant proteins. They contain good supplies of all the essential amino acids including lysine, which is normally lacking in other cereals. Compared with animal proteins, soy proteins contain relatively lesser amounts of the sulfur-containing amino acids (i.e., cysteine and methionine); however, protein digestibility corrected amino acid score (PDCAAS) measurements have shown that soy protein is equivalent in quality in many ways to animal proteins such as milk and meat (Table 13.2) [9]. The protein content of soybean is high (~40 g/100 g), which is twofold higher than meat (20%) and three- to fivefold higher than cereals (7–14%) on a d.w. basis [10, 11].

Albumins and globulins are the major soy proteins. Globulins, which account for up to 80% of soy proteins, consist primarily of the 7S (β-conglycinin) and 11S (glycinin) with molecular weights (MWs) of 140–180 and 320–360 kDa, respectively [12, 13]. Other globulins found in soybean are the 15S proteins, but these are believed to be association products of the 11S proteins. The albumin fraction has an average MW of 26 kDa and contains several different proteins including the Bowman–Birk and Kunitz trypsin inhibitors (KTI). Concerns about the nutritional value of soybean have hovered around the presence of these enzyme inhibitors, which are considered to be antinutritional since they interfere with digestion, absorption, and utilization of proteins. The inhibitory activity of these proteins is, however, destroyed by heating [14].

The rise in the soy food market may be attributed to the many health benefits attributed to soy proteins. In 1999, the U.S. Food and Drug Administration (FDA) approved a health claim for soy proteins, which states that consumption of 25 g of soy protein a day as part of a diet low in saturated fat and

TABLE 13.1 Nutritional Composition of Soybean (per 100 g of Mature Raw Seeds) [7]

Nutrient	Amount
Water (g)	8.54
Energy (kcal)	446
Protein (g)	36.49
Fat (g)	19.94
Saturated fatty acids (g)	2.88
Monounsaturated fatty acids (g)	4.40
Polyunsaturated fatty acids (g)	11.26
Omega-3 fatty acid	1.33
Omega-6 fatty acid	9.93
Carbohydrate (g)	30.16
Fiber (g)	9.30
Ash (g)	4.87
Calcium (mg)	277.00
Iron (mg)	15.70
Magnesium (mg)	280.00
Phosphorus (mg)	704.00
Potassium (mg)	1797.00
Sodium (mg)	2.00
Zinc (mg)	4.89
Copper (mg)	1.66
Manganese (mg)	2.52
Selenium (μg)	17.80
Thiamine (mg)	0.87
Riboflavin (mg)	0.87
Niacin (mg)	1.62
Pyridoxine (mg)	0.38
Pantothenic acid (mg)	0.79
Folate (μg)	375.00
Choline (mg)	115.90
Vitamin C (mg)	6.00
β-Carotene (μg)	13.00
Vitamin A (IU)	22.00
Vitamin E (mg)	0.85
Vitamin K (μg)	47.00
Phytosterol (mg)	161.00

cholesterol may reduce the risk of cardiovascular disease [15]. Furthermore, researchers have found that oligopeptides from soybean protein may have many biological functions such as regulating blood cholesterol levels, angiotensin-converting enzyme inhibition, antithrombotic activity, and antioxidant properties [16–18]. Additionally, it has been reported that soy proteins may be used in treatment of renal diseases in diabetic and nondiabetic patients with nephropathy [19–21].

TABLE 13.2 Protein Digestibility Corrected Amino Acid Score (PDCAAS) for Different Foods [9]

Foods	PDCAAS
Casein	1.0
Egg white	1.0
Soy protein concentrate	0.99
Soy protein isolate	0.92
Beef	0.92
Pea flour	0.69
Kidney beans (canned)	0.68
Rolled oats	0.57
Lentils (canned)	0.52
Peanut meal	0.52
Rice	0.47
Whole wheat	0.42
Wheat gluten	0.25

13.2.1.2. Lipids. Soybean oil is considered to be of high nutritional quality because it contains high amounts of unsaturated fatty acids, such as linoleic acid (LA) (50.11%), oleic acid (22.72%), and linolenic acid (LNA) (6.54%), and lower amounts of the saturated fatty acids (15.25%) [7]. LA (omega-6) and LNA (omega-3) cannot be synthesized by humans and are, therefore, considered to be essential and must be provided by the diet. The ideal omega-6:omega-3 ratio for humans has been extensively discussed. It is generally agreed that there is an imbalance in the consumption pattern in western diets (i.e., overconsumption of omega-6 fatty acids and an underconsumption of omega-3 fatty acids). While the optimal ratio may vary with the disease under consideration, a lower ratio of omega-6:omega-3 fatty acids is more desirable for reducing the risk of many chronic diseases of high prevalence in Western societies [22]. The omega-6:omega-3 ratio for soybean oil is ~7:1.

13.2.1.3. Carbohydrates. Carbohydrates represent the second largest component of soybean. On a dry basis, soybean contains 30–35% carbohydrates, which comprises of both soluble and insoluble carbohydrates. Soluble carbohydrates found in soybean include sucrose (2.5–8.2%), stachyose (1.4–4.1%), raffinose (0.1–0.9%), arabinose (trace), and glucose (trace) [8 and references within]. Major insoluble carbohydrates are cellulose, hemicellulose, and pectin. Unlike other legumes, soybean contains only trace amounts of starch. Due to the relatively high amounts of dietary fiber and very low glycemic index, it is recommended for diabetic, weight control, and hypercholesterolemic patient diets [23, 24]. Furthermore, oligosaccharides (stachyose and raffinose) present in soybean have been recognized as prebiotics and antagonists to colon cancer [25–27].

13.2.2. Minor Components of Soybean

13.2.2.1. Vitamins and Minerals. Soybean contains high amounts of folate (318–375 μg/100 g on d.w. basis) compared with other legumes (10–277 μg/100 g) [28]. It is also a good source of calcium (277 mg/100 g), iron (15.70 mg/100 g), and zinc (4.89 mg/100 g) [7]. Iron bioavailability of soybean is, however, poor but can be improved by supplementing with vitamin C [29]. Zinc bioavailability is relatively good (22–50%), and although calcium bioavailability (~20%) is lower than in milk and green leafy vegetables, it is still reasonably good [30, 31]. The main reason for the lower bioavailability of minerals in soybean is due to the high content of antinutritional factors such as phytate, oxalate, and tannins, which hinder the absorption of minerals in the body [29, 32, 33]. Other vitamins and minerals found in soybean include vitamin A (22 IU/100 g), vitamin K (47 μg/100 g), niacin (1.62 mg/100 g), thiamine (0.87 mg/100 g), riboflavin (0.87 mg/100 g), potassium (1797 mg/100 g), phosphorus (704 mg/100 g), and selenium (17.8 μg/100 g) [7].

13.2.2.2. Phytochemicals. Phytochemicals in soybean of relevance include isoflavones and saponins. Isoflavones have phytoestrogenic properties and have also been shown to be instrumental in preventing oxidative damage in tissue, in preventing cancer, and in blocking tumor promotion [34]. Daidzein, daidzin, genistein, genistin, glycitein, and glycitin are key isoflavones present in soybean. Soy saponins, specifically, subgroup B and E saponins (e.g.,2,3-dihydro-2,5-dihydroxy-6-methyl-4H-pyran-4-one(DMPP)-conjugated saponins), may have antiviral properties [35]. Reported benefits of some of these minor soybean components have contributed to the popularity of soy as a food ingredient [36].

13.3. ALLERGENICITY OF SOYBEAN

At least 17 allergens have been identified in soybean (Table 13.3). These allergens belong to protein families and have conserved structural features in relation to their biological activity, which explains the wide immunochemical cross-recognition observed among members of the legume family.

13.3.1. Glycinin and β-Conglycinin

Soybean glycinin (11S) and β-conglycinin (7S) have been identified as major allergens [49, 57, 58]. These proteins are of particular interest to the food industry because they account for up to 80% of total soybean seed proteins [59], and they are the major determinants of the nutritional quality and functionality of soybean. Glycinin and β-conglycinin are both storage proteins known as globulins belonging to the cupin superfamily [60]. They share conserved sequence motifs comprising identical or chemically similar amino acid residues and a common three-dimensional (3D) conformation.

TABLE 13.3 Identified Soybean Allergens

IgE-Binding Proteins	Allergen Nomenclature	Molecular Weight (kDa)	Family	References
Hydrophobic proteins	Gly m 1: (IUIS)[a]		Lipid transfer protein	37, 38
	Gly m 1.0101	7.5		
	Gly m 1.0102	7.0		
Unknown	Gly m 2 (IUIS)	8.0	Storage protein	39
Profilin	Gly m 3 (IUIS)	14	Profilin	40
SAM22	Gly m 4 (IUIS)	16.6	Pathogenesis-related protein PR-10	41–43
P34	Gly m Bd 30K	34	Protease	44, 45
Unknown Asn-linked glycoprotein	Gly m Bd 28K	26	Unknown	46
β-Conglycinin	Not assigned	140–170	Storage protein	47
Glycinin	Not assigned	320–360	Storage protein	48, 49
2S albumin	Not assigned	12	Prolamin	50
Lectin (agglutinin)	Not assigned	120	Lectin	51
Lipoxygenase	Not assigned	102	Enzyme	52
Kunitz trypsin inhibitor	Not assigned	21	Protease inhibitor	52, 53
Unknown	Not assigned	39	Unknown	50
Unknown	Not assigned	50	Homology to chlorophyll A–B binding protein	54
P22-25	Not assigned	22–25	Unknown	55

[a]International Union of Immunological Societies (IUIS) official allergen nomenclature.

Source: Reference 56—Reprinted from L'Hocine, L., Boye, J.I. (2007). Allergenicity of soybean: new developments in identification of allergenic proteins, cross-reactivities and hypoallergenization technologies. *Critical Reviews in Food Science and Nutrition*, *47*, 127–143, with permission from Taylor & Francis.

Glycinin accounts for over 40% of the total seed globulin [61]. It is a heterogeneous protein with polymorphic subunit composition, which varies among different cultivars [62]. Each of the glycinin subunits (58–69 kDa) can be dissociated under reducing conditions into acidic (A; 31–45 kDa) and basic (B; 18–20 kDa) polypeptide chains [59]. Presently, five major subunits have been characterized [63, 64], namely A1aB2 (G1), A1bB1b (G2), A2B1a (G3), A3B4 (G4), and A5A4B3 (G5). The acidic polypeptide A5 has an exceptionally low MW of 10 kDa. Pedersen and Djurtoft [57] reported immunoglobulin E (IgE) binding to all of the native acidic subunits of glycinin, but little or none to the basic subunits. Other studies report, however, that both chains of soybean glycinin bind IgE from soybean-allergic patients [49]. Zeece et al. [48] reported that the acidic glycinin G1 subunit exhibited a much stronger IgE binding than acidic glycinin G2, even though the sequence homology is similar. Eleven IgE-binding epitopes in glycinin G2 were identified, all located on the exterior of the molecule [49], among which four were immunodominant.

β-Conglycinin is a trimer composed of three major subunits, α, α′, and β, with MWs of 76, 72, and 53 kDa, respectively [65]. Another subunit, named β′, is present only in some soybean varieties [66, 67]. All of the β-conglycinin subunits are glycoproteins and contain 4–5% carbohydrate [68]. Burks et al. [58] reported IgE binding to the three, α, α′, and β, subunits of β-conglycinin. Ogawa et al. [47], however, found that the α′-subunit, which is highly homologous to the α-subunit, had no allergenicity. Different sensitization histories of individual patients may be one of the reasons for the discrepancy.

13.3.2. Soybean Vacuolar Protein (Gly m Bd 30K)

Gly m Bd 30K, also known as P34, is considered by many as the major allergenic protein of soybean [44, 69], since over 65% of soy-sensitive patients react to it [45, 70]. It is a monomeric, insoluble glycoprotein belonging to the soybean 7S fraction [71]. Gly m Bd 30K has a MW of 34 kDa and is classified as an oil body-associated protein [45] and a member of the papain superfamily of cysteine proteases [70]. Sixteen distinct linear epitopes were identified, among which five were considered as immunodominant [70].

13.3.3. Gly m Bd 28K

Tsuji et al. [46] isolated and characterized the allergenic soybean seed protein Gly m Bd 28K. This allergen was an unknown Asn-linked glycoprotein with a MW of 26 kDa. The glycan moiety of Gly m Bd 28K was shown to bind to IgE antibodies in the sera of soybean-sensitive patients at Asn 20 [72]. This allergen is, however, not present in many soybean accessions; in fact, testing showed that 80% of Japanese soybean varieties did not contain Gly m Bd 28K [71].

13.3.4. Kunitz Trypsin Inhibitor (KTI)

The KTI belongs to the group of proteins of the plant defense system, which has inhibitory activity against various proteases. Soybean KTI is classified in the family of antiparallel β-sheet proteins that are highly resistant to chemical and thermal denaturation [73]. Moroz and Yang [74] demonstrated that the serum of a patient with an occupational respiratory disorder had IgE antibodies specific for soybean KTI (21 kDa). Two subsequent studies [52, 53] showed positive cutaneous and IgE response, as well as a specific bronchial response to purified soybean KTI in bakers who were occupationally exposed to soybean flour and with work-related asthma, thus, confirming the soybean KTI to be an occupational inhalant allergen. KTI allergenic epitopes, however, remain unknown.

13.3.5. Soybean Hull Proteins (Gly m 1.0101, Gly m 1.0102, and Gly m 2)

The International Union of Immunological Society (IUIS) Allergen Nomenclature Sub-Committee has identified several allergens in soybean hull. These include isoallergens Gly m 1.0101 also known as Gly m 1A (glycoprotein S2, MW of 7.5 kDa, *pI* 6.8) and Gly m 1.0102 also known as Gly m 1B (protein S1, MW 7.0 kDa, *pI* 6.1–6.2) [39, 75]. These isoallergens were identified as soybean hydrophobic proteins and are classified as lipid transfer proteins [37]. Soybean hull was identified as a source of aeroallergens that caused soybean epidemic asthma in Barcelona, Spain [37]. Using sera of asthmatic epidemic patients from Barcelona, Codina et al. [39, 75] further identified and characterized a third allergen named Gly m 2 (MW of 8 kDa, *pI* 6.0), which was shown along with Gly m 1 (Gly m 1.0101 and Gly m 1.0102) to be partially responsible for the soybean asthma outbreaks. The N-terminal partial sequence of Gly m 2 was found to have 71% homology with a storage protein from the cotyledon of cowpea and 64% homology with a disease response protein from green pea, but it lacked homology with the Gly m 1. Gly m 1.0101, Gly m 1.0102, and Gly m 2 were found to be not responsible for the occupational asthma caused by soybean flour in bakers, where mainly high MW allergens from both flour and hull were involved [76].

13.3.6. Soybean Profilin (Gly m 3)

The soybean allergen Gly m 3 belongs to the profilin family, which is a group of structurally highly conserved plant proteins containing several allergenic proteins [40]. The sequence conservation among profilins is reflected by a highly similar structure, and profilin-specific IgE was shown to cross-react between profilins from pollen and food [77]. They are now recognized as ubiquitous cross-reactive plant allergens. The 3D structure of Gly m 3 remains to be determined; however, epitope mapping with recombinant profilin indicated that IgE binding to rGly m 3 was dependent on the integrity of a conformational structure [40].

13.3.7. Soybean Protein SAM22 (Gly m 4)

Soybean allergen Gly m 4 is the stress-induced 16.6 kDa PR-10 protein SAM22 (starvation-associated message 22) from soybean seed. It belongs to the group of plant defense system proteins also known as pathogenesis-related (PR) proteins. Gly m 4 has a sequence identity of approximately 50% with Bet v 1, the birch allergen that is responsible for IgE binding in more than 95% of birch pollinosis (the prevailing allergic diseases in Northern and Central Europe and North America). Severe oropharyngeal and anaphylactic reactions in birch pollinotics after ingestion of a dietary powder containing 50% of soy protein isolate (SPI) has been reported [41]. Eighty-five percent of those patients had specific IgE against Gly m 4. Clinical investigation further revealed that 96% of patients reacting to Bet v 1 also had Gly m 4-specific IgE, and 64% of these recognized other soy proteins. These studies provide evidence that Gly m 4 is a major soy allergen for birch pollen-allergic patients.

13.3.8. Other Soybean Allergens

A low-molecular-mass soy protein, identified as P22-25, has been found to induce strong antibody responses in calves [55]. The nature of this protein remains unknown, although it was shown to be different from the allergenic P34 described earlier. Another newly identified soybean allergen is the protein P39, which was originally detected in soybean lecithin [50]. P39 is a 39-kDa hydrophobic protein located in the protein storage vacuoles of soybean seeds and is extracted in the oil body-enriched fraction [78]. IgE from several soybean-sensitive individuals recognized recombinant polypeptide fragments derived from the P39 sequence [78].

Lectins are carbohydrate-binding proteins that are predominantly found in plant seeds, particularly those of legumes [79]. Plant lectins have been shown to react unspecifically with the carbohydrate moieties of IgE, inducing allergy-like symptoms [80]. Lectins have MWs of approximately 120 kDa, and minor amounts are present in soybeans. They are composed of four identical subunits of 30 kDa each, which are considered to be potential allergens. Baur et al. [52] demonstrated by radioallergosorbent tests (RASTs) that lipoxidase (lipoxygenase), another protein found in soybean, was a major allergen for 14 bakers suffering from workplace-related respiratory symptoms and sensitized to soybean. Additionally, a 50-kDa protein with homology to chlorophyll A–B binding protein has also been reported to bind to IgE antibodies of patients with soy allergy [54].

13.4. TECHNIQUES TO REMOVE OR REDUCE THE ALLERGENICITY OF SOYBEAN PROTEINS

The effect of processing on the allergenicity of soybean proteins has been studied by various researchers in an attempt to identify ways to remove or reduce soy protein antigenicity. Attempts have also been made using plant breeding techniques to develop hypoallergenic soybean varieties. A brief summary of these studies is provided below.

13.4.1. Heat Treatment

Heating denatures food proteins and may reduce allergenicity by destroying conformational epitopes or by masking linear epitopes through cross-linking reactions (e.g., through hydrogen bonding, sulfhydryl–disulfide, hydrophobic or electrostatic interactions). At the same time, it could also unveil hidden epitopes as a result of unfolding and increase allergenicity. Results on the effect of heat treatment on soybean antigenicity/allergenicity have, therefore, been somewhat conflicting. Shibasaki et al. [81] reported that IgE binding of the 2S soy fraction was enhanced with heating at 80°C, whereas, the IgE binding to other fractions were reduced by 39–75% compared with that of the crude product. Antigenicity of P34 (the immunodominant soybean allergen) was also reported to increase after 5 minutes of boiling but was found to decrease with further heating, and after 30 minutes of heating, no antigenicity was detected [82]. *In vivo* significance of these findings is warranted due to the presence of multiple linear epitopes on the protein. Burks et al. [83] reported similar decreases in soy protein antigenicity after heat treatment at 80 and 120°C for 60 minutes, but they found no significant changes in IgE and immunoglobulin G (IgG) binding when enzyme allergosorbent test (EAST) inhibition was used on soy proteins heated at 100°C for 5–60 minutes [84]. Clearly, further research on the effect of heating under a variety of processing conditions is needed to clarify the effect of thermal processing on the allergenicity of soy proteins. Environmental conditions, processing parameters such as temperature and duration of heat treatment, as well as matrix effects can all influence antigenicity. Well-conducted *in vitro* as well as *in vivo* comparative studies will also be useful to determine the physiological significance of these findings.

13.4.2. Glycation

Glycation involves the conjugation of proteins with reducing sugars through the Maillard reaction. The reaction can block epitopes recognized by antibodies, which could decrease immunogenicity. Evidently, the size of the sugar and the type of cross-links formed will impact the efficiency of the reaction to block allergenic epitopes. Allergenicity of soy protein was found to markedly decrease after glycation with chitosan and galactomannan [85]. The soy protein–chitosan conjugate was found to be more effective in masking the allergenic epitopes in the 34-kDa soy protein allergen (Gly m Bd 30K) than the soy protein–galactomannan conjugate. van de Lagemaat and coworkers [86] also reported that *in vitro* antigenicity of soy proteins as measured by direct ELISA was reduced by glycation with both fructose and commercial fructose oligosaccharides (FOS) (Raftilose® P95, Orafti España S.L., Barcelona, Spain) by up to 90% for some treatments. As with heat treatment, the physiological significance of these studies is unclear. *In vivo* studies with allergic patients will, therefore, be very useful to ascertain the true effects of glycation. This is especially important as some studies have indicated that advanced stages of the Maillard reaction can result in modifications to protein structure that may unveil novel epitopes that can bind IgE antibodies and, as a consequence, could increase allergenicity [87].

13.4.3. Hydrolysis and Fermentation

As with heat treatment, hydrolysis (chemical or enzymatic) can destroy protein conformational structure and could, therefore, degrade antigenic epitopes and modify antigenic response. Franck et al. [88] compared the allergenicity of three commercial soybean-based products (soybean flour, texturized soy proteins, and soymilk) and two infant formulas (one containing intact proteins and the other containing a soy protein hydrolysate). The hydrolyzed infant formula contained proteins with MMs below 28 kDa and showed a lack of allergenicity by immunoblotting. Tsai et al. [89] used skin prick tests, Western blots, and enzyme-linked immunosorbent assay (ELISA) to determine the residual allergenicity of soybean hydrolysates, and no evidence of IgE-specific binding with both skin tests and Western blots were observed, although some residual IgG binding was detected by ELISA. Maldonado et al. [90] suggested that allergenic proteins that are extensively hydrolyzed by acid hydrolysis (greater than 80%) are unlikely to elicit allergenic reactions. The authors, however, caution that while such hydrolyzed formulas may reduce the incidence or delay the appearance of certain atopic symptoms, they may not consistently prevent IgE-mediated allergic reactions.

Studies have also shown a reduction in soy allergenicity as a result of fermentation. In one such study, it was reported that natural or induced fermentation significantly reduced IgE immunoreactivity by up to 89% in soybean meal [91]. The soy products used in this study were, however, tested against the plasma of only six soybean-sensitive donors, and as a result, it is difficult to generalize that fermented or hydrolyzed products are safe for consumption by a larger population of soy-allergic individuals. A study conducted by Hefle et al. [92], in fact, showed that soy sauces made by fermentation of soy protein retained some of their allergenicity (10–30% that of soy flour). Thus, while hydrolysis and fermentation may reduce the allergenicity of soybeans, they may not eliminate the possibility of allergic reactions.

13.4.4. Breeding

Molecular biology and biotechnology tools have made it possible to identify and manipulate phenotypic traits of interest in plants. A wide variety of these techniques have been applied to soybeans to improve their nutritional and functional properties. In the last decade, efforts have also been made to identify and breed for soybean hypoallergenic varieties. Herman et al. [93] used transgene-induced gene slicing to prevent the accumulation of Gly m Bd 30 K (the immunodominant soy protein) in soybean seeds. The authors found no compositional, developmental, structural, or ultrastructural phenotypic differences between the genetically modified seeds and control seeds. Similar studies have been conducted by other workers [94, 95]. These results are promising, and further research may provide an avenue to remove the allergenic proteins in soy. Evidently, the safety and functional performance of such products will need to be evaluated since extensive modifications in protein profile may markedly affect the ability to use these materials in traditional products, and also, total absence of allergenicity might be difficult to achieve.

13.5. IDENTIFYING REPLACEMENTS FOR SOYBEAN AND SOY PRODUCTS IN DIFFERENT FOOD APPLICATIONS

While processing and breeding may be used to reduce the allergenicity of soy protein, there is presently no proven technique that completely eliminates soybean allergenicity. For soy-allergic consumers, the only way to prevent allergic reactions is by avoidance of foods that contain soy proteins. Soybeans and soy ingredients are extensively used in the formulation of different food products [56] (Fig. 13.1). Avoidance of these products requires their identifica-

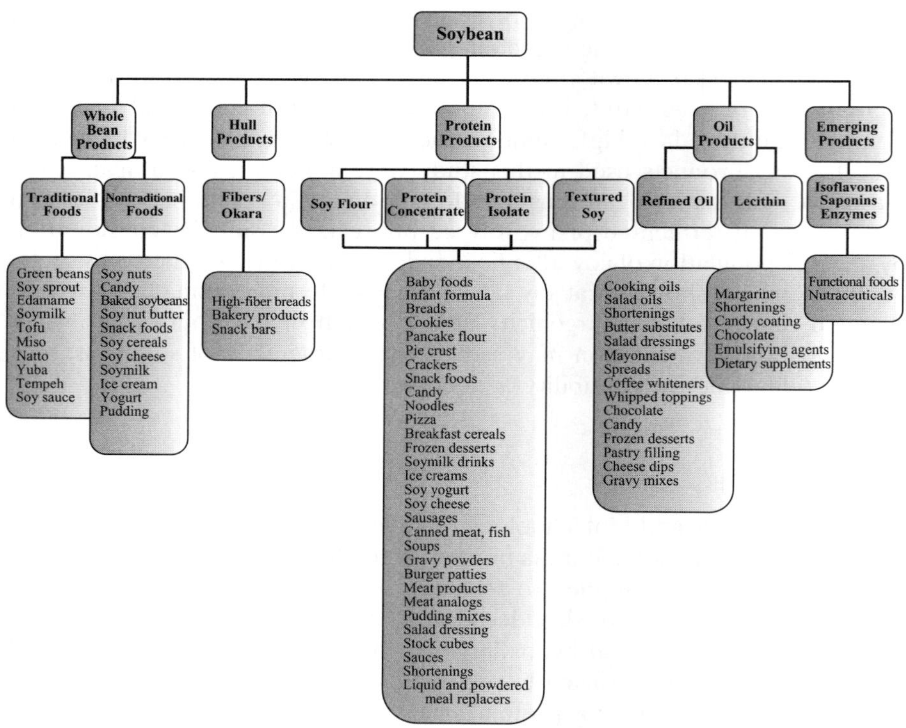

Fig. 13.1 Soybean foods and ingredients (Reference 56—Reprinted from L'Hocine, L., Boye, J.I. (2007). Allergenicity of soybean: new developments in identification of allergenic proteins, cross-reactivities and hypoallergenization technologies. *Critical Reviews in Food Science and Nutrition*, *47*, 127–143, with permission from Taylor & Francis).

tion and the availability of suitable alternative replacements. Commercially available soy foods and soy ingredients include soymilk, tofu, soy sprouts, soy sauce, soy nut butter, soy oil, soy lecithin, edamame, tempeh, natto, miso, soy grits, soy flour, soy protein concentrate (SPC), SPI, and some textured vegetable protein (TVP) and hydrolyzed vegetable protein. Some of these products and their possible replacements are described in greater detail below. In addition to foods, soy ingredients may also be used in personal care products such as soaps, hair care products, body lotions, cleaning agents, and so on. A discussion of these products is outside the scope of this chapter.

13.5.1. Commercially Available Soy Foods and Soy Ingredients

13.5.1.1. Major Soy Foods

13.5.1.1.1. Soybeans and Sprouts. Green soybean and soybean sprouts are two different soybean products that are consumed as a vegetable in many cultures. Soybean sprouts are obtained by germinating soybeans for 5–10 days and are consumed fresh in salads or could be used as a vegetable for cooking. They are normally sold fresh and are kept refrigerated. Green vegetable soybeans, on the other hand, are harvested just before maturity (edamame) and can be cooked and eaten in salads and in soups or as a snack. It is available fresh (in pod or shelled), canned, or frozen.

13.5.1.1.2. Soymilk. Soymilk is the liquid extract obtained after cooking, grinding, and filtering soybeans. The extract has a consistency that is very similar to cow's milk and is frequently used as an alternative to dairy products. There are four major types of soymilk products available (unsweetened, sweetened, flavored, and low fat). Unsweetened soymilk generally contains only water and soybeans. Sweetened soymilk may be sweetened with rice syrup, honey, corn, or barley malt extract. Flavored soymilk may be sweetened or unsweetened, and is often flavored with cocoa, vanilla, carob, or strawberry. Low-fat soymilk may also be sweetened or unsweetened, flavored or unflavored, but usually contains less fat. Soymilk is frequently fortified with vitamins and minerals to increase its nutrient value. Total solid content of soymilk products ranges between 8% and 10% depending on the water:bean ratio used and contains 3–4% protein, ~2% fat, ~3% carbohydrate, and ~0.5% ash [8].

Soymilk can be further processed into a variety of frozen desserts (e.g., ice cream) and fermented products (e.g., soy yogurt and soy cheese) using processes similar to those used in the dairy industry.

13.5.1.1.3. Tofu. Tofu is a curd made from heated soymilk. It is prepared by adding coagulating agents such as glucono-δ-lactone (GDL) or salts (magnesium chloride, calcium chloride, calcium sulfate) to heated soymilk followed by pressing to remove the whey. The final product is a gel with different degrees of hardness depending on the type and amount of coagulant used and the processing method. Tofu has been consumed for years in Asian countries and has only relatively recently found a niche in places such as Europe and North America. Tofu has a soft white texture, which is in some respects similar to cheese. On a wet basis, pressed tofu with a moisture content of about 85% contains 7.8% protein, 4.2% lipid, and 2 mg/g calcium [8, 96].

13.5.1.1.4. Other Soy Foods. Soy sauce and soy paste (miso) are fermented soy products that are frequently used as condiments and seasoning in foods. Soy sauce and miso are made by fermenting soybean with or without other grains (e.g., wheat, rice, barley) with different types of *Aspergillus* sp. Soy sauce is obtained in a liquid form, while miso is a thick paste. Traditionally, these products were used in Asian cuisine, but they have become a mainstay of many modern diets. Tempeh and natto are two other fermented soy foods, but these are less frequently consumed outside of Asia. Tempeh is made by fermenting dehulled and cooked soybeans with *Rhizopus* sp., while natto is fermented with *Baccillus subtilis*.

Soybeans can also be roasted in a manner similar to peanuts. The product obtained has a nutlike flavor and a crunchy texture, which can be consumed as a snack. The product can also be ground to obtain roasted soy flour or roasted soy nut butter. Soynut butter has a texture, flavor, and taste that is similar to peanut butter and is frequently suggested as an alternative to peanut butter. Roasted soybeans are also sometimes ground and used as a substitute to coffee.

13.5.1.2. Major Soy Ingredients

13.5.1.2.1. Soybean Oil. Soybean oil is one of the world's leading vegetable oil for human consumption. It is sold as refined oil and is often labeled as soybean oil or vegetable oil, which could include a mixture of other vegetable oils. Soybean oil is widely used in the manufacture of different foods and in the foodservice industry. It is also frequently used as a salad or cooking oil and in the production of shortening, margarines, mayonnaise, and salad dressings. By-products from the processing of soybean oil are also used to produce mono- and diglycerides and lecithin, which are commonly used as emulsifying agents in foods.

13.5.1.2.2. Soy Flour and Soy Protein Concentrates, Isolates, and Hydrolysates. Traditionally, soybeans have been processed for their oil, leaving behind the soymeal biomass. Significant effort has been made in the last few decades to process this meal into value-added products. Defatted soy flakes (or flour, grits, meal), SPCs, and SPIs are the three major products available from soymeal. Defatted soy flakes contain ~50% protein, while SPC and SPI generally contain at least 65% and 90% protein on a dry basis, respectively. The fat content in these products is quite low (<1%). Full-fat soy flour is also available on the market. It may be steamed and toasted to inactivate enzymes and enzyme inhibitors or unheated that gives an enzyme-active, full-fat soy flour, which is used in the bakery industry to bleach flour.

Soy proteins are widely used in formulated foods, partly because of their nutritional value but especially for their functional properties, which include gelling, foaming, and emulsification, which underlie many food sensory attributes. Food products likely to contain soy ingredients include beverages, nutritional bars, bakery and cereal products, soups, meat products, confectionery, salad dressings, and desserts. A growing use of SPCs and SPIs is in the preparation of texturized food products that are used as meat alternatives or in cereal products. Soy proteins can also be hydrolyzed enzymatically or chemically to produce hydrolyzed vegetable protein (HVP), which is used as a flavor enhancer in many foods.

13.5.1.2.3. Other Soy Ingredients. As indicated above, by-products from soybean processing may also be used in nutraceutical products (e.g., soy dietary fiber, soy hulls, soy isoflavone extracts, vitamin E, and soy phytosterols) and industrial products (e.g., candles, household cleaning agents, fertilizers, adhesives, lubricants, etc.). Suitable precautions need to be taken to identify these when soy-allergic patients are targeted in product development.

13.5.2. Functional and Nutritional Alternatives to Soy Foods and Soy Ingredients

13.5.2.1. Alternative Replacements for Soy Foods. Table 13.4 provides a list of foods that could be used to replace soy foods in the foodservice and food processing industry. While the list is not exhaustive, it provides some examples that could serve as a starting point for the creation of new recipes and food formulations. Different manufacturers use ingredients from varied sources in product development. It is, therefore, important that the ingredient list of the products suggested be verified to ensure that they do not contain soybean in any form.

13.5.2.2. Alternative Replacements for Soy Ingredients: Soy Proteins. Soy protein products deserve special mention because they are used so extensively in food processing. Over the last few decades, soy protein products have been used as nutritional and functional food ingredients in almost every food category [97]. Table 13.5 provides a list of the applications of soy proteins in different foods based on their functional properties.

Identification of appropriate replacements for soybeans represents a significant challenge. Food manufacturers have to ensure that alternative replacements provide adequate nutrition and functionality [97]. Availability and price are other important factors that need to be considered in the selection of replacement ingredients. Animal proteins from milk, eggs, and meat meet most of these criteria and can serve as excellent nutritional and functional alternatives to soy proteins. As these have been studied and discussed extensively in

TABLE 13.4 Nutritional and Functional Alternatives to Soy and Soy Ingredients

Soy Products	Nutritional and Functional Alternatives
Soy nuts	Pea, chickpea, lima bean, kidney bean, navy bean, peanut, almond, cashew, walnut, hazelnut, corn
Soy flour	Cow's milk powder, chickpea flour, pea flour, lentil flour, rice flour, peanut flour, sunflower flour, sesame flour, almond flour
Soy protein concentrate or isolates	Chickpea protein isolate, pea protein isolate, lentil protein isolate, sesame protein isolate, canola protein isolate, rice protein isolates, lupin protein isolate, amaranth leaf protein isolate
Soy protein hydrolysate	Cow's milk protein hydrolysate, pea protein hydrolysate, chickpea protein hydrolysate, lupin protein hydrolysate
Soymilk	Cow's milk, rice milk, almond milk, hemp milk
Tofu	Cheese
Soy ice cream	Cow's milk ice cream, hemp milk ice cream
Soy yogurt	Cow's milk yogurt
Soy oil	Corn oil, peanut oil, sunflower oil, safflower oil, cottonseed oil, canola oil, pumpkin seed oil, olive oil, rice bran oil, grape seed oil, palm oil, coconut oil
Soy margarine	Butter, margarine from palm oil, olive oil, and canola oil
Texturized soy protein	Meat
Soy sprouts	Pea sprouts, chickpea sprouts, mung bean sprouts, kidney bean sprouts, navy bean sprouts
Edamame	Pea pod, fava bean pod
Soy fiber	Rice bran, wheat bran, oat bran, flaxseed meal, psyllium fiber, inulin
Soy phytosterols	Phytosterols from peanut, Brazil nut, pistachio, sunflower seed, and flaxseed

the literature, the present work will specifically focus on other plant protein sources that could be used as replacement for soy proteins. Some of these proteins are commercially available, while others are still in the experimental phase. With the increasing interest of consumers in "healthy" foods, new opportunities may exist for the use of these plant seed proteins in formulating low-calorie, low-cholesterol, and high-density protein items [97].

13.5.2.2.1. Alternative Legume Seeds: Pulses. Legume seeds contain high amounts of protein, starch, and oil. They can be classified into two groups as follows: seeds that are rich in proteins (35% to 40%) and lipids and poor in starch (e.g., soybean, peanut, lupin), and seeds with high amounts of starch and low amounts of lipids (e.g., pea, bean, lentil, chickpea) and are also less rich in protein (20–30%) [103].

As with soybean, most legume seed proteins are storage proteins that accumulate during seed development [104]. The most abundant class of storage

TABLE 13.5 Application of Soy Proteins in Different Foods Based on Functional Properties

Food Applications	Functional Properties	Food Products	References
Meat products	Water and fat absorption, emulsification, gelation, film formation, adhesion, cohesion, flavor binding	Frankfurters, bologna, sausages, loaves, patties, meatballs, restructured meat products, meat analogs	98, 99
Dairy products	Viscosity, foaming, emulsification, aeration, gelation	Cheese, milk, frozen desserts, whipped toppings, yogurt, coffee whiteners	99, 97
Bakery products	Aeration, water-holding capacity, fat absorption, foaming, dough formation, elasticity, adhesion, cohesion, flavor binding	Bread, buns, cakes, rolls, doughnuts, cookies, wafers, crackers, pancakes, pies, muffins, pastas, snack items, breakfast cereals	99–101
Confectionery products	Foaming, emulsifying, aeration, thermoplasticity, crystallization	Candies, caramel, marshmallows, meringue, toffee, fudge	97, 99, 102
Beverages	Solubility, hydration, viscosity, adhesion, cohesion	Instant beverage mixes, soups, gravies	97, 99

proteins in all grain legumes are the globulins generally classified as 7S and 11S globulins according to their sedimentation coefficients (S) [105]. From the nutritional viewpoint, all legume storage proteins contain relatively low quantities of the sulfur-containing amino acids (i.e., methionine, cysteine) and tryptophan, but the amounts of another essential amino acid, lysine, are much greater than in cereal grains [106]. Legume and cereal proteins are, therefore, nutritionally complementary, and essential amino acid deficiencies have traditionally been overcome by integrating legume-based dishes with cereal foods (pasta, rice, bread, etc.) [105].

As with soybean, legume seeds also contain a number of antinutritional compounds such as protease inhibitors, lectins, phytates, tannins, alkaloids, and saponins. The presence and concentration of these antinutritional factors depend on legume type [107]. Heat treatment normally inactivates these antinutritional factors [108]. Once inactivated, antinutritional factors such as protein inhibitors may even play a positive nutritional role, due to their high content of sulfur-containing amino acids relative to the majority of the seed proteins [109]. Moreover, over the last two decades, the presence of certain amounts of some non-nutritional factors, such as phytates, tannins, alkaloids, saponins, and oligosaccharides, have been linked with health-promoting properties, and are at present considered as natural bioactive substances that play an important role in the prevention of cardiovascular diseases and some types of cancer [110], which explains why a reappraisal on their roles and effects is currently taking place [105].

Among the important edible leguminous seeds, pulses like chickpea (*Cicer arietinum*), bean (*Phaseolus vulgaris*), pea (*Pisum sativum*), and lentil (*Lens culinaris*) are increasingly being consumed in many places around the world and are recognized as healthy foods, which could serve as alternatives to soybeans. The nutritional properties of pulses have been investigated extensively [111]. Pulse seeds are high in protein, carbohydrates, and dietary fiber and are a rich source of other nutritional and bioactive components [112, 113]. Sosulski et al. [114] determined the comparative differences of 10 legume proteins as components in imitation milk products. Based on the solubility, physicochemical, and organoleptic properties of the protein isolate and the imitation milk products, the legumes were ranked in the following decreasing order of preference for use as a milk beverage substitute: lima bean = mung bean = pea bean > northern bean = lupin > lentil = soybean > chickpea > field pea > fava bean. For further information on the functional properties of different legumes, the reader is referred to References 115–118. Commercial suppliers of flours and protein concentrates from pulses include Best Cooking Pulses, Parrheim Foods, Nutri-Pea, and Norben Company.

It is important to warn that extensive serological cross-reactivity among legumes has been described [119, 120], although symptomatic hypersensitivity to more than one legume is considered uncommon [58, 121]. Several studies showed that cross-allergenicity among, or multiple allergies to, legumes could be clinically relevant [122, 123]. Dietary habits of people likely influence the food sensitivities found in different geographic areas. Thus, peanut and soybean are the major legumes involved in IgE-mediated hypersensitivity responses in North America, while lentil, chickpea, and pea are the offending legumes in the Mediterranean area and in several Asian countries, like India [124, 125]. These findings emphasize the need for appropriate caution by soy-allergic individuals in the selection of alternative foods.

13.5.2.2.2. Alternative Oilseed Proteins (Rapeseed/Canola). Rapeseed (*Brassica* sp.) is one of the world's major oilseed crops and is the second leading source of protein meal after soybean; it represents 12.4% of world protein meal production [126]. Although China is the world's largest rapeseed producer, Canada is the leader in large-scale production of improved rapeseed varieties, called "canola," which are low in both erucic acid (<2%) and glucosinolate (<30 µmol/g) [127]. After oil extraction, canola meal contains about 38% protein compared with the 44% protein content of soybean meal [128].

Cruciferin and napin (12S globulin and 2S albumin, respectively) are the two major storage canola proteins accounting for 60% and 20%, respectively, of the total protein in mature seeds [129].

Rapeseed/canola meal proteins have a reasonably well-balanced amino acid profile and high biological value [130, 131]. The proteins are rich in lysine and contain adequate quantities of the essential amino acids methionine and cystine, which are the limiting amino acids in most cereal and other oilseed proteins, such as soybean [128]. The protein efficiency ratio of rapeseed/canola meal is 2.64 as compared with 2.19 for soybean [132]. The quality of canola proteins has also been reported to be similar to casein and superior to proteins from other vegetable sources such as pea, and wheat in some applications [133].

The usefulness of rapeseed/canola as a source of food proteins is, however, severely restricted by the presence of undesirable components such as glucosinolates, phytates, and fiber (hull) [134]. Furthermore, commercial oil extraction processes denature the protein, resulting in decreased solubility [135]. These factors have made the use of canola/rapeseed protein impractical in any meaningful way for human food, and hence, the current use of rapeseed meals is generally restricted to animal feed and fertilizer [136].

In the last few decades, many investigations have focused on removing the undesirable components from rapeseed meal [137–141]. Continuous efforts are also being made to improve extraction technologies and to modify canola proteins to obtain desirable functionalities for food applications [130, 135, 142, 143]. As a result, processes have been developed to produce rapeseed/canola protein concentrates and isolates for human consumption, and commercial production is beginning [134, 141, 144–146]. Rapeseed/canola protein products were found to have satisfactory functional properties and could be used in various food formulations [147, 148] (e.g., minced meat products, sausages, canned meat products, etc.). Due to its well-balanced amino acid composition and widespread cultivation globally, rapeseed proteins offer an attractive and promising alternative to soy protein in food formulation.

13.5.2.2.3. Alternative Cereal Grain (Rice). Rice protein could serve as a potentially good alternative to soy proteins. Rice (*Oryza sativa*) plays a prominent role in food consumption worldwide. It is the major food eaten in different parts of Asia, Africa, and South America [149]. Although allergies to rice are reported in the literature, rice is not on the list of the most potent food allergens, and no cross-reactivity with legume allergens has been reported. Rice contains between 6% and 15% protein [150, 151], which can be fractionated to obtain high protein flours, concentrates, and isolates. Several research studies have indicated that the quality of rice proteins is higher than for most cereals [150, 151]. However, the nutritional quality of rice protein is about half that of egg protein [152] and casein [153].

A number of methods have been reported for preparing rice protein isolates [153–158]. The most efficient method of extracting rice endosperm protein that are intended for food application is by using strong alkali conditions similar to those used for the production of other plant protein concentrates and isolates [158]. Production of rice protein isolates with very high purity (~90% protein) is reported [153, 159, 160]. Rice protein flours and concentrates are available commercially and can be used in a variety of products including beverage, meal replacement, and bakery applications. Commercially available rice protein concentrates have purities ranging between 63% and 70%.

13.5.2.2.4. Other Alternatives. Peanuts and tree nuts could be used as alternative sources of protein for soy proteins. Peanut is a legume and is distinctively different from tree nuts (e.g., almond, pecan, hazelnut, macadamia nut, and cashew). Peanuts and tree nuts are, however, often grouped together because of their similar compositional profiles. Peanuts and tree nuts have good amino acid profiles [7, 161], and although they contain a lower amount of protein (8–22%) compared with soybeans (~40%), they provide good sources of mono- and polyunsaturated fatty acids, carbohydrates, vitamins, and minerals. Peanut and tree nut flours and powders are commercially available and could be used in the production of nutritional bars, beverages, bakery products, and other food applications. Unfortunately, peanuts and tree nuts are major allergens and their proteins have been reported to cross-react with soy proteins [162]. Soy-allergic patients, therefore, need to exercise appropriate caution in using peanuts and tree nuts as alternative sources of protein in the diet.

13.6. HIDDEN AND UNINTENTIONAL SOURCES OF SOYBEAN AND SOY PRODUCTS IN FOOD PROCESSING

An allergen is considered hidden when it is unrecognizable or undeclared on the product label [163]. Hidden allergens are one of the key problems linked with the incidence of food allergies and are responsible for the majority of serious allergic reactions to food [164]. Food-allergic individuals are highly dependent on reliable product labeling and on the availability of products not contaminated with allergens [165, 166].

There are many reasons why sensitized individuals may become unknowingly exposed to soy allergens. Soy's versatility and techno-functionality has prompted its wide utilization in a number of applications, which makes it particularly difficult to avoid in food processing. Two particular challenges with soybeans are soybean oil and soy lecithin, and there has been a lot of controversy as to whether these are potential sources of soy proteins.

Soybean oil is used in many applications in the food industry (e.g., in salad dressings, margarine, and baby foods) [167] and was initially thought to be safe for soy-sensitive individuals [168]. Allergenic proteins (28 and 56 kDa) were, however, reported in cold-pressed and deodorized soybean oil [169]. Doke et al. [170] showed that although oxidized soybean oil may not show allergenicity, proteins in soybeans are capable of interacting with oxidized lipid to form products that are allergenic to soybean-sensitive patients. Moneret-Vautrin et al. [171] also reported an unusual soy oil allergy in an allergic infant who reacted to milk, soybeans, and soybean oil emulsion. Thus, the allergenicity of soybean oil depends on its purity, which in turn depends on the extraction process. From the bulk of existing reports, it appears that highly refined soybean oil may not represent a risk of provoking allergic reactions in the overwhelming majority of susceptible people [172], and based on this, U.S. as well as European Union regulators have exempted highly refined vegetable oils from known allergens, such as peanuts and soybeans from their respective new food allergen labeling regulations. Since anecdotal reports of allergic reaction to refined oils still exist [173], the issue continues to deserve further investigation.

The other suspected source of soy allergens is lecithin, which is a by-product from the processing of soybean oil. Lecithin has excellent emulsifying properties and is often used as an ingredient in processed foods, pharmaceutical, cosmetic, and other industrial products. The protein content of lecithin ranges between 100 and 1400 ppm [50] and its potential allergenicity has been an issue of concern. Using EAST, Müller et al. [174] showed that half of the sera of soy-allergic adults used in their study reacted with commercial soy lecithins, which were found to contain residual proteins. Lecithin can be made from egg, egg yolk, soybean, or corn, and on labels of many commercial products containing lecithin, the source of the lecithin is often not clearly stated. Soy lecithin could, therefore, represent an unrecognized source of hidden allergens. Labeling regulations in the European Union, United States, and Canada now require (or will soon be requiring in the case of Canada) that the source of lecithin be stated if it is derived from one of the priority allergens. In Canada and the United States, it is further expected that the new labeling amendments to the Food and Drug Regulations will require the declaration of the sources of lecithin whether or not residual proteins are present. This regulation will also apply to other ingredients such as HVP.

The Canadian Food Inspection Agency further lists the following as possible sources of soy: baby formulas, baked goods and baking mixes (e.g., breads, cookies, cake mixes, doughnuts, pancakes), bean sprouts, beverage mixes (e.g., hot chocolate, lemonade), bread crumbs, cereals, crackers, breaded foods, chili, pastas, stews, taco filling, tamales, canned tuna/minced hams, chewing gum, cooking spray, margarine, vegetable shortening, vegetable oil, diet drinks, imitation milk, dressings, gravies, marinades, frozen desserts, hydrolyzed plant protein (HPP), hydrolyzed soy protein (HSP), HVP, lecithin, monosodium glutamate (MSG) (may contain hydrolyzed protein), processed and prepared meats (e.g., beef, deli, pork, poultry), sauces (e.g., soy, shoyu, tamari, teriyaki, Worcestershire), seafood-based products, fish, seasoning, spices, simulated fish and meat products (e.g., surimi [imitation crab/lobster meat]), simulated bacon bits, snack foods (e.g., candy, chocolate, energy bars, fudge, popcorn, potato chips), soups, broths, soup mixes/stock, spreads, dips, mayonnaise, thickening agents, and vegetarian dishes [175]. Nonfood sources include cosmetics, soaps, craft materials, glycerine, milk substitutes for young animals, pet food, and vitamins.

13.7. SOYBEAN CROSS-REACTING ALLERGENS

When an allergic response is established toward a particular protein, a homologous form of that protein in another substance may also trigger an allergic response (cross-reaction) [176]. Proteins that cross-react with soybean allergens are, therefore, potential sources of allergens for soy-sensitized individuals. Extensive *in vitro* cross-reactivity has been demonstrated among different legumes. As early as 1971, Fries [177], by using a skin test on 30 atopic children with suspected soybean allergy, showed that 90% of patients were sensitive to soybean, 90% to peanut, 80% to green pea, 53% to lima pea, and 43% to string bean. Bernhisel-Broadbent and Sampson [121] in later studies reported that among 69 legume-sensitive patients, 43% showed sensitivity to soybean, 87% to peanut, 22% to green bean, and 26% to pea. Similarly, Eigenmann et al. [162] showed that the removal of soy-binding antibodies, by antigen-affinity chromatography, from the serum of a patient allergic to peanut and soy, resulted in 73% reduction of IgE binding to peanut allergens. Reported clinical coreactivity rates in peanut allergics for soy are low and range only from 1% to 6.5% in placebo-controlled studies [58]. Cross-reactivities between soy and peanut will, however, continue to represent an important clinical challenge, as both share common antigens [178, 179], and both are widely used in food product formulations.

Cross-reactivity between proteins could also occur with materials from different families or species. This type of cross-reaction can be due to structural affinity resulting from homology of linear epitopes and/or similarity of spatial conformation (homology of conformational epitopes). Thus, several authors have reported coexisting clinical soy allergies in patients with cow's milk allergies, ranging from 5% [180] in a prospective study to around 50% in a selected group of patients with cow's milk enterocolitis [181]. Based on a positive RAST/skin test and open or double-blind, placebo-controlled food challenge (DBPCFC), Cantani and Lucenti [182] also reported a 3–4% incidence of soy protein formula allergies in cow's milk-allergic children given soy protein formulas.

13.8. IMPORTANCE OF GOOD MANUFACTURING PRACTICES (GMPs)

13.8.1. Ingredients

Raw materials used in the processing of foods can be a major source of allergen contamination. As a result, it is important that all materials used for

Fig. 13.2 Corn crop in a soybean field.

processing of foods destined for soy-allergic consumers be checked for potential contamination with soy ingredients. This could be a challenge particularly for grains, seeds, and flours as they may be contaminated on the farm (i.e., during production and harvesting) or during transportation and handling. Figure 13.2 shows a picture of a soybean farm in which corn is visibly present. This kind of contamination often happens when crops are rotated on the same farmland and is particularly of concern for soybean and corn crops. Primary processing of grains (cleaning and sorting) can remove contamination of foreign materials, but this is made difficult when the contaminating grain or seed is similar in color and shape to the grain or seed being cleaned. Grain standards vary for different countries, and foreign material ranging from as low as <0.5% to as high as 12% may be permitted in grains destined for the food/feed markets. In the United States as an example, the Food Allergen Labeling and Consumer Protection Act (FALCPA) exempts raw agricultural commodities from allergen labeling. It is, therefore, important that food manufacturers exercise appropriate caution and verify that all incoming materials to be used for production are really allergen free prior to use.

13.8.2. Processing

GMPs need to be put in place for the production of all foods. This is especially important for "allergen-free" foods. Detailed information on GMPs for "allergen-free" food production is provided elsewhere in this book. Deibel et al. [183] reported that inadvertent allergens in foods can result from a number of events. These include misformulation; improper cleanup; cross-contamination by dust or pieces of allergenic food remaining in the processing system; mechanical movement of the allergenic product or raw material that could fall from crossover points from unclosed system or from equipment common to more than one product that was not properly cleaned; product rework or work-in process materials that are added back into the wrong product stream; maintenance tools used on a line producing allergen-containing product and then on a different line; recycling systems; and waste handling. Appropriate allergen management control programs need to, therefore, be implemented when foods intended for allergic consumers are being processed.

13.8.3. Labeling Practices

Confusing or misleading food labeling can pose serious health risks for food-allergic consumers. For the uninformed consumer, the use of technical or uncommon terms for soy and soy derivatives can also be very challenging. Terminologies used for soy that may be less common to the average consumer include the following: edamame, kinako, kouridofu, miso, mono-diglyceride, natto, nimame, okara, soya, soja, vegetable protein, tempeh, soy protein flour or structured protein fiber (SPF), textured soy flour (TSF), textured soy protein (TSP), TVP, HVP, tofu (soybean curds), yuba, and sauces (e.g., shoyu, tamari, teriyaki, Worcestershire) [184]. An investigation on the ability of families of food-allergic children to accurately read food ingredients labels [185] reported that identification of soy was one of the most problematic. Only 22% (6 of 27) of the parents correctly identified soy protein in seven food products. Proper identification of the label was helped by prior instruction by a health professional. There are also some concerns with labeling practices, which arise when manufacturers switch ingredients or modify food composition without making this clear on the label [183, 186]. Additionally, imported food products need to be treated with appropriate precaution as manufacturing and labeling standards may differ from country to country [163].

13.9. CONCLUSION

Soybean has had remarkable growth and gained importance in the last several decades both as food and a source of ingredient for various industrial and health products. They serve as an excellent source of nutrition for humans and are also used extensively in the feed industry. Unfortunately, they are listed as one of the "big eight" allergens, and as such, their labeling is required in many countries. For allergic consumers, avoidance of soybeans may be difficult due to their widespread use. Selection of appropriate ingredients that are free from soybeans and soybean ingredients, the implementation of GMPs, and use of proper labeling are critical to avoid the unintentional presence of soybeans in products targeted for soy-allergic consumers. Extensive studies have been conducted to identify the allergenic proteins in soybean; however, there is limited research on the effect of processing conditions and food matrix on soybean allergenicity. Such studies will be very useful for the food industry, regulatory agencies, and soy-allergic patients. Finally, there is a need for continued research to identify ingredients that could be used as suitable alternatives for soybean and soy ingredients in different food applications.

REFERENCES

1. Freedonia Market Research. (2005). Report on soy products and markets. *Product Code R154-1359*. Cleveland, OH: Freedonia Group.
2. Food and Agricultural Organization of the United Nations. (1995). *Report of FAO Technical Consultation on Food Allergies*, November 13–14, Food and Nutrition Division, Rome, Italy, pp. 1–54.
3. Becker, W., Brasseur, D., Bresson, J.L., Flynn, A., Jackson, A.A., Lagiou, P., Mingrone, G., Moseley, B., Palou, A., Przyrembel, H., Salminen, S., Strobel, S., van Loveren, H. (2004). Opinion of the scientific panel on dietetic products, nutrition and allergies on a request from the commission relating to the evaluation of allergenic foods for labelling purposes (Request nr EFSA-Q-2003-016). *European Food Safety and Authority Journal*, *32*, 1–197.
4. Sicherer, S.H., Sampson, H.A. (2006). Food allergy. *The Journal of Allergy and Clinical Immunology*, *117*, S470–S475.
5. Bindslev-Jensen, C., Briggs, D., Osterballe, M. (2002). Can we determine a threshold level for allergenic foods by statistical analysis of published data in the literature? *Allergy*, *57*, 741–746.
6. Ballmer-Weber, B.K., Holzhauser, I., Scibilia, J., Mittag, D., Zisa, G., Ortolani, C., Oesterballe, M., Poulsen, L.K., Vieths, S., Bindslev-Jensen, C. (2007). Clinical characteristics of soybean allergy in Europe: a double-blind, placebo-controlled food challenge study. *The Journal of Allergy and Clinical Immunology*, *119*, 1489–1496.
7. United States Department of Agriculture (USDA). (2008). Agricultural Research Service Nutrient Database for Standard Reference, Release 21. Available at http://www.nal.usda.gov/fnic/foodcomp/search/
8. Liu, K. (1997). *Soybeans: Chemistry, Technology and Utilization*. New York: Chapman and Hall, pp. 25, 73, 141, 166.
9. Food and Agriculture Organization of the United Nations. (1991). Protein quality evaluation. A Report of Joint FAO/WHO Expert Consultation. *Food and Nutrition Paper 51*. Rome: FAO, p. 39.
10. Alkanhal, M.A. (1995). Nutrient composition of Najdi-camel meat. *Meat Science*, *39*, 71–78.
11. Vaclavik, V.A., Christian, E.W. (2008). *Essentials of Food Science*. New York: Springer Science+Business Media, LLC, p. 83.
12. Wolf, W.J. (1970). *Soybean proteins: their functional*, chemical, and physical properties. *Journal of Agricultural and Food Chemistry*, *18*, 969–976.

13. Kilara, A., Harwalkar, V.R. (1996). Denaturation. In: *Food Proteins Proprieties and Characterization*. Nakai, S., Modler, H.W. (Eds.). New York: VCS Publisher, pp. 71–166.

14. Friedman, M. (1996). Nutritional value of food proteins from different food sources. *Journal of Agricultural and Food Chemistry*, *44*, 6–29.

15. The United States Food and Drug Administration. (1999). Food labeling: health claims; soy protein and coronary heart disease. FDA, Department of Health and Human Services. *Federal Register 64*, 57699–57733. Available at http://www.fda.gov/Food/LabelingNutrition/LabelClaims/HealthClaimsMeetingSignificant-ScientificAgreementSSA/UCM074740.htm.

16. Hori, G., Wang, M.F., Chan, Y.C., Komatsu, T., Wong, Y., Chen, T.H., Yamamoto, K., Nagaoka, S., Yamamoto, S. (2001). Soy protein hydrolyzate with bound phospholipids reduces serum cholesterol levels in hypercholesterolemic adult male volunteers. *Bioscience, Biotechnology, and Biochemistry*, *65*, 72–78.

17. Shin, Z.I., Yu, R., Park, S.A., Chung, D.K., Ahn, C.W., Nam, H.S., Kim, K.S., Lee, H.J. (2001). His-His-Leu, an angiotensin I converting enzyme inhibitory peptide derived from Korean soybean paste, exerts antihypertensive activity *in vivo*. *Journal of Agricultural and Food Chemistry*, *49*, 3004–3009.

18. Gibbs, B.F., Zougman, A., Masse, R., Mulligan, C. (2004). Production and characterization of bioactive peptides from soy hydrolysate and soy-fermented foods. *Food Research International*, *37*, 123–131.

19. Stephenson, T.J., Setchell, K.D., Kendall, C.W., Jenkins, D.J., Anderson, J.W., Fanti, P. (2005). Effect of soy protein-rich diet on renal function in young adults with insulin-dependant diabetes mellitus. *Clinical Nephrology*, *64*, 1–11.

20. Chen, S.T., Peng, S.J., Chen, J.R. (2003). Effect of dietary protein on renal function and lipid metabolism in five-sixths nephrectomized rats. *The British Journal of Nutrition*, *89*, 491–497.

21. Anderson, J.W. (2008). Beneficial effects of soy protein consumption for renal function. *Asia Pacific Journal of Clinical Nutrition*, *17*, 324–328.

22. Simopoulos, A.P. (2002). The importance of the ratio of omega-6/omega-3 essential fatty acids. *Biomedicine and Pharmacotherapy*, *56*, 365–379.

23. Lo, G.S., Goldberg, A.P., Lim, A., Grundhauser, J.J., Anderson, C., Schonfeld, G. (1986). Soy fibre improves lipid and carbohydrate metabolism in primary hyperlipidemic subjects. *Atherosclerosis*, *62*, 239–248.

24. Allison, D.B., Gadbury, G., Schwartz, L.G., Murugesan, R., Kraker, J.L., Heshka, S., Fontaine, K.R., Heymsfield, S.B. (2003). A novel soy-based meal replacement formula for weight loss among obese individuals: a randomized controlled clinical trial. *European Journal of Clinical Nutrition*, *57*, 514–522.

25. Hayakawa, K., Mitzutani, J., Wada, K., Masai, T., Yoshihara, I., Mitsuoka, T. (1990). Effects of soybean oligosaccharides on human faecal microflora. *Microbial Ecology in Health and Disease*, *3*, 293–303.

26. Saito, Y., Takano, T., Rowland, I. (1992). Effects of soybean oligosaccharides on the human gut microflora in *in vitro* culture. *Microbial Ecology in Health and Disease*, *5*, 105–110.

27. Niba, L.L., Niba, S.N. (2003). Role of non-digestible carbohydrates in colon cancer protection. *Nutrition and Food Science*, *33*, 28–33.

28. Rychlik, M., Englert, K., Kapfer, S., Kirchhoff, E. (2007). Folate contents of legumes determined by optimized enzyme treatment and stable isotope dilution assays. *Journal of Food Composition and Analysis*, *20*, 411–419.
29. Davidsson, L., Galan, P., Kastenmayer, P., Cherouvrier, F., Juillerat, M.A., Jercberg, S., Hurrell, R.F. (1994). Iron bioavailability studied in infants: the influence of phytic acid and ascorbic acid in infant formulas based on soy isolate. *Pediatric Research*, *36*, 816–822.
30. Lonnerdal, B., Jayawickrama, L., Lien, E.L. (1999). Effect of reducing the phytate content and of partially hydrolyzing the protein in soy formula on zinc and copper absorption and status in infant rhesus monkeys and rat pups. *The American Journal of Clinical Nutrition*, *69*, 490–496.
31. Kamchan, A., Puwastien, P., Sirichakwal, P.P., Kongkachuichai, R. (2004). *In vitro* calcium bioavailability of vegetables, legumes and seeds. *Journal of Food Composition and Analysis*, *17*, 311–320.
32. Frolich, W. (1995). Bioavailability of micronutrients in a fibre-rich diet, especially related to minerals. *European Journal of Clinical Nutrition*, *49*, 116–122.
33. Messina, M.J. (1999). Legumes and soybeans: overview of their nutritional profiles and health effects. *The American Journal of Clinical Nutrition*, *70*, 439S–450S.
34. Caragay, A.B. (1992). Cancer-preventive foods and ingredients. *Food Technology*, *46*, 65–68.
35. Tsukamoto, C., Shimada, S., Igita, K., Kudou, S., Kokubun, M., Okubo, K., Kitamura, K. (1995). Factors affecting isoflavone content in soybean seeds: changes in isoflavones, saponins, and composition of fatty acids at different temperatures during seed development. *Journal of Agricultural and Food Chemistry*, *43*, 1184–1192.
36. Messina, M., Gardner, C., Barnes, S. (2002). Gaining insight into the health effects of soy but a long way still to go: commentary on the Fourth International Symposium on the role of soy in preventing and treating chronic disease. *The Journal of Nutrition*, *132*, 547S–551S.
37. Gonzalez, R., Polo, F., Zapatero, L., Caravaca, F., Carreira, J. (1992). Purification and characterization of major inhalant allergens from soybean hulls. *Clinical and Experimental Allergy*, *22*, 748–755.
38. Gonzalez, R., Varela, J., Carreira, J., Polo, F. (1995). Soybean hydrophobic protein and soybean hull allergy. *Lancet*, *346*, 48–49.
39. Codina, R., Lockey, R.F., Fernandez-Caldas, E., Rama, R. (1997). Purification and characterization of a soybean hull allergen responsible for the Barcelona asthma outbreaks II. Purification and sequencing of the Gly m 2 allergen. *Clinical and Experimental Allergy*, *27*, 424–430.
40. Rihs, H.P., Chen, Z., Rueff, F., Petersen, A., Rozynek, P., Heimann, H., Baur, X. (1999). IgE binding of the recombinant allergen soybean profilin (r*Gly m* 3) is mediated by conformational epitopes. *The Journal of Allergy and Clinical Immunology*, *104*, 1293–1301.
41. Kleine-Tebbe, J., Vogel, L., Crowell, D.N., Haustein, U.F., Vieths, S. (2002). Severe oral allergy syndrome and anaphylactic reactions caused by a *Bet v* 1-related PR-10 protein in soybean, SAM22. *The Journal of Allergy and Clinical Immunology*, *110*, 797–804.

42. Neudecker, P., Lehmann, K., Rosch, P. (2003). Sequence-specific 1H, 13C and 15N resonance assignments of SAM22, an allergenic stress induced protein from soy bean. *Journal of Biomolecular NMR*, *26*, 191–192.

43. Mittag, D., Vieths, S., Vogel, L., Becker, W.M., Rihs, H.P., Helbling, A., Wuthrich, B., Ballmer-Weber, B.K. (2004). Soybean allergy in patients allergic to birch pollen: clinical investigation and molecular characterization of allergens. *The Journal of Allergy and Clinical Immunology*, *113*, 148–154.

44. Ogawa, T., Bando, N., Tsuji, H., Okajima, K., Nishikawa, K., Sasaoka, K. (1991). Investigation of the IgE binding proteins in soybeans by immunoblotting with the sera of the soybean-sensitive patients with atopic dermatitis. *Journal of Nutritional Science and Vitaminology*, *37*, 555–565.

45. Ogawa, T., Bando, N., Tsuji, H., Nishikawa, K., Kitamura, K. (1993). Identification of the soybean allergenic protein, *Gly m* Bd 30K, with the soybean seed 34-kDa oil-body-associated protein. *Bioscience, Biotechnology, and Biochemistry*, *57*, 1030–1033.

46. Tsuji, H., Bando, N., Hiemori, M., Yamanishi, R., Kimoto, M., Nishikawa, K., Ogawa, T. (1997). Purification and characterization of soybean allergen *Gly m* Bd 28K. *Bioscience, Biotechnology, and Biochemistry*, *61*, 942–947.

47. Ogawa, T., Bando, N., Tsuji, H., Nishikawa, K., Kitamura, K. (1995). α-Subunit of β-conglycinin, an allergenic protein recognised by IgE antibodies of soybean-sensitive patients with atopic dermatitis. *Bioscience, Biotechnology, and Biochemistry*, *59*, 831–833.

48. Zeece, M.G., Beardslee, T.A., Markwell, J.P, Sarath, G. (1999). Identification of an IgE-binding region in soybean acidic glycinin G1. *Food and Agricultural Immunology*, *11*, 83–90.

49. Helm, R.M., Cockrell, G., Connaughton, C., Sampson, H.A., Bannon, G.A., Beilinson, V., Livingstone, D., Nielsen, N.C., Burks, A.W. (2000). A soybean G2 glycinin allergen 1. Identification and characterization. *International Archives of Allergy and Immunology*, *123*, 205–212.

50. Gu, X., Beardslee, T., Zeece, M.G., Sarath, G., Markwell, J. (2001). Identification of IgE-binding proteins in soy lecithin. *International Archives of Allergy and Immunology*, *126*, 218–225.

51. Fu, T.J., Abbott, U.R., Hatzos, C. (2002). Digestibility of food allergens and nonallergenic proteins in simulated gastric fluid and simulated intestinal fluid—a comparative study. *Journal of Agricultural and Food Chemistry*, *50*, 7154–7160.

52. Baur, X., Pau, M., Czuppon, A., Fruhmann, G. (1996). Characterization of soybean allergens causing sensitization of occupationally exposed bakers. *Allergy*, *51*, 326–330.

53. Quirce, S., Fernandez-Nieto, M., Polo, F., Sastre, J. (2002). Soybean trypsin inhibitor is an occupational inhalant allergen. *The Journal of Allergy and Clinical Immunology*, *109*, 178.

54. Codina, R., Ardusso, L., Lockey, R.F., Crisci, C.D., Jaen, C., Bertoya, N.H. (2002). Identification of the soybean hull allergens involved in sensitization to soybean dust in a rural population from Argentina and N-terminal sequence of a major 50 KD allergen. *Clinical and Experimental Allergy*, *32*, 1059–1063.

55. Hessing, M., Bleeker, H., Tsuji, H., Ogawa, T., Vlooswijk, R.A.A. (1996). Comparison of human IgE-binding soya bean allergenic protein *Gly m* 1 with the

antigenicity profiles of calf anti-soya protein sera. *Food and Agricultural Immunology*, *8*, 51–58.

56. L'Hocine, L., Boye, J.I. (2007). Allergenicity of soybean: new developments in identification of allergenic proteins, cross-reactivities and hypoallergenization technologies. *Critical Reviews in Food Science and Nutrition*, *47*, 127–143.
57. Pedersen, H.S., Djurtoft, R. (1989). Antigenic and allergenic properties of acidic and basic peptide chains from glycinin. *Food and Agricultural Immunology*, *1*, 101–109.
58. Burks, A.W. Jr., Brooks, J.R., Sampson, H.A. (1988). Allergenicity of major component proteins of soybean determined by enzyme-linked immunosorbent assay (ELISA) and immunoblotting in children with atopic dermatitis and positive soy challenges. *The Journal of Allergy and Clinical Immunology*, *11*, 1135–1142.
59. Nielson, N.C. (1985). Structure of soy proteins. In: *New Protein Food, V. Seed Storage Proteins*, Altschul, A.M., Wilcke, H.L. (Eds.). Orlando: Academic Press, pp. 27–64.
60. Breiteneder, H., Radauer, C.A. (2004). Classification of plant food allergens. *The Journal of Allergy and Clinical Immunology*, *113*, 821–830.
61. Murphy, P.A., Resurrection, A.P. (1984). Varietal and environmental differences in soybean glycinin and β-conglycinin concentration. *Journal of Agricultural and Food Chemistry*, *32*, 911–915.
62. Mori, T., Utsumi, S., Innaba, H., Kitamura, K., Harada, K. (1981). Differences in subunit composition of glycinin among soybean cultivars. *Journal of Agricultural and Food Chemistry*, *29*, 20–23.
63. Moreira, M.A., Hermodson, M.A., Larkins, B.A., Nielson, N.C. (1979). Partial characterization of the acidic and basic polypeptides of glycinin. *The Journal of Biological Chemistry*, *254*, 9921–9926.
64. Staswick, P.E., Hermodson, M.A., Nielson, N.C. (1981). Identification of the acidic and basic subunit complexes of glycinin. *The Journal of Biological Chemistry*, *256*, 8752–8755.
65. Shattuck-Eidens, D.N., Beachy, R.N. (1985). Degradation of β-conglycinin in early stages of soybean embryogenesis. *Plant Physiology*, *78*, 895–898.
66. Coates, J.B., Medeiros, J.S., Thanh, V.H., Nielson, N.C. (1985). Characterization of the subunits of β-conglycinin. *Archives of Biochemistry and Biophysics*, *243*, 184–189.
67. Morita, S., Fukase, M., Yamaguchi, M., Fukuda, Y., Morita, Y. (1996). Purification, characterization and crystallization of single molecular species of β-conglycinin from soybean seeds. *Bioscience, Biotechnology, and Biochemistry*, *60*, 866–873.
68. Pernollet, J.C., Mosses, J. (1983). Structure and location of legume and cereal seed storage proteins. In: *Seeds Proteins. Annual Proceedings of the Phytochemical Society of Europe*, Daussant, J., Mosse, J., Vaughan, J. (Eds.). London: Academic Press, pp. 155–190.
69. Yaklich, R.W., Helm, R.M., Cockrell, G., Herman, E.M. (1999). Analysis of the distribution of the major soybean seed allergens in a core collection of *Glycine max* accessions. *Crop Science*, *39*, 1444–1447.
70. Helm, R.M., Cockrell, G., Herman, E., Burks, A.W., Sampson, H.A., Bannon, G.A. (1998). Cellular and molecular characterization of a major soybean allergen. *International Archives of Allergy and Immunology*, *117*, 29–37.

71. Bando, N., Tsuji, H., Yamanishi, R., Nio, N., Ogawa, T. (1996). Identification of the glycosylation site of a major soybean allergen, Gly m Bd 30K. *Bioscience, Biotechnology, and Biochemistry*, *60*, 347–348.

72. Hiemori, M., Bando, N., Ogawa, T., Shimada, H., Tsuji, H., Yamanishi, R., Terao, J. (2000). Occurrence of IgE antibody-recognizing N-linked glycan moiety of a soybean allergen, Gly m Bd 28 K. *International Archives of Allergy and Immunology*, *144*, 238–245.

73. Roychaudhuria, R., Sarath, G., Zeece, M.G., Markwell, J. (2004). Stability of the allergenic soybean Kunitz trypsin inhibitor. *Biochimica et Biophysica Acta*, *1699*, 207–212.

74. Moroz, L.A., Yang, W.H. (1980). Kunitz soybean trypsin inhibitor: a specific allergen in food anaphylaxis. *The New England Journal of Medicine*, *302*, 1126–1128.

75. Codina, R., Lockey, R.F., Fernandez-Caldas, E., Rama, R. (1999). Identification of the soybean hull allergens responsible for the Barcelona asthma outbreaks. *International Archives of Allergy and Immunology*, *119*, 69–71.

76. Quirce, S., Polo, F., Figueredo E., Gonzalez, R., Sastre, J. (2000). Occupational asthma caused by soybean flour in bakers: difference with soybean-induced epidemic asthma. *Clinical and Experimental Allergy*, *30*, 839–846.

77. Ramachandran, S., Christensen, H.E., Ishimaru, Y., Dong, C.H., Chao-Ming, W., Cleary, A.L., Chua, N.H. (2000). Profilin plays a role in cell elongation, cell shape maintenance, and flowering in *Arabidopsis*. *Plant Physiology*, *124*, 1637–1647.

78. Xiang, P., Baird, L., Jung, R., Zeece, M.G., Markwell, J., Sarath, G. (2008). P39, a novel soybean protein allergen, belongs to a plant-specific protein family and is present in protein storage vacuoles. *Journal of Agricultural and Food Chemistry*, *56*, 2266–2272.

79. Sharon, N., Lis, H. (1972). Lectins: cell-agglutinating and sugar-specific proteins. *Science*, *177*, 949–959.

80. Shibasaki, M., Sumazaki, R., Isoyama, S., Takita, H. (1992). Interaction of lectins with human IgE: IgE binding property and histamine-releasing activity of twelve plant lectins. *International Archives of Allergy and Immunology*, *98*, 18–25.

81. Shibasaki, M., Suzuki, S., Tajima, S., Nemoto, H., Kuruome, T. (1980). Allergenicity of major component proteins of soybean. *International Archives of Allergy and Immunology*, *61*, 441–448.

82. Wilson, S., Martinez-Villaluenga, C., De Mejia, E.G. (2008). Purification, thermal stability, and antigenicity of the immunodominant soybean allergen P34 in soy cultivars, ingredients, and products. *Journal of Food Science*, *73*, T106–T111.

83. Burks, A.W., Williams, L.W., Helm, R.M, Thresher, W., Brooks, J.R., Sampson, H.A. (1991). Identification of soy protein allergens in patients with atopic dermatitis and positive soy challenges; determination of change in allergenicity after heating or enzyme digestion. *Advances in Experimental Medicine and Biology*, *289*, 295–307.

84. Burks, A.W., Williams, L.W., Thresher, W., Connaughton, C., Cockrell, G., Helm, R.M. (1992). Allergenicity of peanut and soybean extracts altered by chemical or thermal denaturation in patients with atopic dermatitis and positive food challenges. *The Journal of Allergy and Clinical Immunology*, *90*, 889–897.

85. Usui, M., Tamura, H., Nakamura, K., Ogawa, T., Muroshita, M., Azakami, H., Shin, K., Akio, K. (2004). *Enhanced* bactericidal action and masking of allergen structure of soy protein by attachment of chitosan through Maillard-type protein-polysaccharide conjugation. *Nahrung/Food*, *48*, 69–72.

86. van de Lagemaat, J., Silván, J.M., Moreno, F.J., Olano, A., del Castillo, M.D. (2007). *In vitro* glycation and antigenicity of soy proteins. *Food Research International*, *40*, 153–160.

87. Maleki, S.J., Chung, S.Y., Champagne, E.T., Raufman, J.P. (2000). The effects of roasting on the allergenic properties of peanut properties. *The Journal of Allergy and Clinical Immunology*, *106*, 763–768.

88. Franck, P., Moneret-Vautrin, D.A., Dousset, B., Kanny, G., Nabet, P., Guenard-Bilbaut, L., Parisot, L. (2002). The allergenicity of soybean based products is modified by food technologies. *International Archives of Allergy and Immunology*, *128*, 212–219.

89. Tsai, E., Yeung, J., Gold, M., Sussman, G., Perelman, B., Vadas, P. (2003). Study of the allergenicity of plant protein hydrolysates. *Food Allergy and Intolerance*, *4*, 117–126.

90. Maldonado, J., Gil, A., Narbona, E., Molina, J. (1998). Special formula in infant nutrition: a review. *Early Human Development*, *53*, S23–S32.

91. Song, Y.S., Frias, J., Martınez-Villaluenga, C., Vidal-Valverde, C., Gonzalez de Mejia, E. (2008). Immunoreactivity reduction of soybean meal by fermentation, effect on amino acid composition and antigenicity of commercial soy. *Food Chemistry*, *180*, 571–581.

92. Hefle, S.L., Lambrecht, D.M., Nordlee, J.A. (2005). Soy sauce retains allergenicity through the fermentation production process. *The Journal of Allergy and Clinical Immunology*, *115*, S32.

93. Herman, E.M., Helm, R.M., Jung, R., Kinney, A.J. (2003). Genetic modification removes an immunodominant allergen from soybean. *Plant Physiology*, *132*, 36–43.

94. Ogawa, T., Samoto, M., Takahashi, K. (2000). Soybean allergens and hypoallergenic soybean products. *Journal of Nutritional Science and Vitaminology*, *46*, 271–279.

95. Samoto, M., Fukuda, Y., Takahashi, K., Tabuchi, K., Hiemori, M., Tsuji, H., Ogawa, T., Kawamura, Y. (1997). Substantially complete removal of three major allergenic soybean proteins (Gly m Bd 30K, Gly m Bd 28K and the alpha-subunit of conglycinin) from soy protein by using a mutant soybean, Tohoku 124. *Bioscience, Biotechnology, and Biochemistry*, *61*, 2148–2150.

96. Wang, H.L., Swain, E.W., Kwolek, W.F. (1983). Effect of soybean varieties on the yield and quality of tofu. *Cereal Chemistry*, *60*, 245–248.

97. Singh, P., Kumar, R., Sabapathy, S.N., Bawa, A.S. (2008). Functional and edible uses of soy protein products. *Comprehensive Reviews in Food Science and Food Safety*, *7*, 14–28.

98. Schweiger, R.G. (1974). Soy protein concentrates and isolates in comminuted meat systems. *Journal of the American Oil Chemists' Society*, *51*, 192A–194A.

99. Kinsella, J.E. (1979). Functional properties of soy proteins. *Journal of the American Oil Chemists' Society*, *56*, 242–258.

100. Hoover, W.J. (1975). Use of soy proteins in bakery products. *Journal of the American Oil Chemists' Society*, *52*, 267A–269A.
101. Zind, T. (1998). Soy to the world. *Prepared Foods*, *167*, 55–58.
102. Cowan, J.C. (1979). Vegetable proteins in confectionery products. *Journal of the American Oil Chemists' Society*, *56*, 353.
103. Dulau, I., Thebaudin, J.Y. (1998). Functional properties of leguminous protein: applications in food. *Grain Legumes*, *20*, 15–16.
104. Casey, R., Domoney, C., Ellis, N. (1986). Legume storage proteins and their genes. *Oxford Surveys of Plant Molecular and Cell Biology*, *3*, 1–95.
105. Duranti, M. (2006). Grain legume proteins and nutraceutical properties. *Fitoterapia*, *7*, 67–82.
106. Ampe, C., Van Damme, J., de Castro, A., Sampaio, M.J., Van Montagu, M., Vanderkerckhove, J. (1986). The amino acid sequence of the 2S sulphur-rich proteins from seeds of Brazil nut (*Bertholletia excelsa* HBK). *European Journal of Biochemistry*, *159*, 597–604.
107. Sathe, S.K., Salunkhe, D.K. (1984). Technology of removal of unwanted components of dry beans. *Critical Reviews in Food Science and Nutrition*, *21*, 263–287.
108. Vidal-Valverde, C., Frias, J., Estrella, I., Gorospe, N.J., Ruiz, R., Bacon, J. (1994). Effect of processing on some antinutritional factors of lentil. *Journal of Agricultural and Food Chemistry*, *42*, 2291–2295.
109. Ryan, C.A. (1990). Protease inhibitors in plants: genes for improving defenses against insects and pathogens. *Annual Review of Phytopathology*, *28*, 425–449.
110. Muzquiz, M. (2005). Positive impact of pulse consumption on human health. Available at http://www.grainlegumes.com/aep/uses/food_uses/positive_impact_of_pulse_consumption_on_human_health
111. Rochfort, S., Panozzo, J. (2007). Phytochemicals for health, the role of pulses. *Journal of Agricultural and Food Chemistry*, *55*, 7981–7994.
112. Tharanathan, R.N., Mahadevamma, S. (2003). Grain legumes: a boom to human nutrition. *Trends in Food Science and Technology*, *14*, 507–518.
113. Guillon, F., Champ, M.M. (2002). Carbohydrate fractions of legumes: uses in human nutrition and potential for health. *The British Journal of Nutrition*, *88*, S293–S306.
114. Sosulski, F.W., Chakraborty, P., Humbert, E.S. (1978). Legume based imitation and blended milk products. *Canadian Institute of Food Science and Technology Journal*, *11*, 117–123.
115. Paredes-Lopez, O., Ordorica-Falomir, C., Olivares-Vazquez, M.R. (1991). Chickpea protein isolates: physicochemical, functional and nutritional characterization. *Journal of Food Science*, *56*, 726–729.
116. Fernandez-Quintela, A., Macarulla, M.T., Del Barrio, A.S., Martinez, J.A. (1997). Composition and functional properties of protein isolates obtained from commercial legumes grown in northern Spain. *Plant Foods for Human Nutrition*, *51*, 331–342.
117. Clemente, A., Vioque, J., Sanchez-Vioque, R., Pedroche, J., Bautista, J., Millan, F. (1999). Protein quality of chickpea (*Cicer arietinum* L.) protein hydrolysates. *Food Chemistry*, *67*, 269–274.

118. Boye, J.I., Aksay, S., Roufik, S., Ribérea, S., Mondor, M., Farnworth, E., Rajamohamed, S.H. (2010). Comparison of the functional properties of pea, chickpea and lentil protein concentrates processed using ultrafiltration and isoelectric precipitation techniques. *Food Research International*, *43*, 537–546.

119. Barnett, D., Bonham, B., Howden, M.E. (1987). Allergic cross-reactions among legume foods: an *in vitro* study. *The Journal of Allergy and Clinical Immunology*, *79*, 433–438.

120. Szymkiewicz, A., Jedrychowski, L. (1998). Evaluation of immunoreactivity of selected legume seed proteins: short report. *Polish Journal of Food and Nutrition Sciences*, *7*, 539–544.

121. Bernhisel-Broadbent, J., Sampson, H.A. (1989). Cross-allergenicity in the legume botanical family in children with food hypersensitivity. *The Journal of Allergy and Clinical Immunology*, *83*, 435–440.

122. Kalogeromitros, D., Armenaka, M., Galatas, I., Capellou, O., Katsarou, A. (1996). Anaphylaxis induced by lentils. *Annals of Allergy, Asthma & Immunology*, *77*, 480–482.

123. Wensing, M., Knulst, A.C., Piersma, S., O'Kane, F., Knol, E.F., Koppelman, S.J. (2003). Patients with anaphylaxis to pea can have peanut allergy caused by cross-reactive IgE to vicilin (Ara h 1). *The Journal of Allergy and Clinical Immunology*, *111*, 420–424.

124. Pascual, C.Y., Fernandez-Crespo, J., Sanchez-Pastor, S., Padial, A., Diaz-Pena, J.M., Martin-Munoz, F., Martin-Esteban, M. (1999). Allergy to lentils in Mediterranean pediatric patients. *The Journal of Allergy and Clinical Immunology*, *103*, 154–158.

125. Patil, S.P., Niphadkar, P.V., Bapat, M.M. (2001). Chickpea: a major food allergen in the Indian subcontinent and its clinical and immunochemical correlation. *Annals of Allergy, Asthma & Immunology*, *87*, 140–150.

126. Ash, M., Dohlman, E., Simms, D. (2007). Oil crops situation and outlook yearbook. The United States Department of Agriculture Market and Trade Economic Divisions, Economic Research Service, pp. 1–26. Available at http://usda.mannlib.cornell.edu/usda/ers/OCS//2000s/2007/OCS-05-14-2007.pdf

127. Moure, A., Sineiro, J., Domınguez, H., Parajo, J.C. (2006). Functionality of oilseed protein products: a review. *Food Research International*, *39*, 945–963.

128. Tzeng, Y.M., Diosady, L.L., Rubin, L.J. (2006). Preparation of rapeseed protein isolate by sodium hexametaphosphate extraction, ultrafiltration, diafiltration, ion-exchange. *Journal of Food Science*, *53*, 1537–1541.

129. Höglund, A.S., Rödin, J., Larsson, E., Rask, L. (1992). Distribution of napin and cruciferin in developing rapeseed embryo. *Plant Physiology*, *98*, 509–515.

130. Ohlson, R., Anjou, K. (1979). Rapeseed protein products. *Journal of the American Oil Chemists' Society*, *56*, 431–437.

131. Bos, C., Airinei, G., Mariotti, F., Benamouzing, R., Berot, S., Evrard, J., Fenart, E., Tome, D., Gaudichon, C. (2007). The poor digestibility of rapeseed protein is balanced by its very high metabolic utilization in humans. *The Journal of Nutrition*, *137*, 594–600.

132. Delisle, J., Amiot, J., Goulet, G., Simard, C., Brisson, G.J., Jones, J.D. (1984). Nutritive value of protein fractions extracted from soybean, rapeseed, and wheat flours in the rat. *Plant Foods for Human Nutrition*, *34*, 243–251.

133. Sarwar, G.R., Blair, R., Friedman, M., Gumbmann, M.R., Hackler, L.R., Pellett, P.L., Smith, T.K. (1984). Inter and intra-laboratory variability in rat growth assays for estimating protein quality of foods. *Journal of Association of Official Analytical Chemists*, *67*, 976–981.

134. Ghodsvali, A., Haddad, K.M.H., Vosoughi, M., Diosady, L.L. (2005). Preparation of canola protein materials using membrane technology and evaluation of meals functional properties. *Food Research International*, *38*, 223–231.

135. Vioque, J., Sánchez-Vioque, R., Clemente, A., Pedroche, J., Millan, F. (2000). Partially hydrolyzed rapeseed protein isolates with improved functional properties. *Journal of the American Oil Chemists' Society*, *77*, 447–450.

136. Cumby, N., Zhong, Y., Naczk, M., Shahidi, F. (2008). Antioxidant activity and water-holding capacity of canola protein hydrolysates. *Food Chemistry*, *109*, 144–148.

137. Diosday, L.L., Rubin, L.J., Philips, C.R., Naczk, M. (1985). Effect of alcohol-ammonia-water treatment on the glucosinolate content of rapeseed meal. *Canadian Institute of Food Science and Technology Journal*, *18*, 311–315.

138. Dunford, N.T., Temelli, F. (1996). Effect of supercritical CO_2 on the pyrosinase activity and glucosinolate degradation in canola. *Journal of Agricultural and Food Chemistry*, *44*, 2372–2376.

139. Shahidi, F., Gabon, J.E., Rubin, L.J., Naczk, M. (1990). Effect of methanol–ammonia–water treatment on the fate of glucosinolates. *Journal of Agricultural and Food Chemistry*, *38*, 251–255.

140. Shahidi, F., Naczk, M., Rubin, L.J., Diosady, L.L. (1988). A novel processing approach for rapeseed and mustard seed: removal of undesirable constituents by methanol–ammonia. *Journal of Food Protection*, *51*, 743–749.

141. Tzeng, Y.M., Diosady, L.L., Rubin, L.J. (1990). Production of canola protein materials by alkaline extraction, precipitation, and membrane processing. *Journal of Food Science*, *55*, 1147–1151.

142. Lönnerdal, B., Gillberg, L., Törnell B. (1977). Preparation of rapeseed protein isolates: a study of rapeseed protein isolates by molecular sieve chromatography. *Journal of Food Science*, *42*, 75–79.

143. Ismond, M.A.H., Welsh, W.D. (1992). Application of new methodology to canola protein isolation. *Food Chemistry*, *45*, 125–127.

144. Ohlson, R. (1985). Rapeseed. In: *New Protein Foods V. Seed Storage Proteins*, Altschul, A.M., Wilcke, H.L. (Eds.). Orlando, FL: Academic Press, pp. 339–391.

145. Barker, L.D., Green, B.E., Lei, X. (2003). Continuous process for production of oilseed protein isolate. U.S. Patent Application No. 20090076252. Available at http://www.freshpatents.com/-dt20090319ptan20090076252.php

146. Newkirk, R.W., Maenz, D.D., Classen, H.L. (2003). Oilseed processing for protein recovery and phytin removal. International Patent WO 03043438.

147. Igor, S.O., Diosady, L.L., Rubin, L.J. (1993). Catalytic deamidation of canola proteins. *Acta Alimentaria*, *22*, 325–336.

148. Xu, L., Diosady, L.L. (1994). The production of Chinese rapeseed protein isolates by membrane processing. *Journal of the American Oil Chemists' Society*, *71*, 935–939.

149. Athwal, D.L. (1971). Semidwarf rice and wheat in global food needs. *Quarterly Reviews of Biology*, *46*, 1–34.

150. Hegsted, D.M., Juliano, B.O. (1978). Difficulties in assessing the nutritional quality of rice protein. *The Journal of Nutrition*, *104*, 772–781.

151. Juliano, B.O. (1993). *Rice in Human Nutrition*. Rome: Food and Agriculture Organization of the United Nations. Available at http://www.fao.org/inpho/content/documents//vlibrary/t0567e/T0567E00.htm#Contents

152. Murata, K., Nishikaze, M., Tanaka, M. (1977). Nutritional quality of rice protein compared with whole egg protein. *Journal of Nutritional Science and Vitaminology*, *23*, 125–131.

153. Morita, T., Kiriyama, S. (1993). Mass production method for rice protein isolate and nutritional evaluation. *Journal of Food Science*, *58*, 1393–1396, 1406.

154. Shih, F.F., Daigle, K. (1997). Use of enzymes for the separation of protein from rice flour. *Cereal Chemistry*, *74*, 437–441.

155. Shih, F.F., Champagne, E.T., Daigle, K., Zarins, Z. (1999). Use of enzymes in the processing of protein products from rice bran and rice flour. *Nahrung/Food*, *43*, 14–18.

156. Anderson, A., Hettiarachchy, N.S., Ju, Z.Y. (2000). Physicochemical properties of pronase treated rice glutelin. *Journal of the American Oil Chemists' Society*, *78*, 1–6.

157. Paraman, I., Hettiarachchy, N.S., Schaefer, C., Beck, M.I. (2006). Physicochemical properties of rice endosperm proteins extracted by chemical and enzymatic methods. *Cereal Chemistry*, *83*, 663–667.

158. Paraman, I., Hettiarachchy, N.S., Schaefer, C. (2008). Preparation of rice endosperm protein isolate by alkali extraction. *Cereal Chemistry*, *85*, 67–81.

159. Wang, M., Hettiarachchy, N.S., Qi, M., Burks, A.W., Siebenmorgen, T. (1999). Preparation and functional properties of rice bran protein isolate. *Journal of Agricultural and Food Chemistry*, *47*, 411–416.

160. Ju, Z.Y., Hettiarachchy, N.S., Rath, N. (2001). Extraction, denaturation and hydrophobic properties of rice flour proteins. *Journal of Food Science*, *66*, 229–232.

161. Venkatachalam, M., Sathe, S.K. (2006). Chemical composition of selected nut seeds. *Journal of Agricultural and Food Chemistry*, *54*, 4705–4714.

162. Eigenmann, P.A, Burks, A.W., Bannon, G.A., Sampson, H.A. (1996). Identification of unique peanut and soy allergens in sera adsorbed with cross-reacting antibodies. *The Journal of Allergy and Clinical Immunology*, *98*, 969–978.

163. Anibarro, B., Seoane, F.J., Mugica, M.V. (2007). Involvement of hidden allergens in food allergic reactions. *Journal of Investigational Allergology and Clinical Immunology*, *17*, 168–172.

164. Bock, S.A., Muñoz-Furlong, A., Sampson, H.A. (2001). Fatalities due to anaphylactic reactions to foods. *The Journal of Allergy and Clinical Immunology*, *107*, 191–193.

165. Bindslev-Jensen, C., Poulsen, L.K. (1997). Hazards of unintentional/intentional introduction of allergens into foods. *Allergy*, *52*, 1184–1186.

166. Taylor, S.L., Hefle, S.L. (2001). Ingredient and labeling issues associated with allergenic foods. *Allergy*, *56*, 64–69.
167. Kinney, A.J. (1996). Development of genetically engineered soybean oils for food application. *Journal of Food Lipids*, *3*, 273–292.
168. Bush, R.K., Taylor, S.L., Nordlee, J.A., Busse, W.W. (1985). Soybean oil is not allergenic to soybean-sensitive individuals. *The Journal of Allergy and Clinical Immunology*, *76*, 242–245.
169. Montagnon, I., Errahali, Y., Metche, M., Nicolas, J.P., Fremont, S. (2003). Allergenicity of soya oils. *Food Allergy and Intolerance*, *4*, 153–163.
170. Doke, S., Nakamura, R., Torri, S. (1989). Allergenicity of food proteins interacted with oxidized lipids in soybean-sensitive individuals. *Agricultural and Biological Chemistry*, *53*, 231–235.
171. Moneret-Vautrin, D.A., Morisset, M., Flabbee, J., Kanny, G., Kirch, F., Parisot, L. (2002). Unusual soy oil allergy. *Allergy*, *57*, 266–267.
172. Crevel, R.W., Kerkhoff, M.A., Koning, M.M. (2000). Allergenicity of refined vegetable oils. *Food and Chemical Toxicology*, *38*, 385–393.
173. Teuber, S.S., Brown, R.L, Happanen, L.A.D. (1997). Allergenicity of gourmet nut oils processed by different methods. *The Journal of Allergy and Clinical Immunology*, *99*, 502–507.
174. Müller, U., Weber, W., Hoffmann, A., Franke, S., Lange, R., Vieths, S. (1998). Commercial soybean lecithins: a source of hidden allergens? *Zeitschrift-fur Lebensmittel-Untersuchung-und-Forschung-A*, *207*, 341–351.
175. Canadian Food Inspection Agency. Soy—one of the nine most common food allergens. Available at http://www.inspection.gc.ca/english/fssa/labeti/allerg/soye.shtml
176. Sicherer, S.H. (2003). Hidden and cross-reacting food allergens. In: *Food Allergy: Adverse Reactions to Foods and Food Additives*, 3rd edition, Metcalfe, D.D., Sampson, H.A., Simon, R.A. (Eds.). Oxford: Blackwell Publishing, pp. 461–474.
177. Fries, J.H. (1971). Studies on the allergenicity of soybean. *Annals of Allergy*, *29*, 1–7.
178. Beardslee, T.A., Zeece, M.G., Sarath, G., Markwell, J.P. (2000). Soybean glycinin G1 acidic chain shares IgE epitopes with peanut allergen *Ara h* 3. *International Archives of Allergy and Immunology*, *123*, 299–307.
179. Xiang, P., Beardslee, T.A., Zeece, M.G., Markwell, J., Sarath, G. (2002). Identification and analysis of a conserved immunoglobulin E-binding epitope in soybean G1a and G2a and peanut *Ara h* 3 glycinins. *Archives of Biochemistry and Biophysics*, *408*, 51–57.
180. Host, A., Halken, S. (1990). A prospective study of cow milk allergy in Danish infants during the first 3 years of life: clinical course in relation to clinical and immunological type of hypersensitivity reaction. *Allergy*, *45*, 587–596.
181. Burks, A.W., Casteel, H.B., Fiedorek, S.C., Williams, L.W., Pumphrey, C.L. (1994). Prospective oral food challenge study of two soybean protein isolates in patients with possible milk or soy protein enterocolitis. *Pediatric Allergy and Immunology*, *5*, 40–45.

182. Cantani, A., Lucenti, P. (1997). Natural history of soy allergy and/or intolerance in children, and clinical use of soy-protein formulas. *Pediatric Allergy and Immunology*, *8*, 59–74.

183. Deibel, K., Trautman, T., DeBoom, T., Sveum, W.H., Dunaif, G., Scott, V.N., Bernard, D.T. (1997). A comprehensive approach to reducing the risk of allergens in foods. *Journal of Food Protection*, *60*, 436–441.

184. Puglisi, G., Frieri, M. (2007). Update on hidden food allergens and food labeling. *Allergy and Asthma Proceedings*, *28*, 634–639.

185. Joshi, P., Mofidi, S., Sicherer, S.H. (2002). Interpretation of commercial food ingredient labels by parents of food-allergic children. *The Journal of Allergy and Clinical Immunology*, *109*, 920–922.

186. Hugget, A.C., Hischenhuber, C. (1998). Food manufacturing initiatives to protect the allergic customer. *Allergy*, *53*, 89–92.

14

MANUFACTURING A BISCUIT THAT DOES NOT USE MILK, EGGS, OR SOYBEANS

MASAHIRO SHOJI AND TAKAHIDE OBATA

14.1. INTRODUCTION

Recently, food allergy has attracted greater public attention than ever before. Individuals mainly affected by food allergy are young children. Sampson reported that a significant percentage of children (i.e., 6%) is affected by food allergy in the United States, with the most commonly reported allergenic foods for children being milk and eggs, for which the prevalence of allergy is estimated at 2.5% and 1.3%, respectively [1]. Other studies also reported milk and eggs as very common and major food allergens for children across many countries [2, 3].

In the food industry, milk, eggs, and their derivatives are among the most popular ingredients used in food preparations in terms of nutrient supply and processing aids; their inclusion in, for example, bakery, ice cream, soup, fish cake products, however, is not always obvious [4].

Milk and eggs are preferentially used in diet formulations for growing children. As a result, children with allergies to milk and eggs are normally unable to eat many commercially available foods for children. Unfortunately, for these children, the primary demand for food they consume is safety: taste is much less of a priority and is sometimes even overlooked. As a matter of fact, foods given to children with milk and eggs allergies are seldom appetizing

Allergen Management in the Food Industry, Edited by Joyce I. Boye and Samuel Benrejeb Godefroy

Fig. 14.1 Amaranth biscuit for allergic children (launched in 1993).

because milk and eggs contribute much of the taste. Sadly, allergic children sometimes do not enjoy the delight of eating.

Morinaga & Co. Ltd., one of the leading confectionery companies in Japan, manufactures a lot of products for children and recognizes that confectionery plays a significant role as a source of delight and nutrients for children. Additionally, confectioneries serve as a social tool for establishing good relations among children by providing a means for them to eat together.

Accordingly, Morinaga, in pursuing its company vision (i.e., "We offer good health with delight and taste"), has launched a hard confectionery (Marie-type biscuit) named "amaranth biscuit" that is specially prepared for allergic children (Fig. 14.1). The product concept underlying the amaranth biscuit is to provide a safe, tasty, and nourishing confectionery for allergic children.

First, the amaranth biscuit does not use milk, eggs, or soybeans and can therefore be eaten by children with these allergies. Second, the amaranth biscuit is intended to mimic the good taste of the Morinaga "manna" biscuit, a popular Marie-type baby biscuit manufactured since 1930. Third, because children with milk and eggs allergies are often at a disadvantage in terms of protein intake, by formulating with nutritious amaranth, the amaranth biscuit provides protein in a quality and quantity equal to that of the manna biscuit (Table 14.1) [5]. In addition, the amaranth biscuit is likely to provide an opportunity for allergic and nonallergic children to eat the same food together. In this chapter, we would like to share our experience in manufacturing the

TABLE 14.1 Nutritional Facts of the Amaranth Biscuit and Manna Biscuit [5]

Product name	Amaranth Biscuit	Manna Biscuit
Major ingredients	Wheat flour Amaranth powder	Wheat flour Skim milk Milk Eggs
Energy/100 g	427 kcal	432 kcal
Protein/100 g	7.1 g	6.5 g
Amino acid score	28	30
Limiting amino acid	Lysine	Lysine

amaranth biscuit in terms of the management of allergenic ingredients such as milk, eggs, and soybeans.

14.2. ALLERGEN MANAGEMENT INITIATIVE

Generally speaking, the initiative for allergen management in the food industry comes from the need to comply with allergen labeling regulations. Currently, the only effective way to manage food allergy is the complete avoidance of the offending food allergen by encouraging affected individuals to select safe food by means of product labeling. Therefore, regulators at the international and national levels have introduced food labeling policies, which include a mandatory declaration of designated allergenic foods. The intentional incorporation of food requiring mandatory labeling into food products is managed by a positive declaration on the label, so that affected individuals can recognize the presence of an offending food and avoid the concerned product.

If the product contains a food requiring mandatory labeling that is not declared on the product label, the manufacturer is not in compliance with the regulation and will be liable for any incident that results. Accordingly, food companies must take serious action toward allergen management.

The obligatory requirements of allergen management to maintain compliance with regulations are

- to label food requiring mandatory labeling appropriately and
- to prevent cross-contamination of food requiring mandatory labeling with allergenic foods that are not labeled.

The initiative of allergen management for the amaranth biscuit was, nevertheless, distinct because there was no existing allergen labeling regulation in 1993 when the amaranth biscuit was launched. Morinaga was, however, well aware that any failure of allergen management in the amaranth biscuit production would have posed a critical health hazard for allergic individuals; so, strict as well as practical allergen management was needed and of critical importance.

Evidently, the best allergen management is very simple and clear: no allergenic food should be brought into contact with the manufacturing process. This would, however, involve excluding allergenic food from product ingredients and manufacturing in a dedicated production line, requiring heavy investment. The exclusion of allergenic ingredients is workable. However, because the amaranth biscuit is a product for allergic children, who represent only a small market size, it is more practical to consider small batch-scale production in a shared line, with the establishment of an allergen management plan, which considers all processes from product development to product manufacturing. Thus, the management of the amaranth biscuit has become very similar to the principles of management using the Hazard Analysis Critical Control Point (HACCP) system.

14.3. ALLERGEN MANAGEMENT OF THE AMARANTH BISCUIT

14.3.1. Flow of Product Development

Figure 14.2 presents a brief flow diagram for the development of a new product, starting from the formulation of the new product under the designed product concept, the selection of an appropriate production line, and the construction of a practical allergen management plan in order to decide appropriate labeling of the new product, which is very important to help the consumer make the right purchasing decisions, as well as to maintain compliance with regulations and prevent product liability.

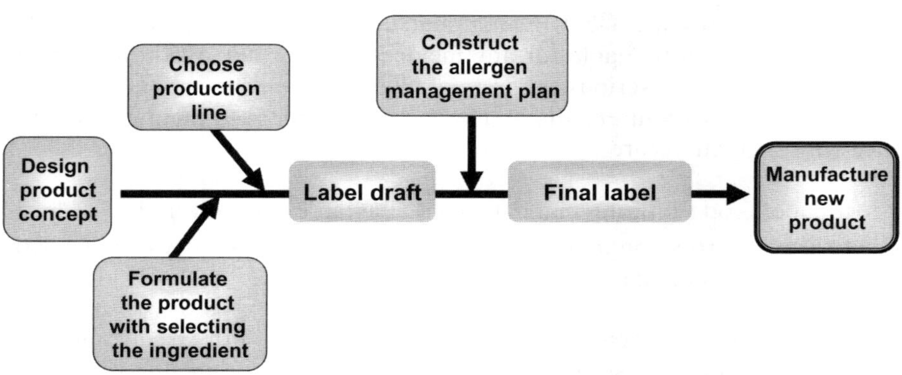

Fig. 14.2 Flow of new product development.

14.3.2. Design Product Concept

The amaranth biscuit was conceived to be a "safe (in terms of food allergy), nourishing, and tasty biscuit" for milk-, egg-, or soybean-allergic children as the biscuit does not use milk, eggs, or soybeans. The concept was founded on food allergen surveys of Japanese allergic children, which indicated two major food allergens (milk and eggs) and one minor allergen (soybean) [6–8]. Accordingly, milk, eggs, and soybeans were eliminated from the ingredients, while wheat, which was not recognized as a significant food allergen, was used. However, a more extensive survey in 2004 unveiled wheat as one of the major food allergens for Japanese children [9]. The manna biscuit, which is popular among Japanese children, was therefore used as a benchmark for the final product.

14.3.3. Formulation and Ingredient Selection

14.3.3.1. Realization of the Product Concept

14.3.3.1.1. A Biscuit That Does Not Use Milk, Eggs, or Soybeans. The ingredients of the amaranth biscuit, from the major ones used in large quantity (e.g., wheat flour, sugar, amaranth powder, and shortening) to minor ones (e.g., baking powder, flavors, and vitamins), were surveyed for their potential food allergen risk.

Wheat Flour. Special care must be taken to avoid contamination with buckwheat, as buckwheat is a significant life-threatening food allergen. In Japan, it is occasionally milled in wheat milling plants; a facility dedicated to wheat is preferred.

Amaranth Powder. Because amaranth cereal is imported from the United States, its potential for soybean contamination during the transportation from the United States was investigated. However, no possibility of the contamination through the warehouse, silo, or transporting vessel was found because amaranth is imported via a route distinct from that of soybean. In addition, it was confirmed that imported amaranth cereal is milled in a plant that does not process buckwheat.

Shortening. Generally, shortening used is made from various oils such as soybean, rapeseed, and/or palm, so the potential contamination of shortening with soybean oil was investigated. Survey of a standard plant revealed potential risk of contamination with soybean oil during the transfer of oil through pipes, storage tanks, and tank lorries. Accordingly, a decision was made to purchase shortening from a factory processing only rapeseed oil and palm oil. In addition, the shortening is supplied in 18-L cans to prevent contamination during transportation. Although highly refined oil is reported to have no allergenicity [10], based on our experiences, potential contamination during storage and transfer was important to keep under consideration.

Minor Ingredients. Food allergens occasionally end up in foods through the use of food additives, for example, flavors, colorants, and processing aids. As milk flavor is commonly used in bakery products to enhance taste, for the amaranth biscuit, a decision was made to use milk flavor not derived from milk. Also, attention was paid to source ingredients that do not contain soy lecithin, a popular food additive. Table 14.2 provides a list of such food ingredients requiring caution. Ingredients, for example, casein/sodium caseinate, lactose, and starch/glucose syrup from wheat repeatedly cause product recall because these ingredients, from which the origins are hardly recognizable, are

TABLE 14.2 Potentially Hazardous Food Ingredients

Origin	Function	Ingredient
Milk	Protein supply	Whey (protein)
	Stabilizer, emulsifier	Sodium caseinate
	Flavor	Casein hydrolysate
	Flavor	Whey hydrolysate
	Fining agent	Casein
	Bulking agent	Lactose
	Bacteriostatic agent	Lactoferrin
Egg	Emulsifier	Lecithin
	Enzyme	Lysozyme
	Fining agent	Egg white
Wheat	Formulation aid	Starch
	Sweetener	Glucose syrup (wheat based)
	Carbohydrate supply	Maltodextrin (wheat based)
	Dough conditioner	Gluten (wheat)
	Flavor	Malt extract
	Flavor	Wheat hydrolysate
	Enzyme	Carboxypeptidase
	Enzyme	Amylase
Soybean	Protein supply	(Soy) protein isolate
	Emulsifier	Lecithin
	Flavor	Soybean hydrolysate
	Enzyme	Amylase
	Antioxidant	Tocopherol
	Foaming agent	Saponin
	Edible oil	Soybean oil
Peanuts	Edible oil	Peanut oil
Buckwheat	Antioxidant	Rutin
Fish	Carrier for vitamin and flavor, fining agent	Fish gelatin

TABLE 14.3 Nutrient Composition of Amaranth Powder and Wheat Flour [5]

	Amaranth Powder	Wheat Flour
Protein	14.9%	7.7%
Fat	8.3%	1.3%
Ash	3.0%	0.4%
Water	9.5%	12.7%
Amino acid score		
Ile	107	118
Leu	72	96
Lys	90	35
Met/Cys	110	52
Phe/Tyr	89	109
Thr	96	81
Trp	155	115
Val	97	102
His	151	129

overlooked in labeling. During formulation or reformulation of the product, ingredients with complex composition like food additives are scrutinized closely in respect to their origins and components.

14.3.3.1.2. Nourishing Biscuit. Elimination of milk and eggs from a wheat-based biscuit results in reduction of the nutritional value, due to the lower protein and poor lysine content of wheat flour. Thus, the new biscuit is fortified by formulating it with a protein- and lysine-rich amaranth powder (Table 14.3) [5]. In addition, the high calcium content of the amaranth powder helps to fortify lost calcium resulting from the elimination of milk.

14.3.3.1.3. Tasty Biscuit. Amaranth biscuit is formulated to attain good taste with limited ingredients. From the nutritional perspective, a high amaranth formulation is preferable to achieve higher protein content (Table 14.3) [5]; however, the final amaranth content was limited to at most 15%, in order to avoid the drawback of the bitter taste of the amaranth.

14.3.3.1.4. Safe Biscuit in Terms of Food Allergy. Due to the packaging label, which declares "does not use milk, eggs, or soybean," the safety of the consumer, namely food-allergic children, is regarded as the first priority. Applying current knowledge and technology, ingredients are carefully selected in formulation and strictly examined in routine production.

14.3.3.2. Ingredient List. In the formulation, an ingredient list is prepared to clarify the original raw materials (Table 14.4). In practice, each ingredient is examined for its components, and then the components are further examined

TABLE 14.4 Ingredient List of the Amaranth Biscuit

	Components		Original Raw Material					
Ingredient	Name	Country of Origin	Name	Country of Origin	Allergen Remark	Supplier Name	Manufacturer Name	Supplier Rank
Wheat flour	Wheat flour	Japan	Wheat	Canada, Australia	Wheat	A Co.	A Co.	Good
Sugar	Sugar	Japan	Sugar cane	Cuba	None	B Inc.	U Co.	Good
Amaranth powder	Amaranth powder	Japan	Amaranth	United States	None	C Co.	C Co.	Good
Shortening	Rapeseed oil	Japan	Rapeseed	United States	None	D Co.	V Co.	Acceptable
	Palm oil	Malaysia	Palm	Malaysia	None			
Starch	Wheat starch	Japan	Wheat	Canada	Wheat	E Ltd.	E Ltd.	Good
	Potato starch	Japan	Potato	Japan	None			
Molasses	Molasses	Japan	Sugar cane	Cuba	None	F Co.	W Inc.	Acceptable
Salt	Sodium chloride	Japan	Seawater	Japan	None	G Inc.	X Ltd.	Acceptable
Baking powder	Ammonium bicarbonate	Japan	Ammonium bicarbonate	Japan	None	H Co.	H Co.	Good
	Sodium bicarbonate	Japan	Sodium bicarbonate	Japan	None			
	Tartaric acid	Japan	Tartaric acid	Japan	None			
Flavors	Flavor without milk derivatives	Japan	Flavor without milk derivatives	Japan	None	I Ltd.	Y Co.	Good
	Dextrin	Japan	Potato	Japan	None			
	Gelatin	Germany	Pork skin	Germany	Gelatin			
Vitamins	Vitamin B_1	Japan	Vitamin B_1	Japan	None	J Co.	Z Co.	Good
	Vitamin B_2	Japan	Vitamin B_2	Japan	None			

until the original raw materials are clarified. This process is done for all listed ingredients. Lastly, the column "allergen remark" in the list is filled if the original raw material is identified as a known allergenic food. In case of an allergic incident, the "allergen remark" helps to identify the potential food allergen (i.e., other than milk, eggs, and soybeans), since allergic individuals occasionally react to more than one allergenic food.

Some ingredients have a complicated composition and, furthermore, sometimes contain an intentionally undeclared substance covered by intellectual property rights, so food allergens are apt to be missed. Therefore, the supplier is encouraged to provide accurate ingredient information by the supply contract or agreement.

Allergen information of imported ingredients is frequently incomplete because of differences in regulations and awareness of food allergens and allergen management levels of exporting countries. Additionally, there may be inadequate communication owing to mileage (i.e., long distance and many operations required to relay the product from the production site). Thus, imported ingredients need to be surveyed much more carefully than domestic ingredients.

14.3.4. Production Line Survey

14.3.4.1. Selection of Production Line. There is a tremendous difference between allergen management for mass production and that for small-batch production. In the case of mass production, the production can be done on a dedicated line, which can exclude allergen cross-contamination; thus, management becomes comparatively simple. On the other hand, small-batch production, as in the case of the amaranth biscuit, needs to share the production with the existing production line. This requires the management system to exclude food allergen cross-contamination during production and proper education of the employees managing the system. When manufacturing in a shared line, a survey of potential allergic contaminants is essential, and this can be carried out by a thorough investigation of the products shared on the same line. The production line of the amaranth biscuit mainly manufactures the manna biscuit, which uses milk- and egg-containing ingredients, but not soybeans. Therefore, the amaranth biscuit production was presumed to have a potential risk for milk and egg contamination.

14.3.4.2. Manufacturing Process

14.3.4.2.1. Scheduled Production. The production line for the amaranth biscuit is shared with several products; accordingly, the amaranth biscuit production is scheduled to minimize the risk of cross-contamination. Normally, scheduled production means manufacturing in the following order: cleaning up the line, manufacturing the nonallergic product, and changeover to the allergenic product. For a complete cleanup of the production line, dismantling of the production machinery and washing the machinery parts are ideal. However, dismantling and washing the machinery require considerable cost; therefore, it is economically not practical for the amaranth biscuit production, since the production size is too small to cover the costs involved. As the second best option, the amaranth biscuit production is scheduled after the production of the mass product (manna biscuit), which therefore requires allergen management mostly targeting for milk and eggs as the potential contaminants from the manna biscuit production.

14.3.4.2.2. Preliminary Survey of the Production Line. The production line was examined for lead time to eliminate remaining allergic contaminants. Practically, after manufacturing the egg-containing biscuit, the line was cleaned as described in Section 14.3.5.1.4, Cleaning and non-egg dough was processed on the line. The amount of egg in the processed dough was examined by an enzyme-linked immunosorbent assay (ELISA) (Section 14.4.1, Analytical method for milk and eggs). Figure 14.3 shows the sampling points and timing of the processed dough, and the amount of egg observed in the dough at different times is presented in Table 14.5.

No egg contamination of dough in the mixer (samples #1 and #6) demonstrates that cleaning of the mixer was successful. Initial samples taken from the dough (sample #3) showed high egg contamination; however, this decreased to an undetectable level (sample #9) during the run. The initial scrap cut from the dough at the start of the operation (sample #4) retained egg contamination from the cutter; accordingly, it was decided not to return to the laminator, and the scrap was disposed of. Using ELISA swab sampling (Section 14.4.3, Sampling), we found out that the major allergen hot spots were the gauge roll pressing the dough, the reciprocating cutter of the cutting machine, and the conveying canvas used for subsequent transport.

Notably, all baked biscuits (samples #5, #8, and #10), even those from the dough containing 1.6-ppm egg (sample #3), showed no detectable egg. During the baking process, biscuit ingredients are exposed to heating, which alters

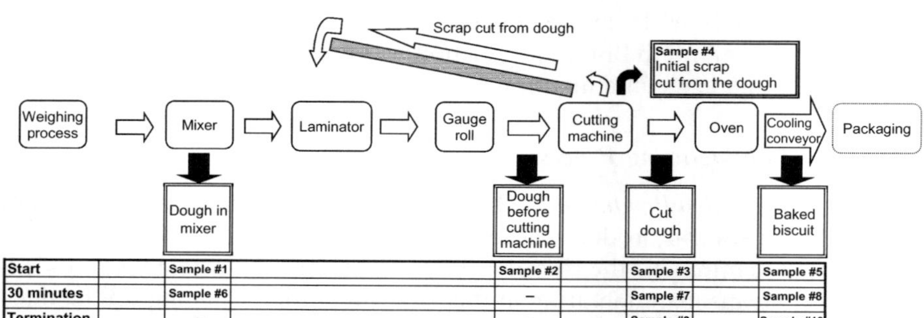

Fig. 14.3 Sampling of dough during the manufacture of the amaranth biscuit.

TABLE 14.5 Egg Concentration in Dough after Changing Dough to Eliminate Egg

	Start		30 Minutes from the Start		Termination	
Dough in mixer	Sample #1	N.D.	Sample #6	N.D.	N.T.	
Dough before cutting machine	Sample #2	0.35 ppm	N.T.		N.T.	
Cutting dough	Sample #3	1.6 ppm	Sample #7	0.55 ppm	Sample #9	N.D.
Initial scrap cutting dough	Sample #4	0.53 ppm				
Baked biscuit	Sample #5	N.D.	Sample #8	N.D.	Sample #10	N.D.

Limit of quantification is 0.1 ppm.
N.D., not detected; N.T., not tested.

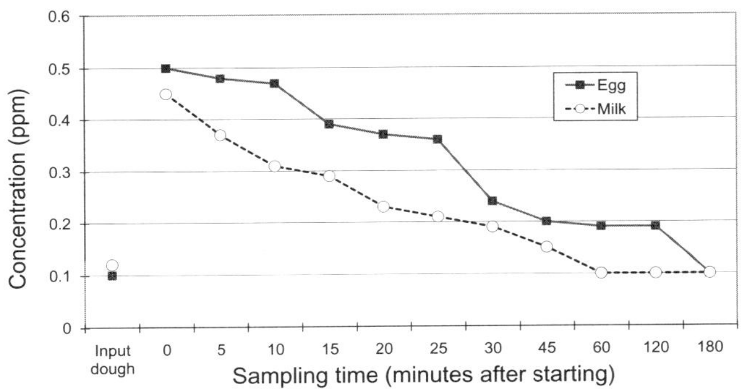

Fig. 14.4 Egg and milk concentration in dough after the start of production.

the intrinsic integrity of existing egg protein molecules by denaturation and aggregation/polymerization in the matrix. This results in the insolubilization of the egg proteins targeted by the ELISA antibody and also the alteration of the molecular structure of the egg proteins, making them unrecognizable by the ELISA antibody. Consequently, the measurement of egg in baked biscuits was unsuccessful (Section 14.4.1, Analytical method for milk and eggs) [11, 12].

In conclusion, the production line for the amaranth biscuit was, therefore, scheduled to be cleaned by the prerun of the amaranth biscuit dough to eliminate contaminants remaining in the machinery. Optimal prerun period of the amaranth production was investigated by mimicking the practical change over from the manna to amaranth biscuit production. As shown in Figure 14.4, milk and egg contamination in amaranth dough gradually decreases with time. From these data, the prerun period was determined.

14.3.4.2.3. Finalized Manufacturing Process. Upon the survey of the production line, the manufacturing process was finalized as follows:

1. After the manna biscuit production, the facility and production line are cleaned (Section 14.3.5.1.4, Cleaning).
2. Amaranth dough is prepared and prerun on the line.
3. Initial cutting dough and initial scrap cutting dough containing milk and egg allergens are disposed safely so as not to mingle with the production process.
4. After the prerun period, the cut dough is sent to the oven.
5. Samples of the cut dough are collected every 10 minutes for product monitoring by ELISA. In the event that milk and/or eggs are detected in the cut dough, the product corresponding to the dough sample is not released for sale (Section 14.3.4.2.4, Product release).
6. Once ingredients are put into the mixer, production should be continuously processed up to the packaging stage in order to minimize cross-contamination during the production. Leaving product in progress, which is susceptible to cross-contamination, is avoided. In case of a breakdown of machinery, any unfinished product is stored in dedicated, lidded, and labeled containers.
7. The packaged amaranth biscuit is labeled with the production time in order to link it to the cut dough sample analyzed.
8. None of the rework material is used, but is disposed. This decision was made because economically, it was not worth the burdensome rework management as well as the product liability risk of misuse.

14.3.4.2.4. Product Release. The final quality control of the amaranth biscuit is verified by checking production documents, such as the input ingredient certificate and production records, as well as by ELISA testing. Although ELISA testing requires significant extra time and labor, it is done because the amaranth biscuit has a potential contamination risk of milk and egg.

If a food-allergic individual should suffer from an imperfect product in terms of milk, egg, or soybean contamination, it absolutely results in the liability of the manufacturer, which causes severe damages, such as recall expense, indemnity to the sufferer, and loss of brand prestige. Thus, it is only after the verification of both product documents and ELISA result that the product can be released for sale. All of the quality control records are kept for future tracing in case of an allergen contamination incident.

14.3.4.3. Validation of the Manufacturing Process. The product from the manufacturing process developed was subsequently validated by a trial production. The allergenic potential of the trial product was performed by passive cutaneous anaphylaxis using guinea pig anti-milk and egg sera [13], as well as by Western blotting using the sera of patients allergic to milk and egg [14],

TABLE 14.6 Label of the Amaranth Biscuit

Wheat flour, Sugar, Amaranth powder, Shortening (Rapeseed oil, Palm oil), Starch (Wheat, Potato), Molasses, Salt, Calcium carbonate, Baking powder (Ammonium bicarbonate, Sodium bicarbonate, Tartaric acid), Flavors, Vitamins

both of which demonstrated negative reactions for milk and eggs. Consequently, the product label was finalized based on the ingredients and product line information (Table 14.6). Since the product label is determined by the results of the above procedures, the label needs to be reviewed after any change in ingredients, formulation, or production process.

14.3.5. Allergen Management Plan

14.3.5.1. Planning for Allergen Management. Allergen management is planned to maintain manufacturing at a level equal to that originally developed. Similar to HACCP [15], the manufacturing process is drawn as a process flow diagram (Fig. 14.5), and an allergen management plan (Table 14.7) is made by reexamining the potential for milk or egg cross-contamination by taking the following into consideration.

14.3.5.1.1. Employee. Amaranth biscuit is manufactured on a shared production line, so prevention of cross-contamination is highly dependent on the employees. Consequently, education of all employees, including temporary staff and contractors involved in production, is essential. Education ranges from fundamental sanitation rules, for example, regular hand washing, and wearing of uniform, cap, gloves, and shoes, to general food allergy knowledge, for example, what is food allergy, how it happens, allergic symptoms, and relation to routine manufacturing (which ingredients contain milk and/or eggs, hot spots for cross-contamination in the production line, etc.).

Regular training sessions are required in order to enlighten employees about food allergy and associated hazards. Effective teaching materials could include

- an article describing how minute amounts of peanuts can threaten the life of allergic individuals,
- pictures of symptoms of allergic children, and
- interview and chats with food-allergic children and their parents.

The objective is to make employees understand the importance of what they must achieve in the production process, by giving them vivid information on food allergy.

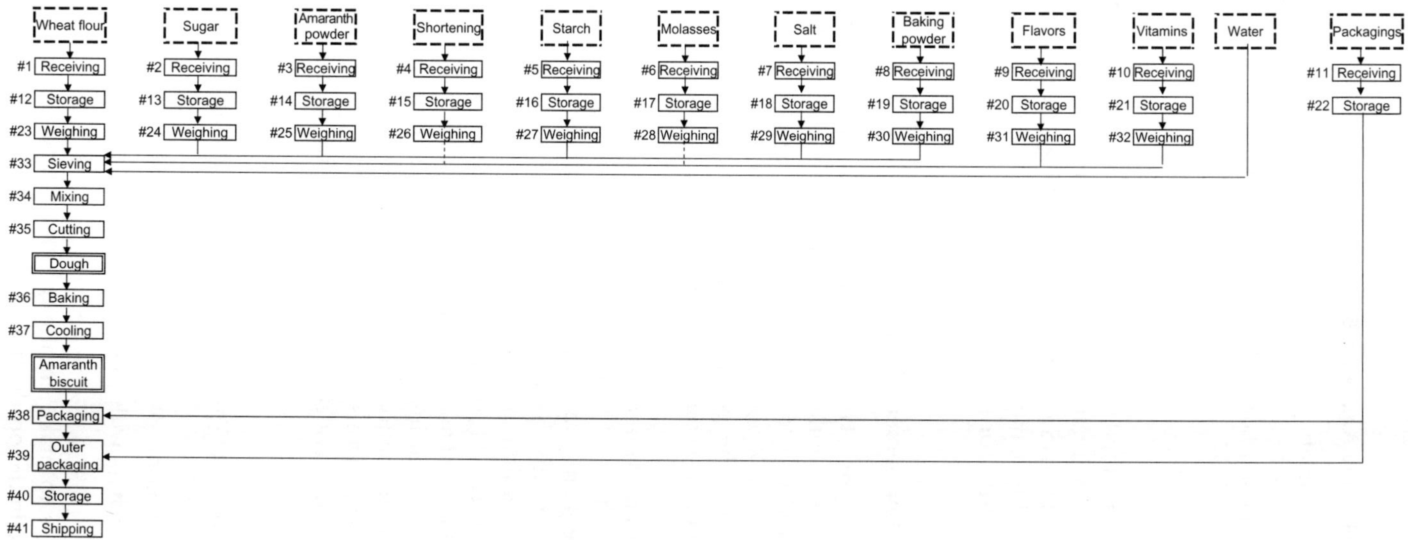

Fig. 14.5 Process flow diagram of the amaranth biscuit.

TABLE 14.7 Allergen Management Plan

Item	Potential for Allergen Incorporation	Control Point	Control Method
Raw materials	#1–#10: presence of milk, eggs, or soybeans	Receiving raw materials	Receiving quality certificate Verify the specification Guarantee by the supplier Verify the information by the supplier Allergen testing of ingredients
Package materials	#11: incorrect labeling and packaging	Receiving packages	Verify the specification
Storage	#12–#21: contamination during storage	Storage	Zoning and segregation of ingredients containing milk, eggs, or soybeans Identify milk- and/or egg-containing ingredients by a mark
	#12–#21: airborne contamination in storage	Storage	Zoning and segregation of ingredients containing eggs, milk, or soybeans Ventilate the storage Regular examination of storage (airborne particle check)
Manufacturing	#23–#32: contamination through mixed utensils	Weighing process	Dedicated utensils
	#23–#32: contamination by weighing machinery	Weighing process	Cleaning of weighing scale Examination of airborne particle
	#23–#32: airborne contamination during weighing	Weighing process	Ventilate weighing area Regular examination of weighing area (airborne particle check)

TABLE 14.7 Continued

Item	Potential for Allergen Incorporation	Control Point	Control Method
	#23–#32: contamination by operator	Weighing process	Uniform and shoes for allergen control zone Air shower, adhesive roller, adhesive sheet
	#23–#32: incorrect formulation	Weighing process	Manual and checklist of the formulation Documentation of input raw materials
	#33: contamination in sieve	Sieving process	Cleaning of sieve Scheduled sieving (after cleaning, amaranth ingredients are sieved first) Allergen testing for contamination
	#33: contamination by operator	Sieving process	Uniform and shoes for allergen control zone Air shower, adhesive roller with mirror, adhesive floor sheet
	#34: contamination in mixer	Mixing process	Cleaning of mixer Scheduled production Allergen testing for cross-contamination
	#34: contamination by operator	Mixing process	Uniform and shoes for allergen control zone Air shower, adhesive roller with mirror, adhesive floor sheet

#35: contamination in laminator, gauge roll, and cutting machine	Cutting process	Cleaning of laminator, gauge roll, and cutting machine Scheduled production
#35: contamination by operator	Cutting process	Uniform and shoes for allergen control zone Air shower, adhesive roller with mirror, adhesive floor sheet
#36: contamination in oven	Baking process	Cleaning of oven Scheduled production
#37: contamination in cooler	Cooling process	Cleaning of cooling equipment Scheduled production
#37: contamination by operator	Cooling process	Uniform and shoes for allergen control zone Air shower, adhesive roller, adhesive sheet
#38: contamination in packaging machine	Packaging process	Cleaning of package machine Scheduled production
#38: contamination by operator	Packaging process	Uniform and shoes for allergen control zone Air shower, adhesive roller with mirror, adhesive floor sheet
#38: incorrect package, incorrect label	Packaging process	Checklist of the packaging Documentation of used label and package (checklist)
#39: incorrect package, incorrect label	Outer packaging process	Checklist of the outer package Documentation of used label and package
#41: presence of milk, eggs, or soybeans in finished product	Dough sample	Allergen testing of periodical dough sample before release

Another point of the training is to change the idea of "clean." On the production site, "clean" traditionally means "no presence of inedible foreign material, for example, hair, stone, or metal," while the presence of milk and/or eggs is regarded as "clean" since these are edible. Altering this idea among the operators involved in the production is essential for successful allergen management.

14.3.5.1.2. Production. The critical points of cross-contamination are the facility, machinery, utensils, and operators in the production line.

Facility. It is necessary to place an air shower, vacuum cleaner, adhesive roller with mirror, and adhesive floor sheet at the facility entrance in order to prevent the entry of external dust. In a bakery site, contamination through airborne particles of milk and eggs is frequently observed. The best prevention of airborne contamination is to separate the room; the second best is to partition the room by a curtain fixed to the ceiling. A standing screen is not effective because airborne contaminants can easily pass over the screen. Powdery ingredients containing milk and eggs are stored separately from other powdery ingredients. Areas releasing airborne milk and eggs particles, such as the weighing area, need to be well ventilated. Ingredients, the product in progress, and the finished product are covered or put into a case, shielding them from potential airborne contamination. The airborne particle in a bakery site is a significant issue even for the operator as they can cause outbreaks of occupational asthma and allergy [16]; therefore, measures to control airborne particle is beneficial to both allergen management and improvement of occupational environment.

Machinery. Machinery is cleaned before production as described in Section 14.3.5.1.4, Cleaning.

Utensils. Utensils such as scoops, trays, racks, and scraper bars are dedicated for amaranth biscuit production and are clearly distinguished by color from usual utensils.

Operator. The operators are presumed as a source of contamination moving around the production area; therefore, the staff should wear a clean uniform, cap, gloves, and shoes during the production, and are encouraged to remove dust from all clothing.

Movement in the Facility. It is preferable to manage movement of the operator, cart, and product in one direction in order to prevent accidental collisions that might bring about cross-contamination.

14.3.5.1.3. Ingredients and Packaging

14.3.5.1.3.1. INGREDIENTS.

Receiving. A written guarantee that the purchased ingredient is free from milk, eggs, and soybeans is demanded in the supplier contract or agreement. On receiving ingredients, an analytical certificate of the batch is required to confirm whether the batch agrees to the stipulated quality, including "free from milk, eggs, and soybeans." These certificates are kept as records.

The purchased ingredients, especially major ingredients, are tested for milk and eggs by quantitative ELISA. Direct testing of the ingredient itself has the advantage of both higher detectability as the contaminant concentration will be higher, and the avoidance of processing effects (Section 14.4.1, Analytical method for milk and eggs).

During the selection of ingredients to be used for the amaranth biscuit, the production sites of the ingredients were field surveyed in order to have a good perspective of the site and also to evaluate the allergen management of the supplier. The field survey is conducted taking into consideration the point mentioned in this section. Depending on the results of the survey, the supplier is ranked as described in the ingredient list (Table 14.4), and the evaluation should be conducted regularly.

On receipt of packaged materials, the materials should be checked to ensure that the package matches the package specification, especially with respect to allergen labeling regulation.

Preventing Contamination in Storage. The milk- and egg-containing ingredients such as for the manna biscuit production are stored in a segregated area of the storage with independent ventilation. In addition, the area where ingredients are opened is separated from the storage area, and unsealed ingredients are never returned to the storage. The condition for nonallergen ingredient storage should be monitored regularly by testing for airborne milk and egg particles (Section 14.4, Milk and egg analysis). Furthermore, ingredients are stored on pallets or plates because the floor is likely to be contaminated.

Preventing Misuse of Ingredients. Ingredients containing milk and/or eggs are marked clearly "milk" and/or "eggs" for careful handling.

14.3.5.1.3.2. PACKAGING. In order to prevent misuse of packaging materials, the packaging used for the amaranth biscuit is stored separately from that of manna. A simple checklist with a picture of the outer wrapper distinguishing the amaranth package from the manna package is placed in the packaging storage area, and is signed before the amaranth packaging is taken out. The operator of the packaging machinery regularly checks and records the package of the finished product.

14.3.5.1.4. Cleaning. Cleaning of the production line consists of two measures:

1. Measure for object of which contaminant can be removed by washing such as utensils and washable parts of machinery like mixing propeller. By washing with warm water containing neutral detergent and rinsing sufficiently, the object is cleaned.
2. Measure for object that is not washable such as hard-to-detach parts of machinery and parts that cannot be washed with water because of rust. Object is cleaned of contaminant as much as possible by using the following techniques:

   - Visible residues remaining in the production machinery are removed by scratching and scrubbing, since incorporation of massive allergenic residues can result in a serious health hazard.
   - Powdery particles are dusted by wiping with dry cloth and vacuum. Moist cloth is also applicable, but imperfect wiping causes permanent contamination by spreading the contaminant on the production machinery. Remaining residues on the contact surface and corner of the machinery are eliminated by the prerun of the production.

Because the cleaning procedure is critical for controlling cross-contamination, a cleaning program, including procedures, timing, staff in charge, verification and records, is documented as standard practice.

14.3.5.1.4.1. CLEANING PROGRAM. Production of the amaranth biscuit is scheduled to follow the manna biscuit production. Immediately after the manna production, the line is cleaned as described in this section and left overnight in order to allow airborne particles scattered during cleaning to settle. The following morning, the machinery is gently cleaned with a dry cloth to remove any particles. Finally, the floor is carefully cleaned. This allows residues and airborne particles from the manna production to be removed from the production facility as much as possible.

14.3.5.1.4.2. CLEANING METHOD. Wet cleaning using water is the best and most complete method; however, several food machinery and parts are made of iron, which is hardly washable by water. Wiping with a moist cloth is used to remove deposits and contaminants stuck on the machinery. However, incomplete wiping results in the spread and permanent adhesion of contaminants to the machinery; thus, cautious wiping is required. Wiping with a dry cloth is suitable for machinery so as not to cause contaminants to stick to the machinery.

Vacuum cleaning of the machinery is preferable so as not to scatter particles or powder; however, it barely removes dough deposits or biscuit fragments firmly retained in the machinery. Compressed air is strong enough to remove dough deposits and biscuit fragments, but scatters particles requiring additional cleaning.

14.3.5.1.4.3. CLEANING PRODUCTION LINE. Initially, all of the troublesome hot spots of allergen contamination and hard-to-clean areas are identified by visual inspection of residues and by testing for allergen contamination. The cleaning method of the production line, that is, wet or dry, is chosen depending on the object to be cleaned, that is, whether the object is biscuit dough or baked biscuit; in principle, wet cleaning is applied for washable objects and objects used before baking, while dry cleaning is used for cleaning objects after baking.

Weighing Scale, Sieving, Vessel, and Mixer. These are washed with detergent and rinsed out. The rotating shaft area of the mixer, especially the hidden part of the socket, is a hot spot.

Laminator, Gauge Roll, Cutting Machine, and Oven. Roll, belt, and conveyor are wiped with a moist cloth. Remaining deposits in washable areas are removed by warm water or by chipping off the deposit by a scraper bar after water spraying. Deposits in unwashable areas are cleaned by compressed air. The cutter of the cutting machine is dismantled and washed with detergent. The steel belt of the oven is cleaned by a band washer. The cutter part and canvas of the conveyor are the most contaminated places, so these must be cleaned carefully.

Cooling Conveyor and Packaging Machine. The cooling conveyor just after the oven, which is a hot spot of broken biscuit and fine biscuit powder, is cleaned by vacuum, and wiped with a dry cloth. The surface of the packaging machine that contacts biscuit is wiped with a dry cloth.

The cleaning result is examined by visual inspection and allergen testing, and the results are recorded. Prior to starting the amaranth biscuit production, the cleaning outcome is verified.

14.3.5.1.5. Record. Work records, for example, input ingredients with the quantity, name of operator, mixing time, and oven temperature, are useful to identify any errors or mistakes in production as well as to provide hints in case of an allergen contamination incident. At the end of the production, the production supervisor checks the work records and approves the release of the batch. Work records of each batch are filed by batch number for tracing.

14.4. MILK AND EGG ANALYSIS

14.4.1. Analytical Method for Milk and Eggs

Two analytical methods have been developed for food allergen analysis: a DNA-based technique and a protein-based technique [17]. However, the DNA-based technique, which amplifies a specific DNA fragment of an allergen gene through polymerase chain reaction, is not suitable for routine testing for the production because of its tedious assay procedure and the requirement of sophisticated instrumentation. The protein-based technique, founded on an immunochemical (antigen–antibody) reaction, consists of an ELISA and a lateral flow assay. ELISA is the most common assay used in routine food analysis due to its quantitative output and high precision. A quantitative measurement gives more information; for example, the amount of allergen contamination in the machinery helps to suggest the origin of the contamination. On the other hand, a lateral flow assay is very simple, rapid, and portable; therefore, it is suitable for on-site testing of the product in progress as well as the cleaning outcome. Table 14.8 presents the usage of ELISA and a lateral flow assay in the amaranth biscuit production.

TABLE 14.8 Usage of ELISA and Lateral Flow Assay during Production

	Ingredient	Work in Process before Baking	Baked Product	Environment of Facility (e.g., Storage)	Cleaning: Machinery, and Facility	Cleaning: Washing after Cleaning
ELISA	◎	◎	X (◎)[a]	◎	◎	◎
Lateral flow	○	○	X	○	○	○
Sampling	Direct	Direct	Direct	Airborne particles	Swab	Direct

[a]Recently baked product can be assayed by ELISA as in Section 14.4.2, Recent development of ELISA.

◎, for generating quantitative data; ○, for generating qualitative data; X, not recommended.

TABLE 14.9 Commercial Quantitative Kits for Milk and Eggs

	ELISA Systems			Neogen		Nippon Meat Packers	
Kit name	–		–	Veratox® for Total Milk Allergen	Veratox® for Egg Allergen	FASTKIT ELISA Ver. II	
Item	Milk	Milk	Egg	Milk	Egg	Milk	Egg
Format	ELISA, Sandwich	ELISA, Sandwich	ELISA, Sandwich	ELISA, Sandwich	ELISA, Sandwich	ELISA, Sandwich	ELISA, Sandwich
Assay range	1.0–10 ppm	2.5–25 ppm	1.0–5.0 ppm	2.5–25 ppm	2.5–25 ppm	0.4–20 ppm	0.4–20 ppm
Analyte	α-Casein	β-Lactoglobulin	Ovalbumin and ovomucoid	Milk protein	Egg white proteins	Multi-antigens (mixture of milk proteins)	Multi-antigens (mixture of egg proteins)
Result reported as	Milk powder	Milk powder	Egg white protein	Total milk	Egg	Milk protein	Egg protein
Incubation time	60 minutes	60 minutes	60 minutes	30 minutes	30 minutes	170 minutes	170 minutes
LOD	–	–	–	–	–	–	–
LOQ	–	–	–	–	–	(1 ppm)	(1 ppm)
Test number/kit	48 wells	48 wells	48 wells	Up to 38 tests	Up to 38 tests	96 wells	96 wells

LOD, limit of detection; LOQ, limit of quantification.

Both methods, however, have the drawback that food processed at high temperature (i.e., by baking, roasting, or retorting) cannot be analyzed, as shown in Table 14.5 [11, 12]. Therefore, ELISA and lateral flow assay are applicable up until the baking process.

14.4.2. Recent Development of ELISA

Allergen analysis of the final product is vital to guarantee whether the declared product labeling is appropriate or not. Furthermore, in an outbreak of food allergen accident, the product in question can be scientifically tested to elucidate the cause and result. Accordingly, we have developed a new ELISA to detect egg allergen even in highly processed form [18, 19] and have diversified this technology to milk, wheat, peanut, and buckwheat ELISAs [20]. After launching these new ELISA kits, the direct analysis of milk and eggs in the final product has been adopted.

Comparisons of the performance of the latest commercial quantitative and qualitative test kits are presented in Tables 14.9 and 14.10, respectively.

14.4.3. Sampling

A test sample should represent the object in terms of homogeneity and location (for swab sampling). Contamination in manufacturing often happens only in a part of the batch; therefore, multiple sampling from different parts,

Morinaga Institute of Biological Science			R-Biopharm		TECRA	Tepnel		
FASPEK Milk	FASPEK Milk	FASPEK Egg	RIDASCREEN® Beta-lactoglobulin	RIDASCREEN® FAST Ei/Egg	Egg	BIOKITS Casein	BIOKITS BLG	BIOKITS EGG
Milk	Milk	Egg	Milk	Egg	Egg	Milk	Milk	Egg
ELISA, Sandwich	ELISA, Sandwich	ELISA, Sandwich	ELISA, Sandwich	ELISA, Sandwich	ELISA, Sandwich	ELISA, Indirect competitive	ELISA, Indirect competitive	ELISA, Sandwich
0.312–20 ppm	0.312–20 ppm	0.312–20 ppm	10–810 ppb	1–27 ppm	0.6–15 ppm	1.6–25 ppm	2.5–40 ppm	0.5–10 ppm
Casein	β-Lactoglobulin	Ovalbumin	β-Lactoglobulin	Egg white proteins (ovalbumin, ovomucoid, ovotransferrin, lysozyme)	Multi-antigens (egg white and yolk proteins)	Casein	β-Lactoglobulin	Ovomucoid
Milk protein	Milk protein	Egg protein	β-Lactoglobulin	Egg white protein	Whole egg protein	Casein	β-Lactoglobulin	Egg white proteins
90 minutes	90 minutes	90 minutes	2 hours 45 minutes	30 minutes	120 minutes	105 minutes	120 minutes	75 minutes
0.078 ppm	0.078 ppm	0.078 ppm	0.2 ppm	0.6 ppm	0.5 ppm	1 ppm	2 ppm	0.1 ppm
0.312 ppm	0.312 ppm	0.312 ppm	5 ppm	1 ppm	0.6 ppm	1.6 ppm	2.5 ppm	0.5 ppm
96 wells	96 wells	96 wells	96 wells	48 wells	96 wells	96 wells	96 wells	96 wells

TABLE 14.10 Commercial Qualitative Kits for Milk and Eggs

	Neogen			Nippon Meat Packers		Morinaga Institute of Biological Science		Tepnel	
Kit name	Alert® for Total Milk Allergen	Alert® for Egg Allergen	Reveal® for Total Milk Allergen	FASTKIT Immunochromato		Nanotrap® Milk	Nanotrap® Egg	BIOKITS RAPID 3-D™	
Item	Milk	Egg	Milk	Milk	Egg	Milk	Egg	Casein	Egg
Format	ELISA, Sandwich		Lateral flow	Lateral flow	Lateral flow	Lateral flow	Lateral flow	Lateral flow	Lateral flow
Detection limit	5 ppm	5 ppm	5 ppm	5 ppm	5 ppm	4 ppm	0.5 ppm	Low level	Low ppm
Analyte	Milk protein	Egg protein	Milk protein	Whole milk proteins	Whole egg proteins	Casein	Ovalbumin	Casein	Ovomucoid
Result reported as	Milk residue	Egg residue	Milk residue	Milk protein	Egg protein	Casein	Ovalbumin	Casein	Egg white protein
Reaction time	30 minutes	30 minutes	5 minutes	15 minutes	15 minutes	15 minutes	15 minutes	10 minutes	10 minutes
Test number/kit	Up to 20 tests	Up to 20 tests	24 tests	20 tests	20 tests	20 tests	20 tests	10 tests	10 tests

locations, and timings can help to minimize the overlooking of allergen contamination. For instance, dough sampling for ELISA is conducted every 10 minutes in the amaranth biscuit production.

Due to the very high sensitivity of ELISA and the lateral flow assay, sample preparation and the assay procedure should be carried out with care to prevent false results due to mishandling. For example, insufficient washing of the sample tube/container/reservoir may generate false-positive results as allergenic protein may remain on the surface of the apparatus. Use of disposable plasticwares and careful washing of the apparatus are, thus, recommended.

Practical areas and materials for samplings are

Airborne Substances. Airborne substances are sampled by a saline-containing cup collecting falling airborne particles. The saline sample is, then, analyzed by ELISA or lateral flow assay.

Received Ingredients. Ingredients are examined by ELISA to obtain quantitative results (Section 14.3.5.1.3, Ingredients and packaging). Allergen testing of ingredients is performed to confirm the batch certificate.

Cleaning of Facilities. Sampling is done by swab method, which involves scraping the surfaces of objects with cotton swab moistened with saline solution, extracting the swab by saline, and analyzing the extract. Sampling by the rinse method is not suitable for biscuit production because of insufficient solubilization from caked residues on the machinery surface.

Washings from Cleaning. When the washing solution is used for ELISA or lateral flow assay sample, it is necessary to consider the composition of the solution. If the washing solution containing concentrated detergents or strong alkaline/acid solutions are directly applied for ELISA/lateral flow assay, the antibody of the assay will be damaged and cannot give correct results. Therefore, the washings must be adjusted to neutral pH, and the rinse water of the final washing must be applied for the assay in order to minimize the effect of the components in the washing.

14.5. CONCLUSION

Given the fact that no allergic incidents have been reported in over a decade of amaranth biscuit sales, it could be said that the management of milk, egg, and soybean allergens in the production of the amaranth biscuit has been successful. In this chapter, food allergen management of the amaranth biscuit, which does not use milk, eggs, or soybeans, has been introduced as a case study. The principle of food allergen management does not differ among food products. The key points requiring management are employee, production facility/machinery, ingredients, labeling (packaging), cleaning, and production record. Once the principles are understood and properly applied, food allergen management tools can be successfully used for the production of allergen-free foods.

REFERENCES

1. Sampson, H.A. (2004). Update on food allergy. *The Journal of Allergy and Clinical Immunology*, *113*, 805–819.
2. Ebisawa, M., Ikematsu, K., Imai, T., Tachimoto H. (2003). Food allergy in Japan. *Allergy and Clinical Immunology International: Journal of the World Allergy Organization*, *15*, 214–217.
3. Dalal, I., Binson, I., Reifan, R., Amitai, Z., Shohat, T., Rahmani, S., Levine, A., Ballin, A., Somekh, E. (2002). Food allergy is a matter of geography after all: sesame as a major cause of severe IgE-mediated food allergic reactions among infants and young children in Israel. *Allergy*, *57*, 362–365.
4. Taylor, S.L., Hefle, S.L. (2001). Ingredient and labeling issues associated with allergenic foods. *Allergy*, *56*, 64–69.
5. Obata, T. (1999). Amaranth bisketto no kaihatsu. (Development of Amaranth biscuit) *Syoku no Kagaku*, *253*, 46–51 (in Japanese).
6. Ministry of Health, Labor, and Welfare. (1992). Nichijyo seikatsu to arerugi you syoujyo (Daily life and food for allergy) (in Japanese).
7. Ministry of Health, Labor, and Welfare. (1993). Atopiseishikkan jittaichosakekka no gaiyou (Summary of Atopic disease survey result) (in Japanese).
8. Ministry of Health, Labor, and Welfare. (1997). Heisei 8 nen syokumotsuarerugi taisakukentouiinkai kenkyu houkokusyo (Research report of Food allergy measure workshop in 1996) (in Japanese).
9. Ebisawa, M. (2009). Arerugi busshitsu wo fukumu syokuhin hyoji seido ni kanshite. (Food labeling regulation for the food containing allergic substance) Report of the Ministry of Health, Labor, and Welfare, May 31, 2004. Available at http://www.mhlw.go.jp/shingi/2004/05/s0531-8a.html (in Japanese).
10. Crevel, R.W., Kerkhoff, M.A., Koning, M.M. (2000). Allergenicity of refined vegetable oils. *Food Chemistry and Toxicology*, *38*, 85–93.
11. Williams, K.M., Westphal, C.D., Shriver-Lake, L.C. (2004). Determination of egg protein in snack food and noodles. *Journal of AOAC International*, *87*, 1485–1491.
12. Shoji, M. (2009). Egg allergen detection. In: *Molecular Biological and Immunological Techniques and Applications for Food Chemists*, Popping, B., Diaz-Amigo, C., Hoenicke, K. (Eds.). Hoboken, NJ: John Wiley & Sons, pp. 311–334.
13. Okada, H. (1975). Morumotto niokeru PCA hanno (PCA reaction of Guinea pig) (in Japanese). In: *Men-eki Jikken Sousahou* (Manual for Immunological Experiments), Migita, T. (Ed.). Kanazawa: Japanese Society for Immunology, pp. 626–629 (in Japanese).
14. Harlow, E., Lane, D. (1988). *Antibodies: A Laboratory Manual*. New York: Cold Spring Harbor Laboratory, pp. 471–510.
15. All Japan Association of Confectionery Manufacturers. (1999). Okashi no eisei kanrimanyuaru (Sanitary management manual for confectionery) (in Japanese).
16. Escudero, C., Quirce, S., Fernandez-Nieto, M., de Migel, J., Cuesta, J., Sastre, J. (2003). Egg white proteins as inhalant allergens associated with baker's asthma. *Allergy*, *58*, 616–620.

17. Poms, R.E., Klein, C.L., Anklam, E. (2004). Methods for allergen analysis in food: a review. *Food Additives Contaminants*, *21*, 1–31.
18. Watanabe, Y., Aburatani, K., Mizumura, T., Sakai, M., Muraoka, S., Mamegosi, S., Honjoh, T. (2005). Novel ELISA for the detection of raw and processed egg using extraction buffer containing a surfactant and a reducing agent. *Journal of Immunological Methods*, *300*, 115–123.
19. Feste, C.K., Lovberg, K. (2007). Extractability, stability, and allergenicity of egg white proteins in differently heat-processed foods. *Journal of AOAC International*, *90*, 427–436.
20. Matsuda, R., Yoshioka, Y., Akiyama, H., Aburatani, K., Watanabe, Y., Matsumoto, T., Morishita, N., Sato, H., Mishima, T., Gamo, R., Kihira, Y., Maitani, T. (2006). Interlaboratory evaluation of two-enzyme-linked immunosorbent assay kits for the detection of egg, milk, wheat, buckwheat, and peanut in foods. *Journal of AOAC International*, *89*, 1600–1608.

PART IV

RISK ASSESSMENT AND RISK MANAGEMENT

15

RISK ASSESSMENT FOR FOOD ALLERGY

RENE W.R. CREVEL

15.1. INTRODUCTION

Food allergy has long been recognized clinically, with numerous reports in the twentieth century medical literature [1–3]. However, perception of its importance as a public health and food safety issue only dates to the last decade of the twentieth century. Early studies by Young et al. [4] in the United Kingdom estimated a population prevalence of 1.4–1.9%, while more recent reports suggest figures of the order of 3.5–4% in the industrialized world [5–10]. The prevalence among children is generally accepted to be higher, of the order of up to 8% [11]. Such figures equate to upward of 10 million affected people in regions such as the European Union (EU) or the United States and, therefore, a significant burden of ill health and reduced quality of life.

Food allergy is a clinically challenging condition. Symptoms range from the very mild, such as a rash or a hive, to the life-threatening, such as anaphylactic shock. They may affect one or several organ systems and may resemble other clinical syndromes. Reactivity within the population varies over five to six orders of magnitude, as expressed by minimum eliciting doses (MEDs) [12], although there is no clear correlation between an individual's threshold and the severity of the reactions that may be experienced. Reactivity can also be

Allergen Management in the Food Industry, Edited by Joyce I. Boye and Samuel Benrejeb Godefroy

modulated by external factors such as exercise, concurrent asthma, and stress. Avoidance of the offending allergen is currently the only treatment for food allergy, and an immediate consequence is that having a member with food allergy affects the whole family and, to a lesser extent, the social circle of the allergic individual. The recommended strategy for patients to manage food allergy, that is, avoidance of all food products containing the allergen, can create a high level of stress and anxiety over food-related activities [13, 14], and stress and anxiety can also occur in people caring for allergic individuals. This situation is made worse by the proliferation of precautionary labels, which purport to warn of the possible presence of allergens that are not part of a product's recipe, but which are increasingly ignored, even by well-informed allergic consumers [15].

Allergic people can find avoidance of specific foods very difficult. Labeling rules generally cover only deliberately added ingredients in prepacked foods. However, food production is a complex process and generally involves the use of shared equipment at all stages from the transport of raw materials, through manufacturing to packaging. Small amounts of allergenic ingredients which are not part of the recipe are therefore an ever-present possibility in finished products and can pose a risk to allergic consumers, even to the extent of occasionally provoking severe reactions [16]. Allergenic ingredients such as peanuts, soy, milk, eggs, and wheat are used extensively in food manufacture, and guaranteeing the total absence of all such constituents from a product when not present as ingredients is often not a practical proposition. Managing the resulting risk on the basis of hazard has led to extensive use of precautionary ("may contain") labeling. Allergic individuals dislike precautionary labeling because there is no consistency in its applications between manufacturers, and it reduces their food choices considerably [17, 18]. Arguably, it also puts them at increased risk not only directly through devaluation of the label [19] but also through adverse effects on nutritional status [20]. Aside from the impact on the allergic consumer, the hazard-based approach to allergen management can also cause problems for the food industry by driving the implementation of measures, which may go beyond what is necessary to make food safe for allergic consumers and may waste significant food resources at a time when these are becoming more scarce [21].

The particular characteristics of food allergy, which make it so challenging to manage clinically, together with the nature of food manufacturing, also mean that elimination of the risk, even in the simplest sense of analytical absence, is not achievable in many circumstances. This risk therefore must be managed in order to minimize it for those affected. Risk management implies risk assessment as a prerequisite. This chapter will discuss the concept of risk assessment and consider how it can be applied to allergenic foods. It will also consider its application to the consequences of food processing, as well as its role in allergen management and describe the various factors currently limiting such risk assessments.

15.2. RISK ASSESSMENT

Risk can be defined as the probability that a hazard will become manifest and is often expressed as a function of the intrinsic hazard and the exposure to that hazard. This definition is sometimes expanded to include severity of the resulting adverse effect. Thus, risk is defined in the International Organization for Standardization/International Electrotechnical Commission (ISO/IEC) Guide 51 as the combination of the probability of occurrence of harm and the severity of that harm [22].

Risk assessment for commonly allergenic foods is solely concerned with establishing the risk of eliciting a reaction; preventing sensitization to such foods is beyond its scope. The purpose of risk assessment is to derive a quantitative estimate of the public health effect in terms of the numbers of people likely to be affected by exposure to a given level of the hazard and the likely consequences of that exposure. This quantitative estimate then forms the basis for possible risk management measures. Thus, the first step in the risk assessment process is the formulation of the risk management question. This will determine what data need to be gathered or generated (Fig. 15.1).

The risk assessment itself begins with hazard identification, followed by hazard characterization, exposure assessment, and risk characterization to scope the overall risk to the population [23–25].

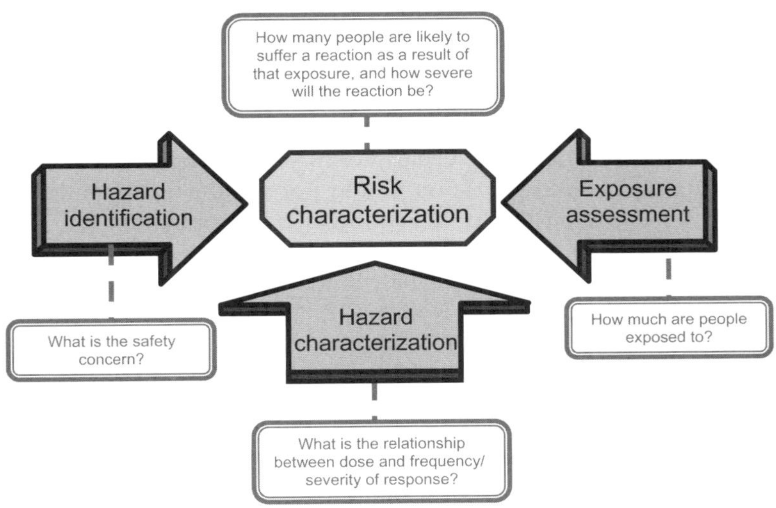

Fig. 15.1 Risk assessment. This diagram illustrates the different components of a risk assessment and how they relate to each other. Sources of data for risk assessment of allergenic foods are indicated.

For allergenic foods, the hazard itself is already defined as their intrinsic allergenicity, in this context their ability to provoke an allergic reaction. However, characterizing this hazard is a more complex task. This complexity originates in the nature of the allergenic components in foods as well as in the response that susceptible individuals make to these allergens. Allergenicity resides in the protein component of foods, and this will almost always be a multiplicity of molecules, some of which will prove to be allergenic, while the majority will not [26]. Allergic individuals often respond to different proteins in a food with different degrees of vigor [27]. Thus, two individuals who react to the same amount of a food with the same symptoms may actually be demonstrating qualitatively very different responses. This clearly poses problems in characterizing the risk from allergenic foods insofar as the responses may differ in terms of their kinetics. This is further complicated because foods are complex materials and some of the nonprotein components can influence the immune response to the proteins [28]. The characteristics and features of immune responses to proteins add to this complexity. The immune system does not recognize whole proteins but areas or structures on the protein known as epitopes, which may be linear stretches of amino acids or discontinuous (conformational) and dependent on the three-dimensional molecular motif of the protein. Thus, while two individuals may react to the same protein, they may be responding to chemically quite distinct structures. The effects observed in individuals undergoing an allergic reaction, therefore, represent a summation of a number of different responses to different structures and molecules, each having its own unique dose–response profile. To characterize the response to an allergenic food requires consideration of the relationship between the amount of food eaten and the clinical response to that food. Since in practice, it is both impracticable and unethical to attempt to derive complete individual dose–response profiles because of the risk to the allergic individual, in this context, characterization is based on the distribution of MEDs (thresholds) in the relevant allergic population. Clearly, critical considerations in this regard include a thorough characterization of the population tested and a careful description of its relationship to the overall allergic population for which the risk assessment is being performed. Only with such knowledge can conclusions about risk in these test populations be confidently generalized to the overall allergic population [29].

Exposure assessment is the next step in risk assessment. Allergic reactions to foods usually occur within a short time of ingestion, ranging from a few minutes to as many as 4 hours, depending on the nature of the food and the extent to which this affects availability of proteins to the immune system. Exposure assessment therefore considers how much allergen may be consumed over a relatively short period of time, such as a meal or serving, rather than over a day, as for conventional toxicological risk assessments. Consideration of exposure needs to take into account the characteristics of allergenic foods and food proteins previously referred to, as well as the effects of other components of the food. Additionally, the effects of digestion need to be considered. Sources of data for food allergy exposure assessment include dietary surveys and analytical measurements, where necessary.

Risk characterization brings together the response characteristics of the population at risk (distribution of MEDs—thresholds, relationship between dose, and severity of reaction), the size of the population at risk, and the extent of exposure. In ideal circumstances, risk assessors could calculate the incidence of reactions and estimate their severity profile for any given level of residual allergen in a food product if people allergic to that food consumed it. In practice, information on all the elements required for a risk assessment remains limited for most allergenic foods.

Studies to describe the reactivity of the allergic population in terms of MEDs (thresholds) are still largely limited to selected populations, such as clinic patients, and further work is needed to map their reactivity onto the general allergic population [29]. It is also limited to certain allergenic foods. Most available information is based on studies undertaken for other purposes, further limiting their value for establishing MEDs because of the type of data generated. Only anecdotal information is available on the relationship between the dose of allergen to which an individual is exposed and the severity of the reaction experienced, as obtaining that information systematically presents considerable ethical difficulties. Further complexities arise from differences in the extent to which allergenic proteins are released from different food matrices, and become available to be recognized by the immune system [30].

Uncertainty also exists over the size of the population at risk, given that epidemiological data are scarce for most regions. In most countries, there is also a surprising lack of detailed information about the incidence of reactions, even severe and fatal ones.

MEDs (thresholds) clearly form a key piece of information for risk assessors. A recent report by the U.S. Food and Drug Administration (FDA) Threshold Working Group [31] identified four ways in which thresholds might be determined: analytically based, statutorily derived, using a safety assessment approach, and using a risk assessment approach. They indicated a preference for the risk assessment approach on the grounds that it was the most scientifically robust, as well as transparent—an important consideration in communicating with other stakeholders. Safety and risk assessment approaches were also recently reviewed by an international workshop [32].

Of the four approaches described by the FDA report, only the safety and risk assessment ones use biological data on the reactivity of allergic patients, generated under controlled conditions. The safety assessment approach applies an uncertainty factor to either the highest dose of allergenic protein observed not to provoke a reaction or the lowest one that does provoke a reaction (depending on what is available). This approach can be valuable in protecting allergic individuals when little or poor quality information is available on population dose distribution. An example of its application is the Voluntary Incidental Trace Allergen Labeling (VITAL) system elaborated under the aegis of the Australian Food and Grocery Council [33]. Clinical studies with quantitative data on reactivity to common allergenic foods were reviewed, and no observed effect levels (NOELs) or lowest observed effect levels (LOELs) were determined. An uncertainty factor of 10 was then applied, and concentrations of allergenic protein above and below which precautionary statements should be applied to products were defined. Such an approach, of course, cannot quantify the residual risk, since it does not take into account exposure or dose distribution. Recently, we proposed an approach based on modeling the distribution of the MEDs of an allergenic food against the proportion of the allergic population responding to that dose, as determined in challenge studies. The goal was to define the expected number of reactions from exposure to specified amounts of allergenic ingredient in a product and use it as a basis for risk assessment [29] and subsequent risk management and communication. Recently, this approach has been applied to published challenge data on peanut [12]. Dose distribution modeling helps to characterize the response to allergens and can be used in a deterministic risk assessment. However, unlike ingredients, the concentration of which always falls within a relatively narrow range, residual allergen content from cross-contact would typically vary in concentration according to production conditions. For instance, concentration in a following product might be expected to be highest immediately after changeover from the product containing the allergenic ingredient. The risk of a reaction to that product by an allergic individual depends not only on the amount of allergen, but also on the probability that the individual would actually encounter and consume a unit of product with an amount of allergen equal to or greater than their MED. This probability will itself depend on the proportion of product units containing that amount of allergen, as well as on the propensity of allergic consumers to use that product. If quantitative data are available on these exposure variables, then a probabilistic risk assessment can be performed, which takes into account those probabilities as well as the uncertainties around them. Figure 15.2 illustrates this approach conceptually, while a more complete description of its application can be found in the paper by Kruizinga et al. [34].

Inevitably, the populations tested in challenge studies contain an element of bias. First, they will tend to exclude individuals who have suffered life-threatening reactions, as these may be less willing to participate and clinicians have been more reluctant to accept them in studies on account of the risks. Perhaps more significantly, participants recruited for challenge studies will tend to come from the more severely affected of the allergic population, other than the category previously mentioned. This happens because volunteers are chosen from the population who attends tertiary referral clinics and who are therefore sufficiently motivated by the impact of their condition to visit a clinical expert. Limitations also exist when using such data as a basis for risk management, because they do not take into account some of the factors that modulate the allergic response, such as alcohol intake and exercise. However, it can be argued that compared with other areas of toxicology, these data present the major advantage of being generated in the species of interest. Currently available data do not permit, however, ready validation of model

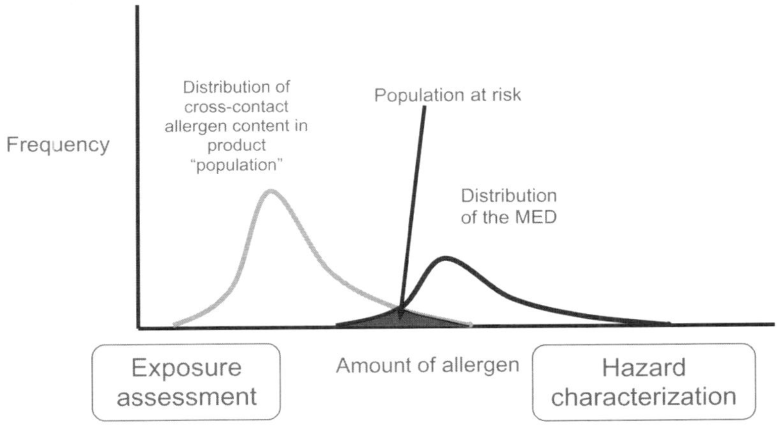

Fig. 15.2 Conceptual illustration of probabilistic risk assessment. An allergic reaction will only occur if the amount of allergen present in a food exceeds the minimum eliciting dose for the person consuming that food. This will clearly depend on the distribution of allergen concentrations in the product as well as the distribution of MEDs in the allergic population. The concept is presented in detail in the work of Spanjersberg and colleagues [108].

predictions. Specifically, data on the incidence of even severe allergic reactions to foods are not reliable, as already discussed, and usable published data on the distribution of undeclared allergen in food remain very limited [35]. Nevertheless, the next step in the development of this approach is to validate the predictions by comparing the number of predicted reactions with the number actually observed.

Despite the challenges that it presents and the uncertainties that attend it, risk assessment methodology offers a practicable way of optimizing outcomes for all the different stakeholders that a hazard-based approach does not.

15.3. CHALLENGES AND LIMITATIONS IN THE ESTABLISHMENT OF MEDS (THRESHOLDS)

As discussed earlier, MEDs and their distribution represent a critical piece of information in characterizing the allergenic hazard from a food. Understanding the uncertainties associated with their determination is therefore needed in order to assess and characterize the risk from allergenic foods.

Food challenges, in particular double-blind, placebo-controlled food challenges (DBPCFCs), remain the only scientifically accepted way to establish or rule out allergy to a food [36] and to determine thresholds of reactivity. Early diagnostic studies featured relatively high starting doses, typically in the range of 250–500 mg of the food for the most sensitive subjects. These were chosen to produce an objective but mild reaction [37–39]. Studies from different sources also differ in critical details, such as challenge procedures, the form of the food used [40, 41], the matrix in which it was presented [42–43], and the weight accorded to subjective and objective manifestations [44–47]. More recently, low-dose challenge adaptations of the DBPCFC have been implemented, which enable the determination of the lowest observed adverse effect level (LOAEL) and no observed adverse effect level (NOAEL) [39, 42, 43, 45, 48–50]. From this perspective, a consensus emerged that in future low-dose challenge studies, each food should be tested in at least 29 patients to ensure adequate statistical power. Clearly, the actual threshold will depend on the criteria used to define a response, with a continuing debate about whether only independently observable ("objective") signs should be considered, or whether symptoms reported by the patient but not capable of independent verification ("subjective") should be taken into account. There is also a debate about how both the threshold (no effect level) and the LOAEL should be derived from the challenge results, but application of the statistical technique known as interval censoring survival analysis (ICSA) can overcome this issue [12]. Threshold studies are performed by titrated food challenges with increasing doses applied at intervals of usually 15–30 minutes, chosen mainly for reasons of practicality. There is currently no agreement on whether the cumulative dose ingested up to the point of reaction, or merely the last discrete dose should be taken as the relevant metric. However, consideration of the rate at which proteins are broken down in the gastrointestinal tract under physiological conditions suggests the cumulative dose to be the most relevant [51].

15.3.1. Factors Affecting the Outcome of Challenge Studies and the Type of Data Generated

As noted by Taylor et al. [37], a wide variety of DBPCFC protocols differing in potentially significant ways have been used. These differences affect both the type of data generated and, therefore, its value for specific purposes, and the extent to which studies can be compared, even when they nominally used the same outcome measures. From the perspective of hazard characterization, the challenge procedure itself, the selection of patients, and the challenge materials can all affect the quality of the data.

15.3.1.1. Challenge Procedure. The precision of threshold estimates depend primarily on the starting dose, dose progression, the time interval between doses, and the way in which placebo and active doses are randomized. Other influential factors include the health status of participants and the avoidance of medications, as well as the scoring of reactions and stop criteria, particularly in the case of subjective reactions. Recent recommendations on challenge protocols [37, 39] include performing placebo and test challenges on different days, starting doses of the order of 10 μg of the suspected food, applying a dose progression ranging from doubling to half or full-log intervals, and setting a time interval between doses of 15–30 minutes, largely for practical reasons. There is a consensus that, principally on grounds of safety, participants should discontinue medications likely to interfere with the outcome (or otherwise be excluded) and, if suffering from asthma, that their condition should be stable. This makes the allergen encounter during a challenge quite different from what might occur in the community where the subjects' health and medication use may vary considerably [52].

15.3.1.2. Patient-Related Criteria. Inclusion and exclusion criteria for low-dose challenge studies will differ according to the purpose for which the data are being generated. A prerequisite in threshold studies is that the subjects are demonstrably still allergic to the food being tested. In several studies, the development of tolerance in previously allergic subjects has been demonstrated, in particular in children with milk and egg allergy [53], which tend to be outgrown. Even up to 20% of peanut-allergic patients may outgrow their food allergy [54]. Where individual thresholds are being estimated, the primary concerns are benefit to and safety of the individual. Scientific studies on thresholds aim to generate data and test hypotheses that can be generalized to the relevant population and, in addition to the patient safety criteria, participant selection must reflect these needs. Participants must therefore be well characterized in relation to the allergic population, particularly in terms of their reactivity.

Population variability encompasses both interindividual variation within an otherwise homogeneous population, and the possible existence of subpopulations with a different distribution of reactivity (e.g., children, people with asthma). Challenge studies show that individual threshold doses can vary by six orders of magnitude or more within a population. With a few exceptions, for example, in Reference 55, MEDs have not been determined in random samples of the population, but instead using groups of food-allergic patients referred to specialist allergy clinics in tertiary care centers. The challenged population will therefore contain individuals who are more reactive than the general allergic population, even though people with a history of severe reactivity have been excluded from most studies. The possible consequences for the distribution of thresholds are illustrated in Figure 15.3. Data are lacking, however, to quantify the relationship between this challenged population and the overall allergic population. Similarly, data are scarce about the existence of subpopulations with different thresholds. One study that looked at possible subpopulation differences in relation to asthma status did not confirm their existence [43].

15.3.1.3. Challenge Materials. The key to success in the DBPCFC is the accurate delivery of a range of doses of the relevant allergenic food in a form that is unrecognizable to the patient. The test material must therefore be well-characterized, and its taste, smell, color, and texture must be masked.

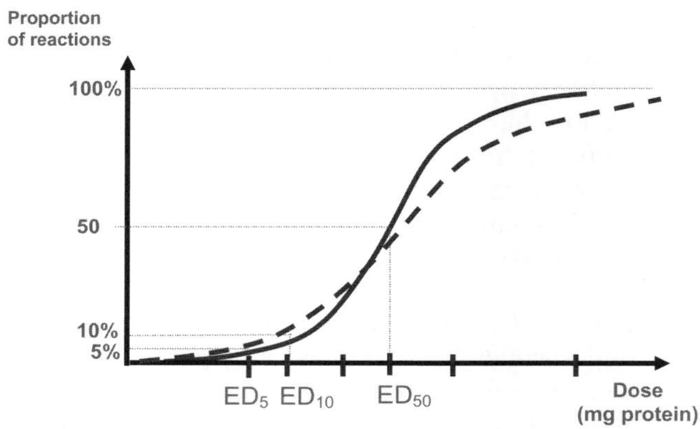

Fig. 15.3 Dose distributions for the clinic and whole allergic populations. The solid line represents the clinic population, which often excludes subjects from both ends of the range of reactivity: individuals who are highly reactive at low doses and those who react at very high doses. The dashed line represents the possible dose distribution across the whole allergic population.

Foods are usually consumed in a processed form; as a result of which they may have undergone a number of treatments any of which could have modified their allergenic potential [56, 57], as will be discussed later. Ideally, in order to provide the greatest margin of safety, the threshold should be determined using the most allergenic form of the food, but in practice, this can only be determined by challenge. Allergic people often differ in their reactivity to individual proteins in foods [e.g. 27, 58, 59], but characterization of challenge materials with regard to their content and profile of allergenic proteins has received little attention. There is, however, evidence for differences between fruits at different stages of ripening [60].

The outcome of DBPCFC can also be affected by the form and matrix in which the material is delivered. The use of capsules resolves issues of blinding but at the expense of eliminating symptoms associated with mucosal contact and possible modifications to the food to achieve encapsulation. Delivery of adequate doses can also be difficult. Recent recommendations thus promote the use of "real" food-type matrices [59]. Key requirements for these foods are as close a match as possible between the challenge matrix and placebo, in terms not only of taste and smell [61, 62], but also of texture [63]. Masking should always aim to maximize the amount of active compound in the matrix, thereby minimizing the amount of material that the patient is required to consume and the probability of nonspecific gastrointestinal symptoms, which could decrease the sensitivity of the test. However, test and placebo materials need not be identical, as long as participants cannot tell which is which. Different matrices can have a profound effect on the kinetics of the clinical reaction, as recently shown in peanut allergy [31, 63]. Other constituents, for example, polyphenols, may also reduce availability. These considerations highlight the need for a more thorough assessment of availability, as well as confirmation of a selection of the doses administered.

In summary, MEDs determined in one study, by one food variety in one state of processing, in one matrix, might not allow final conclusions to be drawn on the "real" threshold level. However, this situation differs a little from others where data must be used to make risk assessments, and where the variability is handled by application of appropriate uncertainty factors. Further studies assessing the impact of all the main variables are needed to circumscribe the degree of uncertainty.

15.4. MOLECULAR, TECHNOLOGICAL, AND PROCESSING CHALLENGES: ASSESSING THE ALLERGENIC RISK FROM FOOD TREATMENTS AND MODIFICATIONS

15.4.1. Allergenicity and Processing: A Reprise

Food processing covers an almost infinitely wide range of activities, which a food or food ingredient can undergo before it is consumed. At its origins, processing aimed either to improve the range or palatability of what could be made out of a raw material (for instance, milling of wheat, cooking of meat) or to preserve perishable foodstuffs (e.g., salting of fish). It now extends to improvements in consumer convenience, generation of useful derivatives, and production of more valuable products [57]. Modern processing has enhanced opportunities considerably to this, for instance, by imparting the properties of one type of food material to another by altering its physical structure (e.g., use of microparticulated protein in a spread [Simplesse]) [64], developing novel ways of using certain foods (e.g., textured soy protein, mycoprotein). Sathe and Sharma [57] classify food processing methods as either thermal (boiling, roasting, frying, blanching, canning, etc.) or nonthermal (washing, filtration, dehydration, spray drying, high-pressure processing, etc.). These processes pose two types of challenge with regard to food allergy. As already discussed, food allergy is an inappropriate immune response to certain food proteins and is mediated at the elicitation stage by the recognition by immunoglobulin E (IgE) antibodies of epitopes on those proteins. Many processing methods may alter those proteins, and the effectiveness with which they may be recognized by antibodies that may have been produced against another form of the same protein (e.g., raw vs. cooked). Processing may also alter the matrix within which those proteins reside. It is unsurprising therefore that food processing can affect allergenicity. It is equally unsurprising that it has proved difficult to develop any useful rules to predict the effect of defined types of processing on the allergenic potential of foods. Thus, one of the challenges that food processing presents is in actually defining the hazardous entity, the risk from which must be characterized. This is well illustrated by the experiments of

Prausnitz and Küstner [1] when they demonstrated the passive transfer to Prausnitz of Küstner's allergy to cooked but not raw fish. Most work since that time has continued to use end points such as IgE antibody binding and other *in vitro* measures. The findings have been extensively reviewed recently [57, 65–68].

However, while it is not possible to relate a specific type of processing to a specific effect on allergenicity, it may be useful to group processes into those that considerably reduce the protein content, those that cleave the proteins to a greater or lesser extent, and those that can alter the conformation of proteins. Edible oil refining and the manufacture of wheat starch are examples of the first. Full oil refining reduces the protein concentration of the final product by around three orders of magnitude or more compared with the original seed oils, such that the oils are very unlikely to provoke allergic reactions [69]. Processes such as fermentation to produce soy sauces, for instance, involve the hydrolysis of much of the protein component of the raw materials and might be expected to result in reduced ability to trigger allergic reactions, compared with the starting materials. However, the extent of any reduction would depend on factors such as process parameters as well as the resistance of individual proteins to hydrolytic cleavage. The differing allergenicity of various hypoallergenic infant formulas illustrates this well [70]. The third category of the process includes the greatest range of techniques and is thus the one for which predictions are most difficult. Numerous examples exist of apparently unchanged allergenicity (microparticulated egg and milk proteins) [64], mango nectar [71], chickpea [72], diminished allergenicity (kiwifruit concentrate) [73], lupine [74], and increased allergenicity (peanuts [75] and wheat [76]). Different processes may affect the same food in different ways. The Ara h 1 content of peanut was reduced by boiling or frying, as was IgE binding to Ara h 2 and Ara h 3, compared with their content and IgE binding in roasted peanuts [77]. Work by Mondoulet and colleagues [78] indicated that the loss of reactivity by boiling was attributable to loss of soluble proteins into the cooking water. They also confirmed the increased reactivity of roasted peanuts. Meyer-Pittroff et al. [79] showed that 19/19 allergic patients, reporting on symptoms after apple consumption, tolerated high-pressure treated apple (600 MPa, 5 minutes) without symptoms in a DBPCFC.

Food processing may also result in the formation of new allergenic entities, not present in the original foods, which do not necessarily trigger reactions in individuals allergic to the native food. The reaction of Küstner [1] to cooked but not raw fish represents one of the first documented examples. The more recent ones include wheat protein isolates that produced an allergic reaction in a patient who could tolerate wheat flour [80].

15.4.2. Processing and Thresholds of Elicitation

Allergy to a food is defined by clinical reactivity to that food, in other words, by the allergic individual suffering an adverse reaction on exposure to or

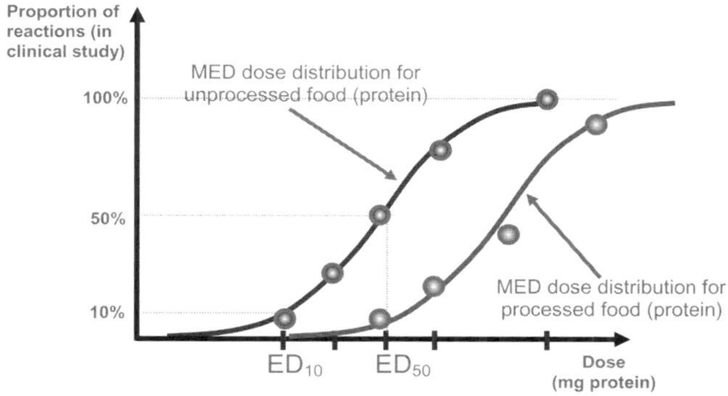

Fig. 15.4 Estimation of the effect of processing on allergenicity as measured by the dose distribution of MEDs. The challenge procedure is in effect a parallel line bioassay, and allergenic potencies can be compared using the ratio of ED_{50}'s.

contact with that food. A diagnosis of food allergy requires a DBPCFC, and to date, no other test has proved able to substitute for it. Similarly, determining thresholds requires DBPCFC with the relevant food in the appropriate form. It follows that, in order to establish whether processing affects the minimum dose of a particular food required to trigger a reaction, data from DBPCFC are needed for both the processed and unprocessed food. Indeed, as already mentioned, the ideal study would investigate the processed and unprocessed food in the same group of well-characterized allergic individuals using foods and processing methods that had been well defined. The result of such a study might resemble Figure 15.4, the effect of processing being represented quantitatively by the ratio of ED_{50}'s, for instance. Few published studies meet these criteria. The question arises, therefore, whether conclusions about changes in thresholds can be drawn from other types of data. These data could include challenge studies in which different forms of the food had been examined in different populations, and the numerous studies using serological methods to investigate alterations in allergenicity, as well as epidemiological data. Two illustrative examples will serve to highlight the possibilities and limitations of the available data.

15.4.2.1. The Effect of Roasting on MED for Hazelnut Protein. Skamstrup-Hansen and colleagues [81] investigated the effect of roasting hazelnuts on the reactivity of individuals, whose hazelnut allergy was the result of cross-reactivity to Bet v 1. The study included 17 patients based in two centers: one in Denmark (Copenhagen) and the other in Switzerland (Zurich). In addition to DBPCFC with hazelnut, masked in a pudding, other clinical assessments included specific IgE (CAP-RadioAllergosorbent Test™ [RAST; Phadia AB] and enzyme allergosorbent test [EAST] inhibition), skin prick tests, and *ex vivo* basophil histamine release. In the DBPCFC, all subjects showed a positive response to challenge with raw hazelnut, while only 5/17 showed a response to the roasted nut. In parallel, median MED increased from 2 to 3g for raw hazelnut to about 7g when they were roasted. This study thus showed a clear effect of processing. However, it also provided information about the value of other measures to determine MEDs. The skin prick test was positive in 16/17 subjects to raw hazelnut, but in only 4/17 to roasted hazelnut. Furthermore, of those four reactions, only one was positive in an individual who was positive to DBPCFC. IgE binding confirmed the skin prick test results, although it was less sensitive with raw hazelnut anyway, possibly due to the lability of the allergen itself. Histamine release showed more promise, with a significant reduction in 9 out of 10 tested, and an estimated 50% reduction in allergenic eliciting potency overall. However, none of the *in vitro* assays was able to discriminate predictively those who would react in DBPCFC and those who would not.

15.4.2.2. The Effect of Roasting on MED for Peanut. No studies equivalent to the Kamstrup-Hansen study exist for peanut. However, comparing the results of two early DBPCFC studies on unroasted peanuts with those of a recent study on roasted peanuts might provide at least the upper limits of a quantitative estimate of the difference in allergenic potency between them. The studies in question are those of May [82], Bock et al. [83], and Wensing et al. [42]. The early studies included a total of 20 patients with a mean age of 9, most of them suffering from asthma, but none of which had experienced anaphylaxis. They were challenged with increasing doses of peanut flour equivalent to 25–2000mg of peanut protein, presented as capsules, except where the patient was under 5, where a pudding-type vehicle was used. In contrast, Wensing and colleagues' study examined adults (mean age 25), 11 of whom had experienced severe symptoms, including three cases of anaphylaxis. Challenge doses ranged from 30μg to 1000mg peanut protein in half-$\log_{10}$ increments, giving a maximum cumulative dose of 1443mg. The peanut flour was incorporated into a pudding type of matrix. The frequency dose distribution of MEDs for these studies is illustrated in Figure 15.5.

Comparison of the amount of allergen that would provoke reactions in half the tested population shows that roasted peanuts appear to be of the order of 100-fold more potent than their unroasted counterparts. However, the actual difference is likely to be considerably less because of differences in the design of the more recent study compared with the earlier ones. The Wensing study included a large proportion of patients who had experienced a severe reaction, whereas severe reactors tended to be excluded from the earlier studies. Critically, capsules were used to administer the allergen in the early studies. This method prevents early warning signs of a reaction in some individuals, resulting in a higher NOAEL. Wensing et al. also used a different basis for determining MEDs, citing both those for "subjective" symptoms and the dose that gave the first "objective" reaction. Consequently, it is difficult to draw conclusions about how much greater the allergenic potency of roasted peanuts is, compared with unroasted ones.

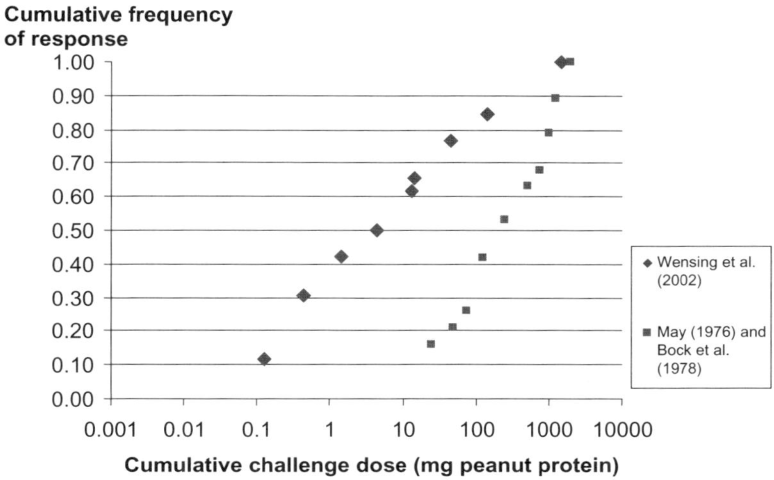

Fig. 15.5 Evaluation of the effect of heating on peanut allergenicity based on challenge results. This evaluation is based on two studies performed on different patients at different times. It illustrates how this type of data may be used, despite its limitations.

15.5. FOODS WITH REDUCED ALLERGENICITY: THE ROLE OF RISK ASSESSMENT

As discussed above, processing can alter the allergenicity of food products. It follows that the techniques used could also be employed specifically to reduce the allergenicity of foods, with a view to widen safe food choices for food-allergic individuals. Critical issues that need to be addressed in this context are how to demonstrate and particularly quantify reduced allergenicity, how to ascertain an absence of potentially adverse unintended effects, and how to maintain the wholesomeness of foods with reduced allergenicity. This section will focus on the first two aspects, since maintaining the wholesomeness of a food is a matter of food technology rather than safety. Some processes, such as edible oil refining or fermentation processes, result in a derived product that is less allergenic than the source material because they remove or destroy most of the protein fraction. However, this reduction in allergenicity is merely a side effect of making the product. This section will not discuss these products but rather products where reduced allergenicity is a primary goal. Early efforts in this field were focused on making infant formula milk products safe for infants with milk allergy. The resulting products, based on different extents of hydrolysis of the milk proteins, achieved a degree of hypoallergenicity, inasmuch as some but not all milk-allergic infants could tolerate them. However, they also illustrated one of the problems that must still be addressed, namely how safe are they for the intended at-risk group for whom they were developed. Quite early on in their development, it became clear that they could cause severe reactions in some individuals [84, 85], making it imperative to find some way of stratifying risk in that population. Initial safety assessment of these products consisted of observing whether they could provoke anaphylaxis in guinea pigs sensitized to milk proteins, when injected intravenously [86], or could generate a response in rabbits using a hyperimmunization protocol [87]. A further criterion, arguably more predictive, was introduced, although it was never formulated as a formal test as it involved human infants. This criterion proposed that an infant formula product could be designated as hypoallergenic if it failed to produce reactions in at least 90% of infants with milk allergy at a statistical confidence level of 95% [88]. Experience with these products indicated that, for milk proteins, peptides below 2 kDa presented very little potential to trigger IgE-mediated reactions, while those between 2 and 5 kDa had only limited ability to do so, compared with the native proteins of milk. Unfortunately, these products serve a very limited purpose and, in particular, provide little general guidance on how to develop foods with reduced allergenicity. To start with, products tended to be somewhat unpalatable [89, 90], and of course, the hydrolytic process destroyed any structure that the proteins provided. Attempts to reduce the allergenicity of several foods have since been attempted, including wheat [91–93], buckwheat [94], rice [95, 96], peach juice and nectar [97], tomatoes [98], apple [99], and soy [100]. Various techniques have been used to achieve this and to estimate changes in allergenicity. Thus, much of the work with wheat has involved treatment of conventional wheat flour with enzymes, while reduction of soy allergenicity involved fermentation of the flour with different types of microorganisms used in food. In contrast, genetic modification techniques, in particular, gene silencing, have been employed with tomatoes and apples, while lye peeling and ultrafiltration were used for peach juice and nectar. Different methods were also used to assess the reduction in allergenic potential ranging from *in vivo* provocation trials (wheat), to skin prick testing (tomato, apple), and various forms of immunochemical assessment, such as immunoblotting (tomato, soy hydrolysates). As already mentioned, only DBPCFC can reliably quantify any reduction in allergenicity and provide data to assess the residual risk. Prior to such evaluation, however, an assessment of possible toxicity needs to have been performed. Depending on the extent of the modifications and the extent to which they could have had unforeseen effects, this assessment may involve extensive animal testing prior to consumption by human volunteers. This may explain why relatively few products resulting from novel techniques, such as genetic modification, have yet been evaluated in this manner. To conclude, production of hypoallergenic foods is undoubtedly feasible, but applications beyond infant formulas have been limited. Considerations that affect the development of such products include the extent to which the modified food differs from its conventional counterpart, which will influence how widely it can be used, as well as demonstration of efficacy (reduction of allergenicity) and safety, in relation to both allergenicity and more general toxicological safety.

15.6. CHALLENGES WITH ALLERGEN DETECTION TOOLS

As previously discussed, risk assessment requires an assessment of exposure and therefore of the amount of allergenic material in a food or food product. Since allergens are the entities that interact with the immune system to provoke allergic reactions, it is logical that they should be targeted as primary analytes. However, while proteins that bind IgE have been identified in a number of foods, this identification may not encompass all of them, and practical constraints make it almost impossible to formally demonstrate a specific protein's involvement in the pathological process by challenge. In practice, therefore, allergen detection assays have usually targeted the whole protein component of an allergenic food, although there are some notable exceptions. Unsurprisingly, immunochemical assays that exploit the specificity and high-affinity interactions of antibodies with protein molecules, such as enzyme-linked immunosorbent assays (ELISAs), have been much favored in allergen analysis [101, 102]. The specificity and sensitivity of ELISA technology, with limits of detection or quantification at low milligram per kilogram level or below, make it a relatively simple tool for allergen detection and quantification, and allow relatively fast and high throughput analysis. It is widely used in food industry laboratories and by official food control bodies to detect and quantify allergens present in food products or commodities. Critical considerations in the development and use of such assays include selection of the material for production of the antibodies (whole food, protein extract, selected purified proteins) and its state (raw, processed, etc.) and possibly the matrix. Processing can alter not only the recognition of the protein(s) of interest but it can also affect how readily it can be extracted from the food for analysis and, in the case of multiple proteins, the profile. Extraction from many matrices is quite limited as well as variable; recovery may also differ between different allergenic proteins and can render the profile of the extracted material quite differently from that of the material used to prepare the detection antibodies. Thus, the accurate quantification that is necessary for risk assessment judgements is often more complex than simply applying a manufacturers' instructions. Instead, understanding the material used to develop the test kit and how it relates to the form of the analyte to be quantified, as well as how effectively it is extracted from the food matrix, is critical. Validation of methods in the individual circumstances of the food manufacturing facility is needed, with appropriate positive controls, such as products manufactured with the allergenic ingredient of interest. So far, ELISA test kits validated for defined matrices include peanut (in cereals, cookies, ice cream, and chocolate; under the auspices of the Association of Analytical Communities [AOAC] and the

European Community Joint Research Centre [EC-JRC]) [103, 104] and hazelnut (in cereals, ice cream, and chocolate; under the auspices of the German Federal Office for Consumer Protection and Food Safety, BVL).

Related antibody-based technologies, which are semiquantitative, include dipsticks and lateral flow devices (LFDs) and are well suited to testing outside the laboratory (e.g., monitoring clean down of food processing lines), where a rapid result is required or qualitative results are needed for only a few samples [105].

Rapid DNA-based tests based on real-time polymerase chain reaction (rt-PCR) may be valuable where no immunochemical assay exists. However, their application to risk assessment is difficult since they do not detect the primary analyte, and it cannot be assumed that the presence of DNA implies the presence of (allergenic) protein, or protein in an allergenic form. Furthermore, quantity of DNA extracted may not relate linearly to amount of protein present, making the value of such results very limited for risk assessment. Short DNA sequences may be detectable in highly processed foods, whereas the protein may no longer be present. Similarly, rt-PCR does not distinguish between different products from the same species: cow's milk and beef, or hen's egg and chicken, are all detected as simply tissues from the same species. Further, PCR detection of DNA from tissues that contain allergens whose content depends on external factors (e.g., temperature, disease, stress) may also be an unreliable method for determining allergen content. On the other hand, antibodies found in commercial ELISAs may no longer detect fragments that still elicit allergic responses.

Mass spectrometric methods offer a potential alternative confirmatory method to the immunochemical and DNA methods since they have the potential to detect protein (and are therefore focused on the hazard itself), to provide information on sequence (giving the species specificity provided by DNA methods), and to detect allergen contamination down to similar levels to those achieved by ELISA and PCR. The automated nature of mass spectrometry experiments and minimum of user involvement naturally lends itself to high-throughput quantification. However, such developments are still at an early stage, with applications to the detection of peanut allergens having been demonstrated only recently [106, 107].

15.7. CHALLENGES AND LIMITATIONS IN DETERMINING ALLERGEN EXPOSURE FOR RISK ASSESSMENT

Exposure in the context of food allergic reactions differs somewhat from exposure as usually considered for toxicological end points. With conventional toxicological end points, long-term exposure is the principal concern, and the metric used is the acceptable or tolerable daily or weekly intake. In contrast, when considering acute allergic reactions to foods, the exposure of concern is that which may occur over a relatively short period, such as a meal occasion.

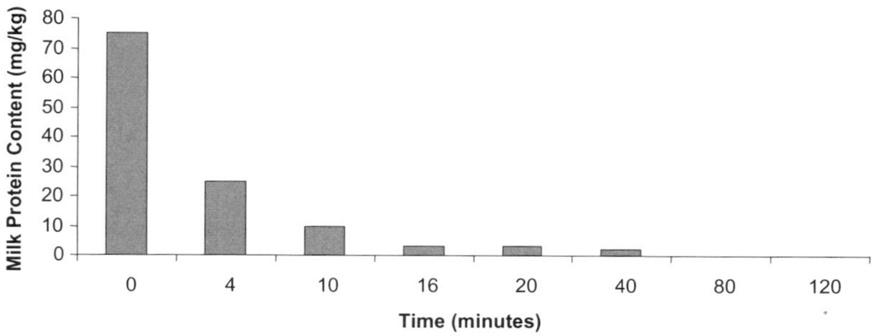

Fig. 15.6 Illustration of the variability of residual allergen content. In this particular illustration, a product that does not contain milk follows one that includes milk in the recipe. The residual milk content is measured in samples taken from the line at different times after changeover.

Assessing allergen exposure requires knowledge of at least two pieces of information. The first is the allergen content of the product of interest, as determined from its concentration (through calculation or analysis) and the amount of the product that will likely be consumed on a relevant occasion. Detecting and measuring allergen in foods has been discussed in the preceding section. Allied with a consideration of allergen content is its distribution among the batch of product made on any one run. Unlike allergen used as an ingredient, allergen present by cross-contact will often be present at a range of concentrations depending, for instance, on how close to the changeover a particular unit of product was produced (Fig. 15.6). Thus, cross-contact allergen will be represented by a distribution, which can be taken into account in the risk assessment if data are available.

The second piece of information required to determine exposure is knowledge of the consumption pattern for the product of interest, preferably in the allergic population if that information is available. Consumption pattern in this context means not only what proportion of the population consumes the product and how frequently, but also how much is generally consumed on any one occasion. Such information may be available from national dietary surveys, such as the Netherlands National Dietary Survey. Again, these data may also contain information about distribution of these intakes, at least in terms of mean and median intakes, as well as the ninetieth or ninety-fifth percentile intakes. They therefore also lend themselves well to a probabilistic approach, as illustrated conceptually in the papers of Spanjersberg et al. [108] and Kruizinga et al. [34].

15.8. LIMITATIONS OF EXISTING ALLERGEN MANAGEMENT PROGRAMS

Allergen management programs aim to control the presence of allergens within products. They ensure that allergenic ingredients are appropriately declared. They also ensure that allergenic materials, which are not ingredients, are excluded so that they do not constitute a risk for allergic individuals, or, if that is not possible, that people at risk are advised of their possible presence. The principles of allergen management have been described in some detail elsewhere [109] and are now accepted to form an integral part of food safety management systems (e.g., ISO 22000:2005). The limitations of allergen management plans can arise from a variety of sources, such as inadequate communications with suppliers or suboptimal design of equipment for easy cleaning. However, since this chapter discusses risk assessment, only limitations related to risk assessment will be discussed here. A critical concept behind all such programs is that they need to consider the presence of allergens at all stages from the design of products, through raw materials and the supply chain to manufacturing and delivery to the final consumer. At each of those stages, this may require a quantitative risk assessment to determine whether management actions have been adequate. Risk assessment is thus critical to allergen management, and the limitations described above in relation to risk assessment limit the efficacy of programs. Less directly, they can also impose additional costs on a manufacturer. For instance, if an accurate risk assessment is not possible, sanitation protocols may be more stringent than they need to be, with more time spent and the line being unavailable for longer. Probably one of the most difficult areas for manufacturers is deciding when to apply precautionary labeling to a product, since this decision encompasses not only the risk assessment, with all its uncertainties, but also a judgement about acceptable or tolerable risk. While it is well accepted that some allergic individuals may react to very small amounts of an allergen, the effect of overuse or abuse of precautionary labeling is much more difficult to quantify. Furthermore, there is a potential tension between the general desirability of limiting precautionary labeling in order to preserve its value and a manufacturer's desire to avoid its products from being involved in provoking allergic reactions.

15.9. CONCLUSIONS

Food allergy is a chronic disease with acute manifestations, which sometimes can be life threatening. The range of doses over which the population of allergic individuals reacts spans up to six orders of magnitude. At an individual level, reactive doses can also vary over time and depending on the cofactors such as exercise, medication, asthma status, stress, and others. These uncertainties have led to a number of consequences that have had an adverse effect on individuals with a food allergy. In the first instance, they have led food manufacturers to make extensive use of precautionary labeling to warn about the possible presence of particular allergens. Second, the general clinical advice to allergic individuals has been to avoid the offending food, without any consideration of their degree of reactivity. However, avoidance can be difficult, as well as inconvenient and deleterious to the quality of life in many respects.

In contrast to hazard-based management of allergens and food allergy, risk-based approaches offer a means to improve the situation both for allergic consumers and food manufacturers, as well as enabling healthcare professionals to provide advice much more tailored to the individual. Sufficient data are now available in the public domain to model population dose distribution of reactivity for some allergenic foods and thus provide quantitative estimates of the population at risk. Allied to estimates of residual allergen content in specific foods and data on their consumption, a full probabilistic quantitative risk assessment can now be developed in a number of cases. Such approaches are of particular value in allergen risk management where they can provide information about the potential effectiveness of different risk reduction measures.

Of course, the application of risk assessment approaches to allergenic foods still presents a number of challenges. Quality of the available data and uncertainty about its applicability are probably the greatest, both with regard to clinical data and data on exposure and allergen content.

REFERENCES

1. Prausnitz, C., Küstner, H. (1921). Studien uber die Uberempfindlichkeit. *Zentralblatt fur Bakteriologie*, *86*, 160–169.
2. Tuft, L., Blumstein, G.I. (1942). Studies in food allergy II. Sensitization to fresh fruits: clinical and experimental observations. *The Journal of Allergy*, *13*, 574–581.
3. Loveless, M.H. (1950). Milk allergy: a survey of its incidence; experiments with a masked ingestion test. *The Journal of Allergy*, *21*, 489–499.
4. Young, E., Stoneham, M.D., Petruckevitch, A., Barton, J., Rona, R. (1994). A population study of food intolerance. *The Lancet*, *343*, 1127–1130.
5. Bruijnzeel-Koomen, C., Ortolani, C., Aas, K., Bindslev-Jensen, C., Bjorksten, B., Moneret-Vautrin, D., Wüthrich, B. (1995). Adverse reactions to food. European Academy of Allergology and Clinical Immunology Subcommittee. *Allergy*, *50*, 623–635.
6. Kanny, G., Moneret-Vautrin, D.A., Flabbee, J., Beaudouin, E., Morisset, M., Thevenin, F. (2001). Population study of food allergy in France. *The Journal of Allergy and Clinical Immunology*, *108*, 133–140.
7. Sicherer, S.H., Munoz-Furlong, A., Sampson, H.A. (2004). Prevalence of seafood allergy in the United States determined by a random telephone survey. *The Journal of Allergy and Clinical Immunology*, *114*, 159–165.
8. Pereira, P., Venter, C., Grundy, J., Clayton, C.B., Arshad, S.H., Dean, T.P. (2005). Prevalence of sensitization to food allergens, reported adverse reaction to foods, food avoidance, and food hypersensitivity among teenagers. *The Journal of Allergy and Clinical Immunology*, *116*, 884–892.
9. Venter, C., Pereira, B., Grundy, J., Clayton, C.B., Arshad, S.H., Dean, T.P. (2006). Prevalence of sensitization reported and objectively assessed food hypersensitivity amongst six-year-old children: a population-based study. *Pediatric Allergy and Immunology*, *17*, 356–363.

10. Rona, R.J., Keil, T., Summers, C., Gislason, D., Zuidmeer, L., Sodergren, E., Sigurdardottir, S.T., Lindner, T., Goldhahn, K., Dahlstrom, J., McBride, D., Madsen, C. (2007). The prevalence of food allergy: a meta-analysis. *The Journal of Allergy and Clinical Immunology*, *120*, 638–646.
11. Sampson, H.A. (2004). Update on food allergy. *The Journal of Allergy and Clinical Immunology*, *113*, 805–819.
12. Taylor, S.L., Crevel, R.W.R., Sheffield, D., Kabourek, J., Baumert, J. (2009). Threshold dose for peanut: risk characterization based upon published results from challenges of peanut-allergic individuals. *Food and Chemical Toxicology*, *47*, 1198–1204.
13. DunnGalvin, A., Dubois, A.E.J., de Blok, B.M.J., Hourihane, J.O. (2007). Child vs maternal perception of HRQoL in food allergy: developmental trajectories and evolution of risk behaviour. *Allergy*, *62*, 70–166.
14. Flokstra-de Blok, B.M.J., DunnGalvin, A., Vlieg-Boerstra, B.J., Oude Elberink, J.N.G., Duiverman, E.J., Hourihane, J.O., Dubois, A.E.J. (2009). Development and validation of the food allergy quality of life questionnaire-adult form (FAQLQ-AF). *Allergy*, *64*, 1209–1217.
15. Muñoz-Furlong, A. (2003). Daily coping strategies for patients and their families. *Pediatrics*, *111*, 1654–1661.
16. Crevel, R.W.R. (2005). Industrial dimensions of food allergy. *The Proceedings of the Nutrition Society*, *64*, 470–474.
17. Sicherer, S.H., Noone, S.A., Munoz-Furlong, A. (2001). The impact of childhood food allergy on quality of life. *Annals of Allergy, Asthma & Immunology*, *87*, 461–464.
18. Avery, N.J., King, R.M., Knight, S., Hourihane, J.O. (2003). Assessment of quality of life in children with peanut allergy. *Pediatric Allergy and Immunology*, *14*, 378–382.
19. Mills, E.N., Valovirta, E., Madsen, C., Taylor, S.L., Vieths, S., Anklam, E., Baumgartner, S., Koch, P., Crevel, R.W., Frewer, L. (2004). Information provision for allergic consumers—where are we going with food allergen labelling? *Allergy*, *59*, 1262–1268.
20. Christie, L., Hine, J., Parker, J.G., Burks, A.W. (2003). Food allergies in children affect nutrient intake and growth. *Journal of the American Dietetic Association*, *102*, 1648–1651.
21. Ambler-Edwards, S., Bailey, K., Kiff, A., Lang, T., Lee, R., Marsden, T., Simons, D., Tibbs, H. (2000). Food futures: rethinking UK strategy. Available at http://www.chathamhouse.org.uk/publications/papers/view/-/id/695/
22. ISO 22000:2005. (2005). Food safety management systems—requirements for any organization in the food chain. International Organisation for Standardisation. Available at http://www.iso.org
23. FAO/WHO. (1995). Application of risk analysis to food standards issues. Report of the Joint FAO/WHO Expert Consultation, March 13–17, World Health Organisation, Geneva, Switzerland. Available at http://www.who.int/foodsafety/publications/micro/en/march1995.pdf
24. FAO/WHO. (1997). Risk management and food safety. Report of a Joint FAO/WHO Consultation. *FAO Food and Nutrition Paper 65*, January 27–31. Rome,

Italy: Food and Agriculture Organisation of the United Nations. Available at http://www.fao.org/docrep/W4982E/W4982E00.htm

25. Codex Alimentarius Commission. (2003). Principles for the risk analysis of foods derived from modern biotechnology. CAC/GL44-2003. Codex Alimentarius Commission, Joint FAO/WHO Food Standards Programme, Food and Agriculture Organisation, Rome. Available at http://www.codexalimentarius.net/download/standards/10007/CXG_044e.pdf
26. Breiteneder, H., Mills, E.N.C. (2005). Molecular properties of food allergens. *The Journal of Allergy and Clinical Immunology*, *115*, 14–23.
27. Peeters, K.A., Koppelman, S.J., Van Hoffen, E., van der Tas, C.W., Hartog Jager, C.F., Penninks, A.H. (2007). Does skin prick test reactivity to purified allergens correlate with clinical severity of peanut allergy? *Clinical and Experimental Allergy*, *37*, 108–115.
28. Dearman, R.J., Alcocer, M.J., Kimber, I. (2007). Influence of plant lipids on immune responses in mice to the major Brazil nut allergen Ber e 1. *Clinical and Experimental Allergy*, *37*, 582–591.
29. Crevel, R.W.R., Briggs, D., Hefle, S.L., Knulst, A.C., Taylor, S.L. (2007). Hazard characterization in food allergen risk assessment: the application of statistical approaches and the use of clinical data. *Food Chemistry and Toxicology*, *45*, 691–701.
30. Grimshaw, K.E.C., King, R.M., Nordlee, J.A., Hefle, S.L., Warner, J.O., Hourihane, J.O. (2003). Presentation of allergen in different food preparations affects the nature of the allergic reaction—a case series. *Clinical and Experimental Allergy*, *33*, 1581–1585.
31. FDA. (2006). Approaches to establish thresholds for major food allergens and for gluten in food. The Threshold Working Group, Department of Health and Human Services, United States Food and Drug Administration. Available at http://www.fda.gov/Food/LabelingNutrition/FoodAllergensLabeling/GuidanceComplianceRegulatoryInformation/ucm106108.htm
32. Madsen, C.B., Hattersley, S., Buck, J., Gendel, S.M., Houben, G.F., Hourihane, J.O., Mackie, A.R., Mills, E.N., Nørhede, P., Taylor, S.L., Crevel, R.W.R. (2009). Approaches to risk assessment in food allergy: report from a workshop developing a framework for assessing the risk from allergenic foods. *Food Chemistry and Toxicology*, *47*, 480–489.
33. Allergen Bureau. (2008). What is voluntary incidental trace allergen labelling (VITAL)? Available at http://www.allergenbureau.net/allergen-guide/vital/
34. Kruizinga, A.G., Briggs, D., Crevel, R.W.R., Knulst, A.C., van den Bosch, L.M.C., Houben, G.F. (2008). Probabilistic risk assessment model for allergens in food: sensitivity analysis of the minimum eliciting dose and food consumption. *Food and Chemical Toxicology*, *46*, 1437–1443.
35. Pele, M., Brohée, M., Anklam, E., Van Hengel, A.J. (2007). Peanut and hazelnut traces in cookies and chocolates: relationship between analytical results and declaration of food allergens on product labels. *Food Additives and Contaminants*, *24*, 1334–1344.
36. Bindslev-Jensen, C., Ballmer-Weber, B.K., Bengtsson, U., Blanco, C., Ebner, C., Hourihane, J., Knulst, A.C., Moneret-Vautrin, D.A., Nekam, K., Niggemann, B., Osterballe, M., Ortolani, C., Ring, J., Schnopp, C., Werfel, T. (2004). Standardization

of food challenges in patients with immediate reactions to foods—position paper from the European Academy of Allergology and Clinical Immunology. *Allergy*, *59*, 690–697.

37. Taylor, S.L., Hefle, S.L., Bindslev-Jensen, C., Bock, S.A., Burks, A.W., Christie, L., Hill, D.J., Host, A., Hourihane, J.O., Lack, G., Metcalfe, D.D., Moneret-Vautrin, D.A., Vadas, P.A., Rance, F., Skrypec, D.J., Trautman, T.A., Malmeheden Yman, I., Zeiger, R.S. (2002). Factors affecting the determination of threshold doses for allergenic foods: how much is too much? *The Journal of Allergy and Clinical Immunology*, *109*, 24–30.
38. Oppenheimer, J.J., Nelson, H.S., Bock, S.A., Christensen, F., Leung, D.Y.M. (1992). Treatment of peanut allergy with rush immunotherapy. *The Journal of Allergy and Clinical Immunology*, *90*, 256–262.
39. Taylor, S.L., Hefle, S.L., Bindslev-Jensen, C., Atkins, F.M., Andre, C., Bruijnzeel-Koomen, C., Burks, A.W., Bush, R.K., Ebisawa, M., Eigenmann, P.A., Host, A., Hourihane, J.O., Isolauri, E., Hill, D.J., Knulst, A., Lack, G., Sampson, H.A., Moneret-Vautrin, D.A., Rance, F., Vadas, P.A., Yunginger, J.W., Zeiger, R.S., Salminen, J.W., Madsen, C., Abbott, P. (2004). A consensus protocol for the determination of the threshold doses for allergenic foods: how much is too much? *Clinical and Experimental Allergy*, *34*, 689–695.
40. Perry, T.T., Matsui, E.C., Conocer-Walker, M.K., Wood, R.A. (2004). Risk of oral food challenges. *The Journal of Allergy and Clinical Immunology*, *114*, 1164–1168.
41. Sampson, H.A., Ho, D.G. (1997). Relationship between food-specific IgE concentrations and the risk of positive food challenges in children and adolescents. *The Journal of Allergy and Clinical Immunology*, *100*, 444–451.
42. Wensing, M., Penninks, A.H., Hefle, S.L., Koppelman, S.J., Bruijnzeel-Koomen, C.A.F.M., Knulst, A.C. (2002). The distribution of individual threshold doses eliciting allergic reactions in a population with peanut allergy. *The Journal of Allergy and Clinical Immunology*, *110*, 915–920.
43. Hourihane, J.O'B., Grimshaw, K.E., Lewis, S.A., Briggs, R.A., Trewin, J.B., King R.M., Kilburn, S.A., Warner, J.O. (2005). Does severity of low-dose, double-blind placebo-controlled food challenges reflect severity of allergic reactions to peanut in the community? *Clinical and Experimental Allergy*, *35*, 1227–1233.
44. Sicherer, S.H., Morrow, E.H., Sampson, H.A. (2000). Dose-response in double-blind, placebo-controlled oral food challenges in children with atopic dermatitis. *The Journal of Allergy and Clinical Immunoloyg*, *105*, 582–586.
45. Wensing, M., Penninks, A.H., Hefle, S.L., Akkerdaas, J.H., van Ree, R., Koppelman, S.J., Bruijnzeel-Koomen, C.A., Knulst, A.C. (2002). The range of minimum provoking doses in hazelnut-allergic patients as determined by double-blind, placebo-controlled food challenges. *Clinical and Experimental Allergy*, *32*, 1757–1762.
46. Bullock, R.J., Barnett, D., Howden, M.E. (2005). Immunologic and clinical response to parenteral immunotherapy in peanut anaphylaxis—a study using IgE and IgG4 immunoblot monitoring. *Allergologia et Immunopathologia*, *33*, 250–256.
47. Osterballe, M., Bindslev-Jensen, C. (2003). Threshold levels in food challenge and specific IgE in patients with egg allergy: is there a relationship? *The Journal of Allergy and Clinical Immunology*, *112*, 196–201.

48. Morisset, M., Moneret-Vautrin, D.A., Kanny, G., Guenard, L., Beaudouin, E., Flabbee, J., Hatahet, R. (2003). Thresholds of clinical reactivity to milk, egg, peanut and sesame in immunoglobulin E-dependent allergies: evaluation by double-blind or single-blind placebo-controlled oral challenges. *Clinical and Experimental Allergy*, *33*, 1046–1051.
49. Flintermann, A.E., Pasmans, S.G., Hoekstra, M.O., Meijer, Y., van Hoffen, E., Knol, E.F., Hefle, S.L., Bruijnzeel-Koomen, C.A., Knulst, A.C. (2006). Determination of no-observed-adverse-effect levels and eliciting doses in a representative group of peanut-sensitized children. *The Journal of Allergy and Clinical Immunology*, *117*, 448–454.
50. Scibilia, J., Pastorello, E.A., Zisa, G., Ottolenghi, A., Bindslev-Jensen, C., Pravettoni, V., Scovena, E., Robino, A., Ortolani, C. (2006). Wheat allergy: a double-blind, placebo-controlled study in adults. *The Journal of Allergy and Clinical Immunology*, *117*, 433–439.
51. Mandalari, G., Adel-Patient, K., Barkholt, V., Bublin, M., Bennett, L., Baro, C., Graser, G., Gaier, S., Ladics G.S., Mierzejewska, D., Vassilopoulou, E., Vissers, Y., Zuidmeer, L., Rigby, N.M., Salt, L., Defernez, M., Mulholland, F., Mackie, A.R., Wickham, M.S., Mills, E.N.C. (2009). *In vitro* digestibility of α-casein and β-lactoglobulin under simulated human gastric and duodenal conditions: a multi-laboratory evaluation. *Regulatory Toxicology and Pharmacology*, *55*, 372–381.
52. Hourihane, J.O., Knulst, A.C. (2005). Thresholds of allergenic proteins in foods. *Toxicology and Applied Pharmacology*, *207*, 152–156.
53. Host, A. (1994). Cow's milk protein allergy and intolerance in infancy: some clinical, epidemiological and immunological aspects. *Pediatric Allergy and Immunology*, *5*, 1–36.
54. Skolnick, H.S., Conocer-Walker, M.K., Koerner, C.B., Sampson, H.A., Burks, W., Wood, R.A. (2001). The natural history of peanut allergy. *The Journal of Allergy and Clinical Immunology*, *107*, 367–374.
55. Zuberbier, T., Edenharter, G., Worm, M., Ehlers, I., Reimann, S., Hantke, T., Roehr, C.C., Bergmann, K.E., Niggemann, B. (2004). Prevalence of adverse reactions to food in Germany—a population study. *Allergy*, *59*, 338–345.
56. Thomas, K., Herouet-Guicheney, C., Ladics, G., Bannon, G., Cockburn, A., Crevel, R., Fitzpatrick, J., Mills, C., Privalle, L., Vieths, S. (2007). Evaluating the effect of food processing on the potential human allergenicity of novel proteins: international workshop report. *Food and Chemical Toxicology*, *45*, 1116–1122.
57. Sathe, S.K., Sharma, G. (2009). Effects of food processing on food allergens. *Molecular Nutrition and Food Research*, *53*, 970–978.
58. Mittag, D., Vieths, S., Vogel, L., Becker, W.M., Rihs, H.P., Helbling, A., Wuthrich, B., Ballmer-Weber, B.K. (2004). Soybean allergy in patients allergic to birch pollen: clinical investigation and molecular characterization of allergens. *The Journal of Allergy and Clinical Immunology*, *113*, 148–154.
59. Crevel, R.W., Ballmer-Weber, B.K., Holzhauser, T., Hourihane, J.O., Knulst, A.C., Mackie, A.R., Timmermans, F., Taylor, S.L. (2008). Thresholds for food allergens and their value to different stakeholders. *Allergy*, *63*, 597–609.
60. Ciardiello, A.M., Giangrieco, I., Tuppo, L., Tamburrini, M., Buccheri, M., Palazzo, P., Bernardi, M.L., Ferrara, R., Mari, A. (2009). Influence of the natural ripening stage, cold storage, and ethylene treatment on the protein and IgE-binding profiles

of green and gold kiwi fruit extracts. *Journal of Agricultural and Food Chemistry*, *57*, 1565–1571.

61. Huijbers, G.B., Colen, A.A.M., Jansen, J.J.N., Kardinaal, A.F.M., Vlieg-Boerstra, B.J., Martens, B.P.M. (1994). Masking foods for food challenge: practical aspects of masking foods for a double-blind, placebo-controlled food challenge. *Journal of the American Dietetic Association*, *94*, 645–649.
62. van Odijk, J., Ahlstedt, S., Bengtsson, U., Borres, M.P., Hulthen, L. (2005). Double-blind placebo-controlled challenges for peanut allergy: the efficiency of blinding procedures and the allergenic activity of peanut availability in the recipes. *Allergy*, *60*, 602–605.
63. Vlieg-Boerstra, B.J., Bijleveld, C.M.A., van der Heide, S., Beusekamp, B.J., Wolt-Plompen, S.A.A., Kukler, J., Brinkman, J., Duiverman, E.J., Dubois, A.E.J. (2004). Development and validation of challenge materials for double-blind, placebo controlled food challenges in children. *The Journal of Allergy and Clinical Immunology*, *113*, 341–346.
64. Sampson, H.A., Cooke, S. (1992). The antigenicity and allergenicity of microparticulated proteins: Simplesse. *Clinical and Experimental Allergy*, *22*, 963–969.
65. Poms, R.E., Anklam, E. (2004). Effect of chemical, physical, and technological processes on the nature of food allergens. *Journal of AOAC International*, *87*, 1466–1474.
66. Hefle, S.L. (1999). Impact of processing on food allergens. *Advances in Experimental Medicine and Biology*, *459*, 107–119.
67. Sathe, S.K., Teuber, S.S., Roux, K.H. (2005). Effects of food processing on the stability of food allergens. *Biotechnology Advances*, *23*, 423–429.
68. Mills, E.N.C., Mackie, A.R. (2008). The impact of processing on allergenicity of food. *Current Opinion in Allergy and Clinical Immunology*, *8*, 249–253.
69. Crevel, R.W.R., Kerkhoff, M.A., Koning, M.M. (2000). Allergenicity of refined vegetable oils. *Food Chemistry and Toxicology*, *38*, 385–393.
70. Kleinman, R.E. (1992). Cow's milk allergy in infancy and hypoallergenic formulas. *The Journal of Pediatrics*, *121*, S116–S121.
71. Dube, M., Zunker, K., Neidhart, S., Carle, R., Steinhart, H., Paschke, A. (2004). Effect of technological processing on the allergenicity of mangoes (*Mangifera indica* L.). *Journal of Agricultural and Food Chemistry*, *52*, 3938–3944.
72. Clemente, A., Vioque, J., Sanchez-Vioque, R., Pedroche, J., Millan, F. (1999). Production of extensive chickpea (*Cicer arietinum* L.) protein hydrolysates with reduced antigenic activity. *Journal of Agricultural and Food Chemistry*, *47*, 3776–3781.
73. Chen, L., Lucas, J.S., Hourihane, J.O., Lindemann, J., Taylor, S.L., Goodman, R.E. (2006). Evaluation of IgE binding to proteins of hardy (*Actinidia arguta*), gold (*Actinidia chinensis*) and green (*Actinidia deliciosa*) kiwifruits and processed hardy kiwifruit concentrate, using sera of individuals with food allergies to green kiwifruit. *Food Chemistry and Toxicology*, *44*, 1100–1107.
74. Alvarez-Alvarez, J., Guillamon, E., Crespo, J.F., Cuadrado, C., Burbano, C., Rodriguez, J., Fernandez, C., Muzquiz, M. (2004). Effects of extrusion, boiling, autoclaving, and microwave heating on lupine allergenicity. *Journal of Agricultural and Food Chemistry*, *52*, 3938–3945.

75. Maleki, S.J., Hurlburt, B.K. (2004). Structural and functional alterations in major peanut allergens caused by thermal processing. *Journal of AOAC International*, *87*, 1475–1479.

76. Simonato, B., Pasini, G., Giannattasio, M., Peruffo, A.D., De Lazzari, F., Curioni, A. (2001). Food allergy to wheat products: the effect of bread baking and *in vitro* digestion on wheat allergenic proteins—a study with bread dough, crumb, and crust. *Journal of Agricultural and Food Chemistry*, *49*, 5668–5673.

77. Beyer, K., Morrow, E., Li, X.M., Bardina, L., Bannon, G.A., Burks, A.W., Sampson, H.A. (2001). Effects of cooking methods on peanut allergenicity. *The Journal of Allergy and Clinical Immunology*, *107*, 1077–1081.

78. Mondoulet, L., Paty, E., Drumare, M.F., Ah-Leung, S., Scheinmann, P., Willemot, R.M., Wal, J.M., Bernard, H. (2005). Influence of thermal processing on the allergenicity of peanut proteins. *Journal of Agricultural and Food Chemistry*, *53*, 4547–4553.

79. Meyer-Pittroff, R., Behrendt, H., Ring, J. (2007). Specific immuno-modulation and therapy by means of high pressure treated allergens. *High Pressure Research*, *27*, 63–67.

80. Leduc, V., Moneret-Vautrin, D.A., Guerin, L., Morisset, M., Kanny, G. (2003). Anaphylaxis to wheat isolates: immunochemical study of a case proved by means of double-blind, placebo-controlled food challenge. *The Journal of Allergy and Clinical Immunology*, *111*, 897–899.

81. Skamstrup-Hansen, K., Ballmer-Weber, B.K., Lüttkopf, D., Skov, P.S., Wüthrich, B., Bindslev-Jensen, C., Vieths, S., Poulsen, L.K. (2003). Roasted hazelnuts—allergenic activity evaluated by double-blind placebo-controlled food challenge. *Allergy*, *58*, 132–138.

82. May, C.D. (1976). Objective clinical and laboratory studies of immediate hypersensitivity reactions to foods in asthmatic children. *The Journal of Allergy and Clinical Immunology*, *58*, 500–515.

83. Bock, S.A., Lee, W.Y., Remigio, L.K., May, C.D. (1978). Studies of hypersensitivity reactions to foods in infants and children. *The Journal of Allergy and Clinical Immunology*, *62*, 327–334.

84. Oldaeus, G., Bradley, C.K., Björkstén, B., Kjellman, N.I. (1992). Allergenicity screening of hypoallergenic milk-based formulas. *The Journal of Allergy and Clinical Immunology*, *90*, 133–135.

85. Hill, D.J., Cameron, D.J.S., Francis, D.E.M., Gonzalez-Andaza, A.M., Hosking, C.S. (1995). Challenge confirmation of late-onset reactions to extensively hydrolyzed formulas in infants with multiple food protein intolerance. *The Journal of Allergy and Clinical Immunology*, *196*, 386–394.

86. Devey, M.E., Anderson, K.J., Coombs, R.R., Henschel, M.J., Coates, M.E. (1976). The modified anaphylaxis hypothesis for cot death: anaphylactic sensitization in guinea-pigs fed cow's milk. *Clinical and Experimental Immunology*, *26*, 542–548.

87. Cordle, C.T., Duska-McEwen, G., Janas, L.M., Malone, W.T., Hirsch, M.A. (1994). Evaluation of the immunogenicity of protein hydrolysate formulas using laboratory animal hyperimmunization. *Pediatric Allergy and Immunology*, *5*, 14–19.

88. American Academy of Pediatrics. Committee on Nutrition. (2000). Hypoallergenic infant formulas. *Pediatrics*, *106*, 346–349.

89. Pedrosa, M., Pascual, C.Y., Larco, J.I., Esteban, M.M. (2006). Palatability of hydrolysates and other substitution formulas for cow's milk-allergic children: a comparative study of taste, smell, and texture evaluated by healthy volunteers. *Journal of Investigational Allergology and Clinical Immunology*, *16*, 351–356.

90. Sawatzki, G., Georgi, G., Kohn, G. (1994). Pitfalls in the design and manufacture of infant formulae. *Acta Paediatrica Supplement*, *402*, 40–45.

91. Watanabe, M., Watanabe, J., Sonoyama, K., Tanabe, S. (2000). Novel method for producing hypoallergenic wheat by enzymatic fragmentation of the constituent allergens and its application to food processing. *Bioscience, Biotechnology, and Biochemistry*, *64*, 2663–2667.

92. Narasaka, S., Endo, Y., Fu, Z., Moriyama, M., Arai, S., Abe, K., Kato, H. (2006). Safety evaluation of hypoallergenic wheat flour using a DNA micro-array. *Bioscience, Biotechnology, and Biochemistry*, *70*, 1464–1470.

93. Tanabe, S. (2008). Analysis of food allergen structures and development of foods for allergic patients. *Bioscience, Biotechnology, and Biochemistry*, *72*, 649–659.

94. Satoh, R., Koyano, S., Takagi, K., Nakamura, R., Teshima, R., Sawada, J. (2008). Immunological characterization and mutational analysis of the recombinant protein BWp16, a major allergen in buckwheat. *Biological and Pharmaceutical Bulletin*, *31*, 1079–1085.

95. Ikezawa, Z., Ikebe, T., Ogura, H., Odajima, H., Kurosaka, F., Komatu, H., Sase, K., Suga, C., Sugiuchi, M., Suguro, H. (1992). Mass trial of hypoallergenic rice (HRS-1) produced by enzymatic digestion in atopic dermatitis with suspected rice allergy. HRS-1 research group. *Acta Dermato-Venereologica Supplementum*, *176*, 108–112.

96. Nakamura, R., Matsuda, T. (1996). Rice allergenic protein and molecular-genetic approach for hypoallergenic rice. *Bioscience, Biotechnology, and Biochemistry*, *60*, 1215–1221.

97. Brenna, O., Pompei, C., Ortolani, C., Pravettoni, V., Farioli, L., Pastorello, E.A. (2000). Technological processes to decrease the allergenicity of peach juice and nectar. *Journal of Agricultural and Food Chemistry*, *48*, 493–497.

98. Lorenz, Y., Enrique, E., Lequynh, L., Fötisch, K., Retzek, M., Biemelt, S., Sonnewald, U., Vieths, S., Scheurer, S. (2006). Skin prick tests reveal stable and heritable reduction of allergenic potency of gene-silenced tomato fruits. *The Journal of Allergy and Clinical Immunology*, *118*, 711–718.

99. Gilissen, L.J., Bolhaar, S.T., Matos, C.I., Rouwendal, G.J., Boone, M.J., Krens, F.A., Zuidmeer, L., Van Leeuwen, A., Akkerdaas, J., Hoffmann-Sommergruber, K., Knulst, A.C., Bosch, D., Van de Weg, W.E., Van Ree, R. (2005). Silencing the major apple allergen Mal d 1 by using the RNA interference approach. *The Journal of Allergy and Clinical Immunology*, *115*, 364–369.

100. Frias, J., Song, Y.S., Martínez-Villaluenga, C., González de Mejia, E., Vidal-Valverde, C. (2008). Immunoreactivity and amino acid content of fermented soybean products. *Journal of Agricultural and Food Chemistry*, *56*, 99–105.

101. Poms, R.E., Klein, C.L., Anklam, E. (2004). Methods for allergen analysis in food: a review. *Food Additives and Contaminants*, *21*, 1–31.

102. Kerbach, S., Alldrick, A.J., Crevel, R.W.R., Dömötör, L., DunnGalvin, A., Mills, E.N.C., Pfaff, S., Poms, R.E., Popping, B., Tömösközi, S. (2009). Managing food

allergens in the food supply chain—viewed from different stakeholder perspectives. *Quality Assurance and Safety of Crops and Foods*, *1*, 50–60.

103. Park, D.L., Coates, S., Brewer, V.A., Garber, E.A.E., Abouzied, M., Johnson, K., Ritter, B., McKenzie, D. (2005). Performance tested method multiple laboratory validation study of ELISA-based assays for the detection of peanuts in food. *Journal of AOAC International*, *88*, 156–160.
104. Poms, R.E., Agazzi, M.-E., Bau, A., Brohee, M., Capelletti, C., Norgaard, J., Anklam, E. (2005). Interlaboratory validation study of five commercial ELISA test kits for the determination of peanut proteins in biscuits and dark chocolate. *Food Additives and Contaminants*, *22*, 104–112.
105. Koppelman, S.J., Hefle, S.L. (2006). *Detecting Allergens in Food*. Cambridge, UK: Woodhead Publishing Limited, pp. 112–113.
106. Shefcheck, K.J., Callahan, J.H., Musser, S.M. (2006). Confirmation of peanut protein using peptide markers in dark chocolate using liquid chromatography-tandem mass spectrometry (LC-MS/MS). *Journal of Agricultural and Food Chemistry*, *54*, 7953–7959.
107. Chassaigne, H., Nørgaard, J.V., Hengel, A.J. (2007). Proteomics-based approach to detect and identify major allergens in processed peanuts by capillary LC-Q-TOF (MS/MS). *Journal of Agricultural and Food Chemistry*, *55*, 4461–4473.
108. Spanjersberg, M.Q., Kruizinga, A.G., Rennen, M.A., Houben, G.F. (2007). Risk assessment and food allergy: the probabilistic model applied to allergens. *Food Chemistry and Toxicology*, *45*, 49–54.
109. Crevel, R.W.R. (2009). Risk management—the principles. In: *Management of Food Allergens*, Coutts, J., Fielder, R. (Eds.). Oxford: Wiley-Blackwell, pp. 83–101.

16

THE CHALLENGES OF PRECAUTIONARY LABELING

FIONA FLEMING, KIRSTEN GRINTER, KIM LEIGHTON,
KEVIN NORMAN, CHRIS PRESTON, AND MARIA SAID

16.1. INTRODUCTION

Managing the risks associated with the presence of allergens in food products is one of the major challenges facing food producers at all stages of the supply chain. Manufacturers grapple with the logistics of factories producing an ever-expanding range of products on shared equipment and the need to provide consistent, accurate, and useful information to allergic consumers and their carers. One of the risk management strategies employed by the food industry for allergen management is to provide consumers with information on the labels of food products to assist them in making informed choices.

In Australia and New Zealand, the requirement for mandatory labeling of allergens present as ingredients came into effect in 2002, ahead of many other countries around the world. Since that time, the issue of what to do about allergenic ingredients that are not intentionally formulated into food products has provided many challenges for both food manufacturers and consumers. In this chapter, we describe the Australian and New Zealand experience of the challenges posed by precautionary labeling as a risk management strategy. We also introduce the voluntary incidental trace allergen labeling (VITAL) process, which has been developed by the Australian and New Zealand food

Allergen Management in the Food Industry, Edited by Joyce I. Boye and
Samuel Benrejeb Godefroy

industry, in response to these challenges in the absence of a regulatory framework.

16.2. PRECAUTIONARY LABELING

16.2.1. What is Precautionary Labeling?

"Advisory" or "precautionary" allergen labeling is a statement on the label of a packaged food to advise the purchaser, or the consumer, that the food has the potential to contain one or more allergens, which are not intentionally added as ingredients in the product. In other words, the purchaser or consumer is being provided with advice at the point of purchase that the product may not be suitable for them, due to the possible presence or unintended inclusion of one or more allergens.

There are a number of ways that an allergen may be unintentionally included in a food, but the most common is due to cross-contact in a manufacturing plant, especially when the same production equipment is used to manufacture a range of products containing different allergens. The term "cross-contact" describes the inadvertent introduction of an allergen into a product that would not intentionally contain that allergen as an ingredient [1]. Cross-contact may occur when the residue of a food allergen is present on food contact surfaces or production machinery or is airborne, and unintentionally becomes incorporated into a product, which is not intended to contain it. This situation could occur where multiple foods are produced in the same facility or on the same processing line or through the use of shared storage and transportation equipment.

While good manufacturing practices (GMPs), such as scheduling products to reduce the risk of allergen cross-contact, are encouraged, they will not always eliminate the potential risk of small quantities of an allergen coming into contact with a food that is not intended to contain it. Even scrupulous cleaning will not completely eliminate all risks of minute traces of food containing allergens remaining trapped in the production equipment.

Many food manufacturers voluntarily use precautionary statements or advisory labeling such as "may contain [an allergen]" or "manufactured in a facility that also processes [an allergen]" to alert consumers that those food products could unintentionally include an allergen.

Precautionary labeling in Australia and New Zealand is currently added at the discretion of each food manufacturer—there is no mandatory or regulatory requirement specifically regarding precautionary labeling. Food manufacturers decide when and how to apply precautionary labeling to their products, and as a result, precautionary labeling is used inconsistently across the Australian and New Zealand food industry. The lack of consistency of precautionary labeling may cause confusion for food-allergic consumers and lead to frustration when they are shopping for safe foods.

In response to this lack of consistency and uncertainty, the Australian Food and Grocery Council (AFGC) launched the Food Industry Guide to Allergen Management and Labelling in June 2007 [2]. This guide provides recommendations as to when it is appropriate for foods to be labeled with a precautionary labeling statement as part of a risk assessment approach known as VITAL, which will be discussed later in the chapter.

16.2.2. The Intent of Precautionary Labeling

The intent of precautionary labeling is to ensure that people with a food allergy, or their carers, are aware of the potential that a food may contain an allergen that is not otherwise declared in the ingredient list. The food-allergic consumer, or their carer, should exercise caution with such foods, and to avoid a possible allergic reaction, it is preferable if they avoid eating the food altogether.

Precautionary labeling may be seen as additional information to the ingredient list that is voluntarily provided by the manufacturer about their product. While the food would be suitable for consumption by the vast majority of the population, it may present a significant risk to people who have a food allergy, given that an allergic reaction can cause extreme health concerns and, potentially, even death. The intent for many food manufacturers in providing this precautionary statement is to help the purchaser make a more informed choice. It may also be perceived to serve as a defense, in the event of legal action against the manufacturer, if an allergic consumer suffered harm after eating such a food. The principle of the legal defense is that the purchaser cannot claim that the manufacturer did not provide advice of the possible presence of the allergen in the food.

16.3. THE CHALLENGES ASSOCIATED WITH PRECAUTIONARY LABELING

The use of precautionary labeling on food product packaging provides challenges for many stakeholders across the food supply chain, including food-allergic and non-food-allergic consumers, food manufacturers, and regulators. These challenges are summarized in Table 16.1 and discussed below. The table also identifies the VITAL response to these challenges to illustrate how they may be addressed.

16.3.1. Non-Food-Allergic Consumers

For non-food-allergic consumers, precautionary statements are potentially just another piece of information on the label of a food package, which may or may not be read, and which could easily be misinterpreted by the consumer. For example, they could consider that the statements may be a poor reflection

TABLE 16.1 The Pros and Cons of Precautionary Labeling

	Precautionary Labeling—Pros	Precautionary Labeling—Cons	Precautionary Labeling—Absence	VITAL Response
Consumers generally	Increased community awareness of the needs of food-allergic consumers	Complicates labels and requires additional space		Declarations are minimized by targeting risks and through statement consistency
Food-allergic consumers	Provided with information to promote safe consumption of foods Increased community awareness of the needs of food-allergic consumers	No information provided as to the rationale for labeling (*why* the allergen might be present), so risks may be misinterpreted Lack of consistency in labeling statements creates uncertainty and confusion, especially "blanket" warnings leading to risk-taking behaviors or unnecessary exclusion of foods from the diet	Potentially life-threatening consequences if allergic reaction is triggered	Systematic, science-based risk safety approach to underlie declarations Consistency in label statements reduces risk-taking behaviors Developed in conjunction with Anaphylaxis Australia to meet the requirements of food-allergic consumers

Food manufacturers	Maintain faith with consumers regarding food safety May provide legal protection if properly implemented Proactive industry response decreases need for regulatory intervention	Costs of establishing and maintaining risk management systems, including testing Loss of consumer confidence or confusion when a precautionary statement appears on a product that did not previously have one	Legal vulnerability Brand and reputation risks Increased likelihood of regulatory intervention Potential for product recalls	Clarity that VITAL cannot prevent all incidents, but it can reduce incidence and severity Best practice risk analysis and management minimizes legal and reputation risks Regulators kept aware of VITAL development Better control over ingredients and systems
Regulators	Properly developed and managed industry self-regulation allows resources to be devoted to other issues Enhanced reputation for safety for Australian food	Poorly managed systems and inconsistent labeling create demand for intervention	Creates demand for intervention	Care taken to ensure consultation development of risk assessment and management tools Scientifically justifiable approach

of the cleaning standards of the food producer or perhaps represent a negative aspect of the food rather than understanding that the precautionary statement conveys important information for the food-allergic consumer.

16.3.2. Food-Allergic Consumers

Food-allergic consumers in Australia and New Zealand initially applauded the amendment to food labeling laws to require mandatory labeling of major food allergens in 2002. In 2003, Food Standards Australia New Zealand (FSANZ) conducted a benchmark quantitative consumer survey on allergen labeling [3]. The results of the survey, released in 2004, found that consumers buying foods for those with food allergy had observed changes to the way allergens were listed on food labels with some 63% of respondents having noticed some specific changes in the form of:

- greater use of "may contain" statements;
- more warnings about nuts;
- the use of "made on same product line or equipment" statement;
- the use of blanket statements such as "contains dairy/seeds/nuts"; and
- greater use of "made in same factory" or "same premises" statements.

While food-allergic individuals and their carers are strongly encouraged by patient support organizations to heed precautionary labels, they sometimes struggle with decision making on food consumption, especially where food choices are limited. The reality is that many allergy sufferers regularly ignore precautionary statements [4].

Parents often claim they have fed their child a particular food a "thousand times" and never had an allergic reaction or any other issue. Children that are fed foods that "may contain [an allergen]" are therefore getting mixed messages about being careful to read food labels, and the relevance of the warning, if they are then told to eat foods that have a "may contain" statement.

Some parents will contact food manufacturers and ask questions as to why a food has a precautionary label and then make an informed decision about consumption depending on what they consider as the level of risk. However, not all allergic consumers (or their carers) gather all of the available information to make an informed decision. We know not all food-allergic consumers contact consumer help lines or check websites to see why a company has used a particular statement. It is unlikely, for example, that a teenager who is out with peers will say "no" to a product, or contact a manufacturer to check on a label statement, when they consider they have consumed a "similar" product before [4]. Information from Anaphylaxis Australia's unpublished survey of 243 members in 2003 indicated that physicians and other health professionals had advised some patients to ignore precautionary labels, with some citing that precautionary statements are only for legal defense and are not a "real" risk [5].

A study conducted in the United States attempted to determine whether food-allergic consumers heed precautionary labels, and whether products labeled as such contained any detectable amounts of the allergen [6]. From this study, two separate groups of parents of children with food allergy were surveyed: one in 2003 (625 parents) and another in 2006 (645 parents). While 85% of parents heeded precautionary labels in 2003, only 75% heeded labels in 2006. It is important to note that parents ignored advisory labels to different degrees, depending on the wording. The decrease in safe food choices has led to an increase in risk-taking behaviors, not only by teenagers but also by parents of allergic children. Manufacturers place specific precautionary labels on *their* product depending on *their* in-house assessment and what they consider to be the level of risk. The study reported that there were no criteria for the use of these statements.

The frustration for the allergic consumer is that each manufacturer conducts their own risk assessment and then uses one of the many precautionary statements they believe will communicate the real level of risk. This was demonstrated in statements used to advise consumers of the risk associated with the presence of peanut in various foods. Researchers found that the wording in the precautionary statements did not correlate with the frequency or amount of peanut that was detected. In fact, peanut was found in more products, and at higher levels, where "shared facilities" was indicated in the precautionary statement than with any other wording. Overall, 7% of products tested (13 of 179) had detectable levels of peanut—in amounts that could cause allergic reactions [7]. Allergic consumers often communicate that they perceive a product with a "shared facility" advisory statement is much less of a risk than one that had an advisory label stating the product is "made on the same line as [allergen] containing food" or "may contain [allergen]."

The lack of precautionary labeling can also create frustration for consumers. A food-allergic individual must make a decision as to whether this is due to the manufacturer having completed a risk assessment, which concluded that precautionary labeling is not required, or because the manufacturer is ignorant of the risk of cross-contact in their product, or ignorant of the risk to some individuals where a cross-contact risk is not declared. Likewise, "blanket labeling," where a manufacturer places a precautionary label listing all allergens despite there being little or no risk of them becoming incorporated into the product, is frustrating as it restricts the diet for the food-allergic individual who makes choices and decisions based on the precautionary statement information.

16.3.3. Food Manufacturers

Cross-contact with allergen sources can occur at any point in the supply chain—from shared harvesting and storage equipment during primary production of ingredients, to shared processing lines and facilities during product manufacture. Any cross-contact could result in small, sporadic amounts of an allergen inadvertently becoming incorporated into a processed food.

Determining which foods in a manufacturing plant are likely to be at risk of containing a particular allergen due to cross-contact can be difficult, time-consuming, and expensive. Furthermore, the allergen may either be present at extremely low levels, or randomly and infrequently present, and therefore difficult to detect by laboratory testing.

Adding a precautionary statement to a product where previously there was not one can have a commercial impact on a product, especially when in a competitive environment, where perhaps similar products do not have precautionary statements. Limiting the diets of food-allergic consumers and shrinking the market for a product requires careful consideration by a company. A "level playing field" for food manufacturers would remove some barriers to using precautionary statements if it was known that competitors are using a similar risk assessment.

In the absence of an industry-recognized standard, or guidance for the determination of the presence of allergens, which may be present in a food product due to cross-contact, some food manufacturers use precautionary labeling as a "blanket" statement on all of their food products. The blanket statements are often used irrespective of whether there was any likelihood of the allergen being present or not. Such blanket statements on food labels have been perceived as providing a means to ensure that food manufacturers were "covered" legally "just in case" an allergen was ever present [5].

An allergen labeling survey conducted in Australia in 2005 [8], consisting of 350 products from various retail food categories, revealed the inconsistent approach to precautionary labeling. In the products surveyed, there were 42 different versions of precautionary statements being declared on products. Of the 42 statements, the following were the key types of statement being used at the time of the survey:

- may contain [an allergen];
- may contain traces of [an allergen];
- manufactured on equipment that processes products containing [an allergen];
- made on equipment that also processes [an allergen];
- manufactured on equipment that also processes products containing [an allergen];
- manufactured in a plant that also processes [an allergen];
- packed in a plant that also pack [an allergen].

What are these messages trying to communicate? Is the consumer to assume that "manufactured on equipment that processes products containing [an allergen]" means the cleaning procedures are inescapably ineffective at removing the allergen? Does "made in a plant that also processes [an allergen]" mean there is a high risk of aerosols, or that human error and the potential for the wrong ingredients to be added is an issue? The consumer is clearly not in a position to accurately interpret the risk without knowing what these messages actually mean and are therefore of arguable use. Without clear guidelines for exactly what the different statements mean, many consumers claim that precautionary labels are of arguable benefit.

The frustration of allergic consumers toward inconsistent use of precautionary labeling is widespread. Statements on labels found in the Australian market place include the following:

- "May contain the occasional nut."
- "May contain peanuts" (on a bag of peanuts with peanut stated in the ingredient list).
- "May contain peanuts, nuts, and other allergens not listed on the label."
- "Carefully baked in a nutty environment."
- "Allergen information: made in a facility that uses dairy."
- "Rice crackers: this product contains wheat, soy, and corn. Manufactured in a facility that may produce products that contain other allergens such as dairy, egg, peanut, and tree nut."
- "The ingredients in this product do not contain peanuts or tree nuts, but this product is made in a factory where nuts are present and may contain peanuts or tree nuts" [4].

This highlights the inconsistency in precautionary labeling and the different approaches taken by different manufacturers to communicating the risk of cross-contact allergens to their consumers.

16.3.3.1. Food Recall. The mandatory labeling requirements for food allergens in the Australia New Zealand Food Standards Code came into effect in 2002. This requires that whenever certain substances are present in foods, they must be declared on the label. Noncompliance may be a result of inadvertently omitting to declare the use of an allergenic ingredient on the label, or the unintentional inclusion of trace levels of an allergenic substance in a product that is not intended to have that allergenic ingredient and, therefore, does not declare the allergen on the label. If the label does not include a precautionary allergen statement, and if the allergen is not included as an ingredient, the manufacturer is faced with the need to make a decision as to whether to conduct a product recall on the grounds of a potential risk to food-allergic consumers, and the risk of prosecution.

In Australia during the period from 2002 to 2005, labeling errors accounted for 47% of food recalls and were due to a variety of errors including failure to identify the presence of allergens. As manufacturers implemented more stringent controls over label design and manufacturing systems, the number of recalls due to labeling faults reduced, and in the period from 2005 to 2009, recalls due to labeling errors averaged less than 30%. While labeling errors, for reasons other than declaring the presence of allergens continued to fall in

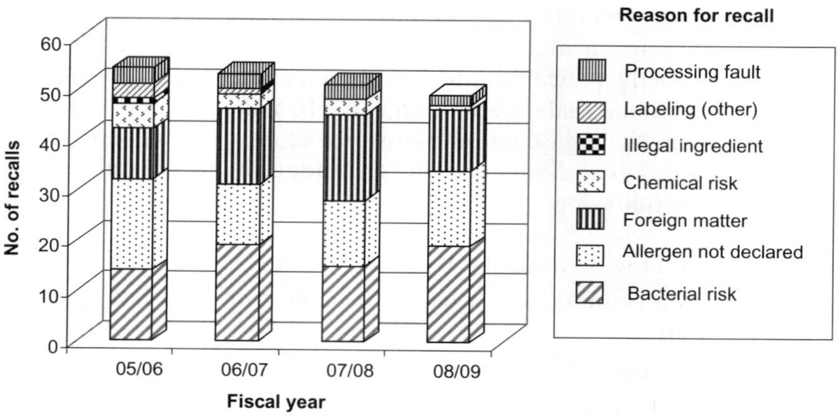

Fig. 16.1 Australian food recalls for the fiscal years 2005/2006 to 2008/2009. Data compiled by the Australian Food and Grocery Council. *Source*: Food Standards Australia New Zealand.

recent years, the failure to declare the presence of allergens remained a problem, and in the financial years 2007/2008 and 2008/2009 accounted entirely for all recalls due to labeling errors. In the fiscal year 2008/2009, failure to declare the presence of allergens (31%) was exceeded only by bacterial contamination (39%) as the reason for undertaking a food recall, as shown in Figure 16.1.

It is not possible to definitively determine what proportion of these recalls were due to inadvertent omission of ingredients from the label, and which were due to the lack of a precautionary statement, which would alert consumers to the potential presence of an allergen that is not contained in the ingredient list. However, consumers who are highly sensitive to allergens such as peanuts are likely to react to very low levels of peanut proteins in food, which may be present from cross-contact sources, and likely to result in customer complaints leading to a recall. The undeclared presence of peanut allergens represented 28% of the recalls, due to failure to declare the presence of allergens in the fiscal years 2005/2006 to 2008/2009, as shown in Figure 16.2.

It may also be possible to draw some assumptions about the level of recalls due to failure to provide a precautionary statement, based on whether the recalled food may normally contain the allergen as an ingredient. For example, foods such as rice crackers or corn chips recalled due to the presence of gluten may be assumed to be due to the inadvertent presence of the allergen, as such products would not use flour-containing gluten. In contrast, failure to declare the presence of egg in a bakery product could be due to the omission of an ingredient from the label, as egg is commonly used in bakery products. Based on such assumptions, the failure to provide an appropriate precautionary statement led to 66% of the allergen-related recalls in this period.

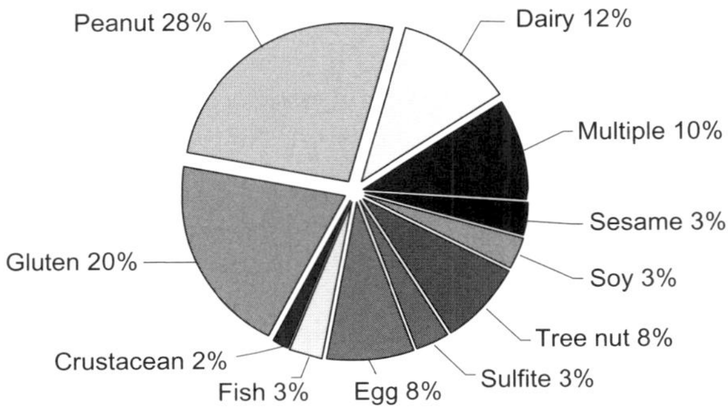

Fig. 16.2 Australian food recalls for the fiscal years 2005/2006 to 2008/2009 showing the percentage of allergenic substances for recalls due to undeclared allergen. Data compiled by the Australian Food and Grocery Council. *Source*: Food Standards Australia New Zealand.

It is estimated that the minimum cost to manufacturers and retailers in conducting a food recall in Australia is $10,000 where the recall has limited distribution in one city only, while the average cost of conducting a food recall with national distribution is estimated at $100,000 for each recall. To avoid a costly recall, food manufacturers may be tempted to include precautionary statements, even when the possibility of cross-contact allergens being present has not been properly assessed.

16.3.3.2. Testing. Testing for food allergens is a valuable tool for use as part of a risk-based approach to allergen management. Test results can provide assurance and verification of critical controls within a comprehensive risk management program.

The most commonly used analytical method for detecting the presence of food allergens is the enzyme-linked immunosorbent assay (ELISA) technique. The sensitivity of ELISA kits currently available is in the low parts per million reporting range (in the order of 10 ppm).

While there are many benefits in using the ELISA method, there are some limitations in using and interpreting results that must be considered. These may include:

- protein extraction for analysis,
- effect of food processing,
- protein structure,
- cross-reactions, and
- availability of kits for different allergens.

It is also important to remember the limitations associated with sampling. As with all food testing methods, results are only representative of the samples tested and cannot be used to categorically prove the absence of allergens in products that have not been tested. This is especially so for particulates, which may be sporadically distributed in a food product.

A number of allergen ELISA test kits are currently available, and international research is being conducted to develop alternative or confirmatory methods for the routine detection of allergens in foods. Rapid ELISA test strips for allergen detection are also becoming more widely available, and in some instances, these may be suitable for allergen screening in-house during production, although these rapid test strips are not as sensitive as current ELISA test kits.

16.3.4. Food Regulators

16.3.4.1. Labeling Regulation. The requirement to declare certain intentionally added allergens in products within the Australian and New Zealand food industry has increased the awareness among food manufacturers of the risks that allergens can also become unintentionally incorporated into a food product. The development of international food standards for labeling allergens was introduced into the Codex Alimentarius [9] in 1999 with a requirement to declare the eight common allergenic classes of ingredients. These are

- cereals containing gluten and their products, namely, wheat, rye, barley, oats, and spelt and their hybridized strains;
- crustacea and their products;
- egg and egg products;
- fish and fish products;
- milk and milk products;
- tree nuts and their products; and
- peanuts and soybeans and their products.

Australia and New Zealand followed the Codex lead with the introduction of a mandatory government regulation under Standard 1.2.3 of the Australia New Zealand Food Standards Code in December 2002. Consistent with the international standards developed by the Codex, the legislation only addresses the mandatory requirement to declare the eight allergenic ingredients, plus the addition of sesame seed in association with tree nuts when added as an ingredient, sub-ingredient, additive, or processing aid [10]. Precautionary statements remain a voluntary option for manufacturers, limited only by the general requirements under Trade Practices legislation that prohibit "false, misleading, and deceptive claims."

16.3.4.2. Legal Implications for Food Manufacturers. Most regulatory schemes governing allergen labeling do not deal with incidental allergens and precautionary labeling. That said, it is not accurate to believe that there is no applicable regulation. There are statutory and court-derived schemes of product liability that make manufacturers liable for defects in their products. This also extends to "defects" in labels that fail to warn of risks that render a product less safe than persons generally are entitled to expect (taking into account all the circumstances, including any warnings in the label).

A duty of care exists under the concept of tort liability due to the fact that the incidental presence of allergens in a food is (or at least should be) within the particular knowledge of the manufacturer, while the consumer has no independent means to assess the risk. This is likely to create a duty of care that requires the manufacturer to take steps to control the allergen or warn of the risk, with any failure to do so potentially attracting liability for any adverse event.

How these liability schemes apply where the affected person is sensitive to a particular food is an open question. The language of the European Union (EU) Product Liability Directive, for example, refers to the degree of safety that "persons generally" are entitled to expect, which may mean that a food is not "defective" under this test if it is safe for persons generally, even if it is unsafe and potentially fatal for persons with a particular food sensitivity. It is certainly true that the test is an objective one relating to the public at large, rather than to an individual or limited class of individuals. However, it might equally be said that the absence of at least some warning about the presence of something in food that might trigger a severe, even fatal, reaction is still a defect. This is because persons generally would expect such a warning, although the consequences of that defect fall on the few rather than the many.

Against this regulatory framework, the Australian food industry has responded to the challenge of precautionary labeling in three distinct, evolutionary ways: the "say nothing" approach, the "blanket statement" approach, and the risk management approach.

The "say nothing" approach was predicated on the belief that statute law does not (typically) require labeling of incidental allergens, and that such laws are a sufficient indication as to the degree of safety that persons generally are entitled to expect. Accordingly, no particular investigation needs to be undertaken into the potential for the incidental introduction of allergens, and no labeling statement is contemplated. Reliance on regulations as a measure of public expectations, though, seems fraught with danger, assuming that the minimum requirement, at some past point in time, is evidence of a sufficient safety requirement some years later. Further, this understanding fails to address the fact that community expectations change and evolve, and that the market circumstances in which products are sold also change and evolve. The "say nothing" approach was in the majority 10 years or so ago. However, given the market prevalence of incidental allergen labeling nowadays, silence is more likely to be interpreted as a positive statement as to the absence of incidental allergens. The "say nothing" approach thus has very little to commend it from a legal perspective and, today, is largely the provenance of those who simply have not considered the risks or who rely on insurance to protect them and their business.

The next stage of precautionary labeling is the "blanket statement" approach, where a "may contain" label warning is made for each allergen that is ever introduced into the manufacturing facility, without any further investigation. At first glance, this seems to be reasonable and is certainly less costly compared with a full risk management approach. However, it fails to address a key issue, which is whether the warning is effective in communicating risks, as judged by consumer behavior. This is important because most statutory liability schemes assess defectiveness by taking into account foreseeable misuse of the product, as well as use according to instructions. In circumstances where almost every food warns against the possible presence of virtually every allergen, it is foreseeable that sensitive consumers, faced with the choice between risk-taking behavior or an unnecessarily restricted diet, will at times opt for the former, and this appears to reflect actual behavior [6]. The "blanket statement" approach thus conveys more an impression of legal protection than of its reality—its efficacy in practice is far from certain.

The legal problems with the "say nothing" and "blanket statement" approaches has led the food industry to develop a third evolution in incidental allergen labeling, based on risk identification, control, and management. The underlying concept is that a precautionary warning statement in a product label is more likely to be legally effective if backed up by a formal risk analysis. The risk analysis must address the whole issue of incidential allergen management, from source ingredients through to manufacturing and process controls, as well as an assessment of the actual risk represented by the level of allergen in the final food. The resulting consumer advice is then appropriate to the risk, such that precautionary statements only appear on labels where the incidental presence of the allergen cannot be avoided, and where the allergen is likely present at a level that poses some real potential for harm.

Further, by developing this risk analysis tool in conjunction with allergy sufferer groups, the expectation is that the warning statements will carry greater credibility with the target audience and actually modify risk-taking behavior (though this, at the present time, is speculative due to the relative recency of analysis tools such as VITAL discussed below).

The mere existence of the risk analysis tool does not, of course, ensure its proper application and use, and it remains likely that there will still be legal issues to be resolved about the adequacy of the tool in its application throughout the food supply. However, the adoption of hazard identification, management and control procedures through the use of the Hazard Analysis Critical Control Point (HACCP) system, along with risk analysis and risk communication, revolutionized food industry thinking on food hygiene. HACCP now enjoys statutory recognition in many jurisdictions and is seen to offer significant improvements in safety and efficiency of production. This risk assessment approach has the potential to do so equally in relation to incidental allergen labeling. It certainly represents current thinking on best practice, and while it may never be possible (or appropriate) to totally exclude liability for allergy incidents, it is the best that a food manufacturer can currently do to protect its customers, and itself.

It is also prudent to note that legal concepts such as duty of care and product liability are not static, but rather evolve according to social, political, and commercial pressures. It is this dynamic process that has driven precautionary labeling through its evolution, and it continues today. The commercial incentive to keep informed and to apply current best practice, such as incidental allergen risk identification and management, exists because any failure to do so has legal implications for manufacturers, their brand owners, and their insurers. It is important to understand that these underlying processes, which have brought voluntary precautionary labeling by food manufacturers to this point, have not stopped, and as our understanding of food-triggered reactions becomes more refined, so too will the challenge facing the food industry.

16.4. EVOLVING A BETTER SYSTEM FOR PRECAUTIONARY LABELING

Precautionary statements should never be used as a substitute for GMP or as a generic disclaimer. Every attempt must be made to eliminate or minimize cross-contact by adhering to GMP.

The food industry is often the target of criticism for the number of different ways that the presence of allergens are declared on a label or packaging, for both mandatory and voluntary declarations. As previously discussed, precautionary allergen labeling is not consistent across the food industry with manufacturers using different risk assessment processes to decide when it is appropriate to precautionary label and when it is not. The same precautionary statement means something different from manufacturer to manufacturer, product to product, and consumer to consumer. A standardized method of assessment and declaration is considered necessary to assist food businesses in presenting allergen information in a way that is more consistent and easily recognized and understood by allergic consumers. A standardized approach will support the allergic consumers' needs by facilitating clear on-pack allergen communication.

In an allergen labeling survey conducted in Australia in 2005 [8], precautionary labeling appeared on 48% of the products surveyed. The use of bold typeface was most commonly used (43% of the time) for the precautionary labeling statement to ensure that the information was emphasized on-pack. The majority of the precautionary statements appeared in conjunction with the ingredient information, usually just below the ingredient list, to facilitate the ease with which the consumer was able to obtain the information. This placement was considered the "normal" place where allergen information is retrieved from a product.

The survey also found that more than 77% of the precautionary allergen statements appearing on-pack had a font size of 1.5 mm or greater, which most consumers considered adequate from a legibility perspective. Additionally, from the survey, 97% of the statements were in distinct contrast from the background pack color. The precautionary labeling statements were predominantly declared in uppercase (73%) for emphasis of information. However, allergic consumer groups considered that a statement in lowercase was easier to read and to obtain the information of concern, and that lowercase was visually more pleasing and legible to the eye.

16.4.1. Benefits of Standardized Precautionary Labeling

It has been recognized within the food industry that a standardized, consistent approach to allergen cross-contact risk assessment is preferred, where one precautionary labeling statement would be universally applied based on the same risk assessment process. A standardized approach facilitates consistent consumer communication, as the cross-contact assessment would be based on the same principles, with the same allergen action levels used to determine if a precautionary labeling outcome is appropriate or not.

Manufacturers must be able to substantiate the assumptions made during the assessment. They must also substantiate the potential risks via allergen cross-contact introduced from the raw materials formulated into the product, or via cross-contact potentially occurring in the manufacturing plant or environment. The only way forward is to use a standardized process approach with one standardized precautionary cross-contact statement. This would ensure that the food industry is communicating with the allergic consumer using one consistent voice.

16.4.2. Protection for Consumers

Protection for the allergic consumer is facilitated if one standardized approach is used to assess and declare allergen cross-contact information. This would mean that if a product carried a precautionary label, then the labeling decision would have been based on the same information as another product carrying a similar precautionary label. An important aspect to understand, however, is that different assumptions can still be made by different manufacturers due to the fact that one standardized approach may have several subjective inputs into the assessment based on the differences between manufacturing facilities, managers, raw material specifications, or experience.

16.5. VITAL: THE AUSTRALIAN RESPONSE TO THE CHALLENGES OF PRECAUTIONARY LABELING

The AFGC's Food Industry Guide to Allergen Management and Labelling [2] recommends that precautionary statements should only be considered as a last resort, only be applied where indicated by the VITAL process, and only when it is not possible to eliminate or control the potential for traces of the allergen(s) in question to be present.

In the absence of regulatory or other international guidance on the issue, Australian and New Zealand food manufacturers identified a need and set about developing a solution, a way forward.

VITAL was developed to provide a standardized allergen risk assessment tool to assist food manufacturers in assessing the potential impact of allergen cross-contact in each of their products and where precautionary labeling is required. It also specifies a standard wording for a precautionary statement—"may be present (allergen)." The aim was to encourage clear, consistent labeling of manufactured food to assist consumers to identify which allergens may be present in the food.

16.5.1. How Was VITAL Developed?

VITAL was developed by a team of food professionals convened by the AFGC's allergen forum. The team reviewed current risk assessment protocols, both food and nonfood, to determine if there was any existing suitable model. The team agreed to adopt an existing risk grid as the basis for the risk assessment process. A literature review was conducted to validate the levels in the existing grid, and a consultant was engaged to review the literature and further refine the grid values. A decision tree [11] and procedure [12] were subsequently developed and refined to assist in the application of the VITAL grid values.

VITAL requires the assessment of likely sources of allergen cross-contact from raw materials and from the processing environment, plus an evaluation of the amount of allergen present and a review of the ability to reduce the allergenic material from all contributing sources. It also provides for ongoing monitoring and verification of the risk assessment process to ensure any changes to the level of risk are acted upon without delay.

The VITAL process can be used for any type of food allergen but was specifically developed for application to the eight generic classes of allergens prescribed in Standard 1.2.3 of the Australian New Zealand Food Standards Code [10], to ensure consistency with mandatory requirements to label allergens present in foods.

In summary the VITAL action levels indicated are the following:

- Action Level 1—Green Zone—precautionary cross-contact statement is not required for the relevant allergen under evaluation.
- Action Level 2—Yellow Zone—precautionary cross-contact statement is required for the relevant allergen using the standard VITAL statement—"may be present."
- Action Level 3—Red Zone—significant levels of the allergen are likely to be present. Labeling of the relevant allergen as an ingredient in the ingredient list is required (rather than use of a precautionary statement) [13].

The VITAL procedure should be followed for each allergen that may be present in the final product due to cross-contact via ingredients or processing. VITAL is not applicable to ingredients intentionally formulated into the product; that is, any allergen added as an ingredient, no matter what the final concentration in the product, must always be labeled in the ingredients list.

VITAL should be used as part of a HACCP-based food safety program when conducting the risk assessment for allergenic hazards. The VITAL process is applicable at all levels of the supply chain, from the farmer to the manufacturer. While it is primarily intended as an aid to determine whether labeling is required on packaged foods, in theory, it could also be used by restaurateurs and caterers.

Unlike the HACCP processes for verification that the presence of microbiological, chemical, and physical hazards lie within acceptable limits, verification of the accuracy of the VITAL risk assessment can be difficult and frustrated by the intermittent and random nature of the occurrence of the allergen in the final food. Reliance on laboratory testing should be used for validation purposes only, as part of the risk assessment process, particularly in relation to ingredients and verification of cleaning processes, as part of a HACCP-based food safety program. It is not expected that testing be used in "real time" as a final quality control step. Only validated test kits should be used for testing.

16.5.1.1. Taking a Line in the Sand. The premise on which VITAL is based is that it is possible to set "action levels" or thresholds. The decision to establish action levels in the absence of internationally agreed thresholds was taken on the basis that the current situation of inconsistent risk assessments could not continue. The approach of sometimes indiscriminate and inconsistent use of precautionary labeling was eroding the value of precautionary labeling as a risk management tool, resulting in allergic consumers taking risks or avoiding foods, thereby reducing their already limited choices.

Throughout the process of developing VITAL, the project team has sought input and feedback from a wide group of stakeholders representing industry and allergic consumers. Feedback on VITAL has been positive and encouraging—the most common sentiment is that "at least we have some numbers (action levels) and a way of doing the risk assessment in a consistent manner." Key among concerns is the robustness of the action level numbers, given there is currently no international consensus on thresholds. However, the action levels are based on the best available information from international literature and a built-in safety margin. It is strongly believed that by following this approach and taking a line in the sand, it provides a positive way forward for consumers and food manufacturers alike. As with any new technology, as new information becomes available, VITAL will continue to improve and evolve.

16.6. CONTINUOUS IMPROVEMENT

Process or manufacturing improvements or changes are continually being executed and can directly impact precautionary labeling statements. This is an important aspect of the manufacturing environment and must be captured formally to assess the potential risks on product manufacture. Hazards that are initially identified may be eliminated, or additional allergen risks could be identified with the introduction of new equipment through capital investment or improvement processes.

The manufacturers' need to provide precautionary labeling statements may reduce over time, as controls or processes that eliminate hazards of incidental allergens are implemented, depending on the complexity of the process and the manufacturing or packing site.

The aim of the manufacturer is to provide clear, consistent, and relevant information so that the allergic consumer has a meaningful and easily identified purchasing decision.

REFERENCES

1. FDA. (2006). Approaches to establish thresholds for major food allergens and for gluten in food. The Threshold Working Group, Department of Health and Human Services, United States Food and Drug Administration. Available at http://www.fda.gov/Food/LabelingNutrition/FoodAllergensLabeling/GuidanceComplianceRegulatoryInformation/ucm106108.htm
2. The Australian Food and Grocery Council. (2007). Food industry guide to allergen management and labelling. Available at http://www.allergenbureau.net/allergen-guide/afgc-allergen-guide/
3. Food Standards Australia New Zealand. (2004). Quantitative consumer survey on allergen labelling: benchmark survey 2003. *Evaluation Report Series No. 7*. Available at http://www.foodstandards.gov.au/_srcfiles/combined_Allergy_Final%20Report_16Feb%20.pdf
4. Said, M. (2008). Anaphylaxis Australia. Personal communication. February.
5. Anaphylaxis Australia Inc. (2003). Summary of AAInc survey of 243 consumers with food allergic consumers to purchase for. Unpublished.
6. Sampson, M.A., Muñoz-Furlong, A., Sicherer, S.H. (2006). Risk-taking and coping strategies of adolescents and young adults with food allergy. *Journal of Allergy and Clinical Immunology*, *117*, 1440–1445.
7. Hefle, S., Furlong, T., Neimann, L., Lemon-Mule, H., Sicherer, S., Taylor, S. (2007). Consumer attitudes and risks associated with packaged foods having advisory labelling regarding the presence of peanuts. *Journal of Allergy and Clinical Immunology*, *120*, 171–176.
8. Australian Food and Grocery Council (AFGC). (2005). Allergen labelling survey. Unpublished.

9. Codex Alimentarius. (2008). General standard for the labelling of prepackaged foods. CODEX STAN 1-1985 Amended 1991-2008, Clause 4.2.1.4. Available at http://www.codexalimentarius.net/web/more_info.jsp?id_sta=32
10. Food Standards Australia New Zealand. (2009). Australia New Zealand Food Standards Code, Standard 1.2.3, Clause 4. Available at http://www.foodstandards.gov.au/_srcfiles/Standard_1_2_3_Warning_Statements_v109.pdf
11. Allergen Bureau. (2007). Voluntary incidental trace allergen labelling (VITAL) decision tree. Available at http://www.allergenbureau.net/downloads/vital/VITAL_Decision_Tree_12_06_07.pdf
12. Allergen Bureau. (2007). Voluntary incidental trace allergen labelling (VITAL) procedure. Available at http://www.allergenbureau.net/downloads/vital/VITAL_Procedure_12_06_07.pdf
13. Allergen Bureau. (2007). Voluntary incidental trace allergen labelling (VITAL) grid. Available at http://www.allergenbureau.net/downloads/allergen-guide/VITAL/VITAL_Grid_02_July_2007.pdf

17

CERTIFICATION PROGRAMS FOR FOODS LABELED AS "FREE FROM" SPECIFIC ALLERGENS

Christine Dupuis and Ferdinand Tchounkeu

17.1. INTRODUCTION TO THE CERTIFICATION PROCESS

To determine whether or not a product, process, or service meets a standard or set of certification program requirements, a compliance assessment must be performed. A compliance assessment performed by the business itself, with its own program requirements, is a first-party assessment (internal audit). A compliance assessment performed by the purchaser of the product, process, or service is a second-party assessment. If an assessment is being done to obtain certification, an independent (third-party) certification body that follows international certification rules can be used to lend greater credibility to the certification.

As well, the third-party certification (TPC) agency can be accredited by an accreditation body (e.g., the Standards Council of Canada [SCC], American National Standards Institute, Accreditation Board [ANAB], United Kingdom Accreditation Service [UKAS], etc.). In Canada, the SCC checks whether or not agencies have the necessary human resources, expertise, and equipment for the assessment work they intend to do. Compliance assessment is a specialized field requiring specific skills, both organizational and in the staff responsible for performing audits (auditors) and evaluating audit files (managers and assessors).

Allergen Management in the Food Industry, Edited by Joyce I. Boye and Samuel Benrejeb Godefroy

Certification agencies (or bodies) can provide various certification programs to show that products, processes, and services are compliant with a current standard or some other recognized document. Accepted tests to prove that a requirement of a standard or certification program has been met are often used to certify products. In the last 10 years, the development of food allergen tests has led to the creation of an important food allergen risk management tool. Allergen-free product certification can therefore be carried out by using a combination of accepted allergen tests and other tools for controlling food manufacturing hazards, such as the Hazard Analysis Critical Control Point (HACCP). Effective management of allergens covered by a certification program can be demonstrated through a certification process similar to those for other products, processes, and services.

Certification can be summarized in five main steps that are the same regardless of the product being certified: application, audit or monitoring visit, decision, certificate delivery, and maintenance or renewal (http://www.bnq.qc.ca/fr/certif/etape_processus.html). The five steps are described briefly below, as they are generally applied by certification agencies for all certification programs, for example, the allergen-free product certification program of the Bureau de normalisation du Québec (BNQ) within the certification program of the Association Québécoise des Allergies Alimentaires (AQAA).

17.2. STEPS IN THE CERTIFICATION PROCESS

17.2.1. Application

A business seeking certification must submit a formal application to the certification agency, with complete information about the products, processes, and services to be certified, as well as information on the business itself (name, address, contacts, etc.). The certification agency may provide an application form to standardize the applications submitted and ensure that all the required information is included. If the business accepts the certification agency's proposed plan, the two parties sign a certification contract with joint obligations to fulfill the requirements of the certification program.

17.2.2. Audit

Usually, before the certification audit, there is a preliminary evaluation of client documentation on how certification program requirements are met. Any specific documentation requirements of the certification program can be checked as well. A preliminary tour of the business location may also take place, to gain familiarity with the facilities and processes. Once it has been determined that any shortcomings found have been resolved and that a certification audit is feasible, the certification audit or visit can then be performed.

The initial certification audit is a critical step in the certification process that requires techniques specific to this kind of audit, if performed in accordance with international rules such as ISO 19011 [1]. ISO 19011 is a standard used mainly in management system audits, but it can also be used for other types of audits, as long as the members of the audit team have the necessary skills. The audit requires all the skills of the auditors, who must be capable of determining whether or not the certification program requirements have been met by the business requesting certification of its products, processes, or services. The auditor or audit team must have adequate knowledge of the certification program requirements, experience with the industry or field, and auditing qualifications to ensure a complete, consistent, and credible assessment. To certify a product as allergen free, the manufacturing, cleanup, and sanitation processes, and all good manufacturing practices are examined, as well as allergen hazard control methods implemented specifically to control hazards associated with the allergens covered by the certification.

The time spent by the audit team on the audit will vary depending on the size of the facility and processes, number of products, number of allergens, and number of employees. The initial audit normally lasts a few days, at the end of which a wrap-up meeting is held to present the findings. Noncompliances may be pointed out, and deadlines may be set for their resolution. A written report is also issued. Following the audit, the audit team may, if all conditions are fulfilled, submit a recommendation for certification to the certification agency, which is the only body that can make the certification decision.

17.2.3. Decision

Once all certification requirements have been fulfilled, certification audit performed, audit report released, and issues (noncompliance or other) resolved, certification agency staff qualified to evaluate certification audit files perform a final assessment and study the recommendations of the auditor responsible. Staff responsible for the final certification decision must not have been involved as auditors for the same audit file. If all requirements of the certification agency are fulfilled, certification is granted and a certificate is issued.

17.2.4. Certificate Delivery

Certificate confirms that the given products, processes, or services meet a standard and fulfill the certification program requirements. It is associated with the use of a certification mark that can be placed on the given products or accompany the given processes or services only if the given products, processes, or services are compliant. The certification period is fixed, normally 2–3 years, and the validity of the certificate is subject to approval mechanisms before certification and following subsequent monitoring afterward. Routine inspections or audits are performed from time to time to ensure that the certification is still valid, that is, that the certification program requirements continue to be met. Amendments can be made to the certificate during the certification period; however, all changes must go through audit and approval mechanisms before a modified certificate is issued.

17.2.5. Maintenance or Renewal

During the certification period, routine inspections or audits must confirm that the requirements of the standard or certification program continue to be met, or ensure that any noncompliance is resolved. The certificate is confirmed as remaining the same, or being amended if products, processes, or services are added or dropped. The certification mark can continue to be used only for the products or services that remain compliant.

Once the certification period has elapsed, the business may renew the certification with the certification agency for another period and continue using the certification mark, or discontinue the certification and stop using the certification mark.

17.3. ENSURING THAT CERTIFIED ALLERGEN-FREE PRODUCTS REMAIN COMPLIANT

To ensure that the certified allergen-free products are compliant, the business must implement an allergen hazard control system or program that meets the requirements of the certification program. Since food allergens are considered a food product safety hazard, the general principles of food hazard control, such as HACCP, may be helpful. The effective use of good food manufacturing practices is also essential. The integration of specific requirements for certified allergen-free products within a larger quality and food hazard management system can therefore be a means of ensuring continued compliance with the allergen-free certification program.

If there is no complete, recognized quality management system (e.g., ISO 9001) [2] or food hazard management system (e.g., ISO 22000 or HACCP), certain management requirements will have to be fulfilled by the business to facilitate continued compliance. Those requirements could involve system management tools such as mandatory internal audits, noncompliance management, and corrective measures.

One of the biggest challenges in the application of HACCP principles to control food allergens lies in the definition of critical control points (CCP); in effect, it is difficult to set a critical limit when the dose required to provoke an allergic reaction (either moderate or serious, i.e., anaphylaxis) is not known. Currently, the control of food allergens in the food industry is done by training of employees, verification of the composition of all raw materials, the use of test kits to detect the presence of allergens (i.e., in ingredients, other nonfood materials used in processing, equipment, and finished product), labeling, and cross-checking to ensure that the right packaging is used for the right product. When analytical test kits are available, they are most frequently used in the context of certification or to support "free-from" labeling. Thus, CCP, as currently defined, allows the maintenance of rigorous control measures that reinforces the safety of the foods produced; it does not, however, represent a real CCP (e.g., as in the case of establishing a limit for pathogenic bacteria) and should therefore be considered as an operational prerequisite program as defined by the requirements of ISO 22000:2005 [3].

17.4. BUSINESS RESPONSE TO NONCOMPLIANCES IDENTIFIED BY THE CERTIFICATION AGENCY

The business must ensure that products certified or to be certified as allergen-free meet the requirements of the certification program. The cause of noncompliance can vary depending on whether it is identified in an initial certification audit or a routine inspection.

In previous experience, most businesses that were discovered in the initial audit to have noncompliances misunderstood the certification program requirements or had not completely implemented those requirements, being too focused on obtaining certification quickly to gain access to the desired market. These businesses responded with action to address an incomplete implementation, a result of underestimating the implementation work required. In routine inspections, noncompliances are often the result of a lapse in applying control measures or a lack in quality management caused by poor management tools. Business response in such cases is usually to "restore" or "readjust" to avoid losing certification and use of the certification mark.

Businesses usually accept findings and agree to rectify noncompliances (thereby improving themselves) when the certification process is credible and audit techniques have been applied properly. If the certification agency does not appear to be credible, competent, and thorough, noncompliances can lead to disputes and misunderstanding with the audited business and, ultimately, a loss in the perceived value of the certification.

17.5. TPC AND HOW TO MAINTAIN AND CONTINUE ALLERGEN-FREE CERTIFICATION

"Allergen-free" certification responds to the needs of a specific group of consumers, that is, consumers with allergies and their family and friends. As long as those needs exist, and suppliers of certified allergen-free food products can meet those needs through certification at a cost that is acceptable to both parties, continued certification is possible.

There are, however, certain overriding factors to maintaining continuity for allergen-free certification. First, the certification must be credible. Second, it must be granted by a third-party agency with proven skills. Consumers must be informed about the differences between certification and the other methods used for highlighting allergen-free products. Self-declarations or claims by a business regarding the absence of allergens in its products, while they may be true, are not proven and verified by an independent third party. The use of the word "certified" is somewhat controversial when the claim has not been verified by an external organization. Consumers must therefore compare the pros and cons of the two types of allergen-free claims, to make informed decisions.

17.6. KEY POINTS TO CONSIDER IN SETTING UP A CERTIFICATION PROGRAM (SPONSORING ORGANIZATION)

First, it is important to make a distinction between a public certification program and a private one. A public certification program requires that specific steps be followed with respect to preestablished guidelines. This generally appeals to a very wide public as the content of the standard is public and available for everyone wishing to consult it. A private program, on the other hand, is conceived according to the sponsoring organization's requirements and its purpose is therefore targeted. The sponsoring organization (i.e., the organization providing the certification) can choose and limit the number of actors who participate in setting up the program requirements. Furthermore, the content of the program is not necessarily available to the general public and, in some instances, could require a confidentiality agreement prior to its consultation.

The information presented in this chapter is a review of information available from the public literature as well as from experience in the private sector. The question of allergen certification is fairly recent, and so there is relatively little information available on allergen certification per se; however, information from other certification programs are of relevance to an allergen certification program and the review will present a summary of the general steps that need to be considered in establishing a certification program irrespective of whether it is public or private.

Prior to launching a certification program, it is important that the sponsoring organization define the skill sets and proficiency levels that the certification will assess and validate, with respect to the presence of allergens in foods. The development effort should include four phases that are all essential to the program's success; these include business analysis, program development, implementation, and management [4].

17.6.1. Business Analysis

To ensure that the program meets the needs of its target clientele, namely the industry, the sponsoring organization should conduct a business analysis to assess the demand and identify the benefits of an allergen-free certification program. The business analysis should include the following components.

17.6.1.1. Internal Analysis. This analysis should cover all the strategic aspects of the program. Strategic advantages might be in the form of distribution benefits or customer franchise for the sponsoring organization. Identifying these advantages should be the first step in any internal analysis, particularly given the great importance attached to it by the industry. A franchise, defined as a contract whereby a company grants to another the right to use its name and trademarks to sell products or services, appears to be more favorably perceived by businesses than a profit-sharing arrangement. In fact, companies are very reluctant to reveal the details of their sales to third parties, for competition reasons, and prefer to pay a lump sum for the right to use the trademark. Similarly, available resources, cost of developing a program, availability of subject matter or domain experts, and so on, should be considered at this stage, with the goal of determining whether the organization possesses all the necessary resources to develop the program.

It is advisable to define the type of program to be established at the outset. The choice made at the beginning to develop a private or public program will determine the basis of the discussions with the various parties who will be invited to participate in its development. This choice of program type determines not only the organizational framework of the development process but also constitutes a key stage in the development of the program's business plan. Indeed, consultations aimed at soliciting comments on the program will be organized differently depending on whether it is private or public. The type of program also determines the level of protection or confidentiality to be accorded to the terms of reference. Regardless of the approach chosen, the business plan must consider each predetermined step in the program development and certification process.

The choice of partners is important for the credibility and reputation of the program. It is essential to join forces with partners who are competent, credible, recognized, and influential in the area and on the subject of food allergies (e.g., government, certification organizations, producers, distributors, and consumer associations). Medical consultants (physicians, nurses, allergists, etc.), manufacturers of allergen detection tests, scientific researchers, and legal experts should be invited to participate in the program development and implementation process. Manufacturers of allergen detection kits, for example, can contribute their expertise concerning detection methodologies and the validation, accuracy, and availability of the tests. Legal experts can examine the legal aspects of applying the program, and they can provide a legal framework for agreements with present and future partners. Generally, the agreements between the sponsoring organization and the companies to be certified must be simple but complete and satisfactory to all the parties concerned. They must necessarily include the duration of the partnership, license fees, legal liability, and partners' rights and obligations (e.g., obligation of manufacturers to carry private civil liability insurance). The legal experts must also examine the implications of the use of the program beyond the established scope (e.g., regional, provincial, national, international, etc.).

Food processors and suppliers are typically subject to a variety of programs and requirements imposed by the government, customers, and distributors. They may, thus, be reluctant to participate in an additional program, even if it is innovative. The involvement of the food industry groups (suppliers, manufacturers, distributors/retailers, etc.) is therefore indispensable to the process. In fact, program participants may act as advocates by encouraging other members of the group to join the program.

The involvement of government officials or authorities in the design of a certification program is very important as it enhances program credibility. Manufacturers are more likely to participate in a program when they know that government authorities have approved it or validated its requirements. Thus, while the sponsoring organization must be selective in identifying partners, it must also remain as broad as possible, so as to represent the target market as a whole.

17.6.1.2. Candidate Profile. The development of a certification program requires careful study and understanding of the needs of the market and of the companies that are potential participants. Their target clientele must also be considered, specifically persons with allergies for whom the products will be targeted. The program should be launched only if it can be demonstrated that the proposed benefits will meet the specific needs of companies and consumers. For example, developing a program to certify products as free of a nonrecognized allergen, or an allergen whose label declaration is not mandatory or required by regulation, although it may be of interest to a few consumers, may be of no importance to the industry. A program with a strong rationale and evidence of success should not be taken at face value unless the proposition is validated through market research of the target clientele.

17.6.1.3. Market Analysis. The "allergen-free" food sector represents a niche market for the agri-food industry. From 1999 to 2003, the market for "allergen-free" foods doubled in the United States, from US$947 million to approximately US$2 billion in retail sales [5]. Trends suggest that this growth will continue, and the category would be an important one in Canada by 2020. As shown in Figure 17.1, the global supply of innovative "allergen-free" products has increased by more than 700%.

Certification programs could help to identify the domain areas or sectors that would be of significant interest and have potential for strong growth for the industry. Figure 17.2 shows, as an example, that over an approximately 10-year period, "gluten-free" foods accounted for more than half of the "allergen-free" foods marketed. This is explained not by the number of individuals who are hypersensitive to gluten or allergic to wheat, but mainly by the fact that this ingredient is present in virtually all foods.

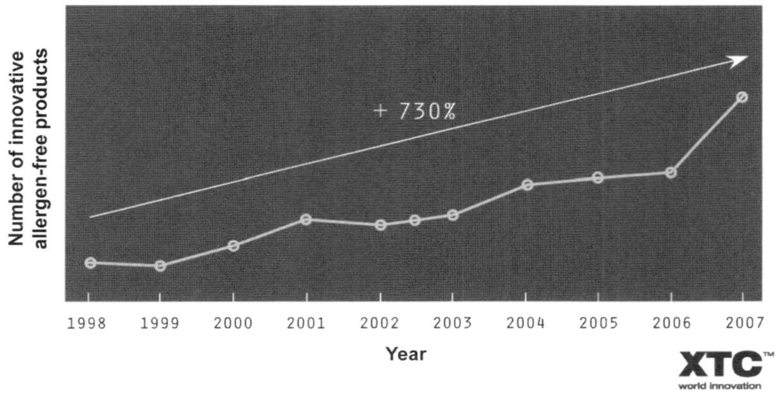

Fig. 17.1 Number of innovative allergen-free food products listed in the XTCSCAN™ database. *Source*: XTC, World Health Food Innovation Panorama, with permission.

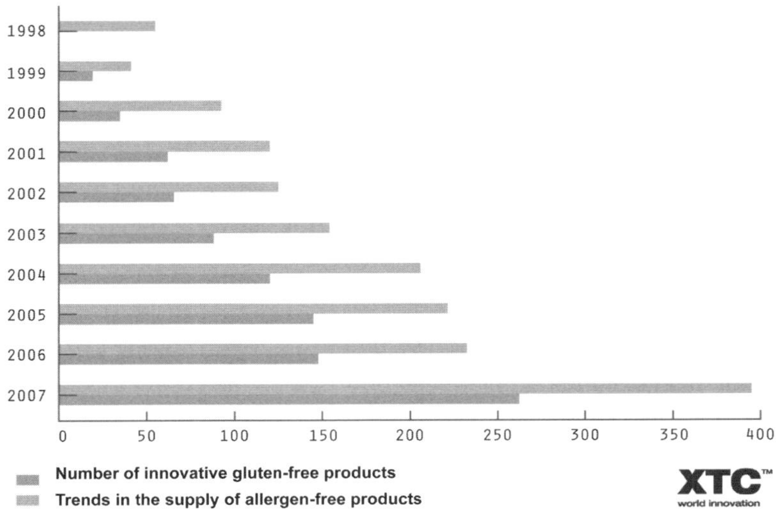

Fig. 17.2 Trends in the global supply of allergen-free products. *Source*: XTC, World Health Food Innovation Panorama, with permission.

Experience has shown that demand for certification often lags a year or so behind market demand for products or services in the targeted domain area. Consequently, when developing a certification program, it is important to conduct a market analysis and identify the sector that will remain viable for the industry.

During the market analysis, steps should be taken to verify whether existing organizations already offer similar certifications in order to assess the potential volume of work and revenues; in fact, there could be major competition with existing organizations. In this case, collaboration with the existing programs in a region or other countries can be considered. Such an approach could also be useful in standardizing practices in the processing industry and even promoting exports. It should be noted that even in the absence of competition, having partners in several countries would help to promote the program and its expansion.

17.6.1.4. Assessment of Requirements. The requirements to be assessed for granting certification must be defined on the onset and must take into account scientific research and findings, as well as industry practices. These requirements must remain rigorous, realistic, and applicable, which is why a multidisciplinary team composed of various partners in the domain area is required at the development phase. This process makes it possible to objectively determine the parameters and skill sets that must be verified during the certification assessment. Manufacturers are more likely to accept and agree to participate in a program whose requirements are clearly defined and applicable to their processes.

17.6.2. Program Development

The program development phase essentially sets the stage for organizing the program design, business plan, assessment tools, and a strategy for learning.

17.6.2.1. Business Plan. A business plan for the certification program should be created on the basis of the business analysis. At a minimum, the business plans should attempt to answer the following five questions:

- What is the program's business value to the organization?
- Do the tangible and intangible benefits exceed the cost of developing and supporting it?
- How will the program be positioned to customers and partners?
- What kind of marketing effort or plan will be pursued to make the program a success with customers?
- What kind of support or organizational resources will be needed to manage the program?

In order to ensure the program's long-term viability, the business plan must consider each predetermined phase and must include the overall long-term vision (development, commercialization, etc.). The business plan should form the basis of support within the organization for such an undertaking and justify the commitment. It should help to directly answer questions raised about the value of the program. Additionally, it should also allow an organization to determine how to proceed and how much to allocate toward the program.

17.6.2.2. Program Design. The program design defines the number of certification tracks and levels that might be offered, specifically which allergens and which products. It also defines the general and minimal requirements that a candidate must meet to obtain certification. The program should also define what prerequisite knowledge candidates must possess ahead of pursuing a certification. For example, it would be difficult to implement the requirements of a certification program guaranteeing that a processed food is free of allergens without having basic knowledge of food hygiene and safety.

And finally, the program design should specify how the certification aligns or articulates with current industry practices without, however, generating excessive costs. This would involve, for example, demonstrating that certification would not require the construction of new manufacturing lines, the purchase of new equipment, very expensive tests, and so on. It must not interfere with the effective operation of other existing programs such as HACCP and organic products, nor be perceived as an additional "burden" or one program too many for the industry.

17.6.2.2.1. *Important Aspects to Consider during the Design*

17.6.2.2.1.1. SYSTEM OR PRODUCT CERTIFICATION. A product certification creates a link between the consumer and the product manufacturing process. More specifically, a certification mark on the product (e.g., a logo) provides direct information to potential consumers. A logo (e.g., the picture of a peanut in a red circle with a diagonal line through it) indicating that the product is peanut free is a clear message to consumers about the manufacturing process, that is, that peanuts are banned in the plant and that the product does not contain any.

A system certification such as the HACCP system does not allow producers to place a logo of any kind on the product. A HACCP certification would, thus, offer some degree of security to the distributor, who has the assurance that the products are manufactured under safe conditions and are fit for consumption. However, although the product is manufactured in a safe environment, the end consumer does not have access to this information at the time of purchase. Since consumers with food allergies are looking for information about the product manufacturing process, a product certification could therefore represent added value; and this must be decided during the program design phase.

In addition, certifying a system with respect to food allergens may, in some cases, raise safety issues for consumers, particularly when the company handles several allergenic ingredients. Consider a plant certified to be free of a specific allergen (e.g., "peanut-free" plant) while other allergens may be present; this may potentially lead to confusion in the minds of less well-informed consumers as they might think that all products manufactured in that plant are completely free of all allergens, thus, exposing themselves to potentially serious allergic reactions.

17.6.2.2.1.2. ALLERGEN DETECTION TESTS. Over the past few years, numerous techniques have been developed to detect various sources of food allergens. Some of these methods are capable of detecting allergenic components with a very low detection limit, less than 1 ppm. The performance of some test kits, however, appears to be highly variable and depends on a variety of factors including the methodology used, extraction conditions, system specificity, and the food matrix analyzed.

For certification purposes, the ideal approach would be to adopt tests whose detection limits are as low as possible, at least for the analyses of the finished products targeted. This approach (with the goal of total absence of the allergen) would make it possible to standardize industry practices and promote exports of certified products in several countries where regulations are different. It would reassure allergic consumers, even those who are hyperallergic, perhaps until such a time as there is clearer information on the amount of allergen dose required to provoke allergic reactions.

The use of detection tests that have been validated and approved by relevant official authorities is therefore recommended. This validation is based on sensitivity, performance, repeatability, and reproducibility criteria, and includes a comparison of the various methods and the effect of food matrices [6]. An interlaboratory validation provides a means of evaluating and comparing the abilities of the tests to determine the presence of allergens in various food matrices. This approach makes it possible to evaluate, for a given confidence level, the ability of the techniques to provide reliable results. Omitting this step could lead to results that are nonreproducible or that differ from one food matrix to another or from one laboratory to another.

The cost of the analyses and of commercially available kits is still high, but is tending to stabilize or decrease. Manufacturers are very reluctant to bear these costs, especially for the purpose of certifying a product to be free of several allergens. For example, a manufacturer who needs to test for four allergens in 20 products should allocate a significant budget for performing these analyses. The certification-sponsoring organization should take these costs into account, as it could constitute an obstacle to manufacturers considering participation in the program.

There is currently no simple system that is capable of simultaneously detecting the majority of allergens in a food product. Only certain models can provide simultaneous multi-allergen analysis [7]. However, the existing models reflect the scientific state of the art and constitute a step toward multi-allergen detection. The certification-sponsoring organization must therefore set up a system for monitoring these scientific advances in order to incorporate into the program new single or multi-allergen detection tests that may be of interest to many manufacturers.

17.6.2.2.1.3. THIRD PARTY CERTIFICATION. In response to competitive pressures, manufacturers of food products are keen to include desirable attributes of their products on the label [8]. These statements may or may not be supported.

To support the credibility of these claims, they often rely on third-party certifiers to demonstrate that their practices comply with applicable standards. In any certification process, it is essential to accord a prominent role to an independent certification organization. The fact that this organization is not directly involved in the production chain (i.e., it is not a farmer, processor, or distributor) ensures the reliability and credibility of the program in the eyes of consumers.

The use of TPC has advantages for processed food distributors compared with other certification systems (first- or second-party certification). First, their responsibility to verify the implementation of the requirements related to the quality and safety of the foods they distribute is minimized since this is largely performed by the third party. Additionally, certification is a marketing tool for manufacturers and distributors, who can proudly point to the involvement of an independent organization, guaranteeing the quality and safety of the foods they sell.

The involvement of a third party in the certification process enhances the level of quality of the products by reducing the risk of noncompliance and losses. This increases efficiency in the supply chain and reduces costs for distributors. An additional reason why TPC must be considered is that many suppliers report that the use of TPC is gradually being demanded by distributors and is even sometimes mandatory [9]. Distributors require manufacturers to have foods supplied certified by a third party as the costs associated with quality and safety are largely transferred to manufacturers, which enables budgets to be allocated to other areas.

17.6.2.2.1.4. USE OF THE CERTIFICATION MARK: CLAIMS AND ADVERTISING. In the food sector, where it is essential to remain competitive because profit margins are thin, advertising plays a major role. In 2005, in the United States for example, the food industry spent US$32 billion on advertising and US$66.5 billion on packaging in order to differentiate its products from those of its competitors [8]. Hence, one sees all kinds of claims, which easily cause confusion among consumers with allergies. Even if certification increases consumer confidence in a product, consumers still find it difficult to distinguish or even make no distinction between certified and noncertified products. Companies participating in the program will tend to resort to the practices of their competitors (unfounded claims, misleading advertising, etc.) in order to improve their competitive position. In view of the fact that most countries have no regulations on this subject, it is important during program development to establish rules by developing, for example, a guide on use of the certification mark. The packaging mock-ups and advertising material prepared by the companies must first be screened by the organization.

17.6.2.3. Creation of Assessment Tools. Based on the established requirements, the program sponsors must develop a plan and assessment tools that will be used to verify the compliance of the manufacturer's practices. Since the requirements are determined by the multidisciplinary team composed of subject matter experts, the questions and assessment tools must be validated by these experts in order to determine whether they truly reflect the competencies and qualifications expected from manufacturers. For example, one of the requirements would be that all participating manufacturers provide proof that the ingredients received from suppliers are free of the targeted allergens. The assessment questionnaire validated by the experts would require, for instance, an allergen-free guarantee letter from the supplier (based on a model letter), the procedure describing the practices in place to prevent cross-contamination, and the detection tests carried out on the batches provided with kits recognized and validated by official agencies.

Actual assessments can be organized by companies that voluntarily agree to participate in order to determine the passing scores. The program sponsors ask voluntary participants to implement the stated requirements and they are then assessed. Passing scores are determined only after the program sponsors have the confidence that the question items and the assessment forms are reflective of manufacturer skills. This approach is important in order to ensure the applicability of the established requirements.

For example, in practice, mills are used to produce several different kinds of flours (wheat, spelt, buckwheat, etc.). It would therefore be difficult, for instance, to obtain buckwheat flour containing no traces of wheat gluten. Requiring the "total" absence of gluten in a buckwheat flour-based product without verifying whether this is feasible could therefore prove impractical for manufacturers relying on existing supplies of buckwheat flour. In addition, depending on the detection limit of the test used (low or high), the analyses may or may not reveal the presence of gluten in the products. In such cases, it is important to validate with voluntary participants the feasibility of such a requirement, by determining as an example, the level of gluten in the buckwheat flour and also determining what tests and detection limit to use.

For certain allergens, this assessment may prove unnecessary. In the case of peanuts for example, a total ban on peanuts in the food and the verification of their absence by analysis with a validated, recognized test with the lowest possible detection limit would be sufficient because such requirements are easily applicable.

17.6.2.4. Learning Support. It stands to reason that any certification program launched should be based on the implementation of available training systems and information tools. A certification program could, however, be launched without these tools provided that the sponsors ensure that, at a minimum, the necessary learning support materials are offered (e.g., manuals, guides, published articles, newsletters, brochures, frequently asked questions, etc.). These support materials can be useful to both manufacturers and allergic and nonallergic consumers.

Candidate companies for certification need support when preparing for the certification exams. Experience shows that candidates often prefer to follow a defined road map when preparing to pass a certification. Model forms, procedures, and so on, are often well received by manufacturers, and the sponsoring organization can develop these tools in order to generate interest in the program. Learning support can be offered by providing certification-related manuals, books, and workshops. Either way, for a certification to be successful among manufacturers, a strategy for learning support should be addressed while defining program parameters and design.

17.6.3. Certification Program Implementation

17.6.3.1. Program Implementation. In order to popularize and ensure the credibility and reputation of a program in the certification marketplace, a number of important aspects need to be further considered as indicated below.

17.6.3.1.1. *Marketing Launch and Customer Communications.* A successful program launch should describe the certification program requirements and its benefits through integrated marketing communications. Through websites, frequently asked questions, press releases, and customer service centers, an organization can foster an understanding about the program and its scope in the mind of companies or help convince them of the need to participate. Potential candidates are most likely to follow up with a range of "what-if" scenarios to assess the potential investment and return. A two-way communication should therefore be enabled through the course of the program, and especially in the early days of the launch.

Very often, consumers have little information about existing certification programs [10]. It is important to plan the communication and information tools in order to reach consumers with allergies and members of the public to whom the certified products are aimed. Although these consumers can contribute to the implementation of such a program, there will still be many who will be unable to participate. The communication plan would thus make it possible to reach the majority of these persons and help inform them about the program and the existence of such products.

17.6.3.1.2. *"Test"—Pilot Program Delivery.* Since the certification sponsor is not a food manufacturer, it must rely on outside companies as partners for implementation of the requirements and the pilot test. One or more companies can be used for this purpose. It is well advised to choose a company that has a number of production sites around the world and to involve several of these sites in order to help promote the program around the world. In the absence of such a partnership, arrangements can be made with various companies around the world, in the country, or in the region where the program is being developed. The choice of a recognized partner is vital to ensure that the integrity of the certification assessment and customer satisfaction are not compromised. The level of preparation of the pilot plants is essential for an effective program launch, since they could be contacted by consumers and other manufacturers.

17.6.3.1.3. *Pilot Program Preparation.* The availability of established procedures and learning methods has a significant impact on adoption of the certification. In conjunction with certified trainers and even publishers, the program sponsors must ensure the availability of training and practical and preparation materials that can be used by potentially interested companies. For a smaller audience, that is, for certification limited by the number of candidates, "e-training" and "e-communication" can help to provide preparation materials in a cost-effective manner to a dispersed audience.

Even with a pilot test, a successful launch, and training and information tools developed by the organization, there are still many other challenges that manufacturers must meet in order to ensure the effective implementation of the program. Companies must have qualified and competent in-house personnel to implement the program. Having people who can read, understand, and interpret the requirements is indispensable to their implementation in the plant. In addition, all procedures must be updated regularly. Companies must conduct a feasibility study, which will make it possible to identify and diagnose potential major problems in terms of existing practices. Requirements must also be adapted according to the context and practices of each company.

Food processors must deal with changes that may significantly impact existing habits or methods. Changes in the physical layout may be necessary in certain cases. For instance, a plant that previously manufactured products containing peanuts or nuts and that wants to completely ban them, with the aim of being certified as peanut free, must halt production and thoroughly clean the equipment, facilities, and so on. The company may also find it necessary to purchase new equipment or build a new plant.

Storage facilities must be segregated so as to avoid any accidental contamination, and procedures must be instituted in the event of accidents observed during handling and storage. Storage of finished products containing the allergen from other production sites could require construction of a new loading dock.

Employees must also adjust to the new measures required by the certification, which may upset their work habits and provoke a hostile reaction. For example, employees may be opposed to restrictions on bringing foods containing the banned allergen into the plant for lunch and to changes affecting use of the cafeteria and vending machines. This could force the manufacturer to build a suitable plant restaurant in order to offer its employees lunch menus not containing the banned allergen(s) and so on, which obviously entails additional expenses.

In the case of a company that has several production sites, logistics must be reorganized, for example, delivering ingredients to a plant site that manufactures certified products before deliveries to other sites where the banned allergen is handled. This requires specific material handling procedures as well as a redeployment of human resources.

Developing new recipes that do not contain the banned allergen is also a major challenge for the industry. Developing foods that do not contain the banned allergenic ingredient while preserving a flavor comparable to that of standard products can be a difficult and even complicated undertaking. For example, for a cake that does not contain wheat, eggs, or milk, the recipe must be formulated in such a way that it is able to preserve a texture and flavor comparable to those of regular products. This is an extremely complicated task since substituting ingredients could cause other problems; for instance, lupine, which would be a good substitute for eggs in baked goods, could cause allergic cross-reactions in persons with allergies to peanuts [6].

Manufacturers often have difficulty obtaining allergen information from suppliers of nonfood raw materials such as lubricants, packaging materials, and cleaning products. The company could also be faced with new situations never previously encountered and not anticipated or covered by the program. An anecdotal example concerns a company that manufactures "nut-free" products and was informed by a supplier of nonfood materials of the presence of hazelnut shells in forklift tires used to improve their adherence to the ground.

The company must also provide training to subcontractors (e.g., service suppliers), which complies with the established policies and procedures. Finished products must be sent to outside accredited laboratories for independent analyses. Semifinished products, even if they possess the primary packaging, cannot be packaged outside the plant and be considered certified. Indeed, a product is certified with respect to the process in place in the plant, and any additional off-site handling of the product can affect the integrity of the primary packaging and expose it to a source of contamination.

17.6.4. Program Management

The program management component of certification ensures that the business value and assessment integrity remain intact for the program. Through ongoing evaluation of market needs and continual program updates, a sponsoring organization can ensure that the program remains relevant over the long run. The program management function also ensures that the program continually evolves to stay current with industry expectations, content, and organizational needs.

17.6.4.1. Monitoring and Maintenance. Assessment results of companies participating in a certification program should be tracked and monitored to ensure that candidates are accurately accredited for their achievements. Comments of participating companies can be solicited and auditors can be evaluated. Additionally, the program could maintain a database of participating companies, which provides them with access only to their own records. After candidates complete the requirements, the program administration should ensure that they receive all the appropriate recognition, such as completion certificates, program logos, support, website promotion, and advertising.

Program maintenance should also include review of requirements to ensure that the content is updated and revised regularly to keep pace with changes in the marketplace. A program that continually updates itself in response to the market or new skill needs reduces the risk of assessing candidates on obsolete content or even of losing credibility. Ongoing task analysis, periodic surveys, focus groups, or working committees help validate the relevance of the program to the workforce.

17.6.4.2. Customer Relationship Management. Besides providing customers with access to a resource for answering questions about the program, customer relationship management includes access to support or knowledge tools for those who meet program requirements and certify. When a manufacturer joins the program, it makes a commitment to the sponsoring organization as much as it does to the certification. If a sponsoring organization views a company that is certified as a loyal customer, affinity programs to strengthen the loyalty will yield benefits in generating advocates within the customer base or the industry.

Creation of a community of practice around the topics of interest for certified companies offers an opportunity for knowledge sharing. Tools such as discussion groups and chat rooms can be enabled for such a community of practice. Additional information items, such as program updates, certification-related news, and learning games can also be offered as a bundle of benefits to certified companies.

17.6.4.3. Marketing. The value and integrity of the program should be demonstrated not only to potential candidates for certification, but also to the distributors who sell their products. Distributors and consumers view certification as a screening tool if they determine that the program is credible. Program administrators should therefore ensure that the marketing message describing the value of the program is presented to potential distributors. This involves presenting the benefits and especially the added value that certification could represent. For example, in presenting the financial and social aspects as added value, the distributor would recognize that it not only generates substantial profits from the products it sells, but also participates in a social action, that of meeting the needs of persons with allergies, which is what most companies are looking for.

17.6.4.4. Evaluation. Program administrators should periodically demonstrate the productivity impact of a certification program on participating partners or customers. At the very minimum, through anecdotal evidence and, preferably, through formal research, the progress toward program goals and distributor satisfaction should be measured for effectiveness. Such data provide valuable input for senior management and useful insights for program direction and decision making.

A few data points for assessing whether a certification program is having the desired impact are as follows:

- Measure the self-reported intent to join or renew membership (loyalty) in the program through surveys to determine whether certified companies respond more favorably than noncertified companies.
- Assess the support calls received by customer service to determine whether the frequency of complaints or claims received from certified customers are lower than those reported by noncertified companies.
- Compare the level of satisfaction ratings garnered by pre- and postsales teams working for certified versus noncertified companies.

17.7. DOCUMENTATION

Documentation plays an essential role, both in the implementation of the requirements of a certification program and in the application of these requirements. It is important to ensure proper documentation since it serves as a reference for everyone within the company and helps standardize employee practices. Documents are intended to be used and must therefore be tailored to the users. They must be adapted to the users' needs, reflect the company's situation, and be simple and easy to read and understand by all.

Since it is the employees who apply the content of the documents, it is recommended that documents be drafted with employee collaboration. Their opinions, judgement, and experiences are extremely useful in ensuring effective application of these documents. Going to the employees' workplace and discussing the content of a document or procedure under development is a way of recognizing their role in the company. Such an approach not only reinforces the employees' feeling of belonging and of being useful to the group, but is helpful in securing acceptance of the document and, consequently, ensuring that it will be applied in practice.

In order to ensure proper documentation, it is also well advised to invite the employees themselves to draft documents, procedures, and so on, which can then be edited and corrected as necessary. In this case, it is important to conserve in the text the routine terms and language that they use on a daily basis. Documentation written by employees or with their participation is better accepted than when it is imposed by superiors.

Personnel often do not have time to read the "superfluous" content of very lengthy documents. Care should therefore be taken to avoid including too much information in the documents, since this makes them unwieldy and difficult to use. Essential points must be summarized and limited to a few pages. In order to improve readability, the font size and style should be considered when drafting documents.

Forms are the documents most frequently used by production employees, whose duties include maintaining production volumes as well as recording the details of equipment malfunctions or breakdowns. These forms must therefore be simple and reserved solely for recording important data.

In the case of small- and medium-sized companies or start-ups, despite the support provided by the program sponsor, there may not be enough personnel available to adapt the program requirements to the company's situation, as well as to develop the relevant procedures. Consultants with relevant expertise can help in the development of such documents. However, document development must take place in collaboration with company personnel in order to ensure that the documents produced are suitable and relevant to the company's situation.

The program sponsor could also provide companies with model documents (procedures, forms, etc.) dealing with program requirements, which could then be simply adapted to suit the context of the company and their lingua franca.

Depending on the company's resources, specific software can be useful to ensure proper documentation and data recording; this simply requires training employees in their use. The advantage of software is the capacity to preserve recorded data for very long time periods, during which the documents remain available for consultation at all times. This also avoids the need to print multiple copies and cuts down on paper waste. The above recommendations should ensure proper documentation, employee involvement, and effective application of program requirements.

17.8. CONCLUSION

Ingestion of a food containing an allergen can trigger an immediate and serious allergic reaction, which can be fatal for some consumers. As people's lives are at stake, manufacturers are obliged to ensure the integrity of their manufacturing process, and it is their responsibility to also ensure that their products are safe for consumers with allergies. Although TPC may be onerous, it provides processors with a high level of confidence compared with an "in-house" program. An external program requires the implementation of very strict criteria to guarantee food safety, which are verified by an outside party, a neutral and independent certifying organization.

It is sometimes difficult for manufacturers to have complete confidence when the product manufactured relies only on internal standards. Additionally, in the event of litigation following the proven ingestion of a contaminated food produced, the food manufacturer will be called on to defend itself and to demonstrate that practices and methods in place were adequate to prevent or avoid such an incident. In the absence of an "external" certification program, manufacturers may also be less rigorous and may make concessions or exemptions in regard to requirements that they themselves may have established. However, when these requirements are imposed by an external source and must be verified by a third party, manufacturers make a greater effort to comply. Furthermore, a certification mark on a product label provides assurance to consumers, who can purchase these products with peace of mind; that is to say, it "makes their life easy" by eliminating the need for the repeated reading of labels and long searches for products suitable to their diet.

The market for "allergen-free" products is a niche market. Companies and distributors are all hoping to get a piece of the pie by various means (logo not accompanied by preventive measures, in-house programs, claims, advertising, etc.). The increase in the prevalence of and demand for "allergen-free" products explains this quest for profit. Certification is the "arbiter" of the situation and could reconcile consumer and industry needs as manufacturers are willing to provide foods that meet the needs of customers and consumers, but not beyond the realm of what is feasible.

In order to facilitate the adoption of a certification program by industry and thus meet the needs of consumers, organizations must, in the course of developing a certification program, rely on the services of various partners who are well-known experts and competent in the field. This approach will reinforce the credibility of the program and ensure its future reputation. Finally, it should be mentioned that allergen-free certification is an evolving concept that cannot remain static. The issue of food allergies is relatively new and therefore requires ongoing monitoring of scientific advances and constantly changing regulations. Thus, an open approach will make it possible to continually update and meet the specific needs of a rapidly growing market.

REFERENCES

1. ISO 19011. (2002). Guidelines for quality and/or environmental management systems auditing. International Organisation for Standardisation. Available at http://www.iso.org
2. ISO 9001. (2008). Guidelines on good quality management practices. International Organisation for Standardisation. Available at http://www.iso.org
3. ISO 22000:2005. (2005). Food safety management systems—requirements for any organization in the food chain. International Organisation for Standardisation. Available at http://www.iso.org
4. Kelly, T., Nanjiani, N. (2004). Certifications as a branding tool: sponsoring a certification programme. In: *The Business Case for E-Learning*. Indianapolis, IN: Cisco Press, pp. 67–74.
5. Claytor, R. (2009). Parents offered new choices in allergen-free food. *Medill Reports Chicago*. Available at http://news.medill.northwestern.edu/chicago/news.aspx?id=130001
6. Sébastien La Vieille. (2008). Allergies alimentaires et étiquetage de précaution. France: Agence française de sécurité sanitaire des aliments, pp. 36, 39, 48.

7. Ben Rejeb, S., Abbott, M., Davies, D., Cléroux, C., Delahaut, P. (2005). Multi-allergen screening immunoassay for the detection of protein markers of peanut and four tree nuts in chocolate. *Food Additives and Contaminants*, *22*, 709–715.
8. Golan, E., Kuchler, F., Krissoff, B. (2007). Do food labels make a difference? ... sometimes. *Amber Waves*. Available at http://www.ers.usda.gov/AmberWaves/November07/Features/FoodLabels.htm
9. Hatanaka, M., Bain, C., Busch, L. (2005). Third-party certification in the global agrifood system. *Food Policy*, *30*, 354–369.
10. Botonaki, A., Polymeros, K., Tsakiridou, E., Mattas, K. (2006). The role of food quality certification on consumers' food choices. *British Food Journal*, *108*, 77–90.

18

EMERGING ALLERGENS AND THE FUTURE

Allaoua Achouri and Joyce I. Boye

18.1. OVERVIEW OF MAJOR FOOD ALLERGENS

Food allergy is a disorder in which ingestion of a small amount of food (mostly naturally occurring proteins or glycoproteins) elicit an abnormal immunologically mediated clinical response. A scientifically based list of foods known to cause severe adverse reactions in hypersensitive individuals was developed and endorsed by the Codex Committee on Food Labelling in 1998 [1]. Consideration was given to those foods most commonly identified as causing medical problems that are seriously debilitating or life threatening, or are associated with increased risk of serious chronic disease [2]. The initial list contained eight major serious allergens (MSAs), namely milk, eggs, soy, wheat, peanuts, fish, shellfish, and tree nuts, which should always be identified on food labels. This list was subsequently revised in Europe, Canada, but not in the United States. Health Canada and the Canadian Food Inspection Agency have jointly identified and included sesame in the list of 10 food allergens in addition to sulfites [3]. Health Canada has subsequently developed criteria derived from the guidance issued by the Joint Food and Agriculture Organization of the United Nations (FAO)/World Health Organization (WHO) Expert Committee on Food Additives (JECFA) on how to expand the list of priority allergens targeted by mandatory labeling on prepackaged foods available for sale in Canada (Health Canada's Food Directorate, September 2009 internal

Allergen Management in the Food Industry, Edited by Joyce I. Boye and Samuel Benrejeb Godefroy

report). The European Commission, by issuing Directives 2000/13/EC and 2003/89/EC, has also widened its allergens list, which have now resulted in a mandatory labeling of 12 major allergenic foods (big eight + sesame, mustard, celery, and sulfites) in all its member states [4]. This list needs to be kept under review and updated in light of changing food practices, emerging clinical observations, and relevant scientific data.

18.2. MINOR AND EMERGING ALLERGENS

Beyond the so-called "big eight" major allergens, which have received the greatest attention, over 170 other foods have been documented in the scientific literature as causing allergic reactions [5]. This number may likely increase due to the following factors: (1) people can become allergic to almost any food product and at any time in life; (2) the pervasive presence of proteins in foods; (3) increased use of conventional and novel protein ingredients in product development; and (4) food globalization and immigration leading to changes in dietary habits. This review will provide a summary of some specific emerging allergens that are not necessarily listed as priority allergens but are becoming the focus of public health experts in many countries, especially as their occurrence and awareness has risen in recent years.

18.2.1. Oilseeds

The seeds of sesame and mustard contain potent allergens that have resulted in their inclusion in the priority allergies in Europe (sesame, mustard) and Canada (sesame, mustard). Other countries have been, however, slow to follow suit. A summary of the allergenic properties of these two seeds are provided below.

18.2.1.1. Sesame Allergy

18.2.1.1.1. Sesame Seed Allergenicity. Sesame is an oilseed plant of the family of Pedaliaceae, originating in the savanna of central Africa spreading to Egypt, India, the Middle East, Asia, the Balkans, Latin America, and the United States [6]. It includes four species of which *Sesamum indicum* is the most important found in commerce. In recent years, the worldwide production and consumption of sesame seeds has increased tremendously due to their high oil, protein, and iron content [7]. Sesame seeds can be consumed as whole seeds, or they can be processed into sesame oil or sesame paste. Sesame seeds and ingredients are frequently used in bakery products, salad dressings, dips, and vegetarian foods. Furthermore, sesame oil has been used in the pharmaceutical and cosmetic industries based on its presumed low antigenicity [8].

Sesame allergy is particularly problematic because while its prevalence is assumed to be low, allergic reactions can be anaphylactic and potentially deadly. In fact, until recently, sesame seed was rarely thought to be allergenic. However, its allergenicity may not be as low as assumed, as increasing use in North America, Australia, many European countries, and Asia (Israel and Japan) has been paralleled by an increase in reported sesame-induced allergic reactions [9–15]. Sesame was found to be a major cause of IgE-mediated food allergy in Israel and was second only to cow's milk as a cause of anaphylaxis [7]. Canada and European countries have identified and included sesame in their allergen labeling lists [3, 4]. Experts suggest that sesame is quickly becoming the ninth most common allergen, and one of the fastest-growing allergies in the world. Unfortunately, sesame allergy remains relatively unknown by the general public and is not as prominent in the public consciousness as peanut [16].

18.2.1.1.1.1 IDENTIFIED ALLERGENS IN SESAME SEED. Studies describing the IgE-binding for sesame revealed that allergic patients recognized a wide range of polypeptides with different molecular weight (MW), indicating the existence of multiple allergens in sesame seeds. To date, at least seven potential sesame allergens have been identified, including Ses i 1 [17, 18], Ses i 2, and Ses i 3 [19] with MWs of 9–14, 7, and 45 kDa, respectively. Two oil body-associated proteins (oleosins) were found to contain sites that bound IgE from sesame-allergic individuals. These oleosins were sequenced and named Ses i 4 (17 kDa) and Ses i 5 (15 kDa), in accordance with the International Union of Immunological Societies (IUIS) Nomenclature [20]. More recently, two additional sesame seed allergens, Ses i 6 (52.2 kDa) and Ses i 7 (56.6 kDa), have been identified by Beyer et al. [21], confirming the allergenic potential of the 11S globulin. Numerous other sesame allergens with MWs of 20–25 and 35 kDa have been observed by two-dimensional electrophoresis and immunoblotting. Further studies are, however, needed to confirm the biological activity of these newly identified allergens (Table 18.1) [19, 22–24]. Cross-reactivity between sesame and other plant seeds, such as soybean and peanut [19], hazelnut and rye grain [25], and, more recently, with walnut [26], have been reported.

18.2.1.1.2. Sesame Oil Allergenicity. Sesame oil is one of the few vegetable oils that is used unrefined and, therefore, may contain trace amounts of proteins that could be hazardous to sesame-allergic individuals. Several reports have claimed the allergenicity of sesame oil, as it is widely used by the food [8, 12, 27, 28], pharmaceutical, and cosmetic industries, and has been shown to cause contact allergic dermatitis [29–32]. Although the allergenicity of sesame oil has been previously documented, the results obtained in these studies have not explicitly demonstrated the direct implication of either sesame lignans (sesamin, sesamolin, and sesamol) [33, 34], residual proteins, or oleosins [20] on sesame oil allergenicity. Further research studies will therefore be useful.

18.2.1.2. Mustard Allergy. Mustard seed (*Brassica* spp.) is an annual, cool season crop that belongs to the Brassicaceae family. Mustard has been used

TABLE 18.1 Identified Sesame Seed Allergens

Allergen Fraction	MW (kDa)	Allergen Nomenclature	Cross-Reactivity	References
2S albumin	9–14	Ses i 1		17, 18
2S albumin	7	Ses i 2		19
7S globulin	45	Ses i 3	PN (*Arah 1*)	19, 24
Oil body-associated proteins				
Oleosin isoform	15	Ses i 4		20
Oleosin isoform	17	Ses i 5		20
11S globulin isoform	52.2	Ses i 6		21, 26
11S globulin isoform	56.6	Ses i 7		
11S subunits				
Acidic subunit	35	Not assigned		22–25
Basic subunit	20–25	Not assigned	H, R, K, PS	

H, hazelnut; R, rye; K, kiwi; PS, poppy seed; PN, peanut.

for culinary purposes, mainly in sauces, as a seasoning in main dishes, or as a traditional ingredient in salad dressing. Mustard used in food is typically a mixture of *Sinapis alba* L. (yellow mustard) and *Brassica juncea* (oriental mustard) [35]. It is most commonly consumed in Europe, the United States, Japan, and India according to local food habits.

Despite its widespread use, mustard allergy has been well documented in clinical and laboratory studies, where several severe anaphylactic reactions have been reported in both adults and children. A large proportion of the documentation on mustard allergy was published in France [36–40], Spain [41–43], Sweden [44], Finland and Italy [45, 46], and India [47]. Concerns about mustard safety for food-allergic consumers led the European Commission to include mustard on the list of 12 priority allergenic ingredients requiring mandatory labeling. The same concerns and documentation of mustard-related anaphylactic reaction is leading Health Canada to consider including mustard as part of its priority allergens.

18.2.1.2.1. Identified Allergens. Two major mustard allergens have been characterized, which are closely related in structure. A major allergen of white mustard, Sin a 1, is characterized as a seed storage protein, belonging to the 2S albumin family, with a MW of 14 kDa [48–53]. Sin a 1 has close similarity to other 2S albumins, such as those isolated from rapeseed, castor bean, and Brazil nut. A second major allergen of the oriental mustard, Bra j 1, has also been identified as a seed storage protein from the 2S albumin family, with a MW of approximately 16 kDa [54, 55]. Recently, a protein of 51 kDa (Sin a 2) was recognized as a third mustard allergen. The allergen dissociates to two

TABLE 18.2 Identified Mustard Allergens

Allergen Fraction	MW (kDa)	Allergen Nomenclature	Cross-Reactivity	References
2S albumin	14	Sin a 1	RS, S, BN, some Brassiceae family species	48–52
2S albumin	16	Bra j 1	RS, S, BN, some Brassiceae family species	54, 55
11S globulin	51	Sin a 2		56, 57

RS, rapeseed; S, sunflower; BN, Brazil nut.

subunits of 36 and 23 kDa, belonging to the seed storage 11S globulin [56, 57]. Identified mustard allergens are shown in Table 18.2.

18.2.2. Pulses

Pulses (bean, pea, chickpea, lentil, lupin) are the dried edible seeds of legumes, which belong to the family Leguminosae. There are over 13,000 known species. Pulses play an important role in the diet and provide a rich source of protein and other nutrients. Chickpea, lentils, and white beans are very common components of the Mediterranean and Middle Eastern diets. Sensitization to pulse legumes has been reported. Crespo et al. [58] and Martinez San Ireneo et al. [59] reported that among Spanish children under 5 years of age, legumes are the fifth most prevalent food allergy.

18.2.2.1. Lupin Allergy. Lupins are colorful plants belonging to the same plant family as peas, lentils, and beans. *Lupinus albus* (white lupin, Mediterranean countries), *Lupinus luteus* (yellow lupin, Central Europe), and *Lupinus angustifolius* (blue lupin, Australia) are the three major species found in foods. Studies have shown that lupin can be successfully incorporated into food products [60–62] as a wheat or soya substitute, and it has been successfully used in biscuits, pasta, pizza, sauces, soups, and diet products [63].

The addition of lupin flour in bakery products was authorized by the European Union (EU) and given clearance for use in the United Kingdom in 1996 by the Advisory Committee on Novel Foods and Processes (ACNFP) [64]. Severe allergic reaction to lupin was, for the first time, reported by Hefle et al. [65] and involved a 5-year-old girl with a known peanut allergy who developed urticaria and angioedema after eating spaghetti fortified with lupin flour. A first case of contact urticaria from lupin was also reported by Gutierrez et al. [66]. Moreover, Matheu et al. [67] reported a case of lupin-induced anaphylaxis with positive skin test results and *in vitro* cross-reactivity with other legumes.

A literature search using the PubMed database on lupin allergy revealed the presence of very limited publications on the issue. However, a clear evidence was found that lupin allergy was a significant, serious, and growing problem, with reports mostly from European countries. A study by *The Lancet* published in April 2005 found lupin allergy to be on the rise in European patients known to be allergic to other legumes, particularly peanut, soya, or pea. Lupin allergy has emerged in Italy [68–70], France [71, 72], Spain [73–76], Norway [77], and in the United Kingdom quite recently [64, 78]. Although, a new directive on food labeling came into force in Europe in November 2004, requiring food manufacturers to specifically list 12 potential allergic ingredients, lupin flour was not included on this list, despite recommendation from the U.K.-based Institute of Food Science and Technology. In Australia, three cases of anaphylaxis, the most severe form of allergic reaction, were reported in the *Medical Journal of Australia* by doctors from the Royal Adelaide Hospital's clinical immunology and allergy unit. Since then, specialists have continued to make requests for mandatory warnings on foods containing lupin [79]. Today, food labeling rules require prepacked food sold in the United Kingdom or the rest of the EU to show clearly on the label if it contains lupin (or if one of its ingredients contains it) (http://www.eatwell.gov.uk).

Although cases of lupin allergy have been published, the molecular identity of lupin allergenic polypeptides has not been completely elucidated. Some IgE-binding proteins with MWs of 65, 43, 38, and 13 kDa have been described. The most distinctly reactive band was the 43-kDa protein [68, 80]. Hefle et al. [65] described the IgE-binding proteins of lupin extract to have an approximate MW of 21 and 35–55 kDa and to be heat stable. Recently, the assessment of potentially allergenic polypeptides in lupin seeds by using two-dimensional (2D) electrophoresis combined with immunoblotting revealed two dominant IgE-reacting polypeptides of lupin seeds: one corresponding to conglutin γ (a basic tetrameric protein with sedimentation coefficient 7S, consisting of disulfide-linked 30- and 17-kDa polypeptides) and the other corresponding to the basic subunits of lupin 11S [67, 81]. More recently, Rojas-Hijazo [75] reported the presence of unsuspected, hidden lupin allergen with MW close to 14 kDa in extracts from cookies, a chicken bouillon cube, and a chicken dehydrated soup.

High cross-reactivity to peanut, soybean, and pea globulins has also been reported in clinical studies by double-blinded, placebo-controlled food challenges (DBPCFCs), radioallergosorbent test inhibition, and immunoblot tests [69, 80, 82–84]. However, the risk of lupin allergy due to allergenic proteins in the lupin seeds themselves is very real and could be much higher than the potential cross-reactivities to other legumes. Further work is required to establish the prevalence and significance of lupin allergy in order to help with risk management [64].

18.2.2.2. Chickpea Allergy. Chickpea is an important component of the Mediterranean diet and a very common food for West Asian and Indian populations. Very few studies exist on the allergenicity of chickpea. A first case of anaphylaxis caused by chickpea was described by Niphadkar et al. [85], where multiple IgE-binding bands with MW between 10 and 70 kDa were detected by immunoblotting. Cases of sensitization to chickpea were also reported, mainly in the pediatric population in Spain. Among Spanish children under 5 years of age, allergy to pulses including chickpea was the fifth most prevalent food allergy as earlier indicated [58].

So far, allergens of chickpea have not been purified or immunochemically characterized. A chickpea 2S albumin was partially purified and characterized using sera from chickpea-sensitive individuals [86]. The first attempt to characterize the allergenicity of raw and boiled chickpea extracts was undertaken by Martinez San Ireneo et al. [59]. The results indicated the presence of multiple heat-stable allergens in both extracts in the molecular range of 10–106 kDa. Later, Patil et al. [87] reported, for the first time, chickpea hypersensitivity cases diagnosed with skin tests, DBPCFCs, and enzyme-linked immunosorbent assay (ELISA) in 59 patients who indicated positive histories of allergy to chickpea. Immunoblot analysis showed that 70-, 64-, 35-, and 26-kDa proteins were the major allergens. Furthermore, reduction of antigenicity was observed in extensive hydrolyzed chickpea proteins obtained by sequential treatment with endo- and exopeptidases for the preparation of specialized hypoallergenic food products [88]. A unique clinical cross-reactivity between latex and chickpea has also been reported [89].

18.2.2.3. Lentil Allergy. Lentils are the second most commonly consumed legumes in the Mediterranean area and in some other Asian countries. They are also the most frequent foods associated with IgE-mediated hypersensitivity reactions, particularly in pediatric patients in the Mediterranean region [90]. Research on lentil allergens is surprisingly rare, and very few cases are found in the literature. Martin et al. [91] reported a first severe case of bronchial asthma induced by inhaling vapors from cooking either lentil or chickpea. Later, Kalogeromitros et al. [92] described another case of an 8-year-old girl who suffered four episodes of anaphylaxis related to lentils from ages 3 to 7 years. The first three involved ingestion of small amounts of cooked lentils, and the fourth episode occurred with inhalation exposure to cooking lentil soup. The patient also suffered a clinically relevant hypersensitivity to chickpea. Allergic reaction to lentils starts in early life, usually below 4 years of age, most frequently with oropharyngeal symptoms or acute urticaria and/or angioedema and severe anaphylaxis in 20% of patients [90]. Whether lentil allergy could be outgrown is so far not confirmed. A study by Ibanez-Sandin et al. [93] is the only available report suggesting that tolerance for the ingestion of this legume could be achieved after 5 years.

Two different types of allergens were identified in boiled lentils by Sanchez-Monge et al. [94]. One of them was Len c 1, with a MW of 12–16 kDa corresponding to γ-vicilin subunits, and the other was a 66-kDa protein designed as Len c 2. More recently, by using more advanced analytical techniques (ion exchange chromatography, reverse phase–high performance liquid chromatography [RP-HPLC], matrix assisted laser desorption and ionization [MALDI] analysis, ELISA inhibition assays, and cDNA cloning), the same research group identified a relevant lentil allergen with an apparent molecular size on sodium dodecyl sulfate–polyacrylamide gel electrophoresis (SDS-PAGE) of approximately 50 kDa [95]. This purified and well-characterized lentil allergen was designed as Len c 1.01 and was recognized by 77% (17/22) of patients. Three genetic variants of this allergen have been detected and have been named Len c 1.0103, Len c 1.0101, and Len c 1.0102, according to the proposal of the International Committee of Allergen Nomenclature. The authors also reported that the proteolytic cleavage of mature vicilin polypeptides of greater than 50–60 kDa can generate an array of related allergens with different molecular sizes ranging between 12 and 35 kDa, and some of these subunits could still retain their IgE-binding capacities. This was the case with the IgE-reactive 16- and 26-kDa bands detected in lentil extracts when tested using pooled sera from lentil allergic patients. The 16-kDa allergenic band, designated as Len c 1.02, corresponded to the previously described Len c 1 (12–16 kDa) allergen by Sanchez-Monge et al. [94].

Many of the lentil allergens remain to be characterized; however, studies suggest that lentil allergens are thermally resistant [93, 96]. Cross-reactivity among lentil, chickpea, and pea has also been reported [87, 90, 93, 97]. Amino acid sequence deduced from Len c 1.0101 (or Len c 1.0102) confirmed its homology to several allergenic vicilins from other plant sources, such as peanut Ara h 1 (50% identity) [98], English walnut Jug r 2 (38% identity) [99], cashew Ana o 1 (32% identity) [100], sesame Ses i 3 (31% identity) [19], and β-subunits of soybean conglycinin (56% identity) [101]. Whether the cross-reactivity between peanut and these legumes has a clinical relevance is still a matter of debate.

18.2.2.4. Pea Allergy. In contrast to chickpeas and lentils, adverse reactions to peas have been rarely reported. An advance literature search using the PubMed database revealed only very few (<6) articles dealing with pea allergenicity.

The first reports investigating the isolation of green pea allergens were published by Malley et al. [102, 103]. A low-MW glycoprotein, a fragment of the albumin fraction, with 1.8 kDa was reported as the major allergen component. Later, Bernhisel-Broadbent et al. [104] described numerous IgE-binding pea proteins with molecular masses between 14 and 70 kDa. It was not until recently that case reports of allergic reactions to pea in groups of Spanish and German patients were described [105, 106]. Several pea protein fractions have been identified and nomenclatured as allergenic components (Table 18.3). Vicilin (44 kDa), Pis s 1, and more specifically its proteolytic fragments, Pis s 1.01 (32 kDa), as well as convicilin (63 kDa), Pis s 2 were the two major identified pea allergens. Additional proteolytic subunits of vicilin (36, 16, and 13 kDa) were also found as relevant IgE-binding pea components. These proteins further showed cross-reactivity with the major lentil allergen Len c 1 [105] and with peanut allergen Ara h 1 [107].

TABLE 18.3 Identified Lupin, Chickpea, Lentil, and Pea Allergens

Legumes	Allergen Fraction	MW (kDa)	Allergen Nomenclature	Cross-Reactivity	References
Lupin (*Lupinus albus*)	11S globulin	62–66	**Lup a**	PN, SB	83
Chickpea (*Cicer arietinum*)	11S globulin	α: 35–40 β: 20–23	Not assigned	L, F, Le	59, 87, 88
	2S albumin	10–12	**Cic a**		86
Lentil (*Lens culinaris*)	7S globulin (vicilin)	~50	**Len c 1.01** (three isoforms)	C, P, SB, PN	90, 93–95
		48.6	Len c1.0101		
		49.7	Len c 1.0102		
		50.0	Len c 1.0103		
		12–16	**Len c 1.02** (initially named Len c 1)		
		26	Not assigned		
Pea (*Pisum sativum*)	Vicilin	44	**Pis s 1**	C, Le, PN	105, 106, 108
		32	**Pis s 1.01** (fragment of Pis s 1)		
	Convicilin	63	**Pis s 2**		
		36, 16, and 13	Not designed (minor allergens)		
	Albumin fragment	1.8	Not assigned		102, 103

α, acidic subunit of 11S globulin; β, basic subunit of 11S globulin; F, fenugreek; C, chickpea; P, pea; SB, soybean; PN, peanut; L, latex; Le, lentil.

Sell et al. [106] studied the influence of different maturation levels of green pea seeds on their allergenicity. Their results demonstrated the allergenic potency of pea at all maturation levels, even with immature seeds. The highest allergenic potency was caused by the albumin fraction as well as the globulin and glutelin fractions. More recently, Szymkiewicz and Jedrychowski [108] investigated the effect of cooking on the immunogenicity of pea proteins using an animal model (Balb/c mice) and revealed the resistance of both vicilin and convicilin allergen epitopes to thermal treatment.

Although serological cross-reactivity among legumes is frequently reported, the clinical relevance of these findings remains to be confirmed. Martínez San Ireneo et al. [109] investigated the cross-reactivity between lentils, chickpeas, peas, white beans, and peanuts and the clinical relevance in Spanish pediatric patients by using ELISA inhibition experiments and oral food challenges. Their results indicated that 50% of the sera identified an allergen with approximately 50 kDa in all three legume extracts (lentil, chickpea, and pea). The oral legume challenges demonstrated that 37 children (69%) were allergic to two or more legumes (median three legumes). The most frequent associations were allergy to lentils and chickpeas (57%); allergy to lentils and peas (54%); and allergy to lentils, chickpeas, and peas (43%), which indicated that clinical allergy to all three legumes is frequent among Spanish children.

18.2.3. Cereal Grains

A number of cereals have been reported to cause allergic reactions in sensitive children and adults by either inhalation or ingestion. These include wheat, rye, barley, oats, maize (corn), and rice. Although cereals form an essential part of daily nutrition, IgE-mediated allergy to cereals is not very frequent in the general population and varies from one region to another. Wheat allergy is, for example, more common in the United Kingdom than allergy to rice, which is in turn more common in Japan and in other Asian countries.

The main IgE-mediated protein fractions in cereals are the water- and salt-soluble proteins (albumins and globulins) [110]. Several studies have also shown that the gluten protein fraction named gliadin in wheat, secalin in rye, and hordein in barley are the most harmful antigens in celiac disease [111, 112]. This causal relation between gluten and celiac disease has been extensively studied, and the causative cereal proteins have been well characterized [113]. This review will mainly focus on some of the emerging or lesser known IgE-mediated cereal allergens. A list of these allergens is presented in Table 18.4.

18.2.3.1. Buckwheat Allergy. Buckwheat (*Fagopyrum esculentum*) is frequently consumed in Asian countries and is now increasingly popular as a health food in North America and Europe. Its flour is gluten-free and serves as a common replacement for patients with celiac disease. Nevertheless, buckwheat is considered to be one of the most important emerging food allergens.

TABLE 18.4 Identified Buckwheat and Corn Allergens

Cereal	Allergen Fraction	MW (kDa)	Allergen Nomenclature	Cross-Reactivity	References
Buckwheat (*Fagopyrum esculentum*)	2S albumin	8–10	Fag e 10kD	R, L	121
		10	Fag t 10kD		122
		14	Fag e 14kD		123
		15	Fag e 2		130
		16–18	Fag e 16kD		124–126
	α-Globulin	19	Fag e 19kD	R, L	127, 128
	11S globulin	22	Fag e 1		129
		24			132, 133
Corn (maize) (*Zea mays*)	By ingestion				
	Albumin	9	Zea m 14	R, P, AP	151, 152
		16			
	Glutenin	27	Zea m 27kD Zein		157
		50	Zea m 50kD Zein	AL	153, 157
	By inhalation				
		35	Zea m 1		154
		55	Zea m 13		
		13.9	ZmTRX h1 (Zea m 25)	W	
		13.0	ZmTRX h2	W	155

R, rice; L, latex; AP, apricot; P, peach; AL, almond; W, wheat.

The first report of buckwheat allergy in the literature dates back to 1909 [114]. Since then, most of the literature on this topic has consisted of case reports involving asthma or allergic rhinitis due to buckwheat pillow exposure [115, 116], occupational buckwheat asthma, atopic dermatitis, urticaria [117, 118], and anaphylaxis [119, 120]. There are, however, only a few epidemiological studies available on occupational buckwheat allergy.

The different allergens identified up to now are water- and salt-soluble proteins (albumins and globulins) and alcohol-soluble proteins (prolamins). Identification and characterization of the major allergens of buckwheat have been studied and reported by several authors. Electrophoretic and immunoblotting studies revealed the presence of several allergenic proteins with a wide range of molecular masses in two buckwheat species (*F. esculentum* and *Fagopyrum tartaricum*). In particular, proteins of 8–10 kDa [121, 122]; 14, 16, and 18 kDa [123–126], and 19 kDa [127, 128] have been reported. Tanaka et al. [124] showed that the 16-kDa protein is one of the most pepsin-resistant buckwheat proteins and is very stable to cooking. The 16- to 19-kDa allergens seem to be related to the α-amylase/trypsin inhibitor family (including 2S albumins). Furthermore, Yoshioka et al. [129, 130] reported, after cDNA isolations and epitope analysis, buckwheat allergenic proteins of 22 and 15 kDa, named Fag e 1 and Fag e 2, respectively. Recently, by studying the distribution of proteins and allergens in graded flours prepared from whole buckwheat grains, Morita et al. [131] reported that the IgE-binding activities of Fag e 1 or Fag e 2 of the inner fractions were weaker than those of the outer fractions. They suggested that the graded-milling procedure could be used to separate the inner fractions without major allergic proteins from the whole buckwheat grains in order to provide less-allergenic buckwheat flours.

Wang et al. [132, 133] also purified and identified a 24-kDa protein from tartary buckwheat as a major allergen with strong IgE-binding activity in buckwheat-allergic patients and thermostable under heating for 20 minutes at 100°C. Several other protein fractions with molecular masses of 34–38 and 69–70 kDa have shown IgE-binding activity. Nakamura et al. [134] reported that Maillard-type glycosylation of Fag e 1 with polysaccharides brought about a drastic reduction of the reactivity against human sera of buckwheat allergy subjects, using immuno-dot blotting, quartz crystal microbalance (QCM) analysis, and ELISA. IgE from buckwheat-allergic individuals has been also reported to cross-react with proteins from latex [135], rice [123, 136], and poppy seed [137].

18.2.3.2. Corn Allergy. Corn (*Zea mays* L.) is a plant belonging to the family of the grasses (Poaceae) and one of the most important cereal crops worldwide. A review of literature on food allergies has indicated that allergy to corn is more common than previously reported [2, 138]. There are increasing reports of allergic reactions to corn including anaphylaxis as a result of eating corn [139–142], corn oil [143], and corn-related foods [144, 145], as well as severe reactions after exposure to cornstarch in surgical gloves [146–149] and handling of corn flour [150].

Corn allergy is caused by proteins residing in the kernels. A salt-soluble lipid transfer protein (LTP) having a MW of 9 kDa was one of the first corn allergens to be identified [151]. LTP is an extremely stable protein, highly resistant to food processing, including heating, but also to gastrointestinal digestion [152]. Salt-soluble corn protein fractions (14 kDa [150] and 16 kDa [152]) belonging to the trypsin/α-amylase inhibitor family also showed IgE inmunoreactivity in studies with the human sera. The latter proteins were, however, heat labile. The aqueous alcohol-soluble prolamins (zeins), constituting about 60–70% of the corn endosperm proteins, also show clinical allergenicity. A potential allergen in the corn alcohol-soluble fraction, with a MW of 50 kDa, has been identified [153]. This protein has high stability to cooking and *in vitro* resistance to both gastric and pancreatic digestion. Sensitization to corn allergy through the respiratory tract (upon inhalation) has also been reported. Maize pollen proteins and their putative function as aeroallergens have been investigated by Petersen et al. [154] in order to determine whether maize contains novel allergenic components, as well as their allergenic potency in comparison with pollen from timothy grass (*Phleum pratense*), a typical representative of the native grasses. Two allergens (Zea m 35 kDa and Zea m 55 kDa), both having high IgE-binding reactivity and sequence homologies of 72% and 70%, respectively, to the corresponding Phl p 1 and Phl p 13 allergens of timothy grass pollen were identified. Additionally, Weichel et al. [155] identified two new thioredoxins of maize, ZMTRX*h*1 (Zea m 25) and ZmTRX*h*2, that are associated with allergic asthma. These two thioredoxins shared 74% and 46% sequence homologies, respectively, with Tri a 25, a wheat thioredoxin.

Many cereal allergens are cross-reactive with grass pollen allergens. This is, however, only an *in vitro* cross-reactivity. Studies on clinically significant cross-reactive allergens of cereals and grasses are limited. Cross-reactivity between corn and soy, peanut, and rice has been previously documented [156]. A 50-kDa maize γ-zein also showed marked cross-reactivity with the major almond protein [157]. IgE-mediated allergies to several other cereals, such as millet [158–161], barley [162], and oat [163], have also been reported.

18.2.4. Fruit and Vegetable Allergens

Allergic reactions to fruits and vegetables are relatively mild and often involve the mouth and the pharynx. Direct contact of the offending food triggers oral and pharyngeal itching, oral papule or blisters, lip irritation and swelling, labial angioedema, and glottic edema, otherwise known as the oral allergy syndrome (OAS). Fortunately, allergens in fruits and some vegetables are not as complicated as other foods, and they can often be destroyed by cooking.

In general, people who are allergic to fruits are much more likely to be allergic to birch pollen. Several clinical associations have been described. The most common recognized correlation between oral reactions to fresh fruits and vegetables and pollen sensitization are birch pollen fruit syndrome, latex–fruit syndrome, and LTP sensitization [164]. Examples of such association are represented in Table 18.5.

Fruit and vegetable OAS are most likely to occur with kiwi, the apple family (apple, pear), the plum family (plum, peach, prune, nectarine, apricot, cherry), the parsley family (carrot, celery, dill, anise, cumin, coriander, caraway), and the potato family (potato, tomato, green pepper). Due to the rising popularity

TABLE 18.5 Fruits and Vegetables Associated with Pollen Sensitization

Allergen	Foods
Ragweed pollen	Melons, bananas, cucumbers, and zucchini
Birch pollen	Apple, carrot, hazelnut, potato, almond group[a]
Mugwort pollen	Celery, apple, kiwi
Latex	Banana, kiwi, avocado, chestnut, tomatoes, mangoes, papayas, figs, aubergines, and cassava

[a]Almond group includes pear, plum, nectarine, cherry, and apricot.

of exotic fruits and the globalization of the food supply, allergy against some specific fruits and vegetables has increasingly moved into the focus of public interest in consuming countries. This review will focus on some emerging allergens in this category and on their postulated birch and mugwort pollen association. Fruit and vegetable allergens to be discussed are briefly represented in Table 18.6.

18.2.4.1. Mango Allergy. Mango (*Mangifera indica* L.) is the most frequently cultivated tropical fruit besides banana. Two allergens with molecular masses of 40 and 30 kDa were characterized as the major allergens Man i 1 and Man i 2 by Paschke et al. [165]. The allergenicity of raw mango fruit and its by-products (puree and nectar) was strongly maintained during technological processing (i.e., heating and mechanical or enzymatic tissue disintegration) [166]. Cross-reactivity to mugwort pollen (Bet v 1) and birch pollen (Art v 1) allergens has been documented [167]. Recently, a third mango profilin allergen with a molecular mass of 14 kDa was identified by Song et al. [168], showing high cross-reactivity with birch pollen profilin Bet v 2.

18.2.4.2. Lychee Allergy. Lychee (*Litchi chinensis* Sonn.) is the most renowned of a group of edible fruits of the family Sapindaceae, native to Southeast Asia, but cultivation is now worldwide. Although it is a delicious fruit, lychee can cause severe anaphylactic reactions [169]. While sensitization to lychee fruit has been described, little information about lychee allergen characterization is available. The first lychee allergen with a MW of about 14 kDa was identified by Far et al. [170] as a profilin, and with nomenclature Lit c 1. Recently, another 28-kDa lychee allergen was identified as a triose-phosphate isomerase [171]. Several other IgE-binding protein bands have been identified in the MW range of 14–94 kDa by immunoblotting. After thermal treatments during fruit preservation (canning), the allergenic potency of the fruit decreased, even though high residual allergenic activity was observed, especially with the 55-kDa protein [172]. Cross-reactivities between lychee and other fruits [173], vegetables [174], pollen [175], and latex [176] have been reported.

18.2.4.3. Grape Allergy. Although allergy to grape is considered rare, several cases in children and a few in adults have been reported. Grape allergy cases are most often displayed as an OAS [177, 178]; however, more severe manifestations, including asthma, exercise-induced anaphylaxis [179, 180], and anaphylactic shock [181,182] have also been reported. Pastorello et al. [183] characterized the allergens of grape and wine. The major allergens were an endochitinase 4A (30 kDa) and an LTP termed Vit v 1 (9 kDa) that was homologous to and cross-reactive with peach LTP. A 24-kDa protein homologous to the cherry thaumatin-like allergen was a minor allergen. The allergenic potency of these three fractions was recently confirmed in an anaphylaxis reported case after consumption of grape and wine [184, 185]. A 28-kDa

TABLE 18.6 Emerging Allergens Most Commonly Associated with Oral Allergy Syndrome

	Allergen Nomenclature	MW (kDa)	Heat Stability	Cross-Reactivity	References
Fruits					
Mango (*Mangifera indica* L.)	Man i 1	40	Stable	Bet v 1 and Art v 1	165
	Man i 2	30	Stable	Bet v 1 and Art v 1	167
	Man i 3	14		Bet v 2	168
Lychee (*Litchi chinensis* Sonn.)	Lit c 1 isomerase	14		A, B, L, Bet v 1, and Bet v 2	170
		28			171
		14–94	Stable		172
Grape (*Vitis vinifera*)	Vit v 1 (LTP)	9		P	183, 184
	Thaumatin	24		C	185, 186
	Endochitinase	30			
Cherry (*Prunus avium*)	Pru av 1	18	Stable	C, P, Bet v 1, Bet v 2	188, 189
	Pru av 2	23	Stable		194, 195
	Pru av 3	9	Stable		196
	Pru av 4	14	Stable		190
Vegetables					
Celery (*Apium graveolens*)	Api g 1	16	Labile	Bet v 1, Bet v 2, C, Ca, A	201
	Api g 1.0101	16.3	Labile		
	Api g 1.0201	17.1	Labile		204
	Api g 4	15	Stable		207
	Api g 5	60			174, 208
	Other bands	35–90			
Tomato (*Lycopersicon esculentum*)	Lyc e 1	14–16	Labile	Bet v 2, C, Ca, Ce, B, Pi, Po, BP	213, 214
	Lyc e 2	60	Stable		215
	Lyc e 3	8–10	Stable		224

A, apple; B, banana; L, latex; P, peach; C, cherry; Ca, carrot; Ce, celery; Pi, pineapple; BP, bell pepper; Po, potato; Bet v 1, mugwort pollen; Art v 1, birch pollen; Bet v 2, birch pollen profilin.

expansin, a 37.5-kDa polygalacturonase-inhibiting protein, a 39-kDa 1,3-glucanase, and a 60-kDa protein have also been identified as minor grape allergens [186]. Cross-reactivity with peach and cherry has been reported [187].

18.2.4.4. Cherry Allergy. Cherry (*Prunus avium*) is one of the most delicious and attractive fruit for both human and bird consumption. Largely produced and consumed in Western Europe, sweet cherry is strongly associated to the birch fruit syndrome. In the North of Europe, birch pollen-sensitized individuals having OAS can tolerate cooked cherry, while in Mediterranean countries, most people are allergic to peach and, therefore, may develop adverse reactions to cherry because of the similarity between the allergens in peach and cherry. The involved allergens are highly stable against thermal processing and digestion and show cross-reactivity with other fruits including apple, peach, apricot, plum, and nuts such as hazelnut and walnut. Four different allergens in cherry have been identified to date. The immunological and biochemical properties of the Pru av 1 (18 kDa) and Pru av 4 (14 kDa) as major cherry allergens have been studied intensively [188–190]. Pru av 1 (formerly Pru a 1) is the first pollen-related food allergen for which a highly resolved solution structure became available [191]. Later, Wiche et al. [192] identified a second putative IgE-binding area on Pru av 1, which is relevant for the clinical phenomenon of pollen-related allergy to fruits. Gruber et al. [193] also reported that the allergenicity of Pru av 1 was reduced by enzymatic polyphenol oxidation and Maillard reaction. Other potential allergens in cherry, Pru av 2, a 23-kD thaumatin-like protein [194, 195], as well as Pru av 3, an LTP showing high stability against thermal processing and digestion [196], have been identified. Recently, Primavesi et al. [197] reported that chemical peeling successfully removed Pru av 3 (LTP) responsible for OAS in patients without pollinosis, leading to the industrial production of cherry hypoallergenic derivatives. Furthermore, the syruping process removed almost all allergenic proteins to which patients with pollinosis were responsive.

18.2.4.5. Celery Allergy. Celery (*Apium graveolens*) is consumed around the world as a vegetable, either for the crisp petiole (leaf stalk), mostly popular in North America, or as a fleshy taproot (called celery root) mainly in Europe and the Mediterranean region. Celery is consumed as fresh salad, as cooked vegetable, in vegetable juices and soups, and as fermented or pickled preserves. Dried celery powder is also widely used as a cheap ingredient for spices. Celery allergy is one of the most common of all food allergies in Western Europe and a frequent cause of food allergy in pollen-sensitized patients [198, 199]. Celery root (or celeriac) is known to contain more allergens than the stalk [200].

Three known celery allergens have been identified so far: (1) Api g 1, a 16-kDa protein, which shares high sequence identity with Bet v 1, and other major allergens in birch pollen-related fruits and vegetables [201–203]—two Api g 1 (isoforms), named Api g 1.0101 (16.3 kDa) and Api g 1.0201 (17.1 kDa), were recently purified and characterized [204, 205]; (2) Api g 4 (15 kDa), a celery profilin [206, 207]; and (3) Api g 5, a 60-kDa glycoprotein belonging to a family of flavin adenine dinucleotide (FAD)-containing oxidases [174, 208]. Jankiewicz et al. [209] reported that a complete loss of antibody binding to Api g 1 can be achieved by cooking and even by short microwave treatment for at least 10 minutes, whereas inactivation of profilin required 30 minutes of microwaving but was not observed after 20 minutes of conventional cooking. This latter treatment led to the presence of allergenic profilin even in the cooking water. Multiple allergen bands in the mass range between 35 and 90 kDa were also recognized by celery-allergic patients as cross-reactive carbohydrate determinants (CCDs). The CCDs containing α1,3-fucose and β1,2-xylose, attached to proteins via N-glycoside linkages, are highly immunogenic in mammals; some celery-allergic patients exclusively display IgE binding to these determinants of MW >45 kDa. However, clinical significance of CCD-specific IgE is still controversial [210, 211]. More recently, protocols for the production and purification of high amounts of the well-defined celery allergens were established to obtain homogenous protein batches [212].

18.2.4.6. Tomato Allergy. Tomato (*Lycopersicon esculentum*) is an important part of the human diet. It is a fruit, although it is often regarded as a vegetable, and is widely consumed, either raw or cooked and used in salads, soups, or sauces. Tomatoes supply one of the most powerful, naturally occurring antioxidants, lycopene, which is linked with lowering the incidence of prostate cancer among Italian men. Tomato consumption has therefore been highly recommended. Tomato fruit is, however, currently recognized as a food of allergenic importance, containing a number of proteins with high allergenic potential.

To date, several tomato allergens have been characterized including Lyc e 1, a profilin protein with MW of 14–16 kDa [213, 214]. Two highly similar genes encoding tomato profilin have also been isolated and designated as allergen Lyc e 1.01 and Lyc e 1.02. Recently, the glycosylated protein β-fructofuranosidase was identified as tomato allergen Lyc e 2, including one isoform with MW of 51 kDa and another of 60 kDa [215]. The prevalence of an IgE sensitization to Lyc e 2 is 17% among tomato-allergic patients. Interestingly, the IgE reactivity is completely determined by the glycan structure of Lyc e 2. Other proteins with capacity to bind IgE have also been described such as Lyc e 3 (nonspecific lipid transfer protein [ns-LTP] [216, 217], glucanase [218], peroxidase [219], chitinase [220], patatin [221], and superoxide dismutase [222], as well as polygalacturonase 2A and pectinesterase [216, 222]).

Plant genetic engineering was used as a tool to generate hypoallergenic tomato fruits by silencing allergenic potency of genes in transgenic tomato plants by means of RNA interference (RNAi). Results indicated that the profilin protein (Lyc e 1) [217] and the ns-LTP Lyc e 3 [223, 224] contents in transgenic tomato fruits was decreased by 10-fold and 10- to 100-fold, respectively, compared with untransformed controls. This marked decrease was sufficient to cause a reduced allergenic reactivity in patients with tomato allergy, providing proof of concept and demonstrating that RNAi can be used to design allergen-reduced food.

High cross-sensitization (very high sequence identity of 76–86%) of Lyc e 1 was observed with other allergenic plant profilins such as Pru av 4 from cherry, Api g 4 from celery, and Bet v 2 from birch pollen [225], as well as Ana c 1 from pineapple, Mus xp 1 from banana [226], and Dau c 4 from carrot [214]. Bell pepper allergen Cap a 2 showed the highest sequence homology (91%) [213]. Allergy to tomato is also linked to allergy to potato since they are related plants.

Limited studies on the prevalence of tomato allergy exist in the literature. Reported prevalence of tomato allergy ranges from 0.35% in several countries throughout the world [227], to about 1.2–1.3% in the U.K. population [228], to higher rates of 1.7–39% in a group of Italian children [229, 230], indicating that tomato is a relevant allergenic food in selected populations. More recently, high sensitization to tomato peel and pulp extracts in the Mediterranean Coast of Spain were also reported [231, 232]. Commercial tomato extracts for skin prick tests to diagnose patients sensitized to tomato have been manufactured and tested efficiently [233].

18.3. PREVALENCE AND INCIDENCE

The true prevalence of adverse food reactions is still unknown. The established prevalence, however, of food allergy by the best available studies appears to be about 2–4% of the adult population and between 6% and 8% in young children, particularly in their first year of life [234]. In clinical surveys, children are more sensitive to food allergies than adults due to their immature gastrointestinal tract and immune systems [2]. Other contributing factors such as decline in breast-feeding in some groups, early introduction of solid foods, and a change in the occurrence of infectious disease may also play a role.

According to literature, the common food allergens and their prevalence are strongly related to population consumption habits and ethnic diversities. This certainly explains the observed variation in figures regarding the prevalence of food allergies across different countries. In this regard, a meta-analysis study was conducted to assess uncertainties in the prevalence of food allergy in different scientific communities. The authors reported a marked heterogeneity in the prevalence of food allergy that could be a result of differences in study design or methodology, or differences between populations, and they therefore recommended the use of standardized methods in data collection [235]. In addition, Mills et al. [236] reported that the reasons underlying the marked variation in the rates of food allergy include wide differences in response rates, problems in reliability and consistency of diagnosis across the different studies, and especially the poor clinical specificity of skin tests and measurement of food-specific IgE [237]. Within this context, in 2005, the EU-funded an integrated project called EuroPrevall (www.europrevall.org) with a large spectrum of 56 partners from 21 different countries (from 19 European countries, Ghana, India, and China) with additional collaborating centers and partners from other countries, to provide knowledge about the prevalence of food allergies, as well as ranking of allergenic foods (food groups) as a function of the number of reactions they provoke both in the overall population and in specific population groups (regarding age and geographic location). Data from such studies will be useful and will help policy makers to address food allergy from a holistic public health perspective.

There is little doubt that the big eight allergenic foods are the most potent allergens; however, additional emerging allergens such as sesame seeds, tropical fruits, spices and condiments, and some pulse proteins (chickpea, lentil, pea) could also be potent allergens [238]. The term "emerging" as used in this review therefore refers only to the increasing prevalence of these allergens, but also the increasing awareness of the allergenic properties of these foods by the public, food industry, media, and regulatory institutions.

Table 18.7 summarizes the estimated prevalence of allergies caused by the well-known "big eight" allergenic foods, as well as other emerging allergens in different countries. It is widely recognized that food allergy prevalence varies by country. For example, in North America, peanut is one of the major allergens; in Scandinavia, fish allergy is predominant; in Europe, celery and poppy seed allergies are quite prevalent; in the Middle East, sesame allergy is not uncommon; and in Japan, rice and buckwheat allergies feature among the prominent allergens.

The prevalence of food allergy appears to have increased over the last decades; however, it remains unclear whether this represents an actual increase in the number of reactions or an increased rate of detection and reporting of these allergic reactions. More research is needed in order to ascertain the true global prevalence of food allergies.

TABLE 18.7 Prevalence of Common and Emerging Food Allergens

Country	Food Allergens[a]	Prevalence (%)	References
Canada	Common allergenic foods	1.5–2	2
	Other priority/emerging allergens: sesame, mustard	NA	
United States	Common allergenic foods	3.5–4	239, 240
	Other priority/emerging allergens: sesame	NA	3, 24
United Kingdom	Common allergenic foods	1.4–1.8	228
	Other priority/emerging allergens: sesame, lupin, mustard, corn, grape	NA	12, 64, 78 186
France	Common allergenic foods	2.1–3.8	241, 242
	Other priority/emerging allergens: sesame, mustard, lupin, celery	NA	8, 38, 71

TABLE 18.7 Continued

Country	Food Allergens[a]	Prevalence (%)	References
Italy	Common allergenic foods	6–8	237
	Other priority/emerging allergens: mustard, lupin, corn, grape, cherry	NA	45, 46, 68–70 197
Spain	Common allergenic foods	4.6	243
	Other priority/emerging allergens: mustard, lupin, chickpea, lentil, pea, tomato	NA	56, 57, 59, 89, 96, 218
Germany	Common allergenic foods	2–3.6	244
	Other priority/emerging allergens: pea, corn, mango, lychee, cherry, grape, celery, tomato	NA	165, 166, 171, 172 209 223
Portugal	Common allergenic foods	3.4–6.9	245
	Other priority/emerging allergens: pollen-related fruits and vegetables	NA	
Israel	Common allergenic foods	1.7	7
	Other priority/emerging allergens		
	Sesame	0.18	14, 16
	Pollen-related fruits and vegetables	NA	
Japan	Common allergenic foods	2–5	246–248
	Other allergens: sesame, buckwheat		131, 134
India	Common allergenic foods	>3	
	Other priority/emerging allergens: mustard, chickpea, lentil, black gram	NA	47, 87 249
Australia	Common allergenic foods	1–2	250–252
	Other priority/emerging allergens: sesame, lupin	NA	79
New Zealand	Common allergenic foods	NA	253–255
	Other priority/emerging allergens: sesame, kiwifruit	NA	
Denmark	Common allergenic foods	2.3–3.2	256–258
	Other priority/emerging allergens: pollen-related fruits and vegetables	NA	259

[a]Common allergens: 90% of allergic reactions are caused by eight foods: milk, eggs, peanuts, tree nuts, fish, shellfish, soy, and wheat.
NA, data not available.

18.4. POTENTIAL IMPACT ON THE FOOD INDUSTRY AND FOR THE CONSUMERS

Adverse reactions to food are of growing concern to both consumers and the food industry. The industrial dimensions of food allergy has recently been reviewed by Crevel [260], reinforcing the fact that food allergy risk management is a shared responsibility between the food manufacturer and the consumers, as well as other stakeholders such as retailers, health professionals, and regulatory bodies.

Crucial information is needed by the food industry for managing risks related to food allergies. This includes expert opinions, scientific knowledge regarding the characteristics of allergenic food and ingredients, and the prevalence of allergic reactions, as well as clinical responses such as eliciting doses. The industry needs such information to improve manufacturing practices and food safety, as well as to support the development of specific products destined for allergic consumers. To fulfill industry needs, it is important to

1. improve current scientific knowledge on eliciting doses for allergenic foods, as a basis for establishing thresholds that are both safe for allergic consumers and feasible for the manufacturer [261];
2. develop and validate scientific criteria for identifying the existing common allergens and assessing the potential allergenicity of novel proteins and foods [262], as well as the phenomenon of cross-reactivity between allergens;
3. establish an allergen list critical for public health, which should be kept under review in light of new emerging allergens; and
4. develop advanced well-validated techniques for the detection of food allergens.

To achieve this, better collaboration among the food industry, regulatory bodies, research centers, and test kit manufacturing companies is needed. To protect allergic consumers, the food manufacturing sector must continue to identify improved methods for food processing and production and set precautions in place to avoid potential cross-contact/cross-contamination (including the development of a hazard analysis and critical control point plan, if possible). They must be aware of current and emerging allergens in order to develop a concise communication strategy with employees, consumers, and suppliers, and be able to seek agreement with their suppliers to request any change in the formulation of ingredients supplied. Additionally, food manufacturers should be in a position to quickly respond by providing appropriate labeling, which is readily understood by the consumer for both current and new emerging allergens, if required [263].

Regulatory agencies have the responsibility to set policies for the handling of allergens, including emerging allergens, and these must be supplemented by guidelines that provide practical advice to individual manufacturing units.

These guidelines should clearly include information on allergen handling at all stages in the product life cycle, from its design, through the sourcing of ingredients to manufacture, labeling, and distribution. Education of the public is also a key component to successfully manage allergies [264]; thus, regulatory agencies need to develop appropriate information tools for parents, allergic individuals, and community groups (e.g., day care and schools) as well as the general public in regard to current and emerging allergens.

18.5. CONCLUSION

Scientific understanding of food allergy in many ways remains incomplete, which makes it difficult to effectively assess and manage risks associated with existing allergens and the potential allergenicity of novel foods. Given the important public policy implications of food allergy research and its use in food safety regulation, firm figures on the prevalence of food allergy, identifying the foods involved, and understanding the socioeconomic impact of these factors on the development of food allergy will be very valuable [236, 265]. In this era of globalization, it is not only populations that migrate but also foods, as people adopt foreign diets and import exotic products [266]. This makes it even more important that we remain watchful for new and emerging allergens. Consumer concerns about allergies have increased considerably in the last decade. At present, the only cure for food allergy is to avoid eating the food responsible for the allergy. Thus, continued awareness and access to high-quality information will be needed to increase our understanding and improve our capacity to manage new and emerging food allergens as they arise.

18.6. USEFUL WEBSITES FOR INFORMATION ON FOOD ALLERGY

Consumer concerns about allergy have increased considerably in the last decade. At present the only cure for food allergy is avoidance of the offending foods. Thus, access to timely and accurate information will make a tremendous contribution to the understanding and management of food allergy. Some useful websites are presented below:

- www.anaphylaxis.org.uk
- www.anaphylaxis.net
- www.foodallergens.info
- Food Allergy and Anaphylaxis Network (www.foodallergy.org)
- Food Allergy and Anaphylaxis Alliance (www.foodallergyalliance.org)
- The Allergen Nomenclature Subcommittee on the IUIS making the official list of peer-reviewed allergens available on the Internet (www.allergen.org)

- The Allergome provides information on allergenic molecules for allergist and immunologist.
- The AllFam database is a resource for classifying allergens into protein families. It is based on the allergen data from the Allergome database, which is currently the most comprehensive collection of allergen data, and on the protein family data from the Pfam database (http://www.meduniwien.ac.at/allergens/allfam/).
- The Food Allergy Research and Resource Program (FARRP) contains a list of publicly known allergens (www.allergenonline.com).
- The PROTALL database contains biochemical and clinical information about plant food allergen (http://foodallergens.ifr.ac.uk).
- EuroPrevall, in 2005, launched an integrated project in partnership with different countries (from 19 European countries, Ghana, India, and China) with additional collaborating centers and partners from the United States, Canada, Australia, and New Zealand to gather complete information on food allergy (www.europrevall.org).
- Anaphylaxis Australia Inc. (www.allergyfacts.org.au)
- Australasian Society of Clinical Immunology and Allergy (www.allergy.org.au)
- http://www.allergy.org.nz/
- www.healthcanada.gc.ca/foodallergies

REFERENCES

1. FAO/WHO, Codex Alimentarius Commission. (1998). *Report of the Twenty-Sixth Session of the Codex Committee on Food Labeling*, Ottawa, Canada, May 26–29, ALINORM 99/22.
2. Zarkadas, M. (1999). Common allergenic foods and their labelling in Canada—a review. *Canadian Journal of Allergy and Clinical Immunology*, *4*, 118–141.
3. Gangur, V., Kelly, C., Navuluri, L. (2005). Sesame allergy: a growing food allergy of global proportions. *Annals of Allergy, Asthma & Immunology*, *95*, 4–11.
4. Chassaigne, H., Brohée, M., Norgaard, J.V., van Hengel, A.J. (2007). Investigation on sequential extraction of peanut allergens for subsequent analysis by ELISA and 2D gel electrophoresis. *Food Chemistry*, *105*, 1671–1681.
5. IFST Position Statement. (1999). Food allergens. *Food Science and Technology Today*, *13*, 163–168.
6. Namiki, M. (2007). Nutraceutical functions of sesame: a review. *Critical Reviews in Food Science and Nutrition*, *47*, 651–673.
7. Dalal, I., Binson, I., Reifen, R., Amitai, Z., Shohat, T., Rahmani, S., Levine, A., Ballin, A., Somekh, E. (2002). Food allergy is a matter of geography after all: sesame as a major cause of severe IgE-mediated food allergic reactions among infants and young children in Israel. *Allergy*, *57*, 362–365.

8. Kanny, G., De Hauteclocque, C., Moneret-Vautrin, D.A. (1996). Sesame seed and sesame seed oil contain masked allergens of growing importance. *Allergy*, *51*, 952–957.

9. Alday, E., Curiel, G., Lopez-Gil, M.L., Carreno, D., Moneo, I. (1996). Occupational hypersensitivity to sesame seeds. *Allergy*, *51*, 69–70.

10. Moneret-Vautrin, D.A., Aaghassian, C., Kanny, G. (1997). Anaphylaxie idiopathique récidivante au sésame. Un allergène alimentaire en expansion. *Revue Française d'Allergologie*, *34*, 487–489.

11. Pajno, G.B., Passalcqua, G., Magazzu, G., Barberio, G., Vita, D., Canonica, W. (2000). Anaphylaxis to sesame. *Allergy*, *55*, 199–201.

12. Perkins, M.S. (2001). Raising awareness of sesame allergy. *Pharmaceutical Journal*, *267*, 757–758.

13. Dalal, I., Binson, I., Levine, A., Somekh, E., Ballin, A., Reifen, R. (2003). The pattern of sesame sensitivity among infants and children. *Pediatric Allergy and Immunology*, *14*, 312–316.

14. Caminiti, L., Vita, D., Passalacqua, G., Arrigo, T., Lombardo, F., Pajno, G.B. (2006). Tahini, a little known sesame containing food, as an unexpected cause of severe allergic reaction. *Journal of Investigational Allergology and Clinical Immunology*, *16*, 308–310.

15. Cohen, A., Goldberg, M., Levy, B., Leshno, M., Katz, Y. (2007). Sesame food allergy and sensitization in children: the natural history and long-term follow-up. *Pediatric Allergy and Immunology*, *18*, 217–223.

16. Vadas, P. (2003). Food allergies. *The Canadian Journal of Diagnosis*, August, 86–94.

17. Pastorello, E.A., Varin, E., Farioli, L., Pravettoni, V., Ortolani, C., Trambaioli, C., Fortunato, D., Giuffrida, M.G., Rivolta, F., Robino, A., Calamari, A.M., Lacava, L., Conti, A. (2001). The major allergen of sesame seeds (*Sesamum indicum*) is a 2S albumin. *Journal of Chromatography. B, Biomedical Sciences and Applications*, *756*, 85–93.

18. Wolff, N., Cogan, U., Admon, A., Dalal, I., Katz, Y., Hodos, N., Karin, N., Yannai, S. (2003). Allergy to sesame in humans is associated primarily with IgE antibody to a 14 kDa 2S albumin precursor. *Food and Chemical Toxicology*, *41*, 1165–1174.

19. Beyer, K., Bardina, L., Grishina, G., Sampson, H.A. (2002). Identification of sesame seed allergens by 2-dimensional proteomics and Edman sequencing: sees storage proteins as common food allergens. *The Journal of Allergy and Clinical Immunology*, *110*, 154–159.

20. Leduc, V., Moneret-Vautrin, D.A., Tzen, J.T.C., Morisset, M., Guerin, L., Kanny, G. (2006). Identification of oleosins as major allergens in sesame seed allergic patients. *Allergy*, *61*, 349–356.

21. Beyer, K., Grishina, G., Bardina, L., Sampson, H.A. (2007). Identification of 2 new sesame seed allergens: Ses i 6 and Ses i 7. *The Journal of Allergy and Clinical Immunology*, *119*, 1554–1556.

22. Kolopp-Sarda, M.N., Moneret-Vautrin, D.A., Gobert, B., Kanny, G., Brodschii, M., Bene, M.C., Faure, G.C. (1997). Specific humoral immune responses in 12 cases of food sensitization to sesame seed. *Clinical and Experimental Allergy*, *27*, 1285–1291.

23. Fremont, S., Zitouni, N., Kanny, G., Veneri, V., Metche, M., Moneret-Vautrin, D.A., Nicolas, J.P. (2002). Allergenicity of some isoforms of white sesame proteins. *Clinical and Experimental Allergy*, *32*, 1211–1215.
24. Navuluri, L., Parvataneni, S., Hassan, H., Birmingham, N.P., Kelly, C., Gangur, V. (2006). Allergic and anaphylactic response to sesame seeds in mice: identification of Ses i 3 and basic subunit of 11S globulins as allergens. *International Archives of Allergy and Immunology*, *140*, 270–276.
25. Vocks, E., Borga, A., Szliska, C., Seifert, H.U., Seifert, B., Burow, G., Borelli, S. (1993). Common allergenic structures in hazelnut, rye grain, sesame seeds, kiwi, and poppy seeds. *Allergy*, *48*, 168–172.
26. Wallowitz, M.L., Chen, R.J.Y., Tzen, J.T.C., Teuber, S.S. (2007). Ses i 6, the sesame 11S globulin, can activate basophils and shows cross-reactivity with walnut in vitro. *Clinical and Experimental Allergy*, *37*, 929–938.
27. Chiu, J.T., Haydik, I.B. (1991). Sesame seed oil anaphylaxis. *The Journal of Allergy and Clinical Immunology*, *88*, 414–415.
28. Stevens, W.J., Ebo, D.G., Bridis, C.H., De Clerck, L.S. (2002). Anaphylaxis to sesame (*Sesamum indicum*) seed and sesame oil. *The Journal of Allergy and Clinical Immunology*, *109*, S217.
29. Neering, H., Vitanyi, B.E., Malten, K.E., Van Djik, E. (1975). Allergens in sesame oil contact dermatitis. *Acta Dermato-Venereologica*, *55*, 311–344.
30. Hayakawa, R., Matsunaga, K., Suzuki, M., Hosokawak, K., Arima, Y., Shin, C.S., Yoshida, M. (1987). Is sesamaol present in sesame oil? *Contact Dermatitis*, *17*, 133–135.
31. Pecquet, C., Leynadier, F., Saiag, P. (1998). Immediate hypersensitivity to sesame in foods and cosmetics. *Contact Dermatitis*, *39*, 313.
32. Naoki, O., Yoshika, Y., Norika, H., Shigeru, K., Akira, K. (2008). Allergic contact dermatitis caused by sesame oil in a topical Chinese medicine, s*hi-un-ko*. *Contact Dermatitis*, *58*, 109.
33. Kubo, Y., Nonaka, S., Yoshida, H. (1986). Contact sensitivity to unsaponifiable substances in sesame oil. *Contact Dermatitis*, *54*, 215–217.
34. Dezfoulian, B., de la Brassinne, M. (2003). Urticaire de contact suite à l'application d'un fond de teint contenant de l'huile de sésame. *Revue Française d'Allergologie et d'Immunologie Clinique*, *43*, 338–341 (in French).
35. Figueroa, J., Blanco, C., Dumpiérrez, A.G., Almeida, L., Ortega, N., Castillo, R., Navarro, L., Pérez, E., Gallego, M.D., Carrillo, T. (2005). Mustard allergy confirmed by double-blind placebo-controlled food challenges: clinical features and cross-reactivity with mugwort pollen and plant-derived foods. *Allergy*, *60*, 48–55.
36. Andre, F., Andre, C., Colin, L., Cacaraci, F., Cavagna, S. (1994). Role of new allergens and of allergens consumption in the increased incidence of food sensitizations in France. *Toxicology and Applied Pharmacology*, *93*, 77–83.
37. Kanny, G., Fremont, S., Talhouarne G., Nicolas, J.P., Moneret-Vautrin, D.A., Incornavia, C., Zanussi, C. (1995). Anaphylaxis to mustard as a masked allergen in chicken dips. *Annals of Allergy, Asthma & Immunology*, *75*, 340–342.
38. Rancé, F., Abbal, M., Dutau, G. (2000). Mustard allergy in children. *Allergy*, *55*, 496–500.
39. Morisset, M., Moneret-Vautrin, D.A., Maadi, F., Fremont, S., Guenard, L., Croizier, A., Kanny, G. (2003). Prospective study of mustard allergy: first study with double-blind placebo-controlled food challenge trials (24 cases). *Allergy*, *58*, 295–299.

40. Rancé, F. (2003). Mustard allergy as a new food allergy. *Allergy*, *58*, 287–288.
41. Vidal, C., Diaz, C., Saez, A., Rodriquez, M., Inglesias, A. (1991). Anaphylaxis to mustard. *Postgraduate Medical Journal*, *67*, 404.
42. Jorro, G., Morales, C., Braso, J.V., Pelaez, A. (1995). Mustard allergy: three cases of systemic reaction to ingestion of mustard sauce. *Journal of Investigational Allergology and Clinical Immunology*, *5*, 54–56.
43. Caballero, T., San-Martín, M.S., Padial, M.A., Contreras, J., Cabañas, R., Barranco, P., López-Serrano, M.C. (2002). Clinical characteristics of patients with mustard hypersensitivity. *Annals of Allergy, Asthma & Immunology*, *89*, 166–171.
44. Widstrom, L., Johnsson, S.G.O. (1986). IgE-mediated anaphylaxis to mustard. *Acta Dermato-Venereologica*, *66*, 70–71.
45. Panconesi, E., Sertoli, A., Fabbri, P., Giorgini, S., Spallanzini, P. (1980). Anaphylatic shock from mustard after ingestion of pizza. *Contact Dermatitis*, *6*, 394–395.
46. Monreal, P., Botey, J., Pea, M., Marin, A., Eseverri, J.L. (1992). Mustard allergy: two analphylatic reactions to ingestion of mustard sauce. *Annals of Allergy, Asthma & Immunology*, *69*, 317–320.
47. Kumar, T.S., Gupta, A., Mehta, N. (2001). A prospective study of mustard allergy in allergic rhinitis cases. Available at http://www.bhj.org/journal/2001_4302_apr01/org_261.htm
48. Menendez-Arias, L., Monsalve, R.I., Gavilanes, J.G., Rodríguez, R. (1987). Molecular and spectroscopic characterization of a low molecular weight seed storage protein from yellow mustard (*Sinapis alba* L.). *International Journal of Biochemistry*, *19*, 899–907.
49. Menendez-Arias, L., Moneo, I., Dominguez, J., Rodriguez, R. (1988). Primary structure of the major allergen of yellow mustard (*Sinapis alba* L.) seed, Sin a 1. *European Journal of Biochemistry/FEBS*, *177*, 159–166.
50. Menendez-Arias, L., Dominguez, J., Moneo, I., Rodriguez, R. (1990). Epitope mapping of the major allergen from yellow mustard seeds, Sin a 1. *Molecular Immunology*, *27*, 143–150.
51. Dominguez, J., Cuevas, M., Urena, V., Munoz, T., Moneo, I. (1990). Purification and characterization of an allergen of mustard seed. *Annals of Allergy, Asthma & Immunology*, *64*, 352–357.
52. Gonzales de la Pena, M.A., Monsalve, R.I., Batanero, E., Villalbe, M., Rodriquez, R. (1996). Expression in *Escherichia coli* of Sin a 1, the major allergen from mustard. *European Journal of Biochemistry/FEBS*, *237*, 827–832.
53. Shim, Y.Y., Wanasundara, J.P.D. (2008). Quantitative detection of allergenic protein Sin a 1 from yellow mustard (*Sinapis alba* L.) seeds using enzyme-linked immunosorbent assay. *Journal of Agricultural and Food Chemistry*, *56*, 1184–1192.
54. Gonzales de la Pena, M.A., Menendez-Arias, L., Monsalve, R.I., Rodriquez, R. (1991). Isolation and characterization of a major allergen from oriental mustard seeds, Bra j 1. *International Archives of Allergy and Applied Immunology*, *96*, 263–270.
55. Monsalve, R.I., Gonzales de la Pena, M.A., Menendez-Arias, L., Lopez-Otin, C., Villalbe, M., Rodriquez, R. (1993). Characterization of a new oriental-mustard (*Brassica juncea*) allergen, Bra j IE: detection of an allergenic epitope. *The Biochemical Journal*, *293*, 625–632.

56. Palomares, O., Cuesta-Herranz, J., Vereda, A., Sirvent, S., Villalba, M., Rodriguez, R. (2005). Isolation and identification of an 11S globulin as a new major allergen in mustard seeds. *Annals of Allergy, Asthma & Immunology*, *94*, 586–592.

57. Palomares, O., Vereda, A., Cuesta-Herranz, J., Villalba, M., Rodriguez, R. (2007). Cloning, sequencing, and recombinant production of Sin a 2, an allergenic 11S globulin from yellow mustard seeds. *The Journal of Allergy and Clinical Immunology*, *119*, 1189–1196.

58. Crespo, J.F., Pascal, C., Burks, A.W., Helm, R.M., Esteban, M.M. (1995). Frequency of food allergy in a pediatric population from Spain. *Pediatric Allergy and Immunology*, *6*, 39–45.

59. Martinez San Ireneo, M., Sandin, M.D.I., Fernandez-Caldas, E., Lizana, F.M., Fletes, M.J.R., Borrego, M.T.L. (2000). Specific IgE levels to *Cicer arietinum* (chickpea) in tolerant and nontolerant children: evaluation of boiled and raw extracts. *International Archives of Allergy and Immunology*, *1231*, 137–143.

60. Pompei, C., Lucisano, M., Ballini, N. (1985). Use of lupin flour in pasta manufacture. *Sciences des Aliments*, *5*, 665–687.

61. Hung, T.V., Papalois, M., Nithianandan, V., Jiang, H.H., Versteeg, K. (1990). Utilisation of lupin seeds for human consumption. In: *Food Pacific Convention Papers*, May 6–10, The Australian Institute of Food Science and Technology/New Zealand Institutes of Food Science and Technology, Gold Coast, Australia, pp. 13–15.

62. Dervas, G., Doxastakis, G., Hadjisavva-Zinoviadi, S., Triantafillakos, N. (1999). Lupin flour addition to wheat flour doughs and effect on rheological properties. *Food Chemistry*, *66*, 67–73.

63. Kyle, W.S.A. (1994). The current and potential uses of lupins for human food. In: *Proceedings of the First Australian Lupin Technical Symposium*, Dracup, M., Palta, J. (Eds.), October 17–21, Western Australian Department of Agriculture, South Perth, Australia, pp. 89–97.

64. Radcliffe, M., Scadding, G., Brown, H.M. (2005). Lupin flour anaphylaxis. *The Lancet*, *365*, 1360.

65. Hefle, S.L., Lemanske, R.F., Bush, R.K. (1994). Adverse reaction to lupine-fortified pasta. *The Journal of Allergy and Clinical Immunology*, *94*, 167–172.

66. Gutierrez, D., Conde, A., Duran, S., Delgado, J., Guardia, P., Martinez, R., Garcia-Cubillana, A., Gonzalez, J., Conde, J. (1997). Contact urticaria from lupin. *Contact Dermatitis*, *36*, 311.

67. Matheu, V., de Barrio, M., Sierra, Z., Gracia-Bara, M.T., Tornero, P., Baeza, M.L. (1999). Lupine-induced anaphylaxis. *Annals of Allergy, Asthma & Immunology*, *83*, 406–408.

68. Novembre, E., Moriondo, M., Bernardini, R., Azzari, C., Rossi, M.E., Vierucci, A. (1999). Lupin allergy in a child. *The Journal of Allergy and Clinical Immunology*, *103*, 1214–1216.

69. Magni, C., Herndl, A., Sironi, E., Scarafoni, A., Ballabio, C., Restani, P., Bernardini, R., Novembre, E., Vierucci, A., Duranti, M. (2005). One- and two-dimensional electrophoretic identification of IgE-binding polypeptides of *Lupinus albus* and other legume seeds. *Journal of Agricultural and Food Chemistry*, *53*, 4567–4571.

70. Rotiroti, G., Skypala, I., Senna, G., Passalacqua, G. (2007). Anaphylaxis due to lupine flour in a celiac patient. *Journal of Investigational Allergology and Clinical Immunology*, *17*, 204–205.

71. Kanny, G., Guerin, L., Moneret-Vautrin, D.A. (2000). Risk of serious acute asthma due to lupine flour associated with peanut allergy. *La Revue de Medecine Interne*, *21*, 191–194.

72. Parisot, L., Aparicio, C., Moneret-Vautrin, D.A., Guerin, L. (2001). Allergy to lupine flour. *Allergy*, *56*, 918–919.

73. Crespo, J.F., Rodriguez, J., Vives, R., James, J.M., Reano, M., Daroca, P., Burbano, C., Muzquiz, M. (2001). Occupational IgE-mediated allergy after exposure to lupine seed flour. *The Journal of Allergy and Clinical Immunology*, *108*, 295–297.

74. Mareno-Ancillo, A., Gil-Adrados, A.C., Dominguez-Noche, C., Cosmes, P.M. (2005). Lupine inhalation induced asthma in a child. *Pediatric Allergy and Immunology*, *16*, 542–544.

75. Rojas-Hijazo, B., Garces, M.M., Cabalero, M.L., Alloza, P., Moneo, I. (2006). Unsuspected lupin allergens hidden in food. *International Archives of Allergy and Immunology*, *141*, 47–50.

76. De las Marinas, D., Cojocariu, Z., Escudero, R., Pardo, N., Sanz, M.L. (2007). Anaphylaxis induced by lupine as a hidden allergen. *Journal of Investigational Allergology and Clinical Immunology*, *17*, 283–284.

77. Lindvik, H., Holden, L., Løvik, M., Cvancarova, M., Halvorsen, R. (2008). Lupin sensitization and clinical allergy in food allergic children in Norway. *Acta Paediatrica*, *97*, 91–95.

78. Shaw, J., Roberts, G., Grimshaw, K., White, S., Hourihane, J.O. (2008). Lupin allergy in peanut-allergic children and teenagers. *Allergy*, *63*, 370–373.

79. Smith, W.B., Gillis, D., Kette, K.E. (2004). Lupin: a new hidden food allergen. *Medical Journal of Australia*, *181*, 219–220.

80. Moneret-Vautrin, D.A., Guérin, L., Kanny, G., Flabbee, J., Fremont, S., Morisset, M. (1999). Cross-allergenicity of peanut and lupine: the risk of lupine allergy in patients allergic to peanuts. *The Journal of Allergy and Clinical Immunology*, *104*, 883–888.

81. Dooper, M.M., Holden, L., Faeste, C.K., Thompson, K.M., Egaas, E. (2007). Monoclonal antibodies against the candidate lupin allergen α-conglutin and β-conglutin. *International Archives of Allergy and Immunology*, *143*, 49–58.

82. Guarneri, F., Guarneri, C., Benvenga, S. (2005). Identification of potentially cross-reactive peanut-lupine proteins by computer-assisted search for amino acid sequence homology. *International Archives of Allergy and Immunology*, *138*, 273–277.

83. Peeters, K.A., Nordiee, J.A., Penninks, A.H., Chen, L., Goodman, R.E., Bruijnzeel-Koomen, C.A., Hefle, S.L., Taylor, S.L., Knulst, A.C. (2007). Lupine allergy: not simply cross-reactivity with peanut or soy. *The Journal of Allergy and Clinical Immunology*, *120*, 647–653.

84. Peeters, K.A., Koppelman, S.J., Penninks, A.H., Lebens, A., Bruijnzeel-Koomen, C.A., Hefle, S.L., Taylor, S.L., van Hoffen, E., Knulst, A.C. (2009). Clinical relevance of sensitization to lupine in peanut-sensitized adults. *Allergy*, *64*, 549–555.

85. Niphadkar, P.V., Patil, S.P., Bapat, M.M. (1996). Chickpea-induced anaphylaxis. *Allergy*, *52*, 115–116.
86. Vioque, J., Vioque, R.S., Clemente, A., Pedroche, J., Bautista, J., Millan, F. (1999). Purification and partial characterization of chickpea 2S albumin. *Journal of Agricultural and Food Chemistry*, *47*, 1405–1409.
87. Patil, S.P., Niphadkar, P.V., Bapat, M.M. (2001). Chickpea: a major food allergen in the Indian subcontinent and its clinical and immunochemical correlation. *Annals of Allergy, Asthma & Immunology*, *87*, 140–145.
88. Clemente, A., Vioque, J., Sanchez-Vioque, R., Pedroche, J., Millan, F. (1999). Production of extensive chickpea (*Cicer arietinum*) protein hydrolysates with reduced antigenic activity. *Journal of Agricultural and Food Chemistry*, *47*, 3776–3781.
89. Ferreira, M.B., Pedro, E., Meneses-Santos, J., Pereira dos Santos, M.C., Palma Carlos, M.L., Bartolome, B., Palma Carlos, A.G. (2004). Latex and chickpea (*Cicer arietinum*) allergy: first description of a new association. *European Annals of Allergy and Clinical Immunology*, *36*, 366–371.
90. Pascual, C.Y., Crespo, J.F., Sanchez-Pastor, S., Padial, A., Diaz-Pena, J.M., Martin-Munoz, F., Martin-Esteban, M. (1999). Allergy to lentils in Mediterranean pediatric patients. *The Journal of Allergy and Clinical Immunology*, *103*, 154–158.
91. Martin, J.A., Compaired, J.A., de la Hoz, B., Quirce, S., Alonso, M.D., Igea, J.M., Losada, E. (1992). Bronchial asthma induced by chickpea and lentil. *Allergy*, *47*, 185–187.
92. Kalogeromitros, D., Armenaka, M., Galatas, I., Capellou, O., Katsarou, A. (1996). Anaphylaxis induced by lentils. *Annals of Allergy, Asthma & Immunology*, *77*, 460–482.
93. Ibanez-Sandin, D., Martinez San Ireneo, M., Maranon Lizana, F., Fernandez-Caldas, E., Alonso Lebrero, E., Laso Borrego, T. (1999). Specific IgE determinations to crude and boiled lentil (*Lens culinaris*) extracts in lentil-sensitive children and controls. *Allergy*, *54*, 1209–1214.
94. Sanchez-Monge, R., Pascual, C.Y., Diaz-Perales, A., Crespo, J.F., Martin-Esteban, M., Salcedo, G. (2000). Isolation and characterization of relevant allergens from boiled lentils. *The Journal of Allergy and Clinical Immunology*, *106*, 955–961.
95. Lopez-Torrejon, G., Salcedo, G., Martin-Esteban, M., Diaz-Perales, A., Pascual, C.Y., Sanchez-Monge, R. (2003). Len c 1, a major allergen and vicilin from lentil seeds: protein isolation and cDNA cloning. *The Journal of Allergy and Clinical Immunology*, *112*, 1208–1215.
96. Martinez San Ireneo, M., Ibanez-Sandin, M.D., Fernandez-Caldas, E., Maranon, F., Munoz Martinez, M.D., Laso Borrego, M.T. (2001). The diagnostic value of crude or boiled extracts to identify tolerant versus nontolerant lentil sensitive children. *Annals of Allergy, Asthma & Immunology*, *86*, 686–690.
97. Martinez San Ireneo, M., Ibanez Sandin, D., Fernandez-Caldas, E. (2000). Hypersensitivity to members of the botanic order Febales (legumes). *Journal of Investigational Allergology and Clinical Immunology*, *10*, 187–199.
98. Shin, D.S., Compadre, C.M., Maleki, S.J., Kopper, R.A., Sampson, H., Huang, S.K., Burks, W., Bannon, G.A. (1998). Biochemical and structural analysis of the IgE binding sites on Ara h 1, an abundant and highly allergenic peanut protein. *The Journal of Biological Chemistry*, *273*, 13753–13759.

99. Teuber, S.S., Jarvis, K.C., Dandekar, A.M., Peterson, W.R., Ansari, A.A. (1999). Identification and cloning of a complementary DNA encoding a vicilin-like protein Jur r 2, from English walnut kernel (*Juglans regia*), a major food allergen. *The Journal of Allergy and Clinical Immunology*, *104*, 1311–1320.

100. Wang, F., Robotham, J.M., Teuber, S.S., Tawde, P., Sathe, S.K., Roux, K.H. (2002). Ana o 1, a cashew (*Anacardium occidentale*) allergen of the vicilin seed storage protein family. *The Journal of Allergy and Clinical Immunology*, *130*, 160–166.

101. Ogawa, T., Bando, N., Tsuji, H., Nishikawa, K., Kitamura, K. (1995). α-Subunit of β-conglycinin, an allergenic protein recognized by IgE antibodies of soybean-sensitive patients with atopic dermatitis. *Bioscience, Biotechnology and Biochemistry*, *59*, 831–833.

102. Malley, A., Baecher, L., Mackler, B., Perlman, F. (1975). The isolation of allergens from the green pea. *The Journal of Allergy and Clinical Immunology*, *56*, 282–290.

103. Malley, A., Baecher, L., Mackler, B., Perlman, F. (1976). Further characterization of a low molecular weight allergen fragment isolated from the green pea. *Clinical and Experimental Immunology*, *25*, 159–164.

104. Bernhisel-Broadbent, J., Sampson, H.A. (1989). Cross-allergenicity in the legume botanical family in children with food hypersensitivity. *The Journal of Allergy and Clinical Immunology*, *83*, 435–440.

105. Sanchez-Mange, R., Lopez-Torrejon, G., Pascual, C.Y., Varela, J., Martin-Esteban, M., Salcedo, G. (2004). Vicilin and convicilin are potential major allergens from pea. *Clinical and Experimental Allergy*, *34*, 1747–1753.

106. Sell, M., Steinhart, H., Paschke, A. (2005). Influence of maturation on the alteration of allergenicity of green pea (*Pisum sativum* L.). *Journal of Agricultural and Food Chemistry*, *53*, 1717–1722.

107. Wensing, M., Knulst, A.C., Piersma, S., O'Kane, F., Knol, E.F., Koppelman, S.J. (2003). Patients with anaphylaxis to pea can have peanut allergy caused by cross-reactive IgE to vicilin (Ara h 1). *The Journal of Allergy and Clinical Immunology*, *111*, 420–424.

108. Szymkiewicz, A., Jedrychowski, L. (2006). Determination of pea proteins allergenicity with the use Balb/c mouse. *Central European Journal of Immunology*, *31*, 63–69.

109. Martínez San Ireneo, M., Ibáñez, M.D., Fernández-Caldas, E., Carnés, J. (2008). *In vitro* and *in vivo* cross-reactivity studies of legume allergy in a Mediterranean population. *International Archives of Allergy and Immunology*, *147*, 222–230.

110. Weiss, W., Huber, G., Engel, K.H., Pethran, A., Dunn, M.J., Gooley, A.A., Görg, A. (1997). Identification and characterization of wheat grain albumin/globulin allergens. *Electrophoresis*, *18*, 826–833.

111. Kieffer, M., Frazer, P.J., Daniels, N.W.R., Coombs, R.R.A. (1982). Wheat gliadin fractions and other cereal antigens reactive with antibodies in the sera of celiac patients. *Clinical and Experimental Immunology*, *50*, 651–660.

112. Vainio, E., Varjonen, E. (1995). Antibody response against wheat, rye, barley, oats and corn—comparison between gluten sensitive patients and monoclonal antigliadin antibodies. *International Archives of Allergy and Applied Immunology*, *106*, 134–138.

113. Vader, W., Kooy, Y., Van Veelen, P., De Ru, A., Harris, D., Benckhuisen, W., Pena, S., Mearin, L., Drijfhout, J.W., Koning, F. (2002). The gluten response in children with celiac disease is directed toward multiple gliadin and glutenin peptides. *Gastroenterology*, *122*, 1729–1737.
114. Smith, H.L. (1909). Buckwheat poisoning with report of a case in man. *Archives of Internal Medecine*, *3*, 350–359.
115. Schumacher, F., Schmid, P., Wuthrich, B. (1993). Sarrazin allergy: a contribution to buckwheat allergy. *Schweizerische Medizinische Wochenschrift*, *123*, 1559–1562.
116. Lee, S.Y., Lee, K.S., Hong, C.H., Lee, K.Y. (2001). Three cases of childhood nocturnal asthma due to buckwheat allergy. *Allergy*, *56*, 763–766.
117. Wieslander, G. (1996). Review on buckwheat allergy. *Allergy*, *51*, 661–665.
118. Stember, R.H. (2006). Buckwheat allergy. *Allergy and Asthma Proceedings*, *27*, 393–395.
119. Schiffner, R., Przybilla, B., Burgdorff, T., Landthaler, M., Stolz, W. (2001). Anaphylaxis to buckwheat. *Allergy*, *56*, 1020–1021.
120. Heffler, E., Guida, G., Badiu, I., Nebiolo, F., Rolla, G. (2007). Anaphylaxis after eating Italian pizza containing buckwheat as the hidden food allergen. *Journal of Investigational Allergology and Clinical Immunology*, *17*, 261–263.
121. Yano, M., Nakamura, R., Hayakawa, S., Toril, S. (1989). Purification and properties of allergenic proteins in buckwheat seed. *Agricultural and Biological Chemistry*, *53*, 2387–2392.
122. Matsumoto, R., Fujino, K., Nagata, Y., Hashiguchi, S., Ito, Y., Aihara, Y., Takahashi, Y., Maeda, K., Sugimura, K. (2004). Molecular characterization of a 10 kDa buckwheat molecule reactive to allergic patient IgE. *Allergy*, *59*, 533–538.
123. Yoshimasu, M.A., Zhang, J.W., Hayakawa, S., Mine, Y. (2000). Electrophoretic and immunochemical characterization of allergenic proteins in buckwheat. *International Archives of Allergy and Immunology*, *123*, 130–136.
124. Tanaka, K., Matsumoto, K., Akasawa, A., Nakajima, T., Nagasu, T., Likura, Y., Saito, H. (2002). Pepsin-resistant 16 kD buckwheat protein is associated with hypersensitivity reactions in patients with buckwheat allergy. *International Archives of Allergy and Immunology*, *129*, 49–56.
125. Koyano, S., Takagi, K., Teshima, R., Sawada, J. (2006). Molecular cloning of cDNA, recombinant protein expression and characterization of a buckwheat 16 kD major allergen. *International Archives of Allergy and Immunology*, *140*, 73–81.
126. Choi, S.Y., Sohn, J.H., Lee, Y.W., Lee, E.K., Hong, C.S., Park, J.W. (2007). Application of the 16 kDa buckwheat 2S storage albumin protein for diagnosis of clinical reactivity. *Annals of Allergy, Asthma & Immunology*, *99*, 254–260.
127. Park, J.W., Kang, D.B., Kim, C.W., Ko, S.H., Yum, H.Y., Kim, K.E., Hong, C.S., Lee, K.Y. (2000). Identification and characterization of the major allergens in buckwheat. *Allergy*, *55*, 1035–1041.
128. Choi, S.Y., Sohn, J.H., Lee, Y.W., Lee, E.K., Hong, C.S., Park, J.W. (2007). Characterization of buckwheat 19 kD allergen and its application for diagnosing clinical reactivity. *International Archives of Allergy and Immunology*, *144*, 267–274.
129. Yoshioka, H., Ohmoto, T., Uriso, A., Mine, Y., Adachi, T. (2004). Expression and epitope analysis of the major allergenic protein Fag e 1 from buckwheat. *Journal of Plant Physiology*, *161*, 761–767.

130. Yoshioka, H., Yasueda, N., Ohmoto, T., Uriso, A., Mine, Y., Yamazaki, R., Adachi, T. (2004). cDNA isolation and epitope analysis of allergenic proteins in autogamous common buckwheat. In: *Proceedings of the 9th International Symposium on Buckwheat*, Iva, F., Vaclav, D., Petra, C., Ivan, H., Vojtech, H., Zdenek, S. (Eds.), August 18–22, University of Agriculture, Prague, Czech Republic, pp. 612–620.

131. Morita, N., Maeda, T., Sai, R., Miyake, K., Yoshioka, H., Urisu, A., Adachi, T. (2006). Studies on distribution of protein and allergen in graded flours prepared from whole buckwheat grains. *Food Research International*, *39*, 782–790.

132. Wang, Z., Zhang, Z., Zhao, Z., Wieslander, G., Norback, D., Kreft, I. (2004). Purification and characterization of a 24 kDa protein from tartary buckwheat seeds. *Bioscience, Biotechnology, and Biochemistry*, *68*, 1409–1413.

133. Wang, Z., Wang, L., Li, Y., Zhang, Z., Wieslander, G., Norback, D. (2006). Cloning, expression, and identification of immunological activity of an allergenic protein in tartary buckwheat. *Bioscience, Biotechnology, and Biochemistry*, *70*, 1195–1199.

134. Nakamura, S., Suzuki, Y., Ishikawa, E., Yakushi, T., Jing, H., Miyamoto, T., Hashizume, K. (2008). Reduction of in vitro allergenicity of buckwheat Fag e 1 through the Maillard-type glycosylation with polysaccharides. *Food Chemistry*, *109*, 538–545.

135. De Maat-Bleeker, F., Stapel, S.O. (1998). Cross-reactivity between buckwheat and latex. *Allergy*, *53*, 538–539.

136. Yamada, K., Urisu, A., Morita, Y., Kondo, Y., Wada, E., Komada, H., Yamada, M., Inagaki, Y., Torii, S. (1995). Immediate hypersensitive reactions to buckwheat ingestion and cross allergenicity between buckwheat and rice antigens in subjects with high levels of IgE antibodies to buckwheat. *Annals of Allergy, Asthma & Immunology*, *75*, 56–61.

137. Oppel, T., Thomas, P., Wollenberg, A. (2006). Cross-sensitization between poppy seed and buckwheat in a food allergic patient with poppy seed anaphylaxis. *International Archives of Allergy and Immunology*, *140*, 170–173.

138. Moneret-Vautrin, D.A., Kanny, G., Beaudouin, E. (1998). L'allergie alimentaire au maïs existe-t-elle? (Does food allergy to corn exist?) *Allergie et Immunologie*, *30*, 230 (in French).

139. Pauls, J.D. (1998). Food-dependent exercise-induced anaphylaxis to corn. *The Journal of Allergy and Clinical Immunology*, *101*, 853–854.

140. Tanaka, L.G., El-Dahr, J.M., Lehrer, S.B. (2001). Double-blind, placebo-controlled corn challenge resulting in anaphylaxis. *The Journal of Allergy and Clinical Immunology*, *107*, 744.

141. Lehrer, S.B., Oberhoff, W.W., Tanaka, L.G., Reese, G., El-Dahr, J.M., Wild, L.G. (2009). Immunoblotting studies of DBPC corn challenged subjects to identify clinically relevant corn allergens. *The Journal of Allergy and Clinical Immunology*, *123*, S183.

142. Gonzalo-Garijo, M.A., Pérez-Calderón, R., Muñoz-Rodríguez, A., Tormo-Molina, R., Silva-Palacios, I. (2004). Hypersensitivity to maize pollen. *Allergy*, *59*, 365–365.

143. Crevel, R.W., Kerkhoff, M.A., Koning, M.M. (2000). Allergenicity of refined vegetable oils. *Food and Chemical Toxicology*, *38*, 385–393.

144. Frisner, H., Rosendal, A., Barkholt, V. (2000). Identification of immunogenic maize proteins in a casein hydrolysate formula. *Pediatric Allergy and Immunology*, *11*, 106–110.

145. Borghesan, F., Borghesan, N. (2005). Maize flour-induced rhinitis. *Allergie et Immunologie*, *37*, 283–284.

146. Seggev, J.S., Mawhinney, T.P., Yunginger, J.W., Braun, S.R. (1990). Anaphylaxis due to cornstarch surgical glove powder. *Annals of Allergy, Asthma & Immunology*, *65*, 152–155.

147. Crippa, M., Pasolini, G. (1997). Allergic reactions due to glove-lubricant-powder in health-care workers. *International Archives of Occupational and Environmental Health*, *70*, 399–402.

148. Dave, J., Wilcox, M.H., Kellett, M. (1999). Glove powder: implications for infection control. *Journal of Hospital Infection*, *42*, 283–285.

149. Liu, W., Nixon, R.L. (2007). Corn contact urticaria in a nurse. *Australasian Journal of Dermatology*, *48*, 130–131.

150. Cristaudo, A., Simonato, B., Pasini, G., De Rocco, M., Curioni, A., Giannattasio, M. (2004). Contact urticaria and protein contact dermatitis from corn in a patient with serum IgE specific for a salt-soluble corn protein of low molecular weight. *Contact Dermatitis*, *51*, 84–87.

151. Pastorello, E.A., Farioli, L., Pravettoni, V., Ispano, M., Scibola, E., Trambaioli, C., Giuffrida, M.G., Ansaloni, R., Godovac-Zimmermann, J., Conti, A., Fortunato, D., Ortolani, C. (2000). The maize major allergen, which is responsible for food-induced allergic reactions, is a lipid transfer protein. *The Journal of Allergy and Clinical Immunology*, *106*, 744–751.

152. Pastorello, E.A., Pompei, C., Pravettoni, V., Farioli, L., Calamari, A.M., Scibilia, J., Robino, A.M., Conti, A., Iametti, S., Fortunato, D., Bonomi, S., Ortolani, C. (2003). Lipid-transfer protein is the major maize allergen maintaining IgE-binding activity after cooking at 100°C, as demonstrated in anaphylactic patients and patients with positive double blind, placebo controlled food challenge results. *The Journal of Allergy and Clinical Immunology*, *112*, 775–783.

153. Pasini, G., Simonata, B., Curioni, A., Vincenzi, S., Cristaudo, A., Santucci, B., Dal Belin Peruffo, A., Giannattasio, M. (2002). IgE-mediated allergy to corn: a 50 kDa protein, belonging to the reduced soluble proteins, is a major allergen. *Allergy*, *57*, 98–106.

154. Petersen, A., Dresselhaus, T., Grobe, K., Becker, W.M. (2006). Proteome analysis of maize pollen for allergy-relevant components. *Proteomics*, *6*, 6317–6325.

155. Weichel, M., Vergoossen, N.J., Bonomi, S., Scibilia, J., Ortolani, C., Ballmer-Weber, B.K., Pastorello, E.A., Crameri, R. (2006). Screening the allergenic repertoires of wheat and maize with sera from double-blind, placebo-controlled food challenge positive patients. *Allergy*, *61*, 128–135.

156. Lehrer, S.B., Reese, G., Malo, J.L., Lahoud, C., Leong-Kee, S., Goldberg, B., Carle, T., Ebisawa, M. (1999). Corn allergens: IgE antibody reactivity and cross-reactivity with rice, soy, and peanut. *International Archives of Allergy and Immunology*, *118*, 298–299.

157. Lee, S.H., Benmoussa, M., Sathe, S.K., Roux, K.H., Teuber, S.S., Hamaker, B.R. (2005). A 50 kDa maize γ-zein has marked cross-reactivity with the almond major protein. *Journal of Agricultural and Food Chemistry*, *53*, 7965–7970.

158. Parker, J.L., Yunginger, J.W., Swedlund, H.A. (1981). Anaphylaxis after ingestion of millet seeds. *The Journal of Allergy and Clinical Immunology*, *67*, 78–80.

159. Olivieri, J., Hauser, C. (1998). Anaphylaxis to millet. *Allergy*, *53*, 109–110.

160. Stuck, B.A., Blum, A., Klimek, L., Hormann, K. (2001). Millet, a potentially life-threatening allergen. *Allergy*, *56*, 350.

161. Bohle, B., Hirt, W., Nachbargauer, P., Ebner, H., Ebner, C. (2003). Allergy to millet: another risk for atopic bird keepers. *Allergy*, *58*, 325–328.

162. Garcia-Casado, G., Crespo, J.F., Rodriguez, J., Salcedo, G. (2001). Isolation and characterization of barley lipid transfer protein and protein Z as beer allergens. *The Journal of Allergy and Clinical Immunology*, *108*, 647–649.

163. Boussault, P., Laut-Labrze1, C., Saubusse, E., Maurice-Tison, S., Perromat, M., Roul, S., Sarrat, A., Taeb, A., Boralevi, F. (2007). Oat sensitization in children with atopic dermatitis: prevalence, risks and associated factors. *Allergy*, *62*, 1251–1256.

164. Pastorello, E.A., Ortolani, C. (2003). Oral allergy syndrome. In: *Food Allergy: Adverse Reactions to Foods and Food Additives*, 3rd edition, Metcalfe, D.D., Sampson, H.A., Simon, R.A. (Eds.). Oxford, UK: Blackwell Publishing, pp. 169–182.

165. Paschke, A., Kinder, H., Zunker, K., Wigotzki, M., Wessbecher, R., Vieluf, D., Steinhart, H. (2001). Characterization of allergens in mango fruit and ripening dependence of the allergenic potency. *Food and Agricultural Immunology*, *13*, 51–61.

166. Dube, M., Zunker, K., Neidhart, S., Carle, R., Steinhart, H., Paschke, A. (2004). Effect of technological processing on the allergenicity of mangoes (*Mangifera indica* L.). *Journal of Agricultural and Food Chemistry*, *52*, 3938–3945.

167. Paschke, A., Kinder, H., Zunker, K., Wigotzki, M., Steinhart, H., Weûbecher, R., Vieluf, I. (2001). Characterization of cross-reacting allergens in mango fruit. *Allergy*, *56*, 237–242.

168. Song, J.J., Zhang, H., Liu, Z., Ran, P. (2007). Mango profilin: cloning, expression and cross-reactivity with birch pollen profilin Bet v 2. *Molecular Biology Reports*, *35*, 231–237.

169. Raap, U., Schaefer, T., Kapp, A., Wedi, B. (2007). Exotic food allergy: anaphylactic reaction to lychee. *Journal of Investigational Allergology and Clinical Immunology*, *17*, 199–201.

170. Far, J., Wuthrich, B., Vieths, S. (1995). Anaphylactic reaction to lychee fruit: evidence for sensitisation to profilin. *Clinical and Experimental Allergy*, *25*, 1018–1023.

171. Hoppe, S., Steinhart, H., Paschke, A. (2006). Identification of a 28 kDa lychee allergen as a triose-phosphate isomerase. *Food and Agricultural Immunology*, *17*, 9–19.

172. Hoppe, S., Neidhart, S., Zunker, K., Hutasingh, P., Carle, R., Steinhart, H., Paschke, A. (2006). The influences of cultivar and thermal processing on the allergenic potency of lychees (*Litchi chinensis* Sonn.). *Food Chemistry*, *96*, 209–219.

173. Saraswat, A., Kumar, B. (2005). Anaphylactic reaction to apple, banana and lychee: what is common between botanically disparate plant families? *International Journal of Dermatology*, *44*, 996–998.

174. Luttkopf, D., Ballmer-Weber, B.K., Wuthrich, B., Vieths, S. (2000). Celery allergens in patients with positive double-blind placebo-controlled food challenge. *The Journal of Allergy and Clinical Immunology*, *106*, 390–399.

175. Song, J.J., Zhang, H.Y., Liu, Z.G., Ran, P.X. (2007). Cloning of the panallergen profilin from lychee fruit and its cross-reactivity with birch pollen profilin Bet v 2. *Food and Agricultural Immunology*, *18*, 129–138.

176. Niggemann, B., Reibel, S., Hipler, C., Wahn, U. (2002). Anaphylactic reaction to lychee in a 12-year-old girl: cross-reactivity to latex. *Pediatric Allergy and Immunology*, *13*, 64–67.

177. Giannoccaro, F., Munno, G., Riva, G., Pugliese, S., Paradiso, M.T., Ferrannini, A. (1998). Oral allergy syndrome to grapes. *Allergy*, *53*, 451–452.

178. Guinnepain, M.T., Rassemont, R., Claude, M.F., Laurent, J. (1998). Oral allergy syndrome (OAS) to grapes. *Allergy*, *53*, 1225–1225.

179. Garcia-Robaina, J.C., de la Torre-Morin, F., Sanchez-Machin, I., Sanchez-Monge, R., Barber, D., Lombardero, M. (2001). Anaphylaxis induced by exercise and wine. *Allergy*, *56*, 357–358.

180. Senna, G., Mistrello, G., Roncarolo, D., Crivellaro, M., Bonadonna, P., Schiappoli, M., Passalacqua, G. (2001). Exercise-induced anaphylaxis to grape. *Allergy*, *56*, 1235–1236.

181. Bircher, A.J., Bigliardi, P., Yilmaz, B. (1999). Anaphylaxis resulting from selective sensitization to Americana grapes. *The Journal of Allergy and Clinical Immunology*, *104*, 1111–1113.

182. Caiaffa, M.F., Tursi, A., Macchia, L. (2003). Grape anaphylaxis. *Journal of Investigational Allergology and Clinical Immunology*, *13*, 211–212.

183. Pastorello, E.A., Farioli, L., Pravettoni, V., Ortolani, C., Fortunato, D., Giuffrida, M.G., Perono Garoffo, L., Calamari, A.M., Brenna, O., Conti, A. (2003). Identification of grape and wine allergens as an endochitinase 4, a lipid-transfer protein, and a thaumatin. *The Journal of Allergy and Clinical Immunology*, *111*, 350–359.

184. Schäd, S.G., Trcka, J., Vieths, S., Scheurer, S., Conti, A., Bröcker, E.B., Trautmann, A. (2005). Wine anaphylaxis in a German patient: IgE-mediated allergy against a lipid transfer protein of grapes. *International Archives of Allergy and Immunology*, *136*, 159–164.

185. Sbornik, M., Rakoski, J., Mempel, M., Ollert, M., Ring, J. (2007). IgE-mediated type-I-allergy against red wine and grapes. *Allergy*, *62*, 1339–1340.

186. Vassilopoulou, E., Zuidmeer, L., Akkerdaas, J., Tassios, I., Rigby, N.R., Mills, E.N.C., van Ree, R., Saxoni-Papageorgiou, P., Papadopoulos, N.G. (2007). Severe immediate allergic reactions to grapes: part of a lipid transfer protein-associated clinical syndrome. *International Archives of Allergy and Immunology*, *143*, 92–102.

187. Anton, E., Jimenez, I., Polo, F., Picans, I., Sanchez, I., Jerez, J. (1997). Immediated hypersensitivity to grape: study of cross-reactivity with other fruits. *Allergy*, *52*, 119.

188. Scheurer, S., Metzner, K., Haustein, D., Vieths, S. (1997). Molecular cloning, expression and characterization of Pru a 1, the major cherry allergen. *Molecular Immunology*, *34*, 619–629.

189. Scheurer, S., Son, D.Y., Boehm, M., Karamloo, F., Franke, S., Hoffmann, A., Haustein, D., Vieths, S. (1999). Cross-reactivity and epitope analysis of Pru a 1, the major cherry allergen. *Molecular Immunology*, *36*, 155–167.
190. Scheurer, S., Pastorello, E.A., Wangorsch, A., Kastner, M., Haustein, D., Vieths, S. (2001). Recombinant allergens Pru av 1 and Pru av 4 and a newly identified lipid transfer protein in the *in vitro* diagnosis of cherry allergy. *The Journal of Allergy and Clinical Immunology*, *107*, 724–731.
191. Neudecker, P., Schweimer, K., Nerkamp, J., Scheurer, S., Vieths, S., Sticht, H., Rosch, P. (2001). Allergic cross-reactivity made visible: solution structure of the major cherry allergen Pru av 1. *The Journal of Biological Chemistry*, *276*, 22756–22763.
192. Wiche, R., Gubesch, M., Konig, H., Fotisch, K., Hoffmann, A., Wangorsch, A., Scheurer, S., Vieths, S. (2005). Molecular basis of pollen-related food allergy: identification of a second cross-reactive IgE epitope on Pru av 1, the major cherry (*Prunus avium*) allergen. *The Biochemical Journal*, *385*, 319–327.
193. Gruber, P., Vieths, S., Wangorsch, A., Nerkamp, J., Hofmann, T. (2004). Maillard reaction and enzymatic browning affect the allergenicity of Pru av 1, the major allergen from cherry (*Prunus avium*). *Journal of Agricultural and Food Chemistry*, *52*, 4002–4007.
194. Inschlag, C., Hoffmann-Sommergruber, K., O'Riordain, G., Ahorn, H., Ebner, C., Scheiner, O., Breiteneder, H. (1998). Biochemical characterisation of Pru a 2, a 23 kD thaumatin-like protein representing a potential major allergen in cherry (*Prunus avium*). *International Archives of Allergy and Immunology*, *116*, 22–28.
195. Pastorello, E.A., Farioli, L., Pravettoni, V., Incorvaia, C., Scibola, E., Ispano, M., Conti, A., Giuffrida, M.G., Fortunato, D., Ansaloni, R., Monza, M., Rivolta, F., Ortolani, C. (1999). Identification of a major allergen from cherry as a thaumatin-like protein. *Giornale Italiano di Allergologiae Immunologia Clinica*, *9*, 24–29.
196. Scheurer, S., Lauer, I., Foetisch, K., Moncin, M.S., Retzek, M., Hartz, C., Enrique, E., Lidholm, J., Cistero-Bahima, A., Vieths, S. (2004). Strong allergenicity of Pru av 3, the lipid transfer protein from cherry, is related to high stability against thermal processing and digestion. *The Journal of Allergy and Clinical Immunology*, *114*, 900–907.
197. Primavesi, L., Brenna, O.V., Pompei, C., Pravettoni, V., Farioli, L., Pastorello, E.A. (2006). Influence of cultivar and processing on cherry (*Prunus avium*) allergenicity. *Journal of Agricultural and Food Chemistry*, *54*, 9930–9935.
198. Wuthrich, B., Stager, J., Johansson, S.G.O. (1990). Celery allergy associated with birch and mugwort pollinosis. *Allergy*, *45*, 566–571.
199. Vieths, S., Jankiewicz, A., Wuthrich, B., Baltes, W. (1995). Immunoblot study of IgE binding allergens in celery roots. *Annals of Allergy, Asthma & Immunology*, *74*, 48–55.
200. Ballmer-Weber, B.K., Vieths, S., Lüttkopf, D., Heuschmann, P., Wüthrich, B. (2000). Celery allergy confirmed by double-blind, placebo-controlled food challenge: a clinical study in 32 subjects with a history of adverse reactions to celery root. *The Journal of Allergy and Clinical Immunology*, *106*, 373–378.
201. Breiteneder, H., Hoffmann-Sommergruber, K., O'Riordain, G., Susani, M., Ahorn, H., Ebner, C., Kraft, D., Scheiner, O. (1995). Molecular characterization of Api g 1, the major allergen of celery (*Apium graveolens*) and its immunological and

structural relationships to a group of 17 kDa tree pollen allergens. *European Journal of Biochemistry/FEBS*, *233*, 484–489.

202. Vieths, S., Luttkopf, D., Reindl, J., Anliker, M.D., Wuthrich, B., Ballmer-Weber, B.K. (2002). Allergens in celery and zucchini. *Allergy*, *57*, 100–105.
203. Schoning, B., Vieths, S., Petersen, A., Baltes, W. (1995). Identification and characterization of allergens related to Bet v 1, the major birch pollen allergen, in apple, cherry, celery and carrot by two-dimensional immunoblotting and microsequencing. *Journal of the Science of Food and Agriculture*, *67*, 431–440.
204. Hoffmann-Sommergruber, K., Ferris, R., Pec, M., Radauer, C., O'Riordain, G., Laimer Da Camara Machado, M., Scheiner, O., Breiteneder, H. (2000). Characterization of Api g 1.0201, a new member of the Api g 1 family of celery allergens. *International Archives of Allergy and Immunology*, *122*, 115–123.
205. Neudecker, P., Lehmann, K., Nerkamp, J., Haase, T., Wangorsch, A., Fotisch, K., Hoffmann, S., Rosch, P., Vieths, S., Scheurer, S. (2003). Mutational epitope analysis of Pru av 1 and Api g 1, the major allergens of cherry (*Prunus avium*) and celery (*Apium graveolens*): correlating IgE reactivity with three-dimensional structure. *The Biochemical Journal*, *376*, 97–107.
206. Vallier, P., DeChamp, C., Valenta, R., Vial, O., Deviller, P. (1992). Purification and characterization of an allergen from celery immunochemically related to an allergen present in several other plant species: identification as a profilin. *Clinical and Experimental Allergy*, *22*, 774–782.
207. Scheurer, S., Wangorsch, A., Haustein, D., Vieths, S. (2000). Cloning of the minor allergen Api g 4 profilin from celery (*Apium graveolens*) and its cross-reactivity with birch pollen profilin Bet v 2. *Clinical and Experimental Allergy*, *30*, 962–971.
208. Bublin, M., Radauer, C., Wilson, I.B., Kraft, D., Scheiner, O., Breiteneder, H., Hoffmann-Sommergruber, K. (2003). Cross-reactive N-glycans of Api g 5, a high molecular weight glycoprotein allergen from celery, are required for immunoglobulin E binding and activation of effector cells from allergic patients. *Journal of the Federation of American Societies for Experimental Biology*, *17*, 1697–1699.
209. Jankiewicz, A., Baltes, W., Bo, K.W., Dehne, L.I., Jamin, A., Hoffmann, A., Haustein, D., Vieths, S. (1997). Influence of food processing on the immunochemical stability of celery allergens. *Journal of the Science of Food and Agriculture*, *75*, 359–370.
210. Aalberse, R.C. (1998). Clinical relevance of carbohydrate allergen epitopes. *Allergy*, *53*, 54–57.
211. Fötisch, K., Altmann, F., Haustein, D., Vieths, S. (1999). Involvement of carbohydrate epitopes in the IgE response of celery-allergic patients. *International Archives of Allergy and Immunology*, *120*, 30–42.
212. Bublin, M., Lauer, I., Oberhuber, C., Alessandri, S., Briza, P., Radauer, C., Himly, M., Breiteneder, H., Vieths, S., Hoffmann-Sommergruber, K. (2008). Production and characterization of an allergen panel for component-resolved diagnosis of celery allergy. *Molecular Nutrition and Food Research*, *52*, S241–S250.
213. Willerroider, M., Fuchs, H., Ballmer-Weber, B.K., Focke, M., Susani, M., Thalhamer, J., Ferreira, F., Wüthrich, B., Scheiner, O., Breiteneder, H., Hoffmann-Sommergruber, K. (2003). Cloning and molecular and immunological characterisation of two new food allergens, Cap a 2 and Lyc e 1, profilins from bell pepper (*Capsicum annuum*)

and tomato (*Lycopersicon esculentum*). *International Archives of Allergy and Immunology*, *131*, 245–255.

214. Westphal, S., Kempf, W., Foetisch, K., Retzek, M., Vieths, S., Scheurer, S. (2004). Tomato profilin Lyc e 1: IgE cross-reactivity and allergenic potency. *Allergy*, *59*, 526–532.

215. Westphal, S., Kolarich, D., Foetisch, K., Lauer, I., Altmann, F., Conti, A., Crespo, J.F., Rodrıguez, J., Enrique, E., Vieths, S., Scheurer, S. (2003). Molecular characterization and allergenic activity of Lyc e 2 (beta-fructofuranosidase), a glycosylated allergen of tomato. *European Journal of Biochemistry/FEBS*, *270*, 1327–1337.

216. Foetisch, K., Son, D.Y., Altmann, F., Aulepp, H., Conti, A., Haustein, D., Vieths, S. (2001). Tomato (*Lycopersicon esculentum*) allergens in pollen-allergic patients. *European Food Research and Technology*, *213*, 259–266.

217. Le, L.Q., Lorenz, Y., Scheurer, S., Fötisch, K., Enrique, E., Bartra, J., Biemelt, S., Vieths, S., Sonnewald, U. (2006). Design of tomato fruits with reduced allergenicity by dsRNAi-mediated inhibition of ns-LTP (Lyc e 3) expression. *Plant Biotechnology Journal*, *4*, 231–242.

218. Palomares, O., Villalba, M., Quiralte, J., Polo, F., Rodriguez, R. (2005). 1, 3-Beta-glucanases as candidates in latex-pollen-vegetable food cross-reactivity. *Clinical and Experimental Allergy*, *35*, 345–351.

219. Weangsripanaval, T., Nomura, N., Moriyama, T., Ohta, N., Ogawa, T. (2003). Identification of suberization-associated anionic peroxidase as a possible allergenic protein from tomato. *Bioscience, Biotechnology, and Biochemistry*, *67*, 1299–1304.

220. Diaz-Perales, A., Collada, C., Blanco, C., Sanchez-Monge, R., Carrillo, T., Aragoncillo, C., Salcedo, G. (1999). Cross-reactions in the latex-fruit syndrome: a relevant role of chitinases but not of complex asparagine-linked glycans. *The Journal of Allergy and Clinical Immunology*, *104*, 681–687.

221. Reche, M., Pascual, C.Y., Vicente, J., Caballero, T., Martın-Munoz, F., Sanchez, S., Martın-Esteban, M. (2001). Tomato allergy in children and young adults: cross-reactivity with latex and potato. *Allergy*, *56*, 1197–1201.

222. Kondo, Y., Urisu, A., Tokuda, R. (2001). Identification and characterization of the allergens in the tomato fruit by immunoblotting. *International Archives of Allergy and Immunology*, *126*, 294–299.

223. Le, L.Q., Mahler, V., Lorenz, Y., Scheurer, S., Biemelt, S., Vieths, S., Sonnewald, U. (2006). Reduced allergenicity of tomato fruits harvested from Lyc e 1-silenced transgenic tomato plants. *The Journal of Allergy and Clinical Immunology*, *118*, 1176–1183.

224. Lorenz, Y., Enrique, E., Le, L.Q., Fotisch, K., Retzek, M., Biemelt, S., Sonnewald, U., Vieths, S., Scheurer, S. (2006). Skin prick tests reveal stable and heritable reduction of allergenic potency of genesilenced tomato fruits. *The Journal of Allergy and Clinical Immunology*, *118*, 711–718.

225. Scheurer, S., Wangorsch, A., Nerkamp, J., Skov, P.S., Ballmer-Weberd, B., Wuthrichd, B., Hausteina, D., Viethsa, S. (2001). Cross-reactivity within the profilin panallergen family investigated by comparison of recombinant profilins from pear (Pyr c 4), cherry (Pru av 4) and celery (Api g 4) with birch pollen profilin Bet v 2. *Journal of Chromatography. B, Biomedical Sciences and Applications*, *756*, 315–325.

226. Reindl, J., Rihs, H.P., Scheurer, S., Wangorsch, A., Haustein, D., Vieth, S. (2002). IgE reactivity to profilin in pollensensitized subjects with adverse reactions to banana and pineapple. *International Archives of Allergy and Immunology*, *128*, 105–114.

227. Woods, R.K., Abramson, M., Bailey, M., Walters, E.H. (2001). International prevalences of reported food allergies and intolerances—comparisons arising from the European community respiratory health survey (ECRHS) 1991–1994. *European Journal of Clinical Nutrition*, *55*, 298–304.

228. Young, E., Stoneham, M.D., Petruckevitch, A., Barton, J., Rona, R. (1994). A population study of food intolerance. *The Lancet*, *343*, 1127–1130.

229. de Martino, M., Novembre, E., Cozza, G., de Marco, A., Bonazza, P., Vierucci, A. (1988). Sensitivity to tomato and peanut allergens in children monosensitized to grass pollen. *Allergy*, *43*, 206–213.

230. Ortolani, C., Ispano, M., Pastorello, E.A., Ansaloni, R., Magri, G.C. (1989). Comparison of results of skin prick tests (with fresh foods and commercial food extracts) and RAST in 100 patients with oral allergy syndrome. *The Journal of Allergy and Clinical Immunology*, *83*, 683–890.

231. Carnés, J., López-Matas, M., Ferrer, A., Larramendi, C.H., Huertas, A.J., Casanovas, M., Fernández-Caldas, E. (2006). Immunochemical characterization of tomato peel and pulp extracts. *The Journal of Allergy and Clinical Immunology*, *117*, S302.

232. Larramendi, C.H., Ferrer, A., Huertas, A.J., Garcıa-Abujeta, J.L., Andreu, C., Tella, R., Cerda, M.T., Bartra, J., Lavın, J.R., Pagan, J.A., Lopez-Matas, M.A., Fernandez-Caldas, E., Carnes, J. (2008). Sensitization to tomato peel and pulp extracts in the Mediterranean Coast of Spain—prevalence and co-sensitization with aeroallergens. *Clinical and Experimental Immunology*, *38*, 169–177.

233. Ferrer, A., Huertas, A.J., Larramendi, C.H., García-Abujeta, J.L., Bartra, J., Lavín, J.R., Andreu, C., Pagán, J.A., López-Matas, M.A., Fernández-Caldas, E., Carnés, J. (2008). Usefulness of manufactured tomato extracts in the diagnosis of tomato sensitization: comparison with the prick-prick method. *Clinical and Molecular Allergy*, *6*, 1–8.

234. Zuidmeer, L., Goldhahn, K., Rona, R.J., Gislason, D., Madsen, C., Summers, C., Sodergren, E., Dahlstrom, J., Lindner, T., Sigurdardottir, S.T., McBride, D., Keil, T. (2008). The prevalence of plant food allergies: a systematic review. *The Journal of Allergy and Clinical Immunology*, *121* (5), 1210–1218.

235. Rona, R.J., Keil, T., Summers, C., Gislason, D., Zuidmeer, L., Sordergren, E., Sigurdardottir, S.T., Lindner, T., Goldhahn, K., Dahlstrom, J., Mcbride, D., Madsen, C. (2007). The prevalence of food allergy: a meta-analysis. *The Journal of Allergy and Clinical Immunology*, *120*, 638–646.

236. Mills, E.N.C., Mackie, A.R., Burney, P., Beyer, K., Frewer, L., Madsen, C., Botjes, E., Crevel, R.W.R., van Ree, R. (2007). The prevalence, cost and basis of food allergy across Europe. *Allergy*, *62*, 717–722.

237. Asero, R., Ballmer-Weber, B.K., Beyer, K., Conti, A., Dubakiene, R., Fernandez-Rivas, M., Hoffmann-Sommergruber, K., Lidholm, J., Mustakov, T., Oude Elberink, J.N., Pumphrey, R.S., Stahl Skov, P., van Ree, R., Vlieg-Boerstra, B.J., Hiller, R., Hourihane, J.O., Kowalski, M., Papadopoulos, N.G., Wal, J.M., Mills, E.N., Vieths, S. (2007). IgE-mediated food allergy diagnosis: current status and new perspectives. *Molecular Nutrition and Food Research*, *51*, 135–147.

238. Ebo, D.G., Stevens, W.J. (2001). IgE-mediated food allergy—extensive review of the literature. *Acta Clinica Belgica*, *56* (4), 234–247.

239. Sicherer, S.H., Munoz-Furlong, A., Sampson, H.A. (2003). Prevalence of peanut and tree nut allergy in the United States determined by means of a random digit telephone survey: a 5 year follow-up study. *The Journal of Allergy and Clinical Immunology*, *112*, 1203–1207.

240. Sicherer, S.H., Munoz-Furlong, A., Sampson, H.A. (2004). Prevalence of seafood allergy in the United States determined by a random telephone survey. *The Journal of Allergy and Clinical Immunology*, *114*, 159–165.

241. Kanny, G., Moneret-Vautrin, D.A., Flabbee, J., Beaudouin, E., Morisset, M., Thevenin, F. (2001). Population study of food allergy in France. *Allergy*, *108*, 133–140.

242. Penard-Morand, C., Raherison, C., Kopferschmitt, C., Caillaud, D., Lavaud, F., Charpin, D., Bousquet, J., Annesi-Maesano, I. (2005). Prevalence of food allergy and its relationship to asthma and allergic rhinitis in schoolchildren. *Allergy*, *60*, 1165–1171.

243. Sandin, A., Annus, T., Bjorksten, B., Nilsson, L., Riikjarv, M.A., van Hage-Hamsten, M., Braback, L. (2005). Prevalence of self-reported food allergy and IgE antibodies to food allergens in Swedish and Estonian school children. *European Journal of Clinical Nutrition*, *59*, 399–403.

244. Zuberbier, T., Edenharter, G., Worm, M., Ehlers, I., Reimann, S., Hantke, T., Roehr, C.C., Bergmann, K.E., Niggemann, B. (2004). Prevalence of adverse reactions to food in Germany—a population study. *Allergy*, *59*, 338–345.

245. Falcao, H., Lunet, N., Lopes, C., Barros, H. (2004). Food hypersensitivity in Portuguese adults. *European Journal of Clinical Nutrition*, *58*, 1621–1625.

246. Mukoyama, T., Nishima, S., Arita, M., Ito, S., Urisu, A., Ebisawa, M., Ogura, H., Kohno, Y., Kondo, N., Shibata, R., Hurusho, M., Mayumi, M., Morikawa, A. (2007). Guidelines for diagnosis and management of pediatric food allergy in Japan. *Allergology International*, *56*, 349–361.

247. Likura, Y., Imai, T. (2001). Prevalence of immediate type food allergy in Japan. *Skin Research*, *43*, 5–9.

248. Yoneyama, K., Ono, A. (2002). Study of food allergy among university students in Japan. *Allergology International*, *51*, 205–208.

249. Kumari, D., Kumar, R., Sridhara, S., Arora, N., Gaur, S.N., Singh, B.P. (2006). Sensitization to blackgram in patients with bronchial asthma and rhinitis: clinical evaluation and characterization of allergens. *Allergy*, *61*, 104–110.

250. Hill, D.S., Hosking, C.S., Zhie, Y.Z., Leung, R., Baratwidjaja, K., Iikura. Y., Iyngkaran, N., Gonzalez-Andaya, A., Wah, L.B., Hsieh, K.H. (1997). The frequency of food allergy in Australia and Asia. *Environmental Toxicology and Pharmacology*, *4*, 101–110.

251. Hill, D.J., Hosking, C.S., Heine, R.G. (1999). Clinical spectrum of food allergy in children in Australia and South-East Asia: identification and targets for treatment. *Annals of Medicine*, *31*, 272–281.

252. Woods, R.K., Thien, F., Raven, J., Walters, E.H., Abramson, M. (2002). Prevalence of food allergies in young adults and their relationship to asthma, nasal allergies, and eczema. *Annals of Allergy, Asthma & Immunology*, *88*, 183–189.

253. Crooks, C., Ameratunga, R., Simmons, G., Jorgensen, P., Wall, C., Brewerton, M., Sinclair, J., Steele, R., Ameratunga, S. (2008). The changing epidemiology of food allergy-implications for New Zealand. *The New Zealand Medical Journal*, *121*, 74–82.

254. Ford, R. (2008). Leading New Zealand paediatrician and allergist response to lack of food allergy data in New Zealand. Available at www.scoop.co.nz/stories/GE0804/S00057.htm

255. Allergy New Zealand. (2008). Frustrated at the lack of allergy services? Join Allergy New Zealand in making submissions to the District Health Boards. Available at http://www.allergy.org.nz/

256. Fuglsang, G., Madsen, C., Saval, P., Østerballe, O. (1993). Prevalence of intolerance to food additives among Danish school children. *Pediatric Allergy and Immunology*, *4*, 123–129.

257. Osterballe, M., Hansen, T.K., Mortz, C.G., Høst, A., Bindslev-Jensen, C. (2004). The prevalence of food hypersensitivity in an unselected population of children and adults. PhD Thesis, Syddansk University, Odense, Denmark.

258. Madsen, C. (2005). Prevalence of food allergy: an overview. *The Proceedings of the Nutrition Society*, *64*, 413–417.

259. Osterballe, M., Hansen, T.K., Mortz, C.G., Bindslev-Jensen, C. (2005). The clinical relevance of sensitization to pollen-related fruits and vegetables in unselected pollen-sensitized adults. *Allergy*, *60*, 218–225.

260. Crevel, R. (2005). Industrial dimensions of food allergy. *The Proceedings of the Nutrition Society*, *64*, 470–474.

261. Taylor, S.L., Hefle, S.L., Bindslev-Jensen, C., Bock, S.A., Burks, A.W., Christie, L., Hill, D.J., Host, A., Hourihane, J.O., Lack, G., Metcalfe, D.D., Moneret-Vautrin, D.A., Vadas, P.A., Rancé, F., Skrypec, D.J., Trauman, T.A., Malmheden Yman, I., Zeiger, R.S. (2002). Factors affecting the determination of threshold doses for allergenic foods: how much is too much? *The Journal of Allergy and Clinical Immunology*, *109*, 24–30.

262. van Putten, M.C., Frewer, L.J., Gilissen, L.J.W.J., Gremmen, B., Peijnenburg, A.A.C.M., Wichers, H.J. (2006). Novel foods and food allergies: a review of the issues. *Trends in Food Science and Technology*, *17*, 289–299.

263. Anandan, C., Sheikh, A. (2005). European develpments in labelling allergenic foods. *British Medical Journal*, *331*, 1155–1156.

264. Reading, D. (2006). Allergy prevention: consumer perspectives. In: *Allergy Matters: New Approaches to Allergy Prevention and Management*, Gilissen, L.J.W.G., Wichers, H.J., Savelkoul, H.F.J., Bogers, R.J. (Eds.). Dordrecht: Springer, pp. 193–199.

265. de Blok, B.M.J., Vlieg-Boerstra, B.J., Oude Elberink, J.N.G., Duiverman, E.J., DunnGalvin, A., Hourihane, J.O., Cornelisse-Vermaat, J.R., Frewer, L., Mills, C., Dubois, A.E.J. (2007). A framework for measuring the social impact of food allergy across Europe: a EuroPrevall state of the art paper. *Allergy*, *62*, 733–737.

266. Hadley, C. (2006). Food allergies on the rise(Determining the prevalence of food allergies, and how quickly it is increasing, is the first step in tackling the problem. *European Molecular Biology Organization*, *7*, 1080–1083.

19

MANAGING RISKS AND PREVENTING FOOD ALLERGY INCIDENTS: A REGULATOR'S PERSPECTIVE

Samuel Benrejeb Godefroy, Sheila Dubois, and Sébastien La Vieille

19.1. INTRODUCTION

Scientific evidence has linked certain food ingredients with severe adverse reactions when ingested by susceptible individuals. These adverse reactions are to be differentiated from adverse reactions that are likely to occur among most members of the population if they are exposed to chemical or microbiological contaminants or to pharmacologically active ingredients in food.

Based on the nomenclature suggested by the European Academy of Allergology and Clinical Immunology (EAACI), *food allergy* is a form of immune-mediated hypersensitivity reaction to food that is reproducible, abnormal, and nonpsychologically mediated. Food allergy involving immunoglobulin E (IgE) antibodies is known as *IgE-mediated food allergy*. When immunologic mechanisms are not involved, we refer to *food intolerance* or *nonallergic food hypersensitivity* [1].

The production of IgE antibodies and a series of interactions between various cell types and chemical mediators are involved in most confirmed cases of food allergy. These cases display the characteristics of IgE-mediated allergy, which is also known as a type 1 hypersensitivity reaction. IgE-mediated allergy produces immediate symptoms, the most severe being anaphylaxis, a

Allergen Management in the Food Industry, Edited by Joyce I. Boye and Samuel Benrejeb Godefroy

hypersensitivity reaction associated with generalized clinical manifestations [2]. Manifestations may also target specific organs, such as the skin (urticaria, atopic dermatitis), the respiratory tract (asthma, rhinitis), and the digestive tract (vomiting, constipation, diarrhea, abdominal pain). Each of these clinical manifestations may precede anaphylaxis; as well, each manifestation can occur alone or in combination with other manifestations, without proceeding to anaphylaxis. Any food that contains protein has the potential to elicit such immediate allergic reactions. More than 170 different foods have been documented to elicit immediate type 1 hypersensitivity reactions [3].

Other foods have been linked with delayed hypersensitivity reactions, in which symptoms may not begin to appear until 24 hours after food ingestion. The mechanisms involved in delayed hypersensitivity reactions remain poorly defined, except perhaps for celiac disease (also known as celiac sprue or gluten-sensitive enteropathy) [4,5]. Celiac disease is an immune-mediated disease, triggered in genetically susceptible individuals by the ingestion of gluten, a general term used to describe storage proteins present in wheat, rye, barley, and their hybridized strains (e.g., triticale) [4,5]. These storage proteins trigger an inflammatory injury in the absorptive surface of the small intestine of susceptible individuals, resulting in malabsorption of protein, fat, carbohydrate, fat-soluble vitamins, folate, and minerals, especially iron and calcium [4–7].

There are also some food ingredients that elicit allergy-like reactions. While the mechanism of action through which this occurs is not fully understood, it does not appear to be immune mediated. Elicitors of such allergy-like reactions include low-molecular-mass chemicals (preservatives, colors, flavors, etc.) such as sulfites. Sulfites are food additives used mainly for preservation purposes. Foods responsible for exposure to sulfites include wines, fruits and vegetables, and vinegars [8, 9]. In terms of adverse reactions associated with sulfites, studies have suggested that 5–10% of people with asthma experience an exacerbation of symptoms within 20 minutes of ingesting sulfites [10, 11]. Other symptoms observed after sulfite ingestion include urticaria, angioedema, rhinoconjutivitis, seizure, and anaphylaxis [10–12].

For the purposes of this chapter, all adverse reactions to food involving an immune-mediated mechanism (either IgE-mediated or non-IgE-mediated), including celiac disease, will be called food allergies. Food intolerances will be defined as food sensitivities that do not involve any immune-mediated mechanism. One example of food intolerance is sulfite sensitivity, another is lactose intolerance. For the most part, food intolerances have less severe manifestations than food allergies. Affected individuals can frequently tolerate some amount of the offending food or food ingredient in their diets. An example to illustrate this fact is the difference between milk allergy and lactose intolerance. Milk-allergic consumers tend to develop systemic and sometimes severe symptoms when ingesting milk, which they cannot tolerate in their diet (or in very small quantities). On the other hand, individuals suffering from lactose intolerance can tolerate larger amounts of milk in their diet. As well, symptoms are localized to the gastrointestinal tract.

The information on regulatory requirements for allergen labeling and other risk management options represents a snapshot of the situation at the time this chapter was written. It is subject to change in the context of a rapidly evolving environment. The information is based on the authors' findings and is not considered endorsed by the regulatory authorities cited in this chapter.

19.2. IMPACTS AND RATIONALE FOR ACTION TO PREVENT FOOD ALLERGY INCIDENTS

Food allergy affects 6–8% of infants and young children and 3.5–4% of adults [13, 14]. Avoidance is currently the only available strategy for preventing potentially serious food allergy-related health problems among susceptible individuals. There is no cure for food allergy.

Systematic avoidance of causal foods can be a substantial burden for allergic individuals. The effort required to avoid food allergens can lead to both stress and social isolation [15]. At the same time, despite great care, accidental exposure can still occur and is sometimes fatal.

Food has been identified as the most common cause of anaphylaxis. Anaphylaxis is the most serious form of allergic reaction, can be life threatening, and often requires emergency room treatment [16]. A study that examined anaphylaxis-related deaths between 1986 and 2000 in Ontario, the most populated Canadian province, identified 63 confirmed deaths due to anaphylaxis; of these, 32 (50%) were food related [17].

Worldwide, the prevalence of celiac disease is estimated to be between 1 in 100 and 1 in 200 [18]. Certain groups are at markedly elevated risk: first-degree relatives of individuals diagnosed with celiac disease have a 10–20% risk of developing the disease; a high prevalence is also observed among those with Down syndrome.

Celiac disease is a lifelong condition. If it is not diagnosed early and treated with a strict gluten-free diet, it can be associated with serious complications, including osteoporosis, lymphoma and other types of malignancies, infertility in both men and women, and a number of autoimmune diseases including insulin-dependent diabetes in children. Moreover, in children, celiac disease can be associated with failure to grow and delayed puberty [4, 6, 7, 19, 20].

A 2008 survey by the Canadian Celiac Association and Health Canada investigated the impacts of the need for lifelong adherence to a strict gluten-free diet. It targeted adults following a gluten-free diet in order to identify both management strategies and emotional impacts of following this diet. A similar study on the impact of a gluten-free diet on adults with celiac disease was published in 2006 [21]. Survey findings of the 2008 study will be reported in both the scientific and the healthcare literature.

The financial and economic costs associated with food allergies can be significant. To the extent that food allergy incidents and the complications of celiac disease are prevented through successful avoidance of exposure to the food allergen(s) of concern, these costs can be reduced.

In a 2007 study from Australia [22], the cost of all types of allergies, including asthma and non-asthma allergies such as food, drug, latex, sting, and bite allergies, contact dermatitis, and anaphylaxis was estimated to be approximately $7200 per person per year in Australian dollars. Burden of disease (disability, premature death) accounted for 73% of costs, reduced productivity for 19% of costs, and health system costs for 4%. The largest share of allergy costs (86%) was borne by the individuals having the allergy themselves, due to the large costs associated with disability and premature death. Nine percent (9%) of costs was borne by the Australian federal government, due to their share of health system and productivity costs. Several groups are currently investigating costs specific to food allergies. One of these is EuroPrevall, a project funded by the European Commission [23].

19.3. RISK MANAGEMENT OPTIONS TO PREVENT AND MANAGE FOOD ALLERGY INCIDENTS

At the present time, it is not possible to cure food allergies. As a result, the emphasis must be on prevention of food allergy incidents, supported by symptom management when accidental exposure does occur. Prevention of food allergy incidents must focus on assisting food-allergic consumers and their caregivers in avoiding exposures to allergens of concern. Such exposure can occur in a range of locations, including in the home, in restaurants, and in school and day care settings. Avoidance of exposure to food allergens of concern requires effort by both food-allergic consumers and their caregivers. It also requires awareness and action by other consumers who sometimes provide food to consumers with food allergies, as well as by the food industry and governmental agencies responsible for food safety.

This chapter will review a range of risk management options aimed at preventing food allergy incidents and the complications of celiac disease. It will discuss risk management options for minimizing exposure to allergens of concern in prepackaged and non-prepacked foods. Many of the examples involving prepackaged food are drawn from current work in two areas: (1) work to revise Canada's regulations on labeling of priority allergens as ingredients and (2) work to review Canada's policy on allergen precautionary labeling. As well, the section on school and day care programs provides summary information for all Canadian provinces and territories.

This chapter will also provide an overview of risk management activities in schools and day care settings. In these latter settings, risk management activities involve both (1) preventive actions to facilitate avoidance of allergen exposure and (2) risk mitigation actions to be implemented if/when accidental exposure does occur. Finally, it provides an overview of a proposed Canadian severe food allergy incident reporting network. While this chapter does not discuss new therapies intended to treat or reverse food allergies, it does provide an overview of recent clinical findings concerning prevention of food allergies through dietary advice for pregnant and breast-feeding women. Industry practices for managing risks associated with allergens in food production settings are not discussed. This topic is addressed in previous chapters.

19.3.1. Product Labeling Initiatives

Avoidance of allergens of concern is the only way that food-allergic consumers can prevent allergic reactions. To avoid foods containing the ingredients to which they are likely to react, food-allergic consumers and their caregivers must rely on information provided on product labels.

The presence of allergens in prepackaged foods can occur in two ways: either as a result of their inclusion in the product formulation (*ingredients*) or where the conditions during processing may have resulted in the presence of an allergen that could not have been avoided or mitigated within reasonable means (*cross-contamination or cross-contact*). Labeling information should allow food-allergic consumers and their caregivers to avoid foods of concern, whether allergens are present as ingredients or potentially present as a result of cross-contamination.

Product labeling can also be used to inform the consumer of the suitability of a particular food product. Such suitability can be because (1) certain ingredients were excluded and/or because (2) special care was taken to control the production process, in a manner that guarantees the absence of the ingredient, and hence an optimum protection for consumers allergic to that specific ingredient (allergen free). These foods are generally considered a targeted source of supply and are sought by food-allergic and gluten-intolerant consumers. Previous chapters of this book describe a range of initiatives intended to make such products available.

Allergen labeling of prepackaged foods is subject to food legislative and regulatory obligations set by each national jurisdiction. The Codex Committee on Food Labelling (CCFL) has considered allergen labeling a priority area, making recommendations that were adopted by the Codex Alimentarius Commission in 1999 [24]. The commission recommended that science-based criteria be used to determine which foods or food ingredients should be placed on a priority list of foods whose presence should always be declared in the list of ingredients on a food label, because of their potential to induce an allergic reaction. The list of priority allergens retained by the Codex Alimentarius Commission included the following foods:

- cereals containing gluten, that is, wheat, rye, barley, oats, spelt, or their hybridized strains and products of these;
- crustaceans and products of these;
- egg and egg products;
- fish and fish products;

- peanuts;
- soybeans;
- milk and milk products;
- tree nuts and nut products; and
- sulfites at concentrations of 10 mg/kg or higher.

Subsequent to the adoption of this list by the Codex Alimentarius Commission, the World Health Organization (WHO) convened a food allergens expert panel, to provide guidance to the Joint Food and Agriculture Organization of the United Nations (FAO)/WHO Expert Committee on Food Additives (JECFA), the committee that advises the Codex Alimentarius Commission on food additives and other chemicals and ingredients in food. This panel was tasked with identifying criteria for amending the Codex list of priority allergenic foods. It recommended that the following criteria be used when considering whether a new substance should be included in the list: (1) the existence of a cause-and-effect relationship, based on positive double-blind, placebo-controlled food challenge (DBPCFC) or unequivocal reports of reactions, including severe symptoms associated with exposure to the food commodity; and (2) prevalence data in children and adults, supported by clinical studies relying on DBPCFC studies would also be appropriate. It was acknowledged that the availability of such data for some foods and in certain regions of the world may represent a challenge.

Building on the Codex list, national food regulatory agencies have used this guidance to develop their own lists of priority allergens, which should be targeted for mandatory listing on labels of foods available for sale in the country/region under their oversight. Canada recently adapted the criteria listed above for use in considering the addition of new allergens to its priority list. These criteria and their use in the assessment of mustard, onions, and garlic are reviewed in a previous chapter of this book.

19.3.2. Ingredient Labeling for Priority Allergens

Several food regulatory agencies have amended their food ingredient labeling regulations for prepackaged foods in order to better address the needs of food-allergic consumers (and their caregivers) for reliable identification of the allergens that they need to avoid. Because of their large number, not all food allergens can be specifically labeled. Rather, the approach is to require labeling of a smaller number of foods and food ingredients that a food regulatory agency has identified as priority allergens.

While the list of priority allergens in each jurisdiction generally builds on the list retained by the Codex Alimentarius Commission in 1999 [24], the specific allergens that are required to be labeled are not the same in all jurisdictions. For example, buckwheat is included in the list of priority allergens developed by Japan but not in the lists of other jurisdictions; lupin and celery are included in the list of the European Union (EU), and mustard was recently added to the Canadian list. Differences are also observed between jurisdictions regarding whether allergen labeling legislation or regulations are worded to reflect the presence of any fraction of an ingredient (e.g., milk) or the presence of any protein fraction of an ingredient (e.g., milk protein). As a result, ingredients that have undergone extensive processing that results in the removal of the protein fraction (e.g., highly refined soy oils or maltodextrins derived from wheat) may require allergen labeling in the first situation (any fraction) but not in the second (any protein fraction). These situations may be subject to scientific opinions and possible exemptions from labeling requirements in different jurisdictions.

These regulations also include specific labeling requirements such as where the allergen name is to appear on the label (in general as part of the list of ingredient or in the immediate vicinity). Another feature of these requirements is an emphasis on use of simple and easy-to-understand language to identify the source of a priority allergen. For example, the words "milk" and "egg" would be required to identify the presence of casein and lysozyme, respectively. Labeling requirements for allergenic ingredients are also generally established so as to remove previous exemptions, for example, those that permitted components of ingredients not to be explicitly identified on labels of prepackaged foods (e.g., when mustard is included in a spice mixture). Such exemptions are being removed in various jurisdictions because they are considered to be a cause of food allergy incidents related to "hidden" sources of allergens.

When imposed by regulation or legislation, labeling requirements help create a uniform environment for labeling of priority allergens as ingredients in prepackaged foods, thereby helping allergic individuals and their caregivers identify products they need to avoid. Such labeling requirements can also help to create a predictable regulatory environment for food processors and importers, including consistency in the enforcement actions taken when problems are identified. To be effective in achieving their intended goals, labeling regulations must be accompanied by compliance and enforcement policies. As an example, provision of up-to-date guidance to industry regarding how to comply with regulations can promote industry compliance, thereby preventing situations in which risk management involves food recalls.

Mislabelling of priority allergens present as ingredients in prepackaged foods can sometimes be an issue. It tends to be discovered in one of two ways: (1) through periodic ingredient verification or (2) through the reporting of allergic reaction incidents among susceptible individuals following the consumption of foods containing the specific priority allergen(s) to which they react. A risk-based approach is desirable for managing such mislabeling. It involves the qualification of the health risk in a given situation, followed by implementation of the relevant risk management action (e.g., food recalls, broadcasting information to consumers on situations of mislabelling of products that cannot be recalled).

Methods to quantify the magnitude of the health risk associated with food allergy incidents are still under development. Current practices rely on (1) existing scientific and medical literature, (2) previous reports of adverse responses, (3) food allergy prevalence information, (4) whether the reported incident was considered life threatening, (5) and an estimate of the amount of consumed allergen that triggered the response. Ongoing development of analytical methods for both the detection and the quantification of undeclared allergens in food is an essential aspect of this work.

The generation of new clinical data in support of the development of no observed adverse effect level (NOAEL) and lowest observed adverse effect level (LOAEL) values will be critical to support risk assessment and risk management activities undertaken by food regulators. For most allergens, data from clinical food challenge studies are still limited and what data are available have somewhat limited usefulness in risk assessment and management work. Most clinical food challenge studies have not been designed to generate data in support of the identification of a NOAEL. One reason for this is that highly sensitive individuals or subpopulations are not eligible to participate in such studies.

Nonetheless, some threshold levels, based on published LOAELs have been used in support of regular risk assessment practices, within the context of incident management (e.g., 0.3–1 mg for egg, 0.02–7.5 mg for tree nuts, etc.). These levels are used as reference values in the qualitative assessment of the risk associated with the presence of undeclared allergens in food; estimated intake levels are compared with the reference levels. It is important to note that a number of other factors need to be taken into consideration when the level of risk is estimated for an allergy incident. This is typically done on an ad hoc basis. Information on such factors, aggravating or mitigating, is obtained from the investigation of the product. Examples of such factors include extent of product distribution, consumers targeted by the product, and presence of an allergen precautionary statement on the label [25]. It is important that food safety regulators worldwide increase their efforts toward harmonization of risk assessment and risk management practices in the context of managing food allergy incidents.

19.3.3. Precautionary Labeling

In the early 1990s, Canadian food regulators were among the first risk managers to (1) identify the need to alert consumers about the possible inadvertent presence of undeclared allergens in prepackaged foods as a result of cross-contamination; (2) select a precautionary labeling statement (also referred to as an advisory labeling statement) as the most appropriate risk communication tool; and (3) define general conditions of use for precautionary statements. Health Canada defined a "food allergen precautionary statement" as "a declaration on the label of a prepackaged food of the possible inadvertent presence of an allergen in the food." Precautionary labeling statements were intended to be used by food manufacturers and importers on a voluntary basis, above and beyond ingredient and nutrition labeling requirements stipulated in the Canadian *Food and Drug Regulations* and other related legislation. Health Canada also indicated that, when used, precautionary statements should aim to (1) alert the consumer to the possible presence of an allergen in a food and (2) prevent the consumption of products labeled with a precautionary statement by persons having a food allergy. The Canadian policy on the use of allergen precautionary labeling statements has been nonprescriptive with respect to the specific wording of such statements. It required only that such statements be truthful, clear, and nonambiguous and are not a substitute for good manufacturing practices.

Since the early 1990s, allergen precautionary labeling statements have proliferated and are a common feature on the labels of prepackaged foods available for sale in Canada and worldwide. A wide variety of wordings are used in these statements. The information provided extends beyond an indication of the possible inadvertent presence of one or more allergens in the food (e.g., "may contain X"; "may be present: X") toward a description of the conditions under which the product was processed and packaged (e.g., "manufactured in the same facility as products containing X"; "packaged in the same facility as products containing X"). Other statement wording simply discloses the presence of the allergen(s) in the manufacturing facility.

The effectiveness of this type of labeling is currently being reexamined in Canada and internationally. Health Canada announced its intent to review its policy on the use of allergen precautionary labeling statements in 2007 [26]. At that time, it indicated that allergen precautionary labeling statements can still be considered as useful tools in mitigating adverse reactions to priority food allergens if they satisfy some general principles:

- Precautionary statements should only be used on the label when, despite all reasonable measures, the inadvertent presence of allergens in food is unavoidable.
- Precautionary labeling must not be used when an allergen or allergen-containing ingredient is deliberately added to a food.
- The use of precautionary labeling where there is no actual risk of an allergen being present should be discouraged, as it would be contrary to the stated objective of enabling a variety of safe and nutritious food choices for consumers with dietary restrictions.

This section will summarize the elements of the evidence being considered during this review.

19.3.3.1. Industry Use of Allergen Precautionary Labeling. In 1994, in response to concerns raised by the Allergy/Asthma Information Association, the Grocery Products Manufacturers of Canada (later renamed Food and Consumer Products of Canada) conducted a survey on the use of precautionary labeling by its members. The survey was repeated in 1997. Response rates were 35% in 1994 and 50% in 1997. While 33% of respondents (10/30) reported using precautionary labeling in 1994, 62% (26/43) did so in 1997. The main reason given in 1994 was that no amount of cleaning and/or training was sufficient to prevent cross-contamination. In 1997, reasons included common processing lines, potency of certain allergens, and inability to source product free of potential cross-contamination. Highlights from the 1997 survey also noted that about half of those who reported use of precautionary labeling also indicated that they planned to phase it out [27]. At the time this chapter was written, Health Canada had initiated a follow-up survey of industry allergen precautionary labeling practices.

In the United States, in a 2006 report to the Senate and the House of Representatives on the Food Allergen Labeling and Consumer Protection Act (FALCPA), the Food and Drug Administration identified 14 reasons why food manufacturing facilities reported using precautionary labeling statements. Many of these were business related (e.g., "recommended or required by the headquarter/legal department/trade association/client firm," "avoid expenses of multiple labels," "keep up with industry's trends and practices") and not directly related to informing consumers about health risks related to possible allergen cross-contamination [28].

Health Canada conducted a survey on the use of allergen precautionary labeling statements on chocolate products and granola bars in conjunction with the presence of specific allergens such as peanut, hazelnut, almond, and Brazil nut in these products. The most commonly observed precautionary statement was "may contain trace(s) of." Levels of allergen protein present in products with this precautionary statement varied widely, from nondetected to 6500 ppm. A level of 6500 ppm peanut protein in a product with a "may contain traces of" precautionary labeling statement is certainly of concern for peanut-allergic consumers [26]. If a 40-g chocolate bar with a precautionary label for peanut were to contain 6500 ppm of peanut protein, this would represent an exposure to 0.26 g (260,000 μg) of peanut protein. In reporting results of a double-blind, placebo food challenge study, Hourihane et al. concluded that as little as 100 μg of peanut protein provoked symptoms in some subjects with peanut allergy [29].

In the United States, Pieretti et al. [30] evaluated the use of statements such as "may contain" in 20,000 unique manufactured products from 99 supermarkets across the United States. Overall, 17% of products contained advisory labeling statements. The highest use was for chocolate candy (54%) and cookies (53%), and the lowest was for canned fish (2%), spices (1%), and baby food (1%). In terms of statement wording, "may contain" was most frequently observed (38%). In 2007, Hefle and colleagues analyzed products bearing allergen advisory statements for the presence of peanut residue. Two of 51 products (4%) with a "may contain" statement had detectable peanut residues in one or both lots analyzed; 7 of 68 products (10%) with a "shared facility" statement had detectable peanut residue in one or both lots [31].

19.3.3.2. Consumer Perceptions and Behaviors. In Canada, Sheth et al. [32] surveyed 766 parents of Canadian children with a confirmed peanut allergy; 583 completed the questionnaire. Responding parents reported that 41% of their children (242) had experienced an accidental exposure at least once in their lifetime. Thirty-four percent (34%) of these exposures were attributed by respondents to product labeling that was complex or incomplete, 32% to failure to read a label, and 6% to ignoring a precautionary labeling statement. In the same survey, some precautionary labeling statements were reported as more likely than others to result in avoidance of a food: 93% of respondents reported that they would avoid a product labeled with "not suitable for people with an X allergy," while 87% indicated that they would avoid a product labeled "may contain X" or "manufactured on same equipment as X." The precautionary statements "may contain traces," "packaged in facility that also processes," and "manufactured in a facility that also processes" were less likely to result in reported avoidance (72%).

Also in Canada, Fenton and Elliott [33] used a qualitative focus group methodology to solicit views about precautionary labeling of commercially packaged food among 10 parents of allergic children recruited through Anaphylaxis Canada. Many of these parents expressed concern that adolescents with food allergies and nonallergic individuals of all ages may not possess the knowledge and experience needed to correctly interpret precautionary labeling statements. Many also indicated that they considered prepackaged foods that do not have precautionary labeling statements as high risk when purchasing decisions are being made by consumers who do not have extensive experience in avoiding food allergens.

In the United States, Hefle et al. conducted surveys during Food Allergy and Anaphylaxis Network (FAAN) patient conferences in 2003 and 2006 [31]. A total of 625 surveys were completed in 2003 and 645 in 2006, most by the parent of a child with a food allergy. Eighty-five percent (85%) of respondents reported that they would "never" purchase a product with an advisory warning in 2003; 75% of respondents did so in 2006. The authors suggested possible explanations for this observed change: parents were noting a proliferation of advisory warnings; they had not noted any reactions to products that had not been previously labeled; possibly, they were presuming that the warnings were for legal rather than health concerns.

Two recent publications looked at how precautionary labeling statements are interpreted by adolescents. In response to open calls for participation made through modalities that included websites of U.S. and Canadian food allergy associations, 174 adolescents and young adults (13–21 years of age) with food allergies completed an anonymous online survey. Seventeen percent (17%) of respondents were considered at high risk for food-related anaphylaxis because they indicated that they (1) will eat a food despite labeling indicating that it "may contain" an allergen of concern and (2) do not always carry their prescribed self-injectable epinephrine [34]. In other work, Akeson et al. [35] studied 15 adolescents aged 13–16 with a history of clinician-diagnosed anaphylaxis and an allergy to peanut or tree nut. Anaphylaxis was typically perceived as "no big deal" by these adolescents. The authors attributed this attitude to the adolescents' inability to remember any severe reaction; only 1 of the 15 could remember having had an anaphylactic reaction.

19.3.3.3. Enhancing the Effectiveness of Allergen Precautionary Labeling. Food-allergic consumers need to know whether each food they consider eating contains a substance to which they are allergic, so that they can take any needed avoidance action. It is not enough to know that a specific prepackaged food does not contain a food allergen as an ingredient. Consumers also need to know if the product may have unintentionally come in contact with the allergen(s) of concern during product processing [36]. Allergic consumers need precautionary labeling information they readily understand and trust, so that they can make informed purchase and serving decisions, either for themselves or for family members/friends/colleagues with food allergies [36]. Such decisions need to effectively manage health risks, without unduly restricting food choice for those with food allergies [16].

A series of guiding principles were developed for use in discussions and decision making related to updating Health Canada's policy on allergen precautionary labeling of prepackaged foods. They are presented below. Some of these principles are based on published research summarized above. Others are based on information received from allergic consumers, allergic consumer associations, industry groups, and other governmental departments. All of these principles are intended to contribute to increased effectiveness of precautionary labeling statements in preventing food allergy incidents. A reduction in such incidents will, in turn, contribute to Health Canada's goal of enhancing the protection of food-allergic consumers, while maximizing choice of safe and nutritious food choices for these consumers with dietary restrictions:

1. To promote consistency of precautionary labeling statements on all prepackaged foods, a small number of wordings would be recommended for use.
2. To establish a consensus on guidelines on good manufacturing practices for allergen handling and to make explicit the conditions of use of food allergen precautionary statements. This work would be led by the industry, with input and support from the government and allergic consumer associations.
3. To establish consensus guidelines to support a standardized approach to conducting allergen-specific risk assessments of potential cross-contamination in prepackaged foods. The ability to estimate both probability of cross-contamination and the likely level of cross-contamination would be among the issues to be addressed as such guidelines are developed. This activity would be led by the industry, with input and support from government and allergic consumer associations.

4. To establish consensus guidelines to support a standardized approach to documentation and on-site storage of the findings of risk assessments of potential cross-contamination with priority allergens in prepackaged foods. These guidelines would explicitly support conduct of risk assessments both (1) when there is a *low probability* that a precautionary statement would be needed and (2) when there is a *high probability* that a precautionary statement would be needed. This activity would also be led by industry, with input and support from government and allergic consumer associations.
5. To develop an education program for consumers, focusing on correct interpretation of precautionary labeling statements. This program would be developed by government, with support from the industry and allergic consumer associations and industry.
6. To establish a mechanism that enables allergic consumers to identify easily and consistently those prepackaged foods for which diligence has been applied in risk assessment and risk management of priority allergens, indicating that precautionary labeling statements are being used judiciously. This information could be disseminated to allergic consumers via several means:
   - symbol on the label,
   - information provided through food allergy associations, and
   - information on a website or at the point of retail sale.
7. To provide a consistent, predictable regulatory environment for industry. It will be important to provide a level playing field between products made domestically, where there is opportunity for inspection to control allergen management practices, and imported foods where such mechanisms are not feasible.

Work on precautionary labeling of priority allergens is also being undertaken by other groups. One example is the work that was done through the voluntary incidental trace allergen labeling (VITAL) initiative developed by the Allergen Bureau [37]. An overview of this initiative and its potential contributions are summarized in a previous chapter of this book. Such work may eventually lead to the development of a code of practice on (1) food allergen handling and (2) appropriate use of allergen precautionary statements. Certification programs associated with the implementation of such code(s) of practice would be useful to ensure compliance of manufacturers to the set requirements.

To the extent that voluntary action by industry does not result in the consistent application of best practices in allergen handling and the use of precautionary statements on prepackaged foods, intervention by the government, for example, through development of labeling standards, may be required.

19.3.4. Allergen-Free Claims

Because considerably more work has been done on gluten-free products than on other allergen-free products, this section of the present chapter focuses on gluten-free products, intended for consumption by individuals with celiac disease and wheat allergy. The focus is specifically on the Canadian regulatory context.

19.3.4.1. Gluten-Free Labeling. The requirements for a gluten-free product according to Division 24 of the Canadian *Food and Drug Regulations* [38] (Foods for Special Dietary Use—B24.018) are as follows: "No person shall label, package, sell or advertise a food in a manner likely to create an impression that it is a gluten-free food unless the food does not contain wheat, including spelt and kamut, or oats, barley, rye or triticale or any part thereof." This regulation was enacted in 1996 and enabled the availability of gluten-free foods for individuals with celiac disease and wheat allergy. This regulation does not specify a threshold level above which compliance action would be taken. Rather, enforcement is based on the currently available detection threshold. When the regulation was enacted in 1996, the detection threshold was 200 ppm of gluten as detected by enzyme-linked immunosorbent assay (ELISA)-based methods. With the development and availability of more sensitive techniques, in 2010, the detection threshold is 20 ppm.

This regulation requires that manufacturers of gluten-free foods ensure that all ingredients comply with the requirements set in the regulations. As well, manufacturers must ensure that cross-contamination in the production facility is prevented, particularly when gluten-containing foods are handled in the facility. This regulation also applies to merchandising outlets (wholesale and retail), which must ensure that gluten-free products are handled in a manner that enables them to remain gluten-free.

Health Canada is currently reviewing its gluten-free regulations. Reasons for this review include (1) the intention to further align its gluten-free regulations with its proposed regulatory amendments for ingredient labeling of allergens, gluten sources, and added sulfites and (2) the fact that gluten-free standard was revised by the Codex Alimentarius Commission in 2008 [24].

One of the considerations in the review of the gluten-free regulations is the use of the same definition for gluten as for other priority food allergens, namely a definition encompassing the "protein notion" for the list of cereal grains excluded from gluten-free foods. A protein-based definition will allow products to make a gluten-free claim as long as the protein fraction of ingredient(s) derived from a cereal grain of concern has been removed and is not present. Such provision would allow, for example, products containing maltodextrin or other sugar-derived ingredients from cereals to bear a gluten-free claim, thereby enabling a greater choice of gluten-free products for people with celiac disease or wheat allergy.

A second consideration is whether oats should be removed from the list of excluded cereal grains. At the same time, specific requirements are being developed for "gluten-free oats," defined as free from protein derived from excluded cereal grains that include wheat, rye, barley, and their hybridized strains (e.g., triticale). The inclusion of oats as an ingredient in gluten-free foods would also be contingent upon the exclusive use of "gluten-free oats" in the manufacturing facility. Finally, in recognition of Health Canada's recent scientific review regarding the introduction of pure, uncontaminated oats into a gluten-free diet [39], Health Canada is considering imposing specific labeling requirements for gluten-free foods that contain pure, uncontaminated oats. This may involve recommending supervision by a health professional when individuals with celiac disease add pure oats to their gluten-free diet.

19.3.5. Policies for the Non-Prepackaged Food Sector

Consumers with food allergies and celiac disease need to know what food they can eat safely. As discussed in a previous section of this chapter, by regulation, the presence of priority allergens as ingredients must be clearly indicated on all prepackaged food sold in Canada. However, these labeling regulations do not apply to non-prepackaged foods [40], which include food sold by restaurants, fast-food outlets and catering establishments, as well as food that that is sold in the same location as it is packaged (e.g., sandwiches made in grocery stores).

While there has been considerable discussion of allergen labeling of prepackaged foods, there has been much less discussion of allergen labeling of non-prepackaged foods. Nonetheless, there are reports that this latter group of foods has been associated with allergic incidents. As an example, in 2006, it was estimated that about 70% of incidents of anaphylactic shock in Europe occurred when people were eating out [41].

A number of factors are believed to be responsible for allergic reactions that occur in restaurants and other foodservice establishments: absence of labeling; insufficient knowledge about ingredients by foodservice staff; errors during food preparation and service including possible cross-contamination; poor communication between customers and foodservice employees. A lack of staff training about food allergies may also be a contributing factor [42–44].

Because there are no regulations concerning allergen labeling of non-prepackaged products, consumers with food allergies typically have to ask for ingredient information at the point of sale, or rely on any written information that has been provided voluntarily by foodservice establishments [44].

There is no official position of the EU concerning allergen labeling for non-prepackaged foods. As well, in the EU, non-prepackaged food products are exempt from most general food labeling requirements. In 2008, the European Commission [41] proposed that some of the requirements for mandatory allergen labeling of prepackaged foods be extended to non-prepackaged foods, including food sold in restaurants and other foodservice establishments. Specifically, it proposed that information on priority allergens present as ingredients either be on display in the establishment or be available to consumers on request. The next step in the process is the discussion of the proposal among EU member states.

While efforts to increase the availability of point of purchase information about ingredients of non-prepackaged foods are not advanced in EU member countries, other initiatives have been undertaken. In 2008, the United Kingdom issued official guidance specifically focused on allergen control information for the non-prepackaged food sector [40]. In 2006, the Switzerland federal government, with the support of their office of federal public health, issued a brochure about allergens in food sold in bulk. This brochure included some educational information for foodservice establishments. As well, in 2009, the New Zealand Food Safety Authority published general guidance on foodservice and catering including some allergen information [45].

In related work, in 2008, the Food Standards Agency of the United Kingdom (UK-FSA) issued a best practices guidance document [40], for voluntary adoption by the non-prepackaged food sector. Its primary aim was to provide guidance to catering establishments regarding how allergen information for non-prepackaged foods should be provided to consumers, for their use in managing allergic risks. The guidance was also considered relevant to businesses providing food in institutional catering operations, such as schools, hospitals, and prisons as well as at corporate events and conferences. Information in the guidance document was organized around three key themes: (1) effective communication with the customer, with suppliers, and between staff; (2) basic training for staff; and (3) accurate, timely, and complete ingredient information. The following is an extract from this document, listing some of its key messages:

> *Effective communication*
> If a customer asks [salesperson] about ingredients in a food:
>
> —never guess
> —if [salesperson] does not know, [he] tries to find out
> —if [salesperson] is unable to provide the information, [he] says so
> —Can [salesperson] provide an alternative food?
>
> Always ensure all relevant staff is advised of any recipe changes or ingredient substitutions.
>
> a) *Basic training for staff*
>
> All staff should receive training on handling allergy requests from their first day in the job and receive refresher training.
>
> There should be an agreed system for dealing with food allergy information requests and all staff should know about this.
>
> b) *Accurate ingredient information*
>
> To know ingredients in the food [salesperson] sells.
>
> To make sure ingredient information is accessible to all staff.
>
> To make sure the ingredient information is up to date.

> If [salesperson] use prepared ingredients, [he] makes sure to know what is in them.

In the United States, a 2007 survey of 100 restaurants and other food establishments found that managers, chefs, and servers believed that their establishments could provide safe meals to allergic patrons. However, 24% of respondents also indicated that they believed that consuming a small amount of allergen would be safe, 35% believed that frying foods would destroy allergens, 54% considered a buffet safe for food-allergic consumers if it was kept "clean," and 25% believed that removing an allergen from a finished meal (e.g., removing nuts) made the dish safe for the allergic consumer. As the authors of this survey concluded, while restaurant personnel expressed a relatively high level of comfort in their ability to provide safe meals for food-allergic consumers, there were deficits in their knowledge about what was needed for such safety, indicating the need for more staff training and greater caution by food-allergic consumers [46, 47].

Also in the United States, allergic consumer associations such as the Food Allergy Initiative (FAI) and FAAN have developed resource material for the foodservice sector that talks about food allergies and how exposure can be minimized for consumers with food allergies. As an example, the document "Welcoming Guests with Food Allergies" (published in 2001 and updated in 2008) was developed by the National Restaurant Association and FAAN [48] with support from FAI. It is a training manual for foodservice professionals that cover topics such as case studies, best practices for allergen control, up-to-date research, allergen food labeling information, and practical strategies for avoiding cross-contamination. Another example is the online training program initiative "Food allergies: challenges and opportunities for food service" which, at the time this chapter was written, was offered free of charge by the Culinary Institute of America [47].

In Canada, the foodservice sector is under provincial jurisdiction. There is no federal regulation specifically regarding labeling of allergens present in non-prepackaged foods. However, the federal rule regarding the labeling of any food, Section 5(1) of Part I of the Food and Drugs Act is applicable. Section 5(1) prohibits the labeling, packaging, treating, processing, selling, or advertising of any food in a manner that misleads or deceives consumers as to the character, value, quantity, composition, merit, or safety of the product.

Because Section 5(1) Food and Drugs Act is very broad and does not explicitly address the concerns of many food-allergic consumers, in 2004, the Association Québécoise des Allergies Alimentaires (AQAA) [44], in collaboration with Health Canada, developed a reference manual for managing allergen risks in non-prepackaged food, for use by restaurateurs and foodservice managers. The initial edition was in French only, but the 2008 edition is in both French and English (http://www.aqaa.qc.ca). The main goal of this reference manual is to help professionals in the foodservice sector understand the importance of taking into account the needs of the growing number of people with food allergies. Of particular interest, this manual includes a "section with seven steps that a manager and team in charge of allergies should gradually deal with to be able to provide safe service to allergic customers" [44]. These steps include

1. training employees
2. appointing food allergy managers
3. establishing policies and procedures
4. choosing specific spaces and equipment
5. creating a safe environment
6. modifying menu presentation
7. creating an allergen chart or a list of ingredients attached to menu.

This manual also provides approximately 20 technical work planning sheets that can be adapted for use with a range of job categories within the foodservice sector [44].

In summary, at the present time, in Canada as in most Western countries, there are no regulations regarding either (1) the labeling of priority allergens in non-prepackaged foods (2) or the provision of such information in other formats by the foodservice sector. The UK-FSA appears to be the only governmental organization with an official position on risk management of food allergens in the retail sector. However, as of January 2010, this position was not supported by regulation that could be used for enforcement purposes.

To date, it appears that most reference tools for risk management of food allergens in the non-prepackaged food sector have been developed through collaborative efforts that included allergic consumer associations.

19.3.6. School and Day Care Programs

In the field of food allergies, it is known that severe incidents occur. Although anaphylactic shock, laryngeal angioedema, and severe acute asthma have the potential to cause death, fatalities are usually avoidable. The risk for the occurrence of such disorders in schools has been stressed by both organizational (e.g., the American Academy of Allergy, Asthma and Immunology [AAAI]) and academic authors [49–51]. Measures must be in place in schools both to reduce the risk of accidental exposure and to respond appropriately when an emergency situation arises. Comprehensive school board policies, standardized school anaphylaxis plans, and greater community support and involvement can supplement good patient management practices and help to avert future episodes of severe food allergy incidents [52].

19.3.6.1. Situation in Canada. In Canada, each province and territory has its own Ministry of Education (Department of Education) that has governmental jurisdiction over schools within its region. Some ministries/departments are responsible for publicly funded schools only, while others are also responsible for privately funded schools and day care centers.

In the paragraphs that follow, the approach of each Canadian provincial/territorial jurisdiction is described briefly. Each description includes an indication of the extent to which a school anaphylaxis prevention and treatment policy was in place at the time this chapter was written.

19.3.6.1.1. Alberta. While Alberta does not have a provincial anaphylaxis policy, it established the *Alberta School Boards Anaphylaxis Policy Advisory* (ASBAPA) to train teachers about how to recognize and treat anaphylactic shock. This work was completed in September 2007. Subsequently, the Alberta Department Education assembled an Allergy and Anaphylaxis Informational Response (AAIR) [53] kit, which was sent to all public and private schools in the province in April 2008. The kit contains a range of anaphylaxis training material, including AAIA's anaphylaxis and asthma brochure, auto-injector trainers, training CDs, and the book *Anaphylaxis in Schools and Other Settings*.

19.3.6.1.2. British Columbia. In September 2007, British Columbia enacted both the *Anaphylactic Protection Order* [54] and the *British Columbia Anaphylactic and Child Safety Framework* [55], which makes it mandatory for all school boards to have emergency plans in place in case a student goes into anaphylactic shock. The *Anaphylaxis Protection Order* issued by the British Columbia Minister of Education requires that every school board establish and maintain policies, procedures, and staff training in order to ensure the safety of children at risk of anaphylaxis in British Columbia schools.

The purpose of the *British Columbia Anaphylactic and Child Safety Framework* is to provide boards of education with a broad overview of the key elements required in district policy, to ensure a consistent appropriate management of anaphylactic incidents that occur in the school setting and throughout the education system. The framework provides guidance and direction to all public schools and is available to all private schools. The framework applies to students as well as to preschool age children participating in early learning programs. All school boards in British Columbia are expected to have a district policy in place that meets the requirements outlined in the *Anaphylaxis Protection* Ministerial Order and the British Columbia Anaphylactic and Child Safety Framework [55].

19.3.6.1.3. Manitoba. Manitoba proclaimed legislation in November 2009 to strengthen the protection for students with life-threatening allergies. Amendments to the Public Schools Act [56] and the Education Administration Act [57] that came into force on November 1, 2009 formalized the requirement that Manitoba school divisions have policies to protect students with life-threatening allergies. This legislation is subsequent to a 2002 provincial directive, which required that by 2004, all community programs, including school divisions, develop local policies for handling life-threatening allergies. In May 2002, a joint letter from three government ministers was sent to school divisions to highlight the finding of the Canadian Society of Allergy and Clinical

Immunology (CSACI) [52] that anaphylaxis was estimated to be a risk for between 1% and 2% of the general population. Since that time, it has been the practice of Manitoba school divisions to have in place anaphylaxis policies that are based on the policy framework outlined in the Manitoba Education, Citizenship and Youth Unified Referral and Intake System (URIS) manual.

19.3.6.1.4. New Brunswick. In 1999, the New Brunswick Department of Education issued a Health Support Services Policy [58] for anaphylaxis stating that "this policy defines standards and procedures required for the provision of health support services to students while they are the responsibility of the public education system, recognizing this responsibility is shared among parents, the public education system and health care providers." This policy was revised in October 2004 and again in December 2008.

19.3.6.1.5. Newfoundland and Labrador. In 2007, a document for informing schools about anaphylaxis [59] was issued by the Government of Newfoundland and Labrador. Topics addressed include how to develop avoidance strategies and what is expected in responding to an emergency situation. An individual support services plan (ISSP) should be in place for all students with life-threatening allergies. There is no provincial anaphylaxis policy.

19.3.6.1.6. Nova Scotia. In 2005, Nova Scotia developed a document for informing public schools about anaphylaxis [60]. It includes information on how to develop avoidance strategies and what can be expected when responding to an emergency situation.

19.3.6.1.7. Ontario. In May 2005, the Ontario government passed Bill 3: An Act to protect anaphylactic pupils, which affects all publicly funded schools in Ontario. Named "Sabrina's law" in honor of an Ontario student who died following an anaphylactic reaction in 2003, the law became effective on January 1, 2006. It requires that every school board establish and maintain an anaphylaxis policy. It also requires that principals develop individual plans for pupils at risk of anaphylaxis [61].

19.3.6.1.8. Prince Edward Island. In 2008, the Minister of Education issued a directive concerning Procedures for Dealing with Life-threatening Allergies. It stated that "the purpose of this Directive is to provide guidance to parents and school personnel concerning procedures for managing students who have life-threatening allergies and are at risk of anaphylaxis" [62].

19.3.6.1.9. Quebec. In Quebec, both private and public schools fall under the jurisdiction of the Ministère de l'Éducation, du Loisir et du Sport. The subsidized day care system (about 1400 day care centers) falls under the Ministère de la Famille et des Aînés and under a separate day care law [63]. Health issues in schools and day care centers are managed by a joint committee within the

Ministère de l'Éducation, du Loisir et du Sport, the Ministère de la Famille et des Aînés, and the Ministère de la Santé et Services Sociaux (Health and Social Services). In 1998, the AQAA published a special newsletter in both French and English that focused on the prevention and management of food allergy and anaphylaxis in schools and day care settings [64]. Information was based on the 1995 CSACI consensus statement [65]. Reviewed by AQAA medical advisors, the newsletter was distributed by public authorities to all elementary schools and day care associations in Quebec. Subsequently, each school board and regional day care association in Quebec has developed its own protocol. Most protocols are similar from region to region and outline the responsibilities of administrators, students, parents, teachers, and others, and provide treatment guidelines.

19.3.6.1.10. Saskatchewan, Northwest Territories, and Nunavut. Saskatchewan, Northwest Territories, and Nunavut have no provincial/territorial policy concerning anaphylaxis; individual schools are responsible for instituting their own policies.

The *Yukon Policy 4003—Administration of Medication to Students* of the *Education Act* [66] explains that "some students have severe and life-threatening illnesses and allergies which may lead to diabetic shock, anaphylaxis and chronic allergy conditions including asthma; there are times that school staff may need to administer medication and/or call for medical support. It is the responsibility of parents to complete and submit a completed copy of the *Administration of Medication Plan* form to the school administrator. This form outlines the roles and responsibilities of parents, school staff, administrators and the student. It also provides authorization to the school to administer the medical plan."

In Canada, only two provinces have mandated through regulation that school boards *must* have an anaphylaxis policy in place: Ontario did so in 2006, Manitoba in 2009. In practice, such plans are proposed by allergic consumer associations (such as Anaphylaxis Canada, Allergy/Asthma Information Association, or AQAA) with technical and medical support by medical associations such as CSACI. Board policies should be flexible enough to allow schools and classrooms to adapt to the needs of individual children and to account for differences in the organizational and physical environment of schools. Each school should develop its own written anaphylaxis plan, which is specific to its environment and complies with the board policy [52].

In a comparative analysis of Canadian school board anaphylaxis management policies published in February 2009, Cicutto et al. [67] reported that the one province with legislation in place at the time the analysis was conducted (i.e., Ontario) had school board policies that are potentially more supportive for managing life-threatening allergies at school [67]. There is some expectation that other Canadian provinces will follow the example of Ontario, and in November 2009, Manitoba enacted similar legislation, creating safe school environments for anaphylactic children.

19.3.6.2. Other International Policies. Strategies to protect children from anaphylaxis in school settings have also been implemented in countries that include the United States, the United Kingdom, and France. In the United States, there is no legislation addressing anaphylaxis in schools. Risk management strategies rely on voluntary cooperation between parents and school. In contrast, regulation has been in place in France since 1999. The United Kingdom has chosen a middle ground between voluntary and legislative initiatives.

In the United States, federal law requires that all children receive a "free and appropriate public education." This means that schools must provide a safe environment for students with life-threatening food allergies. While there were no national food allergy/anaphylaxis guidelines available at the time this chapter was written, many U.S. schools had adopted policies based on the experience and best practices of other schools throughout the country. Practically speaking, parents have to check with their state Department of Education to find out if statewide food allergy management guidelines are in place for public schools. Next, they should determine whether their school or school board has established their own policies, even if there are no statewide guidelines. Most schools want to work with parents to develop a plan that will protect children against severe allergy incidents/anaphylaxis incidents. The school should have an Individualized Health Care Plan (IHP) and an Individualized Emergency Care Plan (IECP) for every student with a life-threatening food allergy. As a last resort, if the school is not cooperative, parents could possibly investigate Section 504 of the Rehabilitation Act of 1973 [68]. Under this law, the school should not be able to refuse admission to a child because he or she has a food allergy. Further, the school should provide services and modify programs to ensure that the child can safely participate in all school activities [69].

In the United Kingdom, the document *Managing Medicines in Schools and Early Years Settings* was published by the Department for Education and Skills and Department of Health in 2005 [70]. It sets out a framework within which local authorities, local health trusts, schools, and early years settings can work together to develop policies to ensure that children requiring medication receive appropriate support. This guidance was designed to help all schools and all early years settings and their employers support individual children with medical needs including food allergies. A range of issues are discussed in the guidance document, including medication policy; role and responsibilities of parents, employer, teachers, and other staff; safety management; healthcare plan; common conditions; and legal framework [71].

In addition, a document *Medical Conditions at School: A Policy Resource Pack* has been compiled by the Medical Conditions at School Group (a group of charities including Asthma UK and The Anaphylaxis Campaign) to complement *Managing Medicines in Schools and Early Years Settings*.

In France, the enactment of safety regulations for food-allergic children in the school setting (public and private) by the French Ministry of Education began in 1993. These regulations were completed and in effect by 1999. They recommended the development of a "Projet d'Accueil Individualisé" document (PAI)—a personalized care project. Practically speaking, schools cannot refuse to register a child with a PAI document (i.e., with a severe anaphylaxis). This document provides information about the child's food allergy or allergies, including their severity and associated risks. It also provides information on symptoms displayed by the child at the beginning of a severe allergic reaction and the specific steps to be taken in case of emergency. Compilation of this type of preventive information requires collaboration between the allergist who establishes the diagnosis, the school physician, teaching staff, parents, and, depending on age, the child [72].

The PAI approach also includes written information about the signs of an allergic emergency and a protocol for use of (1) injectable epinephrine by the intramuscular route, (2) corticosteroids by mouth, and (3) β-agonists by inhalation (as appropriate). This protocol is drafted by the child's allergist or pediatrician. Subsequently, the request for implementation of the management plan is made by the family to the public health physician for the school. The management plan is countersigned by the school physician and the parents, for implementation by school officials [72].

Prior to this 1999 regulation, there was little use of PAIs in French schools. Since then, there has been a significant increase of the use of PAIs for a range of childhood conditions including, for example, diabetes and non-food-related asthma, in addition to severe food allergy. In 2006, 45% of French schools used a PAI [73].

19.4. DIETARY ADVICE DURING PREGNANCY AND LACTATION

The goal of primary prevention in the field of food allergy is to lower the incidence of allergy in the general population by avoiding sensitization of people with a high risk of allergy, for example, those with a family history of atopy. Until the late 2000s, recommendations for the prevention of atopic disease in childhood were based on an antigen avoidance diet during pregnancy and/or lactation. As well, for children with a family history of allergy, the recommendation was to delay the introduction of potentially allergenic foods such as peanuts, tree nuts, and exotic fruits, sometimes until 3 years of age [74]. For children with no family history of allergy, it was recommended that the introduction of solids be delayed until at least 6 months of age [75].

In 1998, following the advice from the Committee on Toxicity of Chemicals in Food (COT), the UK-FSA recommended that "women with a family history of allergy (herself or father or any sibling of the unborn child) *may wish to avoid* eating peanuts and peanut products during pregnancy and breast-feeding and not introduce peanuts into their child's diet before 3 years of age" [76]. In 2008, COT [77] indicated that the available evidence did not indicate clearly whether maternal consumption of peanut during pregnancy or lactation was more likely to increase or decrease the risk of a child's sensitization and allergy to peanut. The UK-FSA subsequently revised their recommendation as follows: "women who would like to eat peanuts or foods containing peanuts during pregnancy and breast-feeding *can choose to do so* as part of a healthy balanced diet, irrespective of whether their child has a family history of allergies. Women should try to exclusively breast-feed their baby for the first 6 months of life. If mothers choose to start giving their baby solid foods before 6 months, they should not introduce peanuts or allergenic foods (such as other nuts, seeds, milk, eggs, wheat, fish, or shellfish) before this time" [40].

This official position of the United Kingdom is consistent with international scientific recommendations published since 2007 concerning exposure to major allergens (not exclusively peanuts) during fetal life and early childhood. For example, until 2004, the recommendations of the American Academy of Pediatrics (AAP) were similar to those of the UK-FSA: to avoid peanut during pregnancy and lactation and to delay introduction of peanut to the diet until 3 years of age. However, in a position statement published in 2008, the AAP indicated there was a lack of evidence regarding whether maternal dietary restrictions during pregnancy play a significant role in the prevention of atopic disease in infants. While noting that more data are needed, they also indicated that it appears that antigen avoidance during lactation does not prevent atopic disease, with the possible exception of atopic eczema. Although solid foods should not be introduced before 4–6 months of age, at the present time, there is no convincing evidence that delaying solids (including foods considered to be highly allergic, such as fish, eggs, and foods containing peanut protein) beyond 4–6 months has a significant protective effect against the development of atopic disease [78].

In 2008, the European Society for Pediatric, Gastroenterology, Hepatology and Nutrition (ESPGHAN) concluded that there was no convincing scientific evidence that avoidance or delayed introduction of potentially allergenic foods reduces allergies, either in infants considering at increased risk for the development of allergy or in those not considered to be at increased risk. They indicated that (1) exclusive or full breast-feeding for about 6 months is a desirable goal and (2) complementary feeding (i.e., solid foods and liquids other than breast milk or infant formula and follow-on formula) should not be introduced before 17 weeks (4 months) or after 26 weeks (6 months) [79].

Also in 2008, the section on Pediatrics, European Academy of Allergology and Clinical Immunology [80] concluded that there is no conclusive evidence for a protective effect of a maternal exclusion diet during pregnancy or lactation. They also indicated that the most effective dietary regimen for primary prevention of food allergies is exclusive breast-feeding for at least 4–6 months, or if breast milk is not available, the use of infant formulas with documented reduced allergenicity for at least 4 months, combined with avoidance of solid foods and cow's milk during the same period [80].

In 1999, Canada published the *National Guidelines for the Childbearing Years*. This guideline indicated that "until the efficacy of a restricted diet during pregnancy and lactation is known, routinely restricting the eating pattern of mother of infants at risk for allergy is not recommended. This could be attempted only when a previous offspring is severely allergic and the mother is strongly motivated to prevent sensitizing her unborn child." In 2005, the Canadian Paediatric Society, Dietitians of Canada, and Health Canada published an opinion on nutrition for healthy term infants. There were two sections to this document; both are available on the Health Canada website.

The first section [81] addresses breast-feeding and indicates the following: "breast-feeding is the optimal method of feeding infants. Encourage exclusive breast-feeding for the first 6 months of life, as breast milk is the best food for optimal growth. Breast-feeding may continue for up to 2 years and beyond."

The second section [82] addresses the introduction of solid food and indicates the following: "introduce nutrient-rich complementary foods at 6 months to meet the infant's increasing nutritional requirements and developmental needs. To prevent iron deficiency, iron-containing foods are recommended as the first foods."

Between 2003 and 2009, a number of authors concluded that eliminating food allergens (particularly peanuts) during pregnancy, lactation, and infancy has failed to prevent IgE-mediated food allergy [75, 83, 84]. In terms of primary prevention of food allergy, it appears that rather than lowering the allergic burden by drastically limiting exposure to food allergens, it may be more effective to induce tolerance to food allergens [85]. However, because a number of factors, including antigen properties, route of exposure, genetics, and age of the host, contribute to the development of oral tolerance [86], each infant should be considered as unique. The final decision should be made by the physician in consultation with the parents.

In 2010, dietary advice to women who are pregnant or breast-feeding can include the following:

1. Allergen avoidance during pregnancy and lactation is not recommended.
2. The most effective dietary regimen for preventing food allergy is exclusive breast-feeding for at least 4–6 months.
3. Although solid foods should not be introduced before 4–6 months of age, there is no scientific reason to further delay their introduction.

As of 2010 a number of important knowledge gaps remain in this area: should solid food be introduced at 4 months or 6 months of age? Should this age be different for infants with and without a family history of food allergy? In terms of food allergy prevention, is there a specific order in which different types of solid food should be introduced? Until what age should exclusive breast-feeding be encouraged? Is there a "best way" to introduce foods with a high potential of allergenicity (e.g., peanuts)? At what age should such foods be introduced?

19.5. FOOD ALLERGY INCIDENT SURVEILLANCE AND REPORTING SYSTEM

Efforts have been made by several groups to establish reporting networks for "severe" food allergy incidents. These incidents are defined as encompassing reactions such as anaphylactic shock, laryngeal angioedema, severe asthma, or a systemic reaction requiring emergency medical treatment. Incident reporting is done by healthcare professionals (ideally physicians) with a specialization or a strong background in treating allergies. Declarations are sent to a coordination center, where the information received is validated by a team of physicians with an expertise in food allergy. It is then entered in a database.

The information in the reporting declaration for each incident must be sufficient to

1. allow for confirmation of the existence of a food allergy (diagnostic elements),
2. provide information concerning patient care (treatment), and
3. facilitate subsequent access to information concerning the food/ingredient that the healthcare professional suspects to be the cause of the reaction.

This section discusses several examples of food allergy incident reporting systems currently in operation. It also reviews the potential benefits of such systems in supporting food allergy incident management and prevention. Finally, it provides an overview of a proposed Canadian network in early stages of development.

19.5.1. French Food Allergy Surveillance Network (Réseau d'allergovigilance)

In 2001, the "Réseau d'allergovigilance" was created by French specialists in Internal Medicine, Clinical Immunology and Allergy at the Centre Hospitalier Universitaire of Nancy (Lorraine, France). Notably, this network is supported by the Agence Française de Sécurité Sanitaire des Aliments (AFSSA)—the agency responsible for conducting food risk assessments in France. The primary goal of this network is to collect data about severe food-related allergic incidents, in support of public health decision making.

Membership in the "Réseau d'allergovigilance" is free and is accessible to all physicians in France with an expertise in clinical allergy. Network coordination is entrusted to three allergy residents (supervised by senior allergists) who review the completeness and accuracy of all submitted incident declarations and one analyst, who is responsible for the quality of electronic data transmission [87].

In January 2009, this network covered approximately 85% of France and included 425 allergists, or approximately one-third of all physicians practicing in France with an expertise in clinical allergy. Forty percent (40%) of network physicians practice in hospitals, while 60% are in private practice (i.e., they operate private medical clinics).

The "Réseau d'allergovigilance" has three goals:

1. To facilitate the reporting of severe cases of food-related incidents, in which there is anaphylactic shock or laryngeal angioedema or acute asthma, or a serious systemic reaction requiring emergency medical treatment;
2. To gather and disseminate information pertaining to new risks (new foods or hidden allergens and any emerging allergen including those resulting from genetically modified organisms [GMOs], once these food products are marketed);
3. To participate in multicenter studies estimating the prevalence of rare or poorly understood food sensitivities especially if these are likely to pose a significant health risk [88].

Physicians who are members of the network can download incident declaration forms from the network website (www.cicbaa.com). Information that must be provided in each declaration includes how the information was gathered (e.g., in a hospital emergency department), age, gender, clinical aspects, suspected foods, emergency measures taken, allergy record (skin tests, bioassays, challenges, etc.), concomitant physical activity or physical exertion, and concomitant use of medications. Descriptive information concerning the food believed responsible for the incident (e.g., the brand) is also requested. The data are reported anonymously, and the patient's identity is known only to the physician who submits the declaration. The information in every declaration is verified by the coordinating physicians. As necessary, further information is requested from the declaring physician (by telephone or *e-mail*).

Information on each reported incident is sent to all members of the network by *e-mail* within 1 week. Summary information is published every second month in a national publication called *Alim'Inter*. Overall statistics are distributed once or twice a year to all members of the network [89].

Benefits associated with the use of such an incident surveillance system include the following:

- The mechanism for submitting a declaration is considered *simple*: the objectives of the network, as well as its operation, are easily understood by individuals who are not public health specialists.
- The system is *acceptable* to the declaring physician, because fortunately, severe food allergy incidents are quite rare, constituting only a small component of a physician's professional practice. Thus, the work involved in reporting does not represent an excessive physician burden.

- The system is *flexible*: replacement physicians can continue to submit declarations when the declaring physician is absent (illness, vacation, etc.). No specific training is required to use the system.
- Implementation of this system is *inexpensive*: declarations are submitted on a voluntary basis by e-mail, and physicians are not remunerated. Costs are associated only with the remuneration of a database manager and acquisition of basic computer equipment with Internet access. The coordinating physicians who validate the reported incidents are salaried public service physicians.
- As with any surveillance system, one of the most important characteristics of the network is its significant *reactivity*. It allows for the rapid retrieval of information if a member physician deems this to be necessary, and it allows for the rapid notification of all members of the network (e.g., in the event of a warning pertaining to improperly labeled food product).
- It is an *ongoing information gathering* system, which makes it possible to conduct selected cross-sectional investigations, insofar as the proposed study includes a limited number of questions (in order to not overburden the declaring physicians).
- One of the network's assets is the *quality of data* collected (validated by the coordinating physicians and including a telephone call or an e-mail request for clarification if necessary).
- It is easy to ensure a *feedback* for data collected by means of an annual statement that is sent to the members of the network by e-mail.

Some of the limitations of this network relate to the fact that it is voluntary, hence, based on the willingness of declaring physicians to participate. The system is also not considered representative of allergists practicing in France. Some regions (Corsica, Ardennes, etc.) are underrepresented, while others are overrepresented (e.g., Lorraine, where the coordination center is located). Because it is often the same physicians (the most motivated/involved) who submit declarations, there may be a significant *selection bias* in the allergy incidents reported to the network.

In light of these limitations, data collected through this network cannot be used to estimate the incidence or the prevalence of severe food allergy incidents at the national level. As well, even though it is a continuous reporting system, for the reasons discussed above, the network does not allow for the reliable identification of time trends. This is because its priority is to focus specifically on severe food anaphylaxis incidents. On the other hand, its ability to rapidly disseminate information may enhance the effectiveness of interventions intended to prevent additional severe food allergy incidents associated with a reported food or food ingredient. This ability suggests that the network has a potentially important role as a public health tool.

19.5.2. Other Food Allergy Incident Surveillance Networks

Similar food allergy incident surveillance networks have been introduced in Sweden [90], Norway [91], and Germany. The latter consists of a unified network of German language partners (including Austria and part of Switzerland) [92, 93].

Although some of these networks are quite new, and there are some differences between them (e.g., in terms of organization, limitations, benefits), they are all designed to contribute to the documentation of severe food allergic incidents. Discussions are currently under way to attempt to standardize the operation across these networks, as well as to lay the foundation for a single European food anaphylaxis network.

19.5.3. Options for a Canadian Severe Food Allergy Incident Surveillance Network

Health Canada is currently investigating options for the development of a Canadian severe food allergy incident reporting network. Anticipated benefits of such a network include many of those discussed above as well as the following:

1. Collection of information specific to Canadians and the Canadian diet. Whereas many cases of food allergy in Western countries are associated with eight food categories (milk, eggs, peanuts, nuts, wheat, fish, crustaceans, soy), the specific food habits of a population are also an important determinant of the food allergies they experience. It is recognized that regional dietary habits determine the food allergies observed more frequently; for example, rice allergies are more frequent in Japan, tomato allergies are more prevalent in Italy, and fish allergies are more common in Scandinavia. Reactions resulting from the consumption of peanuts and/or nuts are particularly common in the United States.

   Since the focus would be severe food allergy incidents, information collected by this network could contribute to (1) the identification of factors associated with the emergence of new food allergens (e.g., allergenicity of "novel" food products, increasing industrial use of an ingredient that was not commonly used in the past); (2) national and international efforts to increase understanding of the etiology of severe allergic reactions to food; and (3) efforts to identify potential opportunities for preventive intervention in occurrence of an allergy incident and the exposure to a hidden allergen in the context of cross-contamination or errors in product labeling. Information can be conveyed very quickly over a network, and, therefore, larger-scale elimination measures would be possible within a very short time frame.
2. If it is implemented from an epidemiological perspective (i.e., organized on a representative basis), this system would allow access to descriptive

information about food allergies in Canada. It could also support targeted cross-sectional investigations, such as those involving a specific allergen or a specific risk management approach to prevention of severe food allergy incidents. Investigations could also focus on specific regions of the country and/or specific population subgroups.

3. Development of such a surveillance network can be expected to provide the opportunity to create an interface across physicians, hospitals, and government organizations responsible for the prevention of severe food allergy incidents.
4. A Canadian surveillance network that focuses on severe allergy incidents would make it possible to gather information about current Canadian procedures for the care and treatment of patients experiencing severe anaphylaxis. It could also contribute to the standardization of diagnostic procedures for severe food-related incidents across Canada.

For all of these reasons, a Canadian Network for Reporting Severe Food Allergy Incidents can be expected to make a significant contribution to the establishment of a relevant, valid evidence base in support of public health and clinical decision making for the prevention of severe food allergy incidents among the Canadian population.

There are also a number challenges associated with the development of a Canadian Allergy Incident Reporting Network. These include the following:

- *Representativeness.* The main limitation of the French "Réseau d'allergovigilance" is the lack of representativeness of the collected data. This suggests that it will be important to consider a Canadian system that is more representative, for example, a network of sentinel physicians. An important consideration will be how to ensure that the number of sentinel physicians participating in the network is sufficient to allow for the reporting of severe incidents that actually occur quite rarely.
- *Cost.* The French system "Réseau d'allergovigilance" is based on the voluntary participation of physicians with an expertise in clinical allergy. The physician coordinators are salaried French hospital employees (at Centre Hospitalier Universitaire de Nancy). For these reasons, the French network is inexpensive to operate. Although the Canadian healthcare system has some elements in common with the French system (notably free access to health care), in most Canadian provinces, physician remuneration is based exclusively on a fee-for-service system. As a result, remuneration for expert physician coordinators would be a network cost, making the operation of a Canadian network more costly than is the case for the French network.
- *Organization.* Canada is a federation in which healthcare delivery is the responsibility of provincial/territorial governments. Each province and territory has specific characteristics in terms of population density, medical

density, and eating habits. Consequently, it may be appropriate to organize the network on a province-by-province basis rather than nationally (centralization of collected data, as is the case in France).

In summary, there are many potential benefits to be derived from the development and implementation of a Canadian severe food allergy incident reporting network. The data generated by this network can be expected to assist clinical experts and policy makers in the development of risk management strategies to protect food-allergic consumers and prevent severe food allergy incidents. It can also be expected to contribute to subsequent evaluation of the effectiveness of those risk management strategies selected for implementation.

19.6. CONCLUSION

For those with food allergies, sensitivities, or intolerances, avoiding specific foods and ingredients is an important health challenge. An allergic individual coming into contact with an undeclared allergen in a food product may have symptoms that develop quickly and rapidly progress from mild to severe, including anaphylactic shock and death. Several national food regulators have developed regulation or legislation with the aim of strengthening allergen labeling requirements and ensuring that the most common food and food ingredients, which can cause life-threatening or severe allergic reactions (as well as gluten sources), are always identified by their common names so that consumers can easily recognize them on food labels. Efforts are still under way to enact these rules and to develop the appropriate enforcement and compliance policies. Analytical detection methods for food allergens will constitute a pillar for the implementation of control strategies, both for industry to comply with such requirements and for government to support its enforcement strategies. Various initiatives are under way to enable improved coordination of efforts in harmonizing labeling requirements and their implementation internationally, where possible [94].

REFERENCES

1. Johansson, S.G., Hourihane, J.O., Bousquet, J., Bruijnzeel-Koomen, C., Dreborg, S., Haahtela, T., Kowalski, M.L., Mygind, N., Ring, J., van Cauwenberge, P., van Hage-Hamsten, M., Wüthrich, B. (2001). A revised nomenclature for allergy. An EAACI position statement from the EAACI nomenclature task force. *Allergy*, *56* (9), 813–824.
2. Johansson, S.G., Bieber, T., Dahl, R., Friedmann, P.S., Lanier, B.Q., Lockey, R.F., Motala, C., Ortega Martell, J.A., Platts-Mills, T.A., Ring, J., Thien, F., Van Cauwenberge, P., Williams, H.C. (2004). Revised nomenclature for allergy for global

use: report of the Nomenclature Review Committee of the World Allergy Organization, October 2003. *The Journal of Allergy and Clinical Immunology, 113* (5), 832–836.

3. Taylor, S.L. (1999). Prospects for the future: emerging problems—food allergens. Available at http://www.fao.org/docrep/meeting/X2670e.htm
4. Briani, C., Samaroo, D., Alaedini, A. (2008). Celiac disease: from gluten to autoimmunity. *Autoimmunity Reviews, 7* (8), 644–650.
5. Kagnoff, M.F. (2007). Celiac disease: pathogenesis of a model immunogenetic disease. *The Journal of Clinical Investigation, 117* (1), 41–49.
6. Zarkadas, M., Scott, F.W., Salminen, J., Ham Pong, A. (1999). Common allergenic foods and their labelling in Canada—a review. *Canadian Journal of Allergy and Clinical Immunology, 4* (3), 118–141.
7. Presutti, R.J., Cangemi, J.R., Cassidy, H.D., Hill, D.A. (2007). Celiac disease. *American Family Physician, 76* (12), 1795–1802.
8. Panconesi, A. (2008). Alcohol and migraine: trigger factor, consumption, mechanisms. A review. *The Journal of Headache and Pain, 9* (1), 19–27.
9. Simon, R.A. (1996). Adverse reactions to food and drug additives. *Immunology and Allergy Clinics of North America, 16* (1), 137–176.
10. Vally, H., Thompson, P.J. (2001). Role of sulfite additives in wine induced asthma: single dose and cumulative dose studies. *Thorax, 56* (10), 763–769.
11. Tarlo, S.M., Sussman, G.L. (1993). Asthma and anaphylactoid reactions to food additives. *Canadian Family Physician Medecin de Famille Canadien, 39*, 1119–1123.
12. Yang, W.H., Purchase, E.C. (1985). Adverse reactions to sulfites. *Canadian Medical Association Journal, 133* (9), 865–867, 880.
13. Grundy, J., Matthews, S., Bateman, B., Dean, T., Arshad, S.H. (2002). Rising prevalence of allergy to peanut in children: data from 2 sequential cohorts. *The Journal of Allergy and Clinical Immunology, 110* (5), 784–789.
14. Sicherer, S.H., Muñoz-Furlong, A., Sampson, H.A. (2003). Prevalence of peanut and tree nut allergy in the United States determined by means of a random digit dial telephone survey: a 5-year follow-up study. *The Journal of Allergy and Clinical Immunology, 112* (6), 1203–1207.
15. de Blok, B.M.J., Dubois, A.E.J., Hourihane, J.O. (2007). Impact of food allergies on quality of life. In: *Managing Allergens in Food*, Mills, C., Wichers, H., Hoffmann-Sommergruber, K. (Eds.). Cambridge, UK: Woodhead Publishing, pp. 47–61.
16. Fernández-Rivas, M., Ballmer-Weber, B. (2007). Food allergy: diagnosis and management. In: *Managing Allergens in Food*, Mills, C., Wichers, H., Hoffmann-Sommergruber, K. (Eds.). Cambridge, UK: Woodhead Publishing, pp. 3–28.
17. Salter, J., Mehra, S., Cairns, J.T., Sussman, G., Vadas, P. (2003). A study of 32 food-induced anaphylaxis deaths in Ontario; 1986-2000. Available at http://www.anaphylaxis.org/content/programs/programs_research_deaths.asp
18. National Institutes of Health. (2005). Consensus Development Conference Statement on Celiac Disease, June 28–30, 2004. *Gastroenterology, 128* (4 Suppl. 1), s1–s9.
19. Frisk, G., Hansson, T., Dahlbom, I., Tuvemo, T. (2008). A unifying hypothesis on the development of type 1 diabetes and celiac disease: gluten consumption may be a shared causative factor. *Medical Hypotheses, 70* (6), 1207–1209.

20. Catassi, C., Kryszak, D., Louis-Jacques, O., Duerksen, D.R., Hill, I., Crowe, S.E., Brown, A.R., Procaccini, N.J., Wonderly, B.A., Hartley, P., Moreci, J., Bennett, N., Horvath, K., Burk, M., Fasano, A. (2007). Detection of celiac disease in primary care: a multicenter case-finding study in North America. *The American Journal of Gastroenterology*, *102* (7), 1454–1460.
21. Zarkadas, M., Cranney, A., Case, S., Molloy, M., Switzer, C., Graham, I.D., Butzner, J.D., Rashid, M., Warren, R.E., Burrows, V. (2006). The impact of a gluten-free diet on adults with coeliac disease: results of a national survey. *Journal of Human Nutrition and Dietetics*, *19* (1), 41–49.
22. ASCIA. (2007). *The Economic Impact of Allergic Disease in Australia: Not to Be Sneezed At*. Access Economics Pty Limited (Eds.). Australia: Australasian Society of Clinical Immunology and Allergy (ASCIA).
23. Mills, E.N., Mackie, A.R., Burney, P., Beyer, K., Frewer, L., Madsen, C., Botjes, E., Crevel, R.W., van Ree, R. (2007). The prevalence, cost and basis of food allergy across Europe. *Allergy*, *62* (7), 717–722.
24. Codex Alimentarius Commission (CAC). (1999). Prospects for the future: emerging problems—food allergens. Available at http://www.fao.org/docrep/meeting/X2670e.htm
25. Health Canada. (2008). Health Canada urges food manufacturers to label priority food allergens, gluten sources and added sulphites in the pre-publication period of the Food Allergen Labelling Regulatory Amendments. Available at http://www.hc-sc.gc.ca/fn-an/label-etiquet/allergen/guide_ligne_direct_indust-eng.php
26. Health Canada. (2007). The use of food allergen precautionary statements on prepackaged foods. Available at http://www.hc-sc.gc.ca/fn-an/label-etiquet/allergen/precaution_label-etiquette-eng.php
27. Food and Consumer Products Manufacturers of Canada (FCPMC). (1997). Survey of food industry on use of allergen labelling.
28. U.S. Food and Drug Administration (FDA). (2009). Food Allergen Labeling and Consumer Protection Act of 2004 (Public Law 108-282, Title II). Available at http://www.fda.gov/Food/LabelingNutrition/FoodAllergensLabeling/GuidanceComplianceRegulatoryInformation/ucm106187.htm
29. Hourihane, J.O., Kilburn, S.A., Nordlee, J.A., Hefle, S.L., Taylor, S.L., Warner, J.O. (1997). An evaluation of the sensitivity of subjects with peanut allergy to very low doses of peanut protein: a randomized, double-blind, placebo-controlled food challenge study. *The Journal of Allergy and Clinical Immunology*, *100* (5), 596–600.
30. Pieretti, M.M., Pacenza, R., Slotkin, T., Sicherer, S.H. (2008). Frequency and language used in allergens advisory labels ("may contain"): a survey of 20,241 commercially available products. *The Journal of Allergy and Clinical Immunology*, February, s182.
31. Hefle, S.L., Furlong, T.J., Niemann, L., Lemon-Mule, H., Sicherer, S., Taylor, S.L. (2007). Consumer attitudes and risks associated with packaged foods having advisory labeling regarding the presence of peanuts. *The Journal of Allergy and Clinical Immunology*, *120* (1), 171–176.
32. Sheth, S.S., Waserman, S., Kagan, R., Alizadehfar, R., Primeau, M., Elliot, S., St Pierre, Y., Turnbu, E., Wickett, R., Joseph, L., Harada, L., Dufresne, C., Allan, M., Allen, M., Godefroy, S., Clarke, A. (2008). Improving food labeling for the allergic consumer. *The Journal of Allergy and Clinical Immunology*, *121* (2 Suppl. 1), s183.

33. Fenton, N.E., Elliott, S.J. (2008). *Consumers Response to Precautionary Statements on Food Labels*. Hamilton, ON, Canada: McMaster University Publishing.

34. Sampson, M.A., Muñoz-Furlong, A., Sicherer, S.H. (2006). Risk-taking and coping strategies of adolescents and young adults with food allergy. *The Journal of Allergy and Clinical Immunology*, *117* (6), 1440–1445.

35. Akeson, N., Worth, A., Sheikh, A. (2007). The psychosocial impact of anaphylaxis on young people and their parents. *Journal of the British Society for Allergy and Clinical Immunology*, *37* (8), 1213–1220.

36. Derr, L.E. (2006). When food is poison: the history, consequences, and limitations of the Food Allergen Labeling and Consumer Protection Act of 2004. *Food and Drug Law Journal*, *61* (1), 65–165.

37. Allergen Bureau. (2009). What is VITAL? Available at http://www.allergenbureau.net/vital/

38. Minister of Justice. (2009). *Canadian Food and Drug Regulations*. Available at http://laws.justice.gc.ca/PDF/Regulation/C/C.R.C.,_c._870.pdf.

39. Pulido, O.M., Gillespie, Z., Zarkadas, M., Dubois, S., Vavasour, E., Rashid, M., Switzer, C., Godefroy, S.B. (2009). Introduction of oats in the diet of individuals with celiac disease: a systematic review. *Advances in Food and Nutrition Research*, *57*, 235–285.

40. Food Standards Agency (UK-FSA). (2008). *The Provision of Allergen Information for Non Pre-Packed Foods*. Food Standard Agency Publication FSA/1226/0108.

41. European Commission MEMO. (2008). Questions and answers on food labelling. Available at http://europa.eu/rapid/pressReleasesAction.do?reference=MEMO/08/64

42. Hefle, S.L., Taylor, S.L. (2002). How much food is too much? Threshold doses for allergenic foods. *Current Allergy and Asthma Reports*, *2* (1), 63–66.

43. Furlong, T.J., DeSimone, J., Sicherer, S.H. (2001). Peanut and tree nut allergic reactions in restaurants and other food establishments. *The Journal of Allergy and Clinical Immunology*, *108* (5), 867–870.

44. Association Québécoise des Allergies Alimentaires (AQAA). (2008). *Food Allergy Management Manual for Restaurants and Food Services*. AQAA and Health Canada Publishing, p. 132.

45. New Zealand Food Safety Authority (NZSA). (2009). *Food Control Plan*. Wellington, New Zealand: New Zealand Food Safety Authority.

46. Ahuja, R., Sicherer, S.H. (2007). Food-allergy management from the perspective of restaurant and food establishment personnel. *Annals of Allergy, Asthma and Immunology*, *98* (4), 344–348.

47. Food Allergy Initiative. (2009). Living with food allergies; eating out. Available at http://www.faiusa.org/?page_ID=7A46E761-D841-0FE4-841B5BC5853D3135

48. Food Allergy & Anaphylaxis Network (FAAN). (2008). *Welcoming Guests with Food Allergies*, Vol. 2. Fairfax, VA: Food Allergy & Anaphylaxis Network.

49. American Academy of Allergy, Asthma and Immunology (AAAAI) Board of Directors. (1998). Anaphylaxis in schools and other childcare settings. *The Journal of Allergy and Clinical Immunology*, *102* (2), 173–176.

50. Nowak-Wegrzyn, A., Isenberg, H., Wood, R.A. (2000). Allergic reactions to foods in the school. *The Journal of Allergy and Clinical Immunology*, *105* (Suppl. 182), 549.

51. Furlong, T.J., Desimone, J., Sicherer, S.H. (2000). Peanut allergic reactions in schools. *The Journal of Allergy and Clinical Immunology*, *105* (Suppl. 182), 550.
52. Canadian Society of Allergy and Clinical Immunology. (2005). Anaphylaxis policies (provinces, territories & school boards). In: *Anaphylaxis in Schools & Other Settings*, pp. 27–30. Canadian Society of Allergy and Clinical Immunology. Available at http://www.allergysafecommunities.ca/assets/anaphylaxisguidelines2ndedition_appendixa%20.
53. Government of Alberta. (2009). Allergy and Anaphylaxis Informational Response (AAIR). Available at http://education.alberta.ca/admin/healthandsafety/aair.aspx
54. British Columbia Minister of Education. (2007). *Anaphylaxis Protection Order*. Victoria, B.C., Canada: Minister of Education.
55. British Columbia Minister of Education. (2007). *British Columbia Anaphylactic and Child Safety Framework*. Victoria, B.C., Canada: Minister of Education.
56. Government of Manitoba. (2009). The Public Schools Act. s. 1 to 58 (parts 1 to 3) and s. 58.1 to 279 (parts 3.1 to 14). Available at http://web2.gov.mb.ca/laws/statutes/ccsm/p250e.php
57. Government of Manitoba. (2009). The Education Administration Act. C.C.S.M. c. E10. Available at http://web2.gov.mb.ca/laws/statutes/ccsm/e010e.php
58. Government of New Brunswick—Department of Education. (1999). *Health Support Services*. Fredericton, NB, Canada: Department of Education.
59. Government of Newfoundland and Labrador—Department of Education. (2007). *Anaphylaxis Policy & Guidelines*. St. John's, NL, Canada: Department of Education.
60. Government of Nova Scotia—Education and Health Promotion. (2005). *Food and Nutrition Policy for Nova Scotia Schools*. Halifax, NS, Canada: Nova Scotia Education and Health Promotion Department.
61. Government of Ontario—Ministry of Education. (2005). Chapter 7. In: *Sabrina's Law*, Available at http://www.e-laws.gov.on.ca/html/statutes/english/elaws_statutes_05s07_e.htm.
62. Prince Edward Island Department of Education and Department of Health. (2006). *Information Handbook on Anaphylaxis*. Charlottetown, PEI, Canada: Department of Education and Department of Health.
63. Gouvernment du Québec. (2007). *Loi sur les services de garde éducatifs à l'enfance*. Québec, QC, Canada: Publication du Ministère de la Famille et des Ainés du Québec.
64. Association Québécoise des Allergies Alimentaires (AQAA). (1998). Prévention des allergies alimentaires à l'école et en service de garde. *Les Mets Sages*, *9* (s-2), 1–8.
65. Gold, M., Sussman, G., Loubser, M. (1996). Anaphylaxis in schools and child care settings. *Canadian Journal of Allergy and Clinical Immunology*, *1*, 4–19.
66. Yukon Education. (2006). *Administration of Medication to Students Policy 4003*. Yukon Education Publications, Records File 3500-34-003.
67. Cicutto, L., Julien, B., Butler, J., Clarke, A., Elliott, S.J., Harada, L., McGhan, S., Nguyen-Luu, N., Stark, D., Vander Leek, T.K., Waserman, S. (2009). A comparative analysis of Canadian school board anaphylaxis management policies. *The Journal of Allergy and Clinical Immunology*, *123* (2 Suppl. 1), s212.

68. U.S. Department of Health and Human Services. (1973). *Your Rights Under Section 504 of the Rehabilitation Act*.

69. Food Allergy Initiative. (2009). Living with food allergies; in school. Available at http://www.faiusa.org/?page=in_school

70. Department of Health. (2005). *Managing Medicines in Schools and Early Years Settings*. Department of Health Publications, UK.

71. The Anaphylaxis Campaign. (2007). *Helping People with Severe Allergies Live Their Lives, Guidance for Schools—Annual Report*. UK.

72. Moneret-Vautrin, D.A., Kanny, G., Morisset, M., Flabbee, J., Guénard, L., Beaudouin, E., Parisot, L. (2001). Food anaphylaxis in schools: evaluation of the management plan and the efficiency of the emergency kit. *Allergy*, *56* (11), 1071–1076.

73. Agence Française de Sécurité Sanitaire des Aliments (AFSSA). (2007). Évaluation de la connaissance et de l'application de la circulaire du 25 juin 2001 relative à la composition des repas servis en restauration scolaire et à la sécurité des aliments dans les établissements publics du second degré (2005–2006). Maisons-Alfort, France: Publications de l'Afssa.

74. Zeiger, R.S. (2003). Food allergen avoidance in the prevention of food allergy in infants and children. *Pediatrics*, *111* (6 Pt 3), 1662–1671.

75. Tarini, B.A., Carroll, A.E., Sox, C.M., Christakis, D.A. (2006). Systematic review of the relationship between early introduction of solid foods to infants and the development of allergic disease. *Archives of Pediatrics & Adolescent Medicine*, *160* (5), 502–507.

76. Committee of the Toxicity of Chemicals in Food, Consumer Products and the Environment (COT). (1998). *COT Report on Peanut Allergy*. UK.

77. Committee of the Toxicity of Chemicals in Food, Consumer Products and the Environment (COT). (2008). *Review of COT Recommendations on Peanut Avoidance*. UK.

78. Greer, F.R., Sicherer, S.H., Burks, A.W., American Academy of Pediatrics Committee on Nutrition, American Academy of Pediatrics Section on Allergy and Immunology. (2008). Effects of early nutritional interventions on the development of atopic disease in infants and children: the role of maternal dietary restriction, breastfeeding, timing of introduction of complementary foods, and hydrolyzed formulas. *Pediatrics*, *121* (1), 183–191.

79. Agostoni, C., Decsi, T., Fewtrell, M., Goulet, O., Kolacek, S., Koletzko, B., Michaelsen, K.F., Moreno, L., Puntis, J., Rigo, J., Shamir, R., Szajewska, H., Turck, D., van Goudoever, J., ESPGHAN Committee on Nutrition. (2008). Complementary feeding: a commentary by the ESPGHAN Committee on Nutrition. *Journal of Pediatric Gastroenterology and Nutrition*, *46* (1), 99–110.

80. Høst, A., Halken, S., Muraro, A., Dreborg, S., Niggemann, B., Aalberse, R., Arshad, S.H., von Berg, A., Carlsen, K.H., Duschén, K., Eigenmann, P.A., Hill, D., Jones, C., Mellon, M., Oldeus, G., Oranje, A., Pascual, C., Prescott, S., Sampson, H., Svartengren, M., Wahn, U., Warner, J.A., Warner, J.O., Vandenplas, Y., Wickman, M., Zeiger, R.S. (2008). Dietary prevention of allergic diseases in infants and small children. *Official Publication of the European Society of Pediatric Allergy and Immunology*, *19* (1), 1–4.

81. Health Canada. (2007). Breastfeeding. Available at http://www.hc-sc.gc.ca/fn-an/pubs/infant-nourrisson/nut_infant_nourrisson_term_3-eng.php

82. Health Canada. (2005). Transition to solid foods. Available at http://www.hc-sc.gc.ca/fn-an/pubs/infant-nourrisson/nut_infant_nourrisson_term_6-eng.php
83. Du Toit, G., Katz, Y., Sasieni, P., Mesher, D., Maleki, S.J., Fisher, H.R., Fox, A.T., Turcanu, V., Amir, T., Zadik-Mnuhin, G., Cohen, A., Livne, I., Lack, G. (2008). Early consumption of peanuts in infancy is associated with a low prevalence of peanut allergy. *The Journal of Allergy and Clinical Immunology*, *122* (5), 984–991.
84. Fox, A.T., Sasieni, P., Du Toit, G., Syed, H., Lack, G. (2009). Household peanut consumption as a risk factor for the development of peanut allergy. *The Journal of Allergy and Clinical Immunology*, *123* (2), 417–423.
85. Rancé, F., Bidat, E., Dutau, G. (2009). The primary prevention of the allergic diseases: the point of view of pediatrist-allergists. *Revue Française d'Allergologie*, *49*, 21–24.
86. Burks, A.W., Laubach, S., Jones, S.M. (2008). Oral tolerance, food allergy, and immunotherapy: implications for future treatment. *The Journal of Allergy and Clinical Immunology*, *121* (6), 1344–1350.
87. Moneret-Vautrin, D.A. (2006). Concept d'allergovigilance: l'activité du Réseau d'allergovigilance. *La Revue de Medecine Interne*, *27* (Suppl. 2), S70–S72.
88. Morisset, M., Moneret-Vautrin, D.A., Kanny, G., Allergo-Vigilance Network. (2005). Prevalence of peanut sensitization in a population of 4,737 subjects—an Allergo-Vigilance Network enquiry carried out in 2002. *European Annals of Allergy and Clinical Immunology*, *37* (2), 54–57.
89. Morisset, M., Boulègue, M., Beaudouin, E., Pirson, F., Rancé, F., Gallen, C., Moneret-Vautrin, D.A. (2003). Anaphylaxie alimentaire sévère et léthale: cas rapportés en 2002 par le réseau d'allergovigilance. *Revue Française d'Allergologie et d'Immunologie Clinique*, *43*, 480–485.
90. Foucard, T., Yman, I.M., Nordvall, L. (2005). Reduced number of fatal and life-threatening reactions to food. Reporting by the medical profession has resulted in effective measures. *Lakartidningen*, *102* (46), 3465–3468.
91. Løvik, M. (2005). Mustard—an important food allergen? *Tidsskrift for den Norske Laegeforening*, *125* (6), 736–738.
92. Hompes, S., Kirschbaum, J., Scherer, K., Treudler, R., Przybilla, B., Henzgen, M., Worm, M., Bircher, A., Merk, H., Kleine-Tebbe, J., Beyer, K., Dickel, H., Bieber, T., Nemat, K., Schäkel, K., Schuster, A., Mahler, V., Hillen, U., Szliska, C., Fuchs, T., Marsch, W.C., Augustin, M., Wedi, B., Fölster-Holst, R., Hunzelmann, N., Rietschel, E., Schirpke, C., Ring, J., Brehler, R., Grabbe, J., Riffelmann, F., Rabe, U., Klimek, L., Friedrichs, F., Rebien, W., Lepp, U., Aberer, W., Reider, N., Huttegger, I., Kinaciyan, T., Hoffmann-Sommergruber, K., Helbling, A., Eigenmann, P., Wiesner, A., Schmid-Grendelmeier, P. (2008). First data of the pilot phase of the anaphylaxis register in German-speaking countries. *Allergo Journal*, *17* (7), 550–555.
93. Worm, M., Hompes, S. (2008). Anaphylaxis: mechanisms and registry. *Allergologie*, *31* (5), 176–181.
94. MoniQA. (2009). Food Allergens working group. Available at http://www.moniqa.org/allergens

INDEX

Note: Page numbers in *italics* refer to figures; those in **bold** to tables.

Allergen Management in the Food Industry, Edited by Joyce I. Boye and Samuel Benrejeb Godefroy